Structural Methods in Inorganic Chemistry

Structural Methods in Inorganic Chemistry

E.A.V. Ebsworth

Vice-Chancellor and Warden
Durham University
Formerly Crum Brown Professor of Chemistry
University of Edinburgh

David W.H. Rankin

Professor of Structural Chemistry
University of Edinburgh

Stephen Cradock

Formerly of the Department of Chemistry
University of Edinburgh

FOREWORD BY
Kenneth Raymond
University of California at Berkeley, California

SECOND EDITION

OXFORD
BLACKWELL SCIENTIFIC PUBLICATIONS
LONDON EDINBURGH BOSTON

MELBOURNE PARIS BERLIN VIENNA

© 1987, 1991 by
Blackwell Scientific Publications
Editorial Offices:
Osney Mead, Oxford OX2 0EL
25 John Street, London WC1N 2BL
23 Ainslie Place, Edinburgh EH3 6AJ
3 Cambridge Center, Cambridge
    Massachusetts 02142, USA
54 University Street, Carlton
    Victoria 3053, Australia

Other Editorial Offices:
Arnette SA
2, rue Casimir-Delavigne
75006 Paris
France

Blackwell Wissenschaft
Meinekestrasse 4
D-1000 Berlin 15
Germany

Blackwell MZV
Feldgasse 13
A-1238 Wien
Austria

First published 1987
Reprinted 1987
ELBS edition 1988
Second edition 1991

Set by Semantic Graphics, Singapore
Printed and bound in Great Britain
by Wm. Clowes Ltd, Beccles

DISTRIBUTORS

Marston Book Services Ltd
PO Box 87
Oxford OX2 0DT
(*Orders*: Tel. 0865 791155
        Fax: 0865 791927
        Telex: 837515)

Australia
Blackwell Scientific Publications
(Australia) Pty Ltd
54 University Street
Carlton, Victoria 3053
(*Orders*: Tel. 03 347-0300)

British Library
Cataloguing in Publication Data

Ebsworth, E.A.V. (Evelyn Algernon Valentine)
1933–
    Structural methods in inorganic chemistry.
    1. Inorganic compounds. Spectroscopy
I. Title   II. Rankin, David W.H.
III. Cradock, Stephen
546

ISBN 0-632-02963-3

Contents

5   Vibrational Spectroscopy, 173

6   Electronic and Photoelectron Spectroscopy, 255

Foreword

It is a pleasure to introduce a revised and updated version of this successful text. Several years ago when I was asked to read the first draft of this book I was impressed with its fresh approach in several key aspects. Simply presenting a coherent approach to the characterization of structure for inorganic compounds is not by itself new. The enormous growth of inorganic chemistry in the last few decades has in large part been the result of improvements in the quality and speed of the characterization of new compounds, including their structure. Many specialist texts exist for various methods of structure determination, but for the student who wishes to have an overview of these techniques and the degree to which they can supplement or interact with one another (or for the instructor who has only one or two semesters to present these topics) it is most useful to have this information presented in a single text. The first thing I liked about the approach of this book was its coherence. The significance of the time scales of physical methods, the relative advantages and disadvantages of those methods, and their use in concert is opened in the first chapter. This theme then continues throughout the body of the text, with individual chapters devoted to nuclear magnetic resonance spectroscopy, electron spin and nuclear quadrupole resonance spectroscopy, rotational and vibrational spectroscopy, and so on. The book ends, however, with another novel feature—the case histories. This is the second aspect that I found extremely valuable and a most useful teaching device. Well-chosen research examples illustrate the use of the techniques (that the student by now has learned about) in real research publications. For the student this adds some of the air of a detective story and makes a strong connection between theoretical topics and the real world of practicing chemists.

Authors routinely ask for initial reviews of their texts. What is not so routine is that they welcome suggested changes or criticisms as well as praise and agreement. I was therefore impressed when rather specific comments I had made about the earliest draft of this book resulted in enthusiastic response from the authors and thanks for the care of my reading! I was even more pleased at the very nice acknowledgment that they gave me in the first edition of this book. I thought that the resultant text was the best available in the field. The revised and updated second edition will be even better. New developments in nuclear magnetic resonance and electronic spectroscopy, the addition of more recent case histories and a general update of the problems at the end of each chapter all represent important additions to the book. As an instructor of graduate and undergraduate inorganic chemistry courses, I offer my congratulations and thanks to the authors for an outstanding textbook.

Kenneth N. Raymond
Berkeley, California
December 7, 1990

Preface

This book has its roots in our experiences of teaching and research in synthetic and structural inorganic chemistry. It is meant for senior undergraduates, graduates and those starting research in inorganic chemistry, and describes some of the enormous range of physical and structural techniques now available. It is, of course, impossible to cover all the fundamental theory necessary for the complete mastering of all these techniques in a single book. Indeed, there are volumes bigger than this one concerned solely with topics to which we devote only a few pages. But we have set out to give a broad view of the methods available, and what they can and cannot reveal. In particular, we have included recently developed techniques that are becoming of increasing importance or seem likely to be used widely in the future. In preparing this second edition, we were aware of the advances made in solid state NMR spectroscopy, and have expanded that section. We have made some reduction in the length of the chapter dealing with rotational spectroscopy. This is an important technique, but the extent to which it is used is limited, and we have concentrated on the more important aspects of the subject. We have also expanded the chapter dealing with electronic spectroscopy, and in particular have given more information about how the spectra of transition metal complexes may be interpreted, in response to the requests of users to provide rather more detail.

Throughout the book our aim is to help the reader to interpret experimental data, to understand the material published in modern journals of inorganic chemistry, and to make decisions about what techniques will be the most useful in solving particular structural problems. We have provided references to more detailed treatments of theory and experimental details, so that the interested reader can find out more about any of the methods we have discussed.

Some points of general importance such as modern advances in computing and other technical developments are discussed in the first chapter, which also includes an account of the plan of the book and a glossary of techniques. The next eight chapters deal with particular techniques, and could be taken in more or less any order. Advanced students would already have met such methods as nuclear magnetic resonance and vibrational spectroscopy, so we do not discuss the basic principles of these in great detail. Group theory, which is relevant to several of the chapters, is described in an appendix. In the final chapter, we move from methods to chemical systems, and discuss a wide range of structural techniques. We regard this chapter as the most important of the book, because all the methods become of value only when they are applied to real chemical problems. When used together, their value is greater than when they are used in isolation. We have therefore extended this chapter in the second edition by the addition of several new case histories, and we have chosen the topics to reflect important developments in inorganic chemistry during the past few years. We hope that readers will find them up-to-date and interesting. In a sense, all the earlier chapters are meant to prepare the reader for this final one, in which the wide experiences of others in tackling structural problems are presented. In this way, the approach to problems which may be encountered later will be learned. We firmly believe that breadth of understanding is vital.

The problems at the end of each chapter except the first have been replaced almost entirely in this edition, and there are many more than in the first edition. Some of the problems are numerical, others involve interpretation of data or some logical analysis and others are meant to start discussion. We have given answers or comments as appropriate for the odd numbered problems and have left the even ones to the students' imagination. In addition, sets of solutions and comments on the even-numbered problems are available to course instructors from the publishers. In this second edition we have also provided throughout the text a series of worked examples. These are related to the case histories, but each one depends on the application of just one experimental technique. They are intended to illustrate the applications of the particular techniques to real chemical problems, and most of them are drawn from recent chemical literature.

We recognize that some of the examples are drawn from our own research or from work we know well, although we have tried to illustrate the various methods with a wide range of examples. Authors with interests in other areas of chemistry might well choose different examples to illustrate the same points. The techniques are, of course, used in organic chemistry as much as in inorganic, although the relative importance of the various methods may be somewhat different. We hope that readers in all levels and in all fields will keep their eyes open for relevant examples, in their own work and as they read journals.

E. A. V. E.
D. W. H. R.
S. C.

Acknowledgements

A book such as this cannot be produced without help from many people. In the preparation of the first edition, Professor R. K. Harris (Durham), Professor A. H. Cowley (University of Texas at Austin) and Dr A. J. Downs (Oxford) provided us with some excellent ideas and examples, and various sections of the manuscript were read and criticized by Dr A. S. F. Boyd (Heriot-Watt), Professor J. R. Durig (University of South Carolina, Columbia), Professor C. D. Garner (Manchester), Professor A. G. Maddock (Cambridge), Dr N. C. Norman (Newcastle), Professor P. C. Ford (University of California at Santa Barbara), and Drs A. J. Blake, R. O. Gould, G. A. Heath, P. R. R. Langridge-Smith, H. E. Robertson, M. Schröder and the late T. A. Stephenson, all of Edinburgh. Dr Boyd also spent a great deal of time running NMR spectra, as did Drs D. Reed and S. G. D. Henderson.

The whole manuscript was painstakingly reviewed by Professor K. N. Raymond (University of California, Berkeley). His comments and criticisms encouraged us in the preparation of a second draft, which incorporated many of his suggested amendments. Professor Raymond then read the entire manuscript again, and to him we owe particular thanks.

There is much new material in this second edition, and we gratefully acknowledge the willing cooperation of Professor B. T. Heaton and Dr J. A. Iggo (Liverpool), Dr R. N. Perutz (York) and Professor J. L. Spencer (Salford) in providing us with data and spectra. Our colleagues in Edinburgh, including Drs A. J. Blake, S. K. Chapman, R. O. Gould, D. Reed, M. Schröder, A. J. Welch and L. J. Yellowlees, have read and corrected parts of the manuscript, and contributed ideas. We have also made many additions and amendments in the light of comments we have received concerning the first edition. In particular we thank Professor T. J. Marks (Northwestern University, Illinois) for his enthusiastic support and the many people around the world who have suggested improvements.

Throughout the years during which this book was being written we were supported by Professor J. J. Turner (Nottingham), who originally proposed the idea. Mr Navin Sullivan of Blackwell Scientific Publications has worked with us on both editions, and has patiently kept our noses to the grindstone. Mrs Jane Gorrie, Mrs Moyra Bain and Mrs Liz Glass typed the original manuscript, and Mrs Charmaine Wilson followed their examples of speed, accuracy and good humor in typing the new material for the second edition.

To all these people we acknowledge our gratitude. We trust that you, the readers, will enjoy using this book and will pass on to us your comments and suggestions for changes and developments.

The following figures are reprinted by permission of VCH Publishers, Inc., 220 East 23rd St, New York, NY 10010: Figure 10.29 from *Chemische Berichte* 1981, Vol. 114, p. 3634, figs 2 and 4; Figure 2.29 from *Angewandt Chemie* (Int. Ed.) 1986,

Vol. 25, p. 1103, figs 1 and 2; Figure 10.30 from *Angewandt Chemie* (Int. Ed.) 1988, Vol. 27, p. 1544, fig. 1; Figure 2.33 from *Stereochemical Applications of Gas-Phase Electron Diffraction*, edited by I. Hargittai and M. Hargittai, figs 14-6 and 14-9.

E. A. V. E.
D. W. H. R.
S. C.

ACKNOWLEDGEMENTS

1 Determining Structures —
How and Why

General introduction — asking questions about structure

In 1954 a young research student began to study the reaction between white phosphorus and iodosilane. The reaction was very complicated, and gave only a small yield of volatile products, containing a number of components. One was recognized to be a known compound, which had already been identified by analysis and molecular weight as SiH_3PI_2. The remainder were separated by laborious fractionation into several different inhomogeneous fractions, and the vapor pressure and density of each were measured, but not much more could be done. Efforts to obtain homogeneous products failed, and after three months' fruitless work he got rid of the solid residue (which blew up like an oil-well when it was exposed to air) and took up a different topic.

However, if he had started his study today, the story would not have ended there. He could have learnt a great deal about what he had made, using physical techniques that are now available generally. As it happens, silyl–phosphorus compounds are very suitable for study by infrared and nuclear magnetic resonance spectroscopy. It would be easy to find out whether likely products such as SiH_3PI_2, $(SiH_3)_2PI$ and $(SiH_3)_3P$ had been formed. Mass spectrometry would provide valuable additional evidence, and a lot of the time wasted in trying to separate pure compounds could have been avoided by monitoring the spectra of fractions; it is easy to find out very quickly whether the composition has changed, even without knowing what the components are. Anyone who reads such papers as Werner's great review [1] in *Liebig's Annalen* of 1910 must be filled with admiration for what could be achieved using classical methods of analysis, cryoscopy, conductivity, and the chemical understanding and intuition of a genius; nowadays we do not have to be so clever to find out what we have made, because we have a range of techniques to tell us about the composition, the groups present, the molecular geometry, the electronic energy-levels, the electron distribution, and much else besides. Modern inorganic chemists must learn about these techniques as part of their training, and this book is meant to help them to acquire the necessary skills and understanding of principles. They can then use their understanding and intuition to tackle systems that are much more difficult than the one we have just described.

This introductory chapter sets the scene, and discusses a few general points that bear on the chapters that follow. The next eight chapters all deal with particular techniques or groups of related methods. When using a particular technique, it may not be essential to know in detail how the instrument works, though such understanding may well help in collecting the best or most useful data. Similarly, it may be possible (particularly using modern methods of processing data) to convert experimental measurements into useful structural information without a *detailed* knowledge of the theoretical basis of the technique. But it is absolutely vital for chemists to be able to interpret the structural information they obtain. In this book we have therefore put the major emphasis on such interpretation. As far as we can, we present spectra or other experimental information to illustrate the points we make. We do of course explain the physical principles of each technique, but we have tried to avoid going into the theory in great detail; we provide references to sources where anyone who wants to will be able to find more formal and rigorous treatments. There are many books that deal with the theory of structural methods in more or less depth; in order to keep this book to a reasonable size we have concentrated on the aspects of the methods that are of greatest concern to the practising inorganic chemist – those primarily related to the interpretation of spectroscopic and structural data. The basic principles of symmetry and of the simple use of Group Theory in the analysis of spectra are presented in outline in the Appendix.

The last chapter is different from Chapters 2–9, which deal with techniques. It is based on compounds or groups of related compounds, and consists of a detailed explanation of what has been learnt about the structures of some 20 inorganic systems by using all appropriate physical methods. It draws on what has been derived and explained in Chapters 2–9, but from the point of view of the chemist who has a compound and wants to know as much as possible about it, rather than from the point of view of someone with an expensive instrument who wants to find a use for it.

Inorganic chemists have many different questions to ask about different types of system, and it is not possible to explain how to try to answer all of them in a single text. We have not set out to discuss questions associated with surfaces or with heterogeneous systems; our book is written from the point of view of the chemist who has to deal with well-defined chemical species, though a good deal of what is described would also be useful to the solid-state scientist. The principles of structure determination apply equally to organic compounds too, although the relative importance of the techniques may be different.

The young research student of 1954 is now one of the authors of this book, and still retains an interest in silicon chemistry. Inevitably, we have drawn on our own experience in selecting examples to illustrate the application of structural methods, but we hope that users of the book will understand the principles of gaining useful chemical information using these methods, and will see how to apply them to their own problems.

1.2   Answering questions about structure

Identifying and characterizing the products of a chemical reaction can be a long and messy process, involving the parallel use of techniques of separation and of recognition. This book is not about methods of separation, but the physical methods we are writing about include methods of recognition which, properly used, can shorten and focus the process of separating and purifying products. The questions we may try to answer depend on the system we are studying, and upon the progress of separation and isolation. If we are dealing with reactions that lead to products that have not been made before, we can consider the following eight questions in order of increasing subtlety:

(a) Does our material contain any known compound that we can identify?
(b) Is it pure?
(c) What functional groups does it contain?
(d) What is its composition and (if appropriate) molecular weight?
(e) How are the functional groups linked together – what is the pattern of connectivity?
(f) What is the molecular symmetry?

CHAPTER 1: DETERMINING STRUCTURES

(g) What are the bond lengths (or internuclear distances) and angles?

(h) What can we say about its electronic structure?

We can ask the first four of these questions about mixtures; the later ones only mean anything in relation to a single species. The techniques we can use to try to answer some or all of them will depend on the kind of material we are working with. In Table 1.1 we have correlated some of these questions with the techniques discussed

Table 1.1 Uses of some physical techniques in determining structures

Information sought	Phase of sample		
	Gas	Liquid or solution	Solid
Fingerprint[a]	IR *Microwave* Mass spec. UV/vis *UPS*	IR, R NMR Mass spec.[b] UV/vis	IR, R Powder X-ray diffraction Mass spec.[b] UV/vis
Groups present	IR Mass spec. NMR	IR, R Mass spec.[b] NMR	IR, R Mass spec.[b] *NMR *Mössbauer*
Molecular symmetry	IR[c] *Microwave* **ED*	IR, R NMR	IR, R *NMR ***NQR* Single crystal X-ray diffraction
Bond lengths and angles	IR, R[d] *Microwave* **ED*	****EXAFS* LCMNR	Single crystal X-ray diffraction ***Neutron diffraction
Electronic structure[e]	UV/vis *UPS ESR	UV/vis ESR **XPS*	UV/vis *UPS *XPS *Mössbauer ***NQR* Single crystal X-ray and neutron diffraction

This table is not exhaustive, but it indicates the uses of methods discussed or treated in depth in this book. The meanings of abbreviations are given in Section 1.6.

Key: Italic type, method of limited importance in this context; no asterisk, equipment available in most laboratories; *, equipment available in some laboratories; **, equipment only available in a few laboratories; ***, very expensive specialized equipment needed.
[a]Almost any spectroscopic technique can be used to recognize a known species; this table only lists some important applications.
[b]The sample may be in a condensed phase, but the resulting ions are always studied in the gas phase.
[c]The shapes of bands from small molecules may be characteristic.
[d]Using vibration–rotation structure.
[e]Deductions about electronic structure can be made from many spectroscopic measurements such as chemical shifts (NMR), force constants and band intensities (IR/R), and bond lengths (microwave). This list only mentions methods that give direct information about electronic energy-levels or charge densities.

elsewhere in this book. The table includes an assessment of the importance of each technique for answering each question, and also shows whether the necessary equipment is widely available.

Most spectroscopic methods can be used to provide fingerprints of particular compounds. For example, complicated organic molecules give infrared spectra that cannot usually be assigned in detail but which are nonetheless most useful, without assignment, as fingerprints. Some solid phases give X-ray powder patterns that are well-defined and characteristic, and these may be used to identify a material without any detailed analysis. Small molecules, produced in high temperature reactions in the gas phase, may give characteristic band-patterns in ultraviolet photoelectron spectra, and so may be identified specifically in complicated mixtures. The ^{57}Fe Mössbauer spectra of minerals may give characteristic magnetic splitting patterns that can be used without analysis to identify known species. Indeed, complicated spectra may be hard to analyze in detail, but they are often much more useful than simple spectra for straightforward identification. In the same sort of way, it is often possible to follow the weakening of bands due to impurities as invaluable indicators of the progress of purification. Here, though, we must be careful. Impurities do not necessarily give infrared bands or NMR peaks that can be resolved from those of the products we are trying to purify. If we are certain that by using a particular technique we can detect the impurity we are trying to remove, we have no problems; but there is no spectroscopic technique that can be relied on as a general test for purity. ^{31}P NMR spectroscopy, for instance, will not tell us if a sample of $P(OMe)_3$ contains 90% MeOH. The best we can do is to use more than one method and to make sure that each type of spectrum contains no bands other than those due to the pure product.

When we come to more detailed questions about molecular composition, functional groups present, molecular weight, and molecular geometry, the methods we use depend on the phase and type of sample. Elemental analysis is important, but it is outside the scope of this book. There are now reliable ways of doing quantitative analysis for almost every element, and results are both more reliable and more precise than they were thirty years ago. Normally this kind of work is done in specialist laboratories, using methods based on combustion or hydrolysis followed by weighing, titration, or some physical technique such as atomic absorption spectroscopy. In modern chemistry, relatively few inorganic compounds are characterized by complete elemental analysis, and it is unwise to rely exclusively on carbon and hydrogen analyses to characterize a new compound. Nature seems to have an annoying habit of finding an impurity with C and H content similar to that of the product sought. Moreover, analysis is almost always destructive, and can be expensive.

To determine the molecular weight of a volatile compound, we can use mass spectrometry as well as the old Dumas method for the direct weighing of a known volume of vapor. Modern methods of ionization make it possible to obtain mass spectra from virtually all compounds, whether volatile in the conventional sense or not, and by adjustment of the ionizing voltage it is very often possible to identify the molecular ion and so to measure the molecular weight. The compound present need not even be pure: mass spectrometry can give the molecular weights of individual compounds, if necessary with great precision. In contrast, measurement of vapor density gives an average molecular weight for the whole of the sample studied. Molecular weights in solution are normally determined using colligative properties, particularly osmotic pressure. They may not be very precise, but they can be most useful in helping to decide whether a compound is associated or dissociated in solution, particularly through the use of several different solvents.

We can usually get a good idea of the functional groups present in a compound by using vibrational spectroscopy. The correlation patterns that have been drawn up for organic functional groups can be extended to organometallic systems, and there are similar correlations for purely inorganic species. Where a compound contains only heavy atoms, it may be necessary to rely on low-frequency vibrations. We may then find that bands are sometimes weak and frequency patterns less well-defined, but there is a lot of background information available and it is almost always possible to learn something useful. If the compound contains suitable nuclei, we may be able to discover a great deal from NMR spectra, or perhaps (if it contains unpaired electrons) from ESR spectra. We may even be able to determine stereochemical relationships between groups using these techniques.

If we want precise measurements of bond lengths and angles, our choice of method is more limited. The geometries of many simple molecules in the gas phase can be determined very accurately by microwave spectroscopy or electron diffraction, but complicated molecules of low symmetry are not suitable for study by these methods. Furthermore, we must usually know what we are studying before we can determine its structure. If a material is suitable for single crystal X-ray diffraction, these limitations no longer hold: very complicated structures can be determined, and we do not always have to know what we are studying before we determine the precise molecular geometry. In suitable cases we can answer all questions about connectivity, symmetry and geometry at once by crystallography. In such cases the most serious uncertainty may be whether or not the crystal studied is typical of the sample as a whole. There are plenty of examples where after much effort at crystallization the single crystal obtained turned out to be of a product of decomposition, oxidation or hydrolysis, or of a minor impurity in the compound studied.

But when we begin to look in detail at the information we obtain from any of the methods mentioned, we find that much of it is limited. X-ray crystallography seems to answer all our questions, but it does so for species in very specific situations. It gives us a frozen snapshot of a molecule or ion in the form it adopts under the pressure of the forces exerted by its neighbors. In solids many molecules take up different conformations or even have structures which are completely different from those existing in solution or in the freedom of the gaseous phase. Microwave spectroscopy gives us information about the rotation of molecules in whichever vibrational states are populated at the temperature of the experiment: we see separate sets of lines for each. Electron diffraction tells us about the distances between pairs of atoms in molecules, but these are averaged over all populated states; while these distances lead in principle to a complete structure, they are obviously affected by vibrations, and so we need to know something about the vibrations of the molecule before we can make a reliable interpretation of the radial distribution curve from an electron diffraction experiment. Both microwave spectroscopy and electron diffraction are affected by molecular vibrations, but in different ways. The internuclear distances we obtain from these and other methods do not represent exactly the same physical parameters, and different types of calculation lead to values with slightly different physical significance. Relationships between some of these definitions of distances for gas or solution phase molecules are shown in Fig. 1.1. The differences emphasize that molecules are not static and rigid like most molecular models. We may think of them in this way, but it is

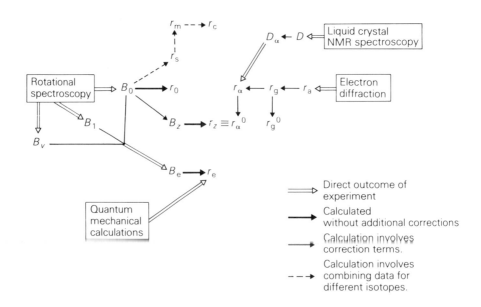

Fig 1.1 Relationships between various definitions of inter-atomic distances for molecules in the gas or solution phase. These distances may be measured directly by diffraction experiments or derived from rotation constants (B) or dipolar couplings (D). The correction terms depend on vibrational properties, and their calculation requires knowledge of the harmonic and/or anharmonic force field.

The distances are defined as follows:

r_e	Equilibrium structure distance	Geometrically consistent
r_α	Distance between average positions of atoms, at temperature of experiment	Geometrically consistent
r_α^0	Distance between average positions of atoms at absolute zero temperature	Geometrically consistent
r_g	Average distance between atoms, $\langle r \rangle$	Geometrically inconsistent
r_a	Apparent distance from electron diffraction experiment, $<1/r>^{-1}$	Geometrically inconsistent
r_0	Distance in structure derived from ground-state rotation constants, B_0	Inconsistent with everything else; depends on isotope
r_s	Distance obtained by isotopic substitution of two atoms concerned, or from structure derived by multiple isotopic substitution	Geometrically consistent, but exact physical significance of distance cannot be expressed simply
r_c, r_m	Distances derived by allowing for effects of isotopic substitution on geometric parameters in calculation of structure by multiple isotopic substitution	Geometrically consistent, and r_c is close to r_e
r_z	Distance calculated from B_z, which is B_0 corrected for harmonic vibrations of molecule	Identical to r_α^0; difference in name reflects different experimental origin
r_{av}	Name given to distances based on both ED data and rotational constants	Identical in meaning to r_α^0 and r_z

All these distances refer to nuclear positions, which are what we obtain from electron diffraction, liquid crystal NMR and rotation constants. Neutron diffraction experiments also

CHAPTER 1: DETERMINING STRUCTURES

wrong to do so. They are dynamic things. They twist and turn, they vibrate and rotate; they may exchange electrons, single atoms or even groups with other molecules, and groups within a single molecule may change partners. If we are to understand the structure of a particular molecule, we must study it by as wide a range of methods as we can.

1.3 Plan of the book

The physical methods discussed in detail in Chapters 2−9 can conveniently be classified by the type of physical process involved, as methods based on ionization, on emission or absorption of radiation (which may also involve ionization), or on diffraction. There are a very large number of spectroscopic methods based on the absorption or emission of radiation. Those considered in detail in this book are electronic (including ultra-violet and X-ray photoelectron), vibrational, rotational, ESR, NMR, NQR and Mössbauer spectroscopy. Others are referred to, and more are mentioned in Section 1.6, which provides a quick guide to the many confusing synonyms and acronyms used to describe structural methods. The energy spectrum (see Fig. 1.2) embraces an enormous range of atomic and molecular energies, and the techniques used vary as widely. Each spectroscopic method depends on the appropriate selection-rules; these are given, but not derived, in the appropriate chapters.

We start in Chapter 2 with nuclear magnetic resonance spectroscopy (NMR), which involves transitions of very low energy, and which has become of great importance to inorganic chemists since about 1960. It can provide enormously detailed information about the groups present in molecules or ions if they contain suitable nuclei, and it may open a window for us to find out about the rates of processes that are fast at room temperature and that would otherwise be hard to study. Electron spin resonance (ESR) spectroscopy may provide a detailed picture of the energy-levels involving unpaired electrons; it tells us nothing about materials in which all the electrons are paired, but it can give us insights into the electronic structures of free radicals which have no parallel in our understanding of the electronic structures of diamagnetic species. It is discussed in Chapter 3, together with nuclear quadrupole resonance spectroscopy (NQR). NQR is only relevant to the study of compounds containing quadrupolar nuclei, and is of limited value for some of them, but it has important applications in the identification

---

give nuclear positions, but X-ray diffraction actually finds the positions of the centers of electron distribution around the nuclei. Distances between these positions are then of the same type as r_α distances from electron diffraction. Distances of the same type as r_g are often derived by applying corrections for thermal motions (vibrations); these are not geometrically consistent.

It is also possible to extrapolate from B_0 rotation constants and rotation constants B_v for some other vibrational state v to arrive at 'equilibrium rotation constants' B_e; the r_e structure may then be calculated in some cases.

Quantum mechanical calculations generally give equilibrium structures, r_e.

For further details, see A.G. Robiette, in *Molecular Structure by Diffraction Methods*, Vol. 1, p. 160, *Specialist Periodical Reports*, The Chemical Society, London (1973).

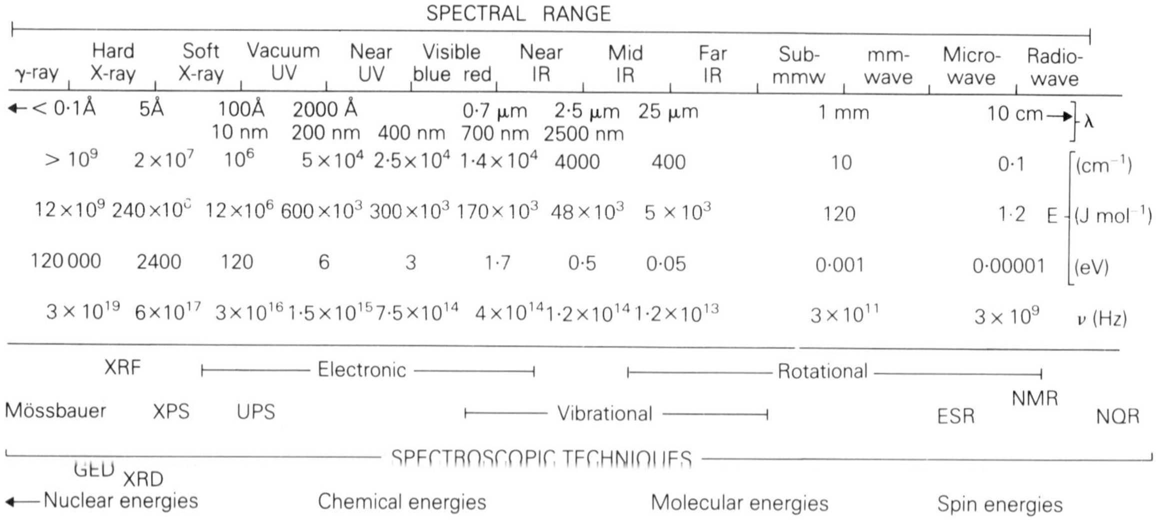

SPECTRAL RANGE

γ-ray	Hard X-ray	Soft X-ray	Vacuum UV	Near UV	Visible blue red	Near IR	Mid IR	Far IR	Sub-mmw	mm-wave	Micro-wave	Radio-wave	
← <0.1Å	5Å	100Å	2000 Å		0.7 μm	2.5 μm	25 μm			1 mm		10 cm →	λ
		10 nm	200 nm	400 nm	700 nm	2500 nm							
> 10^9	2×10^7	10^6	5×10^4	2.5×10^4	1.4×10^4	4000	400		10		0.1		(cm^{-1})
12×10^9	240×10^6	12×10^6	600×10^3	300×10^3	170×10^3	48×10^3	5×10^3		120		1.2		E (J mol^{-1})
120 000	2400	120	6	3	1.7	0.5	0.05		0.001		0.00001		(eV)
3×10^{19}	6×10^{17}	3×10^{16}	1.5×10^{15}	7.5×10^{14}	4×10^{14}	1.2×10^{14}	1.2×10^{13}		3×10^{11}		3×10^9		ν (Hz)

XRF ├── Electronic ──┤ ├── Rotational ──┤

Mössbauer   XPS   UPS   ├── Vibrational ──┤   ESR   NMR   NQR

GED   XRD ── SPECTROSCOPIC TECHNIQUES ──

←— Nuclear energies   Chemical energies   Molecular energies   Spin energies

Fig 1.2 Spectrum of electromagnetic radiation, showing the relationship between important spectroscopic techniques and the energy and wavelength of the radiation. Note that energies are often quoted in cm^{-1}, eV and Hz, none of which is, strictly speaking, a unit of energy.

of non-equivalent sites in solids and in understanding the electronic structures of compounds of the halogens or of nitrogen.

Rotational spectroscopy, used mainly to obtain bond lengths and angles for simple species in the gas phase, is described in Chapter 4. Vibrational spectroscopy, discussed in Chapter 5, can be approached through infrared or Raman methods. The information obtained by these is complementary though different. We can learn about molecular symmetry, and where it is possible to resolve and interpret vibration−rotation structure we may be able to derive bond lengths and angles; we can also learn about the resistance of bonds to stretching and bending (molecular force fields) by making calculations based on the observed spectra. In addition, techniques such as resonance Raman spectroscopy can be used to tease out particular parts of complicated Raman spectra and to excite high overtones of particular vibrations. But by far the most familiar use of vibrational spectroscopy is the qualitative analysis of compounds or mixtures through their infrared spectra.

Electronic spectroscopy is discussed in Chapter 6. It provides information about the differences between molecular energy-levels, and so leads to an understanding of the electronic structures of compounds; it is therefore closely connected with quantum mechanical calculations. Similarly, ultraviolet photoelectron spectroscopy provides information about the electronic energy-levels of molecules by giving ionization potentials. It is most useful for studying small molecules in the gas phase, where it can provide a clear and simple picture of orbital energy-levels, and also give information about the vibrations of molecular ions which have been ionized from particular molecular energy-levels. We learn less from the UPS spectra of complex molecules, where the results are often hard or impossible to interpret without the help of detailed calculations based on theory, but at worst we can determine the lowest ionization energy of a molecule, and that in itself can sometimes be a very important property. X-ray photoelectron spectroscopy gives similar information about the core levels of atoms in molecules. Most such work has been with solids, where the effects of local charges can be difficult to allow for. However, the technique has proved valuable for qualitative and quantitative analysis, it can be used to identify species like O^{2-} on metal surfaces after the adsorption of non-metallic oxides such as NO, and it may help to characterize the oxidation-state in which a particular element is present in a sample.

Chapter 7 deals with Mössbauer spectroscopy, which is only useful for solids and for a few nuclides (of which the most important so far are ^{57}Fe and ^{119}Sn). It is important as it can be used to characterize oxidation-states, coordination and local symmetry, as well as identifying species, and it also affords information about the effects of electric field gradients and magnetic fields at Mössbauer-active nuclei.

Diffraction methods are considered in Chapter 8. They differ from spectroscopic methods in that the incident radiation or particles do not undergo any energy change. Instead, there is scattering, and the techniques involve interpretation of an interference pattern in terms of the relative positions of atoms. We have also included Extended X-ray Absorption Fine Structure (EXAFS) spectroscopy in this chapter. This is not a diffraction technique, but the manner of interpretation of the data is very like that used in gas diffraction studies, and it does not logically belong in any other chapter.

Mass spectrometry, discussed in Chapter 9, is of great importance to inorganic chemists for the identification of known species through recognized fragmentation patterns, and it also provides a way of identifying a molecular species through its exact molecular weight. No other technique can measure molecular weights accurately enough to allow us to distinguish between, say, O_2 and N_2H_4. The details of fragmentation patterns can be used to derive ionization and bond energies and are of great importance to the ion physicist. They can also tell us about how the atoms and groups in a molecule are linked together. In the past, inorganic chemists have made less use of mass spectrometry than might have been expected, because many inorganic compounds are very hard to vaporize and ionize. In recent years, the development of new methods of ionization has gone far to reduce this disadvantage, and it seems very likely that mass spectrometry will prove of growing importance to inorganic and organometallic chemists.

We are sorry that we have been unable to include much discussion of the techniques available for quantum-mechanical calculations. The importance and practical value of calculations based on different levels of approximation has grown dramatically with the growth and power of computing facilities, and such calculations may be almost essential in interpreting or understanding spectroscopic measurements. It would need a separate book to explain the principles, advantages, snags, uses and limitations of the different methods available; we hope that somebody will write one.

1.4   Timescales

The various physical techniques we may use to study inorganic species depend on a variety of processes. The conclusions we may draw about structures are related to the timescales associated with these processes, and it is important for us to understand these if we are to avoid making erroneous deductions. In relation to any one type of experiment, there are in fact four different times for us to consider: the time during which a quantum of radiation or a particle can interact with a molecule; the lifetime of any excited state of the molecule; the minimum lifetime that the species being studied must have to allow it to be seen as a distinct species; and the total time of an experiment in which the species is observed, which may be several days, or as little as 10^{-12} s. Before we consider these further we must look at the timescales of typical molecular processes, so that we can relate them to timescales associated with structural techniques. Typical vibrational frequencies are of the order of 10^{13} to 10^{14} Hz, while rotational frequencies are around 10^{10} to 10^{12} Hz. The inversion of ammonia has a rate of about 10^{11} Hz at room temperature, while the corresponding rate for phosphine is 10^{3} Hz. The inversion rate for methane is 10^{-15} Hz – so any one molecule inverts, on average, once every 100 million years! Pseudorotation in PF_5, which switches axial and equatorial fluorine atoms, has a rate of about 10^{5} Hz at room temperature, while the rate for PCl_5 is 10^{-4} Hz.

The time during which radiation can interact with a molecule is essentially the time taken for photons to pass by the molecule or relevant part of it. Particles used in a diffraction experiment move at rather less than the speed of light, but in all these cases we are concerned with interaction times of 10^{-16} to 10^{-19} s. This is very much less than the time taken for molecular vibrations, rotations or rearrangements, and so each particle or photon "sees" a molecule with an instantaneous structure, and in a fixed electronic, vibrational and rotational state.

In some spectroscopic techniques using electromagnetic radiation this molecular state is changed, although in techniques such as NMR or ESR it is the nuclear or electronic spin state that is changed: in diffraction experiments there is no equivalent phenomenon. In these processes, there is an effective lifetime associated with the conversion of the upper, excited state to the lower one. This lifetime, called the relaxation time, is in general short if the energy gap is large, and longer if the gap is small, but there is no simple direct relationship. Thus relaxation times associated with low-energy techniques such as NMR are comparatively long, often many seconds, while those associated with higher-energy techniques are correspondingly shorter. Typical transition frequencies for some spectroscopic techniques are given in Table 1.2, with typical relaxation times.

When relaxation times are short, the Uncertainty Principle becomes important, because the lifetime τ of an excited state and the uncertainty in its energy, ΔE, are related by $\tau \Delta E \approx \hbar$. Spectra consist of lines representing transitions, and if the uncertainty in the upper state becomes large, these lines may be broadened, so that

Table 1.2 Timescales associated with some spectroscopic techniques

Technique	Energy of excited state (Hz)	Typical relaxation time (seconds)	Typical linewidth (Hz)
NMR (solution)	10^{8}	10	10^{-1}
ESR (solution)	10^{10}	10^{-5}	10^{5}
Rotational spectroscopy (gas)	10^{11}	10^{-4}	10^{4}
Vibrational spectroscopy (gas)	10^{14}	10^{-8}	10^{8}
Electronic spectroscopy (solution)	10^{16}	10^{-15}	10^{15}
Mössbauer spectroscopy (solid)	10^{19}	10^{-8}	10^{8}

resolution is lost, and in extreme cases the whole spectrum may become just a single, extremely broad hump. The constant $\hbar$ is very small, about 10^{-34} J s, but for electronic spectra of transition metal complexes in solution, relaxation times are typically around 10^{-15} s, so that ΔE is of the order of 10^{-19} J molecule^{-1}, or 60 kJ mol^{-1}. This is comparable with the transition energies involved. Most such electronic spectra therefore consist of a few broad bands, and much potentially useful information is lost.

It is important to realize that the relaxation times may depend on some factors which are properties of the atom or molecule itself, and on others which are related to its environment. Thus rotational spectra of gases have linewidths (related to the rotational relaxation times) which depend on the mean times between collisions for the molecules, which in turn depend on the gas pressure. In liquids, the collisional lifetimes are much shorter, and so rotational energy is effectively unquantized. On the other hand, if the probability of collisions is reduced, as in a molecular beam, we can increase the relaxation time, reduce linewidths, and so improve resolution. Of course, the relaxation time only defines a minimum width of spectral lines, which may be broadened by other experimental factors.

We can see from Table 1.2 that relaxation times in NMR experiments are relatively long, typically of the order of seconds or tens of seconds, so line broadening is very much less than in most other spectra. However, even this degree of line broadening can cause problems, because the range of energies covered in an NMR spectrum is very small. This is particularly true when relaxation times are much less than a second, as they can be when quadrupolar nuclei are involved. In such cases lines may be so broad that they cannot be detected at all.

In most spectroscopic experiments we observe the absorption of energy as the excited state is formed. Sometimes, particularly in the case of Fourier transform NMR, we observe the subsequent relaxation of the excited state. As this may take place over several seconds, it is possible that nuclei in one molecule are excited, but that they then change into something else by chemical reaction, or rearrange internally, while we are watching. In such a case the experiment can tell us about the relative rates of the exchange or reaction process and the relaxation. For this sort of rate, around 1 Hz, information is rather difficult to obtain by other methods. The technique, as applied in NMR saturation transfer studies, is described in more detail in Section 2.15.

The third sort of time we must consider is related to the frequency scale of the spectrum. This is best considered by discussion of NMR spectra. Suppose that we have a compound with two chemically distinct types of methyl group, and that these give rise to two proton NMR resonances 100 Hz apart. Then let us suppose that these methyl groups exchange positions at a rate much greater than 100 Hz. In our spectrum we will observe just one average peak, rather than two distinct ones. Thus if the lifetime of the molecule between rearrangements is $\gg 0.01$ s we see two resonances, and if the lifetime is $\ll 0.01$ s we see just one, while for intermediate lifetimes we may see broad lines. Thus we can use NMR to study exchange processes with rates of 10 to 10^6 Hz. This subject is described in greater detail in Section 2.15.

The same arguments can apply to other forms of spectroscopy. However, transition energies are much larger than those involved in NMR; a typical vibrational spectrum, for example, may cover hundreds or thousands of wavenumbers, and 1 cm^{-1} corresponds to about 10^{10} Hz. In this case, even with the highest attainable resolution, ca. 10^{-4} cm^{-1}, only exchange rates of more than 10^6 Hz can be studied. Hence exchange studies using techniques other than NMR are relatively rare. In Section 10.10 we describe a rather unusual example, in which a lower rate limit for an electron transfer process is provided by an NMR experiment, and an upper limit is provided by a Mössbauer spectroscopy experiment.

Finally, we must take account of the total time during which we acquire our experimental data. In a diffraction experiment, each electron, neutron or X-ray photon is diffracted by a molecule or crystal in around 10^{-18} s, and so sees an effectively frozen instantaneous structure. But the experiment involves accumulating data from many photons or particles, and each sees a molecule at a different stage of its vibrational motion. For gases we have to consider the effects of rotation as well. Thus some diatomic molecules, for example, will be seen to have an interatomic distance less than the mean, and some greater. A radial distribution curve would show a broad, Gaussian distribution of distances. The distance between the end atoms of a linear, triatomic molecule such as OCS would equal the sum of the bond lengths while the molecule was in its average, unbent position, but at all other times this distance would be shortened, as the molecule underwent its bending vibration. This is the origin of the shrinkage effect in gas phase electron diffraction, and of important vibrational effects in crystallography (see Chapter 8). In contrast to this, if the molecule were studied by microwave spectroscopy, the relaxation time would be *ca.* 10^{-4} s, during which the molecule would execute many (10^9) vibrations. The spectrum would therefore consist of transitions characteristic of vibrational states, which would enable us to determine the average structure in the ground state, and possibly also in excited states. The average ground state structure is, of course, truly linear.

The effects of averaging over vibrations just described are inevitable because our diffraction experiment lasts much longer than the time taken for a vibration to occur. Similarly, if the experiment lasts longer than some chemical reaction or exchange process, we can only expect to collect data characteristic of a mixture. Thus if a compound A isomerizes to form an equilibrium mixture of A and B, with a lifetime of one minute, and we take an hour to record an infrared spectrum, we will see bands attributable to both A and B, superimposed. But if we could start with pure A and obtain a spectrum in one second, we would see almost pure A. In some circumstances, we may want to study a species that lasts for a very short time indeed, and techniques such as flash photolysis (Section 5.16) have been developed for such purposes.

1.5   Modern technology and the development of instruments

The technological advances of recent years have had a dramatic impact on almost all the experimental techniques described in this book. Cheap and reliable computers have simplified life from the operator's point of view, while allowing much more sophisticated experiments to be performed. Indeed, in much of the routine apparatus of a modern chemical laboratory a computer is an *essential* component − as in Fourier transform NMR and infrared spectrometers, or in diffractometers. High energy sources, such as lasers (which are generally available) and synchrotrons and high flux neutron sources (which are national or international facilities) have opened up new areas of research. Similarly, the scope of mass spectrometry has been greatly extended by the development of powerful new methods for obtaining ions in the gas phase from involatile materials.

In this book we are primarily concerned with the interpretation of experimental data, and to a much lesser extent with instrumental matters, which change rapidly as technology progresses. Here we present some general observations on these technological developments, and discuss matters that may be relevant to several of the techniques described in subsequent chapters.

<table>
<tr><td>1.5.1</td><td>

Sources

</td></tr>
</table>

1.5.1 **Sources**

In order to acquire a simple absorption spectrum, we require a source of radiation, a sample, and a detector. The nature of the source must of course depend on the type of spectrum required. The extremely low energy radiation involved in NMR experiments is derived from a coil and a radio-frequency generator, while the extremely high energies required for a Mössbauer experiment are obtained by γ-rays emitted by a radioactive isotope. Radiation with intermediate energies used in electronic, vibrational and rotational spectroscopy can be emitted from various sorts of "lamps".

In these experiments it is obviously desirable to have an intense source, as the size of sample required or the time taken to perform an experiment can thereby be minimized. It is also important to have radiation that is as monochromatic as possible, and this can be achieved by using a broad-band source with some monochromating device, or by using a narrow-band source whose output frequency can be varied. If the source is also coherent (that is, its output radiation is in phase) or polarized, then we may be able to perform special experiments utilizing these properties.

There are now many kinds of lasers available, producing radiation with energies stretching from the far infrared region, through the infrared and visible and into the ultraviolet region, but excluding microwaves at one extreme and X-rays at the other. The radiation produced by these lasers is generally intense, monochromatic, coherent, polarized and directional. The total power output varies enormously from one type of laser to another, but it is always possible to deliver a high flux of photons with a particular energy (very narrow linewidth) to a particular place. The probability of a molecule absorbing a photon is correspondingly high, and the probability of absorbing *two* quanta increases as the square of the flux. Thus lasers, which can be valuable in any form of spectroscopy, have allowed double quantum methods to be developed. These may involve two identical quanta, or two different ones, using two lasers. It is in such experiments that the coherence of the source can be utilized.

The output frequency of some lasers can be tuned, and these can be used in a conventional way to scan a spectrum. One application of this is in the use of a diode laser to obtain very high resolution infrared spectra. The resolution is superb (10^{-4} cm^{-1}) but the range of tuning is very limited for a particular diode, so that spectra can only be obtained over a wide range by using many diodes. In other cases, the laser can only produce a single, fixed frequency, and this obviously limits the range of applications. But such lasers are still valuable. In Raman spectroscopy, for example, all that is required is an intense, monochromatic source with an energy somewhere in the visible or ultraviolet region. For colored samples, the frequency of the source should fall outside the absorption bands in the spectrum; in resonance Raman spectroscopy, it is necessary to use a frequency that falls within an absorption band of the species under study. In this case several laser frequencies may need to be available to cover all possibilities, but again the precise frequency is immaterial.

Another difference between lasers is that some produce continuous output, whereas others are pulsed. Both have advantages for particular experiments. Continuous output is useful in Raman spectroscopy, for example, but pulsed lasers can be used in experiments involving short-lived species. The short pulse of radiation can be used to produce an unstable chemical species, or an unstable excited state, and its subsequent behavior can then be investigated. The observation of the species may well involve the use of a second laser. Thus we may be able to obtain an excited electronic state using a laser operating in the visible or ultraviolet regions of the spectrum, and then use an infrared laser to study its vibrational spectrum.

Another increasingly important source of radiation is the electron storage ring, which is essentially a very large and exceedingly expensive light bulb, producing synchrotron radiation. It consists of an evacuated ring, which may be some tens of metres in diameter, into which pulses of high energy electrons are introduced from an accelerator. The electrons are deflected by a series of magnets so that they move round and round the ring, and as they are being continuously accelerated in this way, they emit radiation tangentially to the ring. The radiation is polarized and is also pulsed, as at any one point in the ring, radiation is only emitted as each bunch of electrons goes by. It is possible to arrange that the pulses are so close together that we can regard the storage ring as a continuous source.

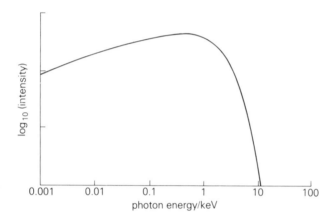

Fig 1.3 Typical spectrum, showing the output of synchrotron radiation from a storage ring.

The output of a synchrotron storage ring is illustrated by Fig. 1.3. The range of energies produced is enormous, with useful radiation over about eight orders of magnitude, and over most of that range the intensity is very high. Thus the attractions of the synchrotron lie in the strength of the source at a chosen energy, the fact that the output lies only in the plane of the ring and so is partially collimated, and in the almost unlimited choice of energies. In the X-ray region, the high intensity compared with that attainable using conventional X-ray tubes enables data to be collected very quickly, and very small crystals to be studied (see Section 8.5). The choice of energies (wavelengths) allows investigation of phenomena associated with absorption of X-rays by a particular element, and this may help in solving complex structures. The absorption spectrum in the vicinity of an absorption edge is also utilized in Extended X-ray Absorption Fine Structure (EXAFS) studies, which can yield structural information about surfaces, liquids and solutions, as well as amorphous and crystalline solids (Section 8.9). The range of elements having suitable absorptions is limited by the fall-off of output shown at the high energy end of Fig. 1.3. Some improvement can be made by using devices called wigglers and undulators, which cause the electron beam to turn with a smaller radius than normal. The upper limit of frequencies emitted thus rises, and it is also possible to superimpose output from several wiggles, and so to gain intensity.

In the ultraviolet region, the main benefit is in the unlimited range of energies available, as here the choice of lasers is limited. However, in the visible region, tunable lasers are often preferable, while in the infrared region, sources such as a glowbar are just as powerful, and tunable lasers are inherently monochromatic. In the far infrared, or millimeter wave region, the storage ring can offer perhaps one order of magnitude more intensity than a conventional source. In the microwave region, the storage ring output is no longer intense enough to be of importance.

One other characteristic of synchrotron radiation can also be used in spectroscopic studies. If just one bunch of electrons circulates, the output radiation consists of pulses, a few hundred picoseconds in length, separated by a few hundred nanoseconds. Thus time-dependent studies of short-lived species are possible; and given the wide range of energies that are output, all sorts of sophisticated multiple resonance experiments are also feasible.

Techniques involving the use of neutrons, such as neutron diffraction (Section 8.7) and inelastic neutron scattering (Section 5.3), have been limited by the low fluxes of velocity-selected and collimated neutrons available from fission reactors. A recent development is the construction of purpose-built neutron sources involving processes such as spallation. Such facilities are so expensive that they can be constructed only by national or international organizations, but they have made experiments easier and faster.

Several new techniques for generating gas-phase ions from involatile samples are now commercially available in mass spectrometers. This development has greatly extended the range of applicability of the technique, which is discussed in Chapter 9.

1.5.2   Detectors

The recent developments in sources have been accompanied by advances in the detectors used, but as yet there has been nothing quite so dramatic as the advent of the laser source. New devices using semiconductors and semiconductor junctions ("diodes") to detect radiation, particularly in the infrared region, have appeared, and special detectors for particles and radiation with sensitivity, noise level and response time especially suited to particular experiments have been developed. But the old familiar devices, such as thermocouples in the IR, photomultipliers in the visible/UV, ionization detectors for X-rays and photographic plates, are still widely used.

One area of development that holds great promise is that of position-sensitive electronic detection. Here a linear or 2-dimensional array of semiconductor devices gives a signal depending on the radiation or particle flux at each position, allowing an electronic analog of photographic recording. At its simplest this might allow simultaneous recording of radiation over a range of wavelengths after dispersion using a prism or grating, while it is also possible to record an entire diffraction pattern from a diffraction experiment using such a device. One particularly exciting development of this kind is in X-ray crystallography. In a conventional diffractometer, the intensity of each reflection is measured in turn, and collection of a complete data set may take hours or days. By using position-sensitive detectors this total time can be reduced to minutes or even seconds, and by using a powerful synchrotron source as well, the possibility of studying structural changes that occur over a period of a few seconds or less has been opened up.

Electronics and computers

Much more important than the development of new detectors as such has been the development of the electronics for handling the signals produced by detectors. Above all, the microprocessor, essentially a computer on a single multi-element silicon "chip", has allowed a vast increase in the capacity and sophistication of signal handling, and there is at present no sign of any slackening in the rate of increase. In the past the ideal signal from a detector was a steady voltage which could be used to drive a mechanical system such as a servo-mechanism controlling a pen, whose deflection was a measure of the size of the signal. Nowadays the signal from a detector is more likely to be converted into digital form so that it can be transmitted, stored, retrieved and manipulated in the form of a string of binary digits. The sheer speed at which data can be

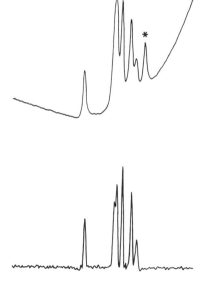

Fig 1.4 NMR spectrum, before and after computer manipulation to improve its appearance. Resolution has been improved by line-narrowing, the baseline has been flattened, and a peak due to an impurity (marked *) has been removed.

collected today is very hard to comprehend; thousands or even hundreds of thousands of eight-bit or sixteen-bit numbers can be recorded per second, so that detectors with very short response times are needed to allow full advantage to be taken of the data-handling capabilities of the electronics. The full advantage of such systems is shown best in instruments using Fourier transform techniques (see Subsection 1.5.4 below) and in experiments with very short timescales, such as those involving pulsed lasers, but microprocessor technology can also be useful in manipulating more conventional spectra.

Some important areas where even conventional spectra can benefit from microprocessor involvement are:

spectrum storage and retrieval;
spectrum "massage" (removal of backgrounds, of noise and of unwanted features, and modification of line shape − see Fig. 1.4). Note that this cannot give any new information. It may take some features more obvious to the eye, but the effects are really only cosmetic;

CHAPTER 1: DETERMINING STRUCTURES

spectrum comparison, subtraction and ratioing;

spectrum simulation;

spectrum reduction (conversion to a list of peak positions and intensities, for instance).

Many commercial instruments now incorporate a microprocessor both to allow such processing of spectra and to control the actual acquisition of the spectra, using time-delay signals to control power pulses or grating movement, logic signals to control filter changes, and so on. In Fourier transform instruments (see below) the same microprocessor also performs the mathematical operations required to generate the spectrum from the detector output.

1.5.4	**Fourier transform techniques**

The simplest way of recording a spectrum is to use a tunable monochromatic source, whose output is passed through a sample onto a detector. The absorption spectrum of the sample is then given by plotting I/I_0 (the ratio of the transmitted intensity to the source intensity) against the source frequency. However, few monochromatic sources that can be tuned over a wide range are available, even now, and most spectra are recorded using broad-band sources, whose output contains all frequencies of interest. There is then the problem of how to separate the information about I/I_0 at each frequency from the overall transmission $\Sigma I/\Sigma I_0$. The usual method over most regions of the spectrum, at least in the range from UV to IR, is to use a physical device, such as a refracting prism or diffraction grating, which separates rays of different frequency in a spatial sense. The transmitted spectrum $I(v)$ can then be recorded all at once using a position-sensitive detector such as a photographic plate. Alternatively, the spatial separation can be converted into a time-domain separation by scanning, that is moving the detector or the dispersing element so that rays of successive frequencies are detected in sequence. Photographic plates are not suitable for recording IR radiation, and scanning prism or grating spectrophotometers have been used almost exclusively in this region until fairly recently.

There is, however, another way of separating the information according to frequency, which involves the interference phenomena that are also responsible for the dispersive effect of a diffraction grating. It can be shown that if interference effects can be made to alter the amplitude of the total radiation reaching the detector as a function of *time*, a purely mathematical manipulation, called a Fourier transform, will convert this record of amplitudes $A(t)$ into the desired spectrum $I(v)$. The process is also called Fourier inversion, as there is an inverse relationship between time and frequency. As mentioned above in Subsection 1.5.3, the Fourier transformation is carried out using a computer, which may also be used for other functions such as controlling the operation of the instrument and storage and comparison of spectra.

Fourier transform (FT) techniques are now of great importance in two areas: NMR spectroscopy and IR spectroscopy. The physical basis of the interference effects is different in the two areas. In NMR, the interference is generated during the relaxation of nuclear spins which have been aligned in a strong magnetic field by a short pulse of radiation of a chosen frequency, v_0, close to but not coinciding with the expected resonance frequency of the nuclei in the sample (v_s). The interference between v_0 and v_s gives beats of frequency $v_0 - v_s$. Fourier transformation of such beat patterns then generates a spectrum $I(v)$ whose zero frequency corresponds to v_0, *i.e.* a spectrum of

$v_0 - v_s$. More details of the application of FT techniques in NMR will be found in Chapter 2, and there is a clear account of the basis of the method in the review [ref. 2] cited in that chapter.

In FT infrared spectroscopy, the interference effects are generated by recombination of beams that have travelled different distances, in a Michaelson interferometer. The interference is between two sets of rays of frequency v whose paths differ by a distance x. If x is zero or a whole number of wavelengths, *constructive* interference occurs, whereas if x is an odd number of half-wavelengths, *destructive* interference occurs. The path difference x is scanned and the amplitude record $A(x)$ is then converted by the Fourier transformation into a spectrum $I(v)$. More details of the application of FT techniques in IR spectroscopy will be found in Chapter 5 and in the book [ref. 6] cited therein.

There is in principle no reason why the same methods should not be used in the visible and UV regions as in the IR, except that increasing mechanical precision is required as the frequency v rises: precision in placing and orientating the mirrors to the order of one quarter of the wavelength is necessary. A few commercial FT instruments are now available that can cover the visible region and part of the UV region as well as the IR. FT techniques have been applied in the microwave region as well; the basis of the method is more akin to that used in NMR than that used in the IR. No commercial instruments are available.

We should also note that diffraction methods all depend on interference effects, and that obtaining structural information from the observed patterns therefore involves Fourier transformation. For data obtained for gases or liquids the process is exactly analogous to that involved in FT NMR or IR spectroscopy, but for crystal data a three-dimensional Fourier transform is required, to generate a three-dimensional map of the positions of atoms in the crystal.

In all applications of FT techniques there are certain inescapable rules, which may at first seem arbitrary or even perverse. A careful consideration of the inverse nature of frequency and time will, however, show how they arise. First, the *resolution* (δv) obtainable in the frequency spectrum $I(v)$ is governed by the total extent over which data is collected in the time-domain, or the maximum path difference in the interferometer, t_{max} or x_{max}:

$$\delta v \propto \frac{1}{t_{max}}, \text{ or } \frac{1}{x_{max}}. \tag{1.1}$$

Secondly, the total extent of the frequency spectrum, v_{max}, is governed by the sampling interval, or data-point separation, in the recording of $A(t)$ or $A(x)$, δt or δx:

$$v_{max} \propto \frac{1}{\delta t}, \text{ or } \frac{1}{\delta x}. \tag{1.2}$$

Thus to extend the frequency-range covered by our final spectrum we must collect data points more closely-spaced, while to improve the resolution of the spectrum we must collect data points over a wider range of times or path-differences. Conversely, if we know that all bands in our spectrum are likely to be broad, we need only collect data points over a short range without degrading the spectrum, while if we are only interested in low frequencies we can use widely-spaced data points.

CHAPTER 1: DETERMINING STRUCTURES

Finally, we should note that each point in the amplitude record $A(t)$ contains information about every point in the spectrum $I(v)$. If $A(t)$ can be recorded rapidly, as is the case in FT NMR experiments, the Fourier transform method may enable spectra to be recorded much more rapidly than by direct observation of absorption.

1.6   Glossary of spectroscopic and structural techniques

Technique	Description	Application	Ref.
AAS Atomic absorption spectroscopy	Electronic spectra of atoms in flames	Elemental analysis	[2]
AES Auger electron spectroscopy	Velocity spectra of secondary photoelectrons ejected from solids after absorption of high-energy radiation	Study of solid surfaces, giving identity of atoms and information about their local environments	[3]
CARS Coherent anti-Stokes Raman spectroscopy	Non-linear two-photon process giving very intense Raman spectra with intensity-patterns and selection-rules different from those for conventional Raman spectra	Studying species in flames, biological systems, obtaining rotational and vibration–rotation Raman spectra	[4]
CD Circular dichroism	Study of difference in absorption coefficients for left and right circularly polarized light	Studying optically active species, particularly complexes of transition metals	[5] Sect. 6.8
CEMS Conversion electron Mössbauer spectroscopy	Velocity spectra of back-scattered electrons ejected from thin films or surfaces by internal conversion of Mössbauer-excited nuclei	Study of surfaces and thin films by Mössbauer techniques, with some possibility of depth resolution	Sect. 7.6
Chemical ionization mass spectrometry	see Mass spectrometry		
CIDEP/CIDNP Chemically-induced dynamic electron/nuclear polarization	Generation of products or intermediates in chemical or photochemical reactions with non-equilibrium distribution of electron or nuclear spins, thus creating abnormal ESR or NMR signal	Study of transient intermediates	[6]
COSY Correlation spectroscopy	2-Dimensional NMR technique	Gives information about couplings between nuclei	Sect. 2.11.4
Coulomb explosion	Mass spectral fragmentation pattern of molecular beams of ions that fragment spontaneously after multiple electron-loss by collision with electrons in thin foils	Geometries of molecular ions deduced from joint energy and angular distribution of fragments	[7]

Technique	Description	Application	Ref.
CP-MAS(S) Cross polarization magic angle (sample) spinning	Method for narrowing lines in NMR spectra of nuclei, often of low natural abundance, in solids using pulsed heteronuclear double resonance combined with magic angle spinning (*q.v.*)	Important for ^{13}C NMR spectra of solids; useful for many other nuclei	Sect. 2.14
CRAMPS Combined rotation and multi-pulse spectroscopy	Method for narrowing lines in NMR spectra of abundant nuclei in the solid phase	Important for ^1H and ^{19}F NMR of solids	Sect. 2.14
DANTE Delays alternating with nutations for tailored excitation	Used in NMR to obtain resonances due to *one set* of chemically equivalent nuclei	Simplifies complex spectra, and used to study exchange processes	Sect. 2.11.3
DEPT Distortionless enhancement by polarization transfer	Enhances intensities in some NMR experiments; used to select resonances from only those nuclei coupled to 3, 2, 1 or no protons	Distinguishes ^{13}C resonances in CH_3, CH_2, CH groups *etc.*	Sect. 2.11.2 and 2.11.3
2D-NMR 2-Dimensional nuclear magnetic resonance spectroscopy	Extension of NMR, presenting resonances as function of two frequency parameters	Useful in analyzing complicated 1D-NMR spectra, and determining connectivity in unknown structures	Sect. 2.11.4
ED Electron diffraction	Diffraction of electrons, usually by gases	Gives distances between atoms	Chap. 8
EDX (sometimes EDAX) Energy dispersive X-ray spectroscopy	X-ray fluorescence (*q.v.*) of solid samples when bombarded with electrons	Used in electron microscopes, often in conjunction with EELS, as analytical tool	
EELS Electron energy-loss spectroscopy	Inelastic scattering of electrons from gas or molecular beam samples	Electronic or vibrational energy spectra. Particularly important in studies of surfaces by electron microscopy	Sect. 5.3.4
ELDOR Electron−electron double resonance	ESR double resonance experiment	Measurement of hyperfine splitting constants in complex spectra	Sect. 3.5
Electronic spectroscopy	Absorption or emission of radiation, usually in visible and UV regions, accompanying changes in electronic state	Electronic energy spectra, which may contain vibrational and rotational detail for both upper and lower states	Chap. 6
Electron impact mass spectrometry	see Mass spectrometry		

Technique	Description	Application	Ref.
EM Electron microscopy including SEM (scanning), TEM (transmission), STEM (scanning transmission) and HREM (high resolution)	Use of electrons as radiation to obtain micrographs of surfaces	Reveals extremely fine detail (resolution almost to molecular level) of surface structure and defects	[8]
ENDOR Electron−nuclear double resonance	Observation of ESR spectrum while irradiating nuclear spin transitions	Simplification and interpretation of spectra	Sect. 8.5
EPR Electron paramagnetic resonance spectroscopy	Another name for electron spin resonance spectroscopy (*q.v.*), particularly applied to radicals in gas phase or in molecular beams		
ESCA Electron spectroscopy for chemical analysis	Another name for X-ray photoelectron spectroscopy (*q.v.*)		
ESR Electron spin resonance spectroscopy	Observation of transitions between energy levels associated with unpaired electrons in magnetic field	Study of any species having at least one unpaired electron	Chap. 3
EXAFS Extended X-ray absorption fine structure	X-ray absorption spectra in region of absorption edge of particular element	Gives distribution of inter-atomic distances around atoms of that element − for solids, liquids, solutions, surfaces *etc.*	Sect. 8.9
EXSY Exchange Spectroscopy	2-Dimensional NMR technique, the same as NOSEY (*q. v.*)	Gives information about rates of exchange reactions	Sect. 2.15.2
Far infrared spectroscopy	IR spectroscopy below 200 cm^{-1}	Study of deformation modes involving heavy atoms, stretching modes involving very heavy atoms, torsional modes, *etc.*	Chap. 5
Far infrared laser magnetic resonance	see Laser magnetic resonance		
Fast atom bombardment (FAB) mass spectrometry	see Mass spectrometry		
Field desorption mass spectrometry	see Mass spectrometry		
Fluorescence spectroscopy	Analysis of fluorescence from samples irradiated by visible or UV light − see also LIF and electronic spectroscopy		Chap. 6

Technique	Description	Application	Ref.
FTIR Fourier transform infrared spectroscopy	Infrared spectroscopy, using Fourier transform spectrometer		Sect. 5.4
GED Gas-phase electron diffraction	see Electron diffraction		
HETCOR (HCOR) Heteronuclear correlation spectroscopy	Heteronuclear 2-dimensional NMR technique; related to COSY (q. v.)	Gives information about heteronuclear couplings	Sect 2.11.4
HOESY Heteronuclear Overhauser effect spectroscopy	Heteronuclear variant of NOESY (q. v.)	Gives information about proximity of nuclei in space, or exchange processes	Sect. 2.11.4
HOHAHA Homonuclear Hartmann–Hahn experiment	2-Dimensional NMR technique. The same as TOCSY (q. v.)	Correlates all spins in any NMR spin system	Sect. 2.11.4
Hyper Raman spectroscopy	Scattering under irradiation by giant-pulse laser with low irradiance	Selection-rules are different from those for both IR and conventional Raman, so it can reveal transitions forbidden in both	[4]
ICR Ion cyclotron resonance spectroscopy	Mass analysis using cyclo- tron resonance frequency of ion in magnetic field	Very sensitive mass spectro- metric detection; prepara- tion of pure ion samples for further spectroscopy; study of ion–molecule reactions	[9]
INADEQUATE Incredible natural abundance double quantum experiment	NMR technique for obser- vation of homonuclear satellites of low-abundance spinning isotopes	Enables pattern of linkages between nuclei to be built up	Sect. 2.11.3 and 2.11.4
INDOR Inter-nuclear double resonance	Double resonance NMR technique	Gives information about one set of nuclei while observing resonances of another set	Sect. 2.10.2
INEPT Insensitive nuclei enhancement by polarization transfer	Superseded by DEPT (q.v.)		
INS Inelastic neutron scattering	Energy-loss spectroscopy using neutrons – cf. EELS	Vibrational energy spectra	Sect. 5.3.4
Inverse Raman spectroscopy	Stimulated absorption under simultaneous irradiation by giant-pulse laser beam and continuum	Study of short-lived species (spectrum obtained in 10^{-8}-10^{-11} s). Experimentally very difficult	[4]

Technique	Description	Application	Ref.
IR Infrared spectroscopy	Absorption or emission of radiation in IR region, usually associated with changes in vibrational state	Vibrational and vibration–rotational energy spectra	Chap. 5
Laser desorption mass spectrometry	see Mass spectrometry		
LCNMR Liquid crystal nuclear magnetic resonance spectroscopy	NMR using liquid crystals as solvents	Gives geometrical information about solute	Sect. 2.13
LEED Low energy electron diffraction	Diffraction of low-energy electrons (< 100 keV)	Study of surfaces	[3] and [10]
LIF Laser-induced fluorescence	Fluorescence spectroscopy using laser source for excitation. If source laser is tunable, extra information is obtainable about *excitation spectrum*	Simplification of electronic spectra	Sect. 6.7.1
LMR, LZS Laser magnetic resonance spectroscopy or laser Zeeman spectroscopy	Use of magnetic fields to tune energy levels of free radicals into resonance with fixed-frequency IR or far-IR laser, whose absorption is detected	Detection, identification and detailed spectroscopy of free radical species	[11]
MAS(S) Magic angle (sample) spinning (also called MAR)	Technique for narrowing NMR lines due to solids. CP-MASS is same plus cross-polarization	Enables high resolution NMR spectra of solids to be obtained	Sect. 2.14
Mass spectrometry	Analysis by mass and charge of molecular ions and their breakdown products. Often prefixed by a term describing means of obtaining ions	Gives molecular weights, and fragmentation patterns may help in structure (connectivity) elucidation. Can also give bond dissociation energies *etc.*	Chap. 9
Matrix isolation	Preparation of sample of material in frozen matrix of chemically-inert substance for ease of subsequent spectroscopic examination by IR, Raman, ESR or electronic spectroscopy	Study of reactive, unstable and short-lived species; also allows study of IR spectra of isolated molecules of stable species without rotational bands	Sect. 5.16.3
MCD Magnetic circular dichroism	Study of difference in absorption coefficients for left and right circularly polarized light of sample in magnetic field	Can reveal electronic transitions not resolved by conventional electronic spectroscopy, and provide information about electronic and magnetic properties of excited states and ground states	[12], Sect. 6.8

Technique	Description	Application	Ref.
Microwave spectroscopy	Absorption spectroscopy of gaseous samples in 3-60 GHz region $(0.1-2.0 \text{ cm}^{-1})$ of electromagnetic spectrum	Pure rotational spectra of gaseous molecules, low-J values	Chap. 4
MIKES Mass-analyzed ion kinetic energy spectroscopy	Mass spectrum showing all daughters of a particular ion	Detailed study of fragmentation processes in mass spectrometer	Sect. 9.4
Millimetre wave spectroscopy	Absorption spectroscopy of gaseous sample in 60-600 GHz region (5-0.5 mm wavelength)	Pure rotational spectra of gaseous molecules, to higher J values than in microwave	Chap. 4
MODR Microwave optical double resonance spectroscopy	Simultaneous irradiation of sample in microwave and IR regions, using changes in signal in one region to identify absorption in other region as second frequency is scanned	Detailed analysis of rotational spectra in vibrationally-excited states of small molecules	[12]
Mössbauer spectroscopy	Absorption of nuclear gamma rays by nuclei	Gives information about chemical environment (e.g. oxidation state) of, and magnetic interactions and electric field gradient at, nuclei	Chap. 7
MPI (REMPI) Multiphoton ionization spectroscopy (resonance enhanced)	Detection of formation of ions following absorption of several photons. May use one or two excitation frequencies, one of which is tunable	Can detect and characterize intermediate excited states. Used as species-specific ionization mechanism for mass spectrometry	[13]
MS/MS Multi-stage mass spectrometry	Two mass spectrometers, used to obtain and select particular ion, which is then excited further, so that its fragmentation can be studied	Selective study of ions, and powerful analytical tool	Sect. 9.4
Multiphoton induced fluorescence spectroscopy	Study of fluorescence from very highly-excited electronic states following one- or two-color multiphoton excitation	Characterization of highly-excited states	[11]
Neutron diffraction	Diffraction of neutrons, mainly by crystalline solids	Structure determination, particularly useful for locating hydrogen atoms	Sect. 8.7
NGR Nuclear gamma resonance spectroscopy	Another name for Mössbauer spectroscopy $(q.v.)$		
NMDR Nuclear magnetic double resonance spectroscopy	As NMR, but irradiating with two radiofrequencies	Many applications, but particularly useful for simplifying very complex NMR spectra	Sect. 2.10

Technique	Description	Application	Ref.
NMR Nuclear magnetic resonance spectroscopy	Observation of transitions between energy levels associated with nuclear spins in magnetic field	Many applications, particularly in identification of known and unknown compounds	Chap. 2
NOE (Nuclear Overhauser Effect) difference spectroscopy	Very low power double resonance NMR experiment	Information about groups of protons that are close in space, in small molecules	Sect. 2.10.6
NOESY Nuclear Overhauser spectroscopy	2-Dimensional NMR technique	Information about groups of protons that are close in space to each other in large molecules	Sect. 2.11.4
NQR Nuclear quadrupole resonance spectroscopy	Involves energy levels arising from nuclear quadrupole moments of nuclei with spin of 1 or more	Information about environments of quadrupolar nuclei in solids	Chap. 3
ODR (OODR) Optical double resonance spectroscopy	Observation of changes in absorption or emission in IR or visible/UV spectrum as a second intense frequency is tuned across another part of spectrum	Assignment of rotational levels of vibrationally- or electronically-excited molecules	[11]
ORD Optical rotary dispersion	Study of way in which rotation of plane of polarization of polarized light varies with wavelength	Study of optically active species and their electronic structures	[5] Sect. 6.8
PE, PES Photoelectron spectroscopy	General name for spectroscopic methods in which electrons are ejected as result of absorption of radiation and their energy-distribution is analyzed. See also Auger spectroscopy, ultraviolet photoelectron spectroscopy, X-ray photoelectron spectroscopy, conversion electron Mössbauer spectroscopy	Energies of occupied electronic levels (UVPE); identity and environment of atoms in sample (Auger, XPS, CEMS)	Chap. 6
Photoionization mass spectrometry	see Mass spectrometry		
Raman spectroscopy	Observation of emitted radiation from sample irradiated by intense monochromatic light in visible or UV regions	Rotational and vibrational spectra of sample, with different selection rules from IR absorption/ emission	Chap. 4 and 5
RIKES Raman-induced Kerr-effect spectroscopy	Coherent two-color Raman experiment in which rotation of plane of polarization of one beam is monitored as frequency of other beam is altered	Enhanced sensitivity, with normal Raman selection rules	[14]

Technique	Description	Application	Ref.
ROESY Rotating frame Nuclear Overhauser Effect spectroscopy	2-Dimensional NMR technique	As NOESY (*q. v.*) but applied to smaller molecules	
Rotational Raman spectroscopy	see Raman spectroscopy		Chap. 4
Rotational spectroscopy	Study of transitions involving changes in rotational quantum numbers of gas molecules, using IR, microwave, millimeter wave, sub-millimeter wave and Raman techniques (*q.v.*)	Can give moments of inertia of molecules, and hence molecular structures	Chap. 4
RRS Resonance Raman spectroscopy	Raman spectroscopy (*q.v.*) performed with an irradiation frequency that is absorbed by sample	Enhanced signal, often in one or a few bands, compared with normal Raman spectrum. May include short or long progressions in one or more vibrational frequencies	Sect. 5.17
SERS Surface-enhanced Raman spectroscopy	Raman spectroscopy of adsorbed species on surfaces		[15]
SEXAFS Surface extended X-ray absorption fine structure	Same as EXAFS	Applied to surfaces	
SIMS Secondary ion mass spectrometry	see Mass spectrometry		
Stimulated Raman spectroscopy	Scattering under irradiation by giant-pulse laser with high irradiance	Spectra differ much from conventional Raman: often only one mode excited; special angular dependence of scattering; lines may be very intense	[4]
TOCSY Total correlation spectroscopy	2-Dimensional NMR technique. Same as HOHAHA	Correlates all spins in any NMR spin system	[16]
TRRR, TR³	Resonance Raman of electronically excited states produced by pulsed laser excitation	Vibration frequencies and excitation profiles for electronically-excited states	[17]
UVPE or UPS ultraviolet photoelectron spectroscopy	Velocity analysis of photo-electrons emitted from sample irradiated by monochromatic UV source	Energies of occupied electronic levels; band shapes may indicate bonding/non-bonding/antibonding character of levels	Chap. 6
UV/visible spectroscopy	Absorption or emission in UV or visible regions due to electronic transitions	Energies of electronic transitions; may also contain resolved vibrational or rotational detail	Chap. 6

Technique	Description	Application	Ref.
Vibrational spectroscopy	Study of changes in vibrational states, most often by IR absorption or Raman scattering ($q.v.$)	Many applications, including identification of known compounds and of functional groups in unknown compounds	Chap. 5
Visible spectroscopy	Absorption or emission by colored samples in visible region due to electronic transitions	Energies of electronic transitions; may also contain resolved vibrational or rotational detail	Chap. 6
XANES X-ray absorption near edge structure	Essentially same as EXAFS	As for EXAFS: also gives information about high-energy electronic transitions	Sect. 8.9
X-ray diffraction	Diffraction of X-rays, usually by crystalline solids	Gives precise structural information (atom positions) and some information about atomic motions	Chap. 8
X-ray diffuse scattering	Diffraction of X-rays by disordered or partially ordered systems	Gives geometrical information, but often hard to interpret	Sect. 8.4
XRF X-ray fluorescence spectroscopy	Emission of characteristic X-rays from solid samples excited by bombardment with electron beam	Detection and measurement of atomic composition of samples	[18]
XPS X-ray photoelectron spectroscopy	Velocity analysis of photoelectrons emitted from sample irradiated by X-ray source	Energies of occupied valence and core electronic levels; identities and environments of constituent atoms in sample	Chap. 6

References

1 A. Werner, *Liebig's Annalen der Chemie*, **375**, 1 (1910).
2 K.C. Thompson and R.J. Reynolds, *Atomic Absorption, Fluorescence and Flame Emission Spectroscopy*, Charles Griffin and Co., London (1978).
3 F.C. Tompkins, *Chemisorption of Gases on Metals*, Academic Press, New York (1978).
4 D.A. Long, *Raman Spectroscopy*, McGraw-Hill, New York (1977).
5 R.D. Gillard, in *Physical Methods in Advanced Inorganic Chemistry* (eds H.A.O. Hill and P. Day), Interscience, London (1968), p. 167.
6 J.K.S. Wan, *Adv. Photochem.*, **12**, 283 (1980).
7 D.S. Gemmell, *Chem. Rev.*, **79**, 233 (1979).
8 I.M. Watt, *The Principles and Practice of Electron Microscopy*, Cambridge University Press, Cambridge (1985).
9 L.R. Thorne, V.G. Anicich and W.T. Huntress, *Chem. Phys. Letters*, **98**, 162 (1983).
10 H.E. Bauer, R.W. Schmid and H. Seiler, *J. Phys. E*, **15**, 277 (1982).
11 J.M. Hollas, *High Resolution Spectroscopy*, Butterworths, London (1982).
12 P.J. Stephen, *Ann. Rev. Phys. Chem.*, **25**, 201 (1974).
13 D.H. Parker, *Ultrasensitive Laser Spectroscopy*, Academic Press, New York (1983), chap. 4.
14 M.D. Levenson, *Physics Today*, **44** (May 1977). H.W. Schroetter, *NATO Adv. Study Inst. Ser., Ser. C*, **93**, 603 (1982).
15 R.K. Chang and T.E. Furtak (eds), *Surface Enhanced Raman Spectroscopy*, Plenum Press, New York (1982).
16 A. Bax and D.G. Davis, *J. Magn. Reson.*, **65**, 335 (1985).
17 P.C. Bradley, N. Kress, B.A. Hornberger, R.F. Dallinger and W.H. Woodruff, *J. Am. Chem. Soc.*, **103**, 7441 (1981).
18 E.P. Botin, *Introduction to X-ray Spectrometric Analysis*, Plenum Press, New York (1978).

2 Nuclear Magnetic Resonance Spectroscopy

2.1 Introduction

It is probably safe to say that, of all the structural and spectroscopic methods described in this book, none has had the impact of nuclear magnetic resonance (NMR) spectroscopy. In the early days, studies were limited mainly to ^{1}H, with ^{19}F, ^{31}P and ^{11}B playing minor roles, but even so this method was unique, in that it was possible to count spinning nuclei by looking at multiplet coupling patterns. Moreover, the spectra enabled quantitative measurements of nuclear concentrations to be made, and chemical shifts and coupling constants gave information about chemical environments of nuclei.

The introduction of Fourier transform methods around 1970 enabled many more nuclei to be observed, and double resonance methods, particularly noise decoupling, greatly simplified spectra of complex systems, particularly for ^{13}C in organic compounds. Relaxation could also now be readily studied. Since then we have seen dramatic improvements in computer technology, giving easier acquisition and better presentation of spectra. There have also been major advances in electronics and probe design, leading to vast gains in signal to noise ratios, and in magnet technology, so that high field superconducting magnets are now commonplace, with valuable consequences for spectral dispersion and sensitivity.

Thus small samples and very complicated systems can be studied, and developments have enabled complex spectra to be simplified. For example, it is possible to run separate ^{13}C spectra showing resonances arising from carbon atoms bonded to three hydrogens, or to two, or one, or none (see Section 2.11.3). This can be of great help in assigning complex spectra, and 2-dimensional spectra (Section 2.11.4) may also be useful in this respect. Another development enables an insensitive nucleus to be studied effectively using the sensitivity of some other nucleus to which it is coupled. Very important developments which have also been made in NMR spectroscopy of solids (Section 2.14) extend the usefulness of the technique even further.

With such a diverse range of methods and applications of NMR, it is clear that it is not possible, in a book of this size, to go into the theory in detail, and many applications, such as those in metallurgy and medicine, are not even mentioned. Our aim is to show what information about the structures of inorganic compounds can be obtained, using the routine NMR methods, and the more esoteric tricks devised for us by the specialists. It is very unlikely that any reader of this book will have no experience at all of NMR. So for the simpler aspects of the subject which are covered in the early part of the chapter, up to Section 2.7, we have avoided lengthy descriptions of the theoretical basis of the method, and of instrumentation. In the latter part of the chapter many different types of NMR experiment are described, and it would take a great deal of space to explain the theory behind all these techniques. Throughout the chapter, therefore, we have aimed to provide enough of the theory to enable you to understand what is going on, and plenty of specific references to books and reviews that expand particular topics in detail. There is also a more general bibliography at the end of this chapter. Most of these books go into the theory and experimental aspects in much greater depth than we can here.

It may seem that there is a bias in this chapter in favor of the novel, the obscure, and the bizarre. But the importance of a topic is not proportional to the space given to it. Of course, it is of more general relevance to know that a nucleus coupled to n equivalent spin 1/2 nuclei will have an $(n + 1)$ line multiplet resonance than to understand the niceties of triple resonance. But both are described here, so that when faced with an apparently intractable structural problem, you may know how to assemble as much relevant information as possible. What appears now to be obscure and irrelevant may soon become routine. In our own laboratory, our whole approach to studying ^{13}C, ^{29}Si and some other nuclei was revolutionized when we first started to use the DEPT multipulse sequence, which until then had seemed to be of only minor importance. In many cases we observed spectra, where previously there was nothing but noise!

Although some of the latest ideas have mainly been applied to problems in organic chemistry or biochemistry, we have included them here, in the belief that once the principles are understood, and the information to be obtained has been illustrated, the applications to inorganic systems will follow.

2.2 The nuclear magnetic resonance phenomenon

The property of a nucleus known as its spin is the basis of nuclear magnetic resonance spectroscopy. For every isotope of every element there is a ground state nuclear spin quantum number, I, which has a value of $n/2$, where n is an integer. Isotopes having atomic and mass numbers that are both even (e.g. ^{12}C, ^{28}Si, ^{56}Fe) have $I = 0$, and these nuclei have no NMR spectra. Isotopes with odd atomic number but even mass number (e.g. ^{2}H, ^{10}B, ^{14}N) have n even, while those with odd mass numbers (many examples, such as ^{1}H, ^{13}C, ^{19}F, ^{55}Mn) have n odd. Some of this last group have values of $I = 1/2$, and these are the nuclei most commonly studied by NMR, but spin quantum numbers of 3/2, 5/2, 7/2 and 9/2 also occur frequently.

When I is non-zero, the nucleus has a magnetic moment, μ, given by the equation:

$$\mu = \gamma \hbar \, [I(I + 1)]^{1/2} \tag{2.1}$$

where γ is the magnetogyric ratio, a constant characteristic of the particular isotope. If $I > 1/2$ the nucleus also has a nuclear quadrupole moment. In the presence of a strong magnetic field, B_0, the spin axis orientation is quantized, with magnetic quantum numbers m taking values of $I, I - 1, I - 2 \ldots -I$. Irradiation at an appropriate (radio) frequency causes transitions, with a selection rule $\Delta m = -1$, and it is these transitions that are observed in nuclear magnetic resonance spectroscopy.

For the simplest liquids, solutions and gases under normal conditions, the stationary state energies, E, are given by:

$$E = -h \sum_A \nu_A m_A + h \sum_{A<B} J_{AB} m_A m_B. \tag{2.2}$$

Here ν_A is a frequency characteristic of nucleus A, called the Larmor frequency, and is related to the nuclear magnetogyric ratio γ, the field B_0 and its shielding constant, σ_A, by:

$$\nu_A = \gamma_A/2\pi B_0(1 - \sigma_A) \tag{2.3}$$

Differences between shielding constants for nuclei are called chemical shifts, and these are normally quoted relative to a standard reference compound for each isotope. They are discussed further in Section 2.4.

In Eqn (2.2) J_{AB} is the spin–spin coupling constant for the nuclei A and B. Coupling constants (discussed in Section 2.7) are several orders of magnitude smaller than Larmor frequencies, but couplings between nuclei separated by one, two or three bonds are normally large enough to be observed. Thus the frequencies of transitions between energy levels observed in NMR experiments provide information primarily about the nuclei directly involved, but also about neighboring spinning nuclei.

The equation for the stationary state energies given above is only valid when the coupling constants are small compared with the chemical shifts, $v_A - v_B$. In general, the calculation of the theoretical spectrum for a collection of magnetic nuclei in a magnetic field involves solving the Schrödinger equation:

$$\hat{\mathcal{H}}\psi = E\psi \tag{2.4}$$

where ψ is the nuclear spin wavefunction and $\hat{\mathcal{H}}$ is the Hamiltonian operator. This operator may be written as a function of magnetogyric ratios, shielding constants, coupling constants, and the field B_0. Diagonalization techniques yield eigenvalues and eigenvectors, and so the energies (frequencies) and probabilities (relative intensities) of permitted transitions can be obtained. The analyses are usually performed with the aid of standard computer programs, which are widely available. Libraries of NMR programs are maintained by the Quantum Chemistry Program Exchange and the UK Science and Engineering Research Council. There are also excellent spectral analysis programs designed to work on the computers of most modern NMR spectrometers.

Analysis of spectra, however achieved, yields chemical shifts and coupling constants. These are parameters which are related to the chemical structure of the system, and are discussed in Sections 2.4 and 2.7. But there are other measurable NMR parameters which may be useful to chemists, relating to the rate of relaxation to equilibrium after disturbance. For a nucleus having two possible spin states, there is a Boltzmann distribution between the two levels, so that the ratio of populations is $\exp(-\Delta E/kT)$, where ΔE is the energy difference between the levels, k the Boltzmann constant, and T the absolute temperature. For nuclei under typical NMR conditions, the population difference is a few parts per million, and radiation at the resonance frequency would reduce it. The absorption signal (the intensity of which is proportional to the population difference) would soon disappear, were it not that there are relaxation mechanisms, whereby the spin system exchanges energy with its surroundings. The relaxation time T_1 characterizes this process, and a second relaxation time T_2 relates to redistribution of energy between spins of different nuclei. A small T_2 leads to broad lines, and consequent uncertainty in the determination of resonance frequencies. For mobile fluids T_1 and T_2 are effectively equal. These relaxation times may be measured fairly easily, and their chemical significance is discussed in Section 2.9.

In general the most useful nuclei for NMR observation are those for which the spin quantum number, I, is 1/2. Some of these are listed in Table 2.1, together with natural abundances, magnetogyric ratios, approximate resonance frequencies in a field in which the proton resonance of SiMe$_4$ is at exactly 100 MHz, and relative receptivities. This last quantity is a useful guide to the relative signal strengths obtainable from solutions of different elements at equal concentrations and is proportional to $\gamma^3 NI(I + 1)$, where γ is the magnetogyric ratio, N the natural abundance, and I the nuclear spin

Table 2.1 NMR properties of some spin 1/2 nuclei

Isotope	Natural abundance/%	Magnetogyric ratio[a] /10^7 rad $T^{-1}s^{-1}$	Relative NMR frequency/MHz	Relative receptivity
^{1}H	99.985	26.7519	100.0	1.00
^{3}H	—	28.535	106.7	—
^{3}He	0.00013	−20.380	76.2	5.8×10^{-7}
^{13}C	1.11	6.7283	25.1	1.8×10^{-4}
^{15}N	0.37	−2.712	10.1	3.9×10^{-6}
^{19}F	100.0	25.181	94.1	8.3×10^{-1}
^{29}Si	4.7	−5.3188	19.9	3.7×10^{-4}
^{31}P	100.0	10.841	40.5	6.6×10^{-2}
^{57}Fe	2.2	0.8661	3.2	7.4×10^{-7}
^{77}Se	7.6	5.12	19.1	5.3×10^{-4}
^{89}Y	100.0	−1.3155	4.9	1.2×10^{-4}
^{103}Rh	100.0	−0.846	3.2	3.2×10^{-5}
^{107}Ag	51.8	−1.087	4.0	3.5×10^{-5}
^{109}Ag	48.2	−1.250	4.7	4.9×10^{-5}
^{111}Cd	12.8	−5.6926	21.2	1.2×10^{-3}
^{113}Cd	12.3	−5.9550	22.2	1.3×10^{-3}
^{117}Sn^b	7.6	−9.578	35.6	3.5×10^{-3}
^{119}Sn	8.6	−10.021	37.3	4.5×10^{-3}
^{125}Te^b	7.0	−8.498	31.5	2.2×10^{-3}
^{129}Xe	26.4	−7.441	27.8	5.7×10^{-3}
^{169}Tm	100.0	−2.21	8.3	5.7×10^{-4}
^{171}Yb	14.3	4.712	17.6	7.8×10^{-4}
^{183}W	14.4	1.120	4.2	1.1×10^{-5}
^{187}Os	1.6	0.616	2.3	2.0×10^{-7}
^{195}Pt	33.8	5.768	21.4	3.4×10^{-3}
^{199}Hg	16.8	4.8154	17.9	9.8×10^{-4}
^{203}Tl	29.5	15.436	57.1	5.7×10^{-2}
^{205}Tl	70.5	15.589	57.6	1.4×10^{-1}
^{207}Pb	22.6	5.540	20.9	2.0×10^{-3}

^{a} G.H. Fuller, *J. Phys. Chem. Ref. Data* **5**, 835 (1976). ^{b} Other spin 1/2 isotopes also exist.

quantum number. For those elements which have several isotopes with $I = 1/2$ some less important ones are not included in the table, and radioactive isotopes (except ^{3}H) are also excluded.

For nuclei with $I > 1/2$, the nuclear electric quadrupole moments, Q, are also important, because large moments shorten the T_2 relaxation times greatly, and so broaden the NMR resonances. Thus the quadrupolar nuclei, some of which are listed in Table 2.2, are generally less suitable for NMR studies than those with $I = 1/2$, but nevertheless most of those in Table 2.2 have been studied, and some, such as ^{2}H, ^{11}B, ^{14}N and ^{17}O, are widely and routinely observed.

There is full discussion of the applications of NMR to all the elements less commonly observed in Ref. [1]. This is an essential reference book for those interested in applying NMR widely in inorganic chemistry.

2.3   Experimental methods

2.3.1   NMR spectrometers

The essential components of an NMR spectrometer are a powerful magnet, providing a magnetic field into which the sample is placed, a radiofrequency transmitter, receiver, and possibly decoupler, and some recording device. The magnet has to provide a very stable field, which is also extremely homogeneous in the region of the sample. Some simple and cheap spectrometers use permanent magnets, giving fields of 1-2 tesla, but more often electromagnets are used, with field strengths typically in the range 1.8-2.3 T. Stability is maintained by locking the field to the resonance frequency of a nucleus not under investigation. The deuterium resonance of a solvent is most commonly used, but an external sample, kept permanently in the magnetic field, is useful in some circumstances. For very high fields, up to about 13 T, superconducting magnets are used, with a large coil of superconducting material cooled by liquid helium. In general, higher fields give better sensitivity, at least for light nuclei, and wider line separation, which often greatly simplifies the procedure of spectral analysis.

The probe, at the center of the magnetic field, is an electrical device with a sample holder, and one or more coils. The radiofrequency for the nucleus being observed is

Table 2.2 NMR properties of some quadrupolar nuclei[a]

Isotope	Spin	Natural abundance/ %	Magnetogyric ratio[b]/ 10^7 rad $T^{-1}s^{-1}$	Relative NMR frequency/ MHz	Relative receptivity	Quadrupole moment[b]/ $10^{-28}m^2$
^{2}H[c]	1	0.015	4.1066	15.4	1.5×10^{-6}	2.8×10^{-3}
^{6}Li	1	7.4	3.9371	14.7	6.3×10^{-4}	-8×10^{-4}
^{7}Li	3/2	92.6	10.3975	38.9	2.7×10^{-1}	-4×10^{-2}
^{9}Be	3/2	100.0	-3.7596	14.1	1.4×10^{-2}	5×10^{-2}
^{10}B	3	19.6	2.8746	10.7	3.9×10^{-3}	8.5×10^{-2}
^{11}B	3/2	80.4	8.5843	32.1	1.3×10^{-1}	4.1×10^{-2}
^{14}N[c]	1	99.6	1.9338	7.2	1.0×10^{-3}	1×10^{-2}
^{17}O	5/2	0.037	-3.6279	13.6	1.1×10^{-5}	-2.6×10^{-2}
^{23}Na	3/2	100.0	7.0801	26.5	9.3×10^{-2}	1×10^{-1}
^{25}Mg	5/2	10.1	-1.639	6.1	2.7×10^{-4}	2.2×10^{-1}
^{27}Al	5/2	100.0	6.9760	26.1	2.1×10^{-1}	1.5×10^{-1}
^{33}S	3/2	0.76	2.055	7.7	1.7×10^{-5}	-5.5×10^{-2}
^{35}Cl	3/2	75.5	2.6240	9.8	3.6×10^{-3}	-1×10^{-1}
^{37}Cl	3/2	24.5	2.1842	8.2	6.7×10^{-4}	-7.9×10^{-2}
^{39}K[d]	3/2	93.1	1.2498	4.7	4.8×10^{-4}	4.9×10^{-2}
^{43}Ca	7/2	0.15	-1.8025	6.7	8.7×10^{-6}	2×10^{-1e}
^{45}Sc	7/2	100.0	6.5081	24.3	3.0×10^{-1}	-2.2×10^{-1}
^{47}Ti	5/2	7.3	-1.5105	5.6	1.5×10^{-4}	2.9×10^{-1}
^{49}Ti	7/2	5.5	-1.5109	5.6	2.1×10^{-4}	2.4×10^{-1}
^{51}V[d]	7/2	99.8	7.0453	26.3	3.8×10^{-1}	-5×10^{-2}
^{53}Cr	3/2	9.6	-1.512	5.7	8.6×10^{-5}	3×10^{-2}
^{55}Mn	5/2	100.0	6.608	24.7	1.8×10^{-1}	4×10^{-1}
^{59}Co	7/2	100.0	6.317	23.6	2.8×10^{-1}	3.8×10^{-1}
^{61}Ni	3/2	1.2	-2.394	8.9	4.1×10^{-5}	1.6×10^{-1}
^{63}Cu	3/2	69.1	7.0974	26.5	6.5×10^{-2}	-2.1×10^{-1}
^{65}Cu	3/2	30.9	7.6031	28.4	3.6×10^{-2}	-2.0×10^{-1}
^{67}Zn	5/2	4.1	1.6768	6.3	1.2×10^{-4}	1.6×10^{-1}
^{69}Ga	3/2	60.4	6.4323	24.0	4.2×10^{-2}	1.9×10^{-1}
^{71}Ga	3/2	39.6	8.1731	30.6	5.7×10^{-2}	1.2×10^{-1}
^{73}Ge	9/2	7.8	-0.9357	3.5	1.1×10^{-4}	-1.8×10^{-1}
^{75}As	3/2	100.0	4.595	17.2	2.5×10^{-2}	2.9×10^{-1}
^{79}Br	3/2	50.5	6.7228	25.1	4.0×10^{-2}	3.7×10^{-1}
^{81}Br	3/2	49.5	7.2468	27.1	4.9×10^{-2}	3.1×10^{-1}

CHAPTER 2: NUCLEAR MAGNETIC RESONANCE SPECTROSCOPY

Table 2.2 NMR properties of some quadrupolar nuclei (cont'd)

Isotope	Spin	Natural abundance/ %	Magnetogyric ratio[b]/ 10^7 rad T^{-1}s^{-1}	Relative NMR frequency/ MHz	Relative receptivity	Quadrupole moment[b]/ 10^{-28}m^2
^{87}Rb[d]	3/2	27.9	8.7807	32.8	4.9×10^{-2}	1.3×10^{-1}
^{87}Sr	9/2	7.0	-1.163	4.3	1.9×10^{-4}	3×10^{-1}
^{91}Zr	5/2	11.2	-2.4959	9.3	1.1×10^{-3}	-2.1×10^{-1f}
^{93}Nb	9/2	100.0	6.564	24.5	4.9×10^{-1}	-2.2×10^{-1}
^{95}Mo	5/2	15.7	1.750	6.5	5.1×10^{-4}	$\pm 1.2 \times 10^{-1}$
^{97}Mo	5/2	9.5	-1.787	6.7	3.3×10^{-4}	± 1.1
^{99}Ru	5/2	12.7	1.234^g	4.6	1.5×10^{-4}	7.6×10^{-2}
^{101}Ru	5/2	17.1	1.383^g	5.2	2.8×10^{-4}	4.4×10^{-1}
^{105}Pd	5/2	22.2	-1.23	4.6	2.5×10^{-4}	8×10^{-1}
^{115}In[d]	9/2	95.7	5.8908	22.0	3.4×10^{-1}	8.3×10^{-1}
^{121}Sb	5/2	57.3	6.4355	24.0	9.3×10^{-2}	-2.8×10^{-1}
^{123}Sb	7/2	42.7	3.4848	13.0	2.0×10^{-2}	-3.6×10^{-1}
^{127}I	5/2	100.0	5.3817	20.1	9.5×10^{-2}	-7.9×10^{-1}
^{131}Xe[c]	3/2	21.2	2.206	8.2	5.9×10^{-4}	-1.2×10^{-1}
^{133}Cs	7/2	100.0	3.5277	13.2	4.8×10^{-2}	-3×10^{-3}
^{137}Ba[d]	3/2	11.3	2.988	11.1	7.9×10^{-4}	2.8×10^{-1}
^{139}La	7/2	99.9	3.801	14.2	6.0×10^{-2}	2.2×10^{-1}
^{177}Hf	7/2	18.5	1.081	4.0	2.6×10^{-4}	4.5
^{179}Hf	9/2	13.8	-0.679	2.5	7.4×10^{-5}	5.1
^{181}Ta	7/2	99.99	3.22	12.0	3.7×10^{-2}	3
^{185}Re	5/2	37.1	6.077	22.7	5.1×10^{-2}	2.3
^{187}Re	5/2	62.9	6.138	22.9	8.8×10^{-2}	2.2
^{189}Os[c]	3/2	16.1	2.096	7.8	3.9×10^{-4}	8×10^{-1}
^{191}Ir	3/2	37.3	0.4643	1.7	9.8×10^{-6}	1.1
^{193}Ir	3/2	62.7	0.5054	1.9	2.1×10^{-5}	1.0
^{197}Au	3/2	100.0	0.4625	1.7	2.6×10^{-5}	5.9×10^{-1}
^{201}Hg[c]	3/2	13.2	-1.7776	6.6	1.9×10^{-4}	4.4×10^{-1}
^{209}Bi	9/2	100.0	4.2342	16.2	1.4×10^{-1}	-3.8×10^{-1}

[a] In general, radioactive nuclei, the rare gases, and the lanthanides are omitted. All the lanthanides except cerium have potentially useful NMR isotopes. [b] G.H. Fuller, *J. Phys. Chem. Ref. Data*, **5**, 835 (1976), except where otherwise stated. [c] A spin 1/2 isotope also exists. [d] Other less important quadrupolar isotopes also exist. [e] R. Neumann, F. Träger, J. Kowalski and G. zu Putlitz, *Z. Physik*, **A279**, 249 (1976). [f] S. Büttgenbach, R. Dicke, H. Gebauer, R. Kuhnen and F. Träber, *Z. Physik*, **A286**, 125 (1978). [g] C. Brévard and P. Granger, *J. Chem. Phys.*, **75**, 4175 (1981).

applied to the sample via the transmitter coil, and at resonance a voltage is induced in the receiver coil, which detects sample magnetization. Another frequency has to be transmitted and received to operate the field/frequency lock, and in multiple resonance experiments yet another radiofrequency must be applied, via a decoupling coil. However, one coil may perform several of these functions. There is also a need to spin the sample tube, to get optimum field homogeneity, and this is usually achieved by using a sample holder which acts as a turbine, driven by compressed air. In most cases, the probe is complicated still further by equipment used to raise, lower and control the temperature.

Finally, the spectrometer must have some means of recording spectra. Normally they may be displayed on an oscilloscope or viewing screen, and plotted using a pen recorder, but in most modern instruments the primary recording device is a computer.

2.3.2 Continuous wave spectra

As an NMR spectrum is a plot of nuclear magnetism against frequency, the simplest and most obvious way to observe the spectrum is to sweep the frequency (or field) slowly through the appropriate range. A low power must be used to avoid saturation, a decrease in the height of peaks caused by reduction of the difference in population between energy levels. The peaks obtained by this method (known as the continuous wave method) have a Lorentzian line shape, with a width dependent on the relaxation time, T_2, from the excited state, as given by Eqn (2.5), in which $I(v)$ is the intensity at frequency v, and v_0 is the frequency of the line origin. Lines may possibly be broadened by other factors, and they have an area or integrated intensity which is proportional to the number of resonant nuclei.

$$I(v) = \frac{2T_2}{1 + 4\pi^2 T_2^2 (v_0 - v)^2}$$
(2.5)

2.3.3 Fourier transform spectra

Continuous wave spectroscopy is inefficient, as the spectral range is scanned slowly, and it is much quicker to monitor all the frequencies simultaneously. This can be achieved by applying a short pulse of energy at a frequency close to the resonances of interest. Typically, a pulse of a few microseconds can be used to excite all nuclei having resonances within a range of 50 kHz. The duration of the pulse affects both the range of frequencies that can be excited, and the direction of the resultant nuclear magnetization. The magnetization, or total magnetic moment, of a sample is the sum of all individual nuclear moments. In classical mechanics interaction of a magnetic field with a magnetic moment induces a torque, and so the nuclear moments precess about the applied field direction with a characteristic frequency. This is the Larmor frequency, which also appears in the equation for energy levels, Eqn (2.2). It is often necessary to consider nuclear magnetization in a frame of reference which rotates at the Larmor frequency. In the present case, the radio frequency magnetic field associated with the pulse, B_1, lies at right angles to the static spectrometer field, B_0, as shown in Fig. 2.1. The second field, which we take to be along the x axis, causes the magnetization to precess about this axis, so that it is no longer parallel to the applied field. After a 90° pulse it is perpendicular to the field, along the y axis, while after a 180° pulse it is

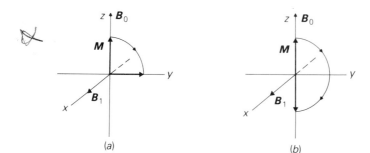

Fig 2.1 Effects of (a) a 90° pulse and (b) a 180° pulse on the nuclear magnetization, M, in the rotating frame of reference.

CHAPTER 2: NUCLEAR MAGNETIC RESONANCE SPECTROSCOPY

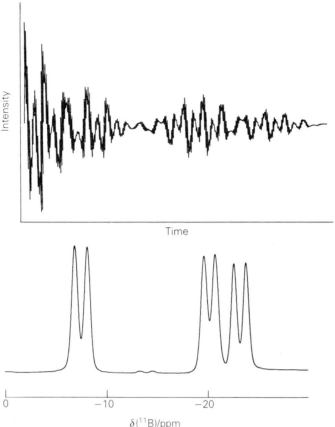

Fig 2.2 Free induction decay and NMR spectrum obtained from it by Fourier transformation.

parallel to the field, but inverted. The nuclei then relax towards equilibrium, and during the relaxation period they induce voltages in the receiver coil, giving the 'time domain spectrum' or 'free induction decay' (normally referred to as FID), and this is stored in the spectrometer's computer. Fourier transformation of the FID (a computer operation taking a few seconds) yields the 'frequency domain', or normal, spectrum, as illustrated in Fig. 2.2.

In practice a pulse of a few microseconds is followed by a data collection time of a few seconds or less, and rapid repetition of the pulse–acquisition sequence is thus possible, averaging out the noise and so giving great improvements in signal-to-noise ratio. Typically, in the time taken to scan a continuous wave spectrum, 100 FID's may be accumulated, giving a ten-fold gain in signal-to-noise ratio (the gain is the square root of the number of scans averaged), and data collection may continue for as long as is desired. Thus the method is almost essential for nuclei whose sensitivity is low, and pulse excitation is now preferred to swept excitation for all nuclei, except for routine proton work.

As a computer is needed to control the Fourier transform spectrometer, and to collect and transform the data, it may also be used to manipulate, compare, tidy, level, add, subtract or simulate spectra, or to control the more complicated pulse sequences described in Section 2.11. A full account of the effects of the computer processes on the appearance, resolution, signal-to-noise ratios *etc.* of the spectra obtained is given in Ref. [2], while the principles and practice of Fourier transform NMR spectroscopy are discussed fully in Refs [3] and [4]. Shorter accounts (but fuller than the present one) are given in standard texts on NMR listed in the bibliography.

2.4　Information from chemical shifts

2.4.1　General principles

Although it is necessary to understand the basic theory of NMR and the operation of spectrometers in order to know what spectra may usefully be obtained, the prime concern of the practising inorganic chemist is the interpretation of spectra. The first stage of this is the determination of the fundamental parameters, chemical shifts and coupling constants, from the line positions and intensities.

In atoms and molecules the nuclei are screened to some extent from the magnetic field, B_0, by electrons, so that the net field effective at a nucleus is:

$$B_{\text{eff}} = B_0(1 - \sigma) \tag{2.6}$$

where σ is the shielding constant. Thus each chemically distinct nucleus is associated with a characteristic resonance frequency, so that for decaborane 14, for example, the ^{11}B spectrum has four distinct resonances (Fig. 2.3). The difference in screening

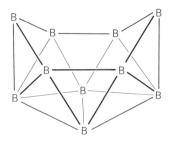

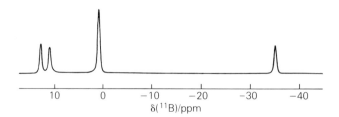

Fig 2.3　^{11}B NMR spectrum of $B_{10}H_{14}$, with all couplings to protons removed. The boron cage structure is as shown, so four equivalent nuclei give rise to the largest peak, and three pairs of equivalent nuclei give the three smaller peaks.

constants for two nuclei is the chemical shift (δ) between them, and shifts are normally reported relative to a standard for the isotope concerned. Lists of recommended standards are given in Ref. [1]. The shifts are quoted in parts per million (ppm) of the observation frequency, and the convention is now that a shift to high frequency is positive, so that a decrease in shielding increases the chemical shift. The τ scale, formerly used for protons, was positive to *low* frequency, with the standard ($SiMe_4$) at 10 ppm. With old chemical shift data always check what convention has been used.

A group of nuclei must have the same chemical shift if they are *chemically equivalent*, i.e. they have the same chemical environment, as a consequence of the symmetry properties of the molecule. Thus, for example, the protons in $SiBrClH_2$ (2.I) are equivalent, being related by a plane of symmetry, but the phosphorus nuclei in *cis-*$[PtBrCl(PPh_3)_2]$ (2.II) are non-equivalent, as one occupies a position *trans* to Br,

　CHAPTER 2: NUCLEAR MAGNETIC RESONANCE SPECTROSCOPY

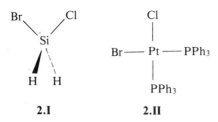

2.I   2.II

while the other is *trans* to Cl. Chemically distinct nuclei may by chance have the same chemical shift, and there is a need to be aware of the confusion that this may cause. It should also be noted that a group of nuclei which are not strictly related by symmetry may be effectively equivalent, due to some averaging process which is fast on the NMR time scale (i.e. the rate of exchange between non-equivalent sites is much greater than the difference between the frequencies of their associated resonances). Thus, the hydrogen atoms of a methyl group are usually equivalent, due to rapid rotation, and the axial and equatorial fluorine atoms of PF_5 are equivalent at room temperature, due to the exchange process known as pseudorotation. The use of NMR spectroscopy to study exchange processes is discussed in Section 2.15.

The precise reproduction of experimental chemical shifts by theoretical calculations is not yet possible, and a chemist interested in determining the composition or structure of a material will work by analogy, and by using empirical rules, which may, of course, have a sound theoretical basis. The shielding constant, σ, has so-called diamagnetic and paramagnetic components, σ_d and σ_p, where σ_d depends on the electron distribution in the electronic ground state, whereas σ_p depends on excited states as well. The paramagnetic term is zero for electrons in s orbitals, and so for ^1H the diamagnetic term is normally dominant, but for heavier nuclei the paramagnetic term predominates.

2.4.2   Proton chemical shifts

For protons, increasing the $1s$ orbital electron density increases the shielding. Thus an electronegative neighbor will reduce the $1s$ density and the shielding, and the chemical shift will move to high frequency. This effect is illustrated for some simple methyl, silyl and germyl compounds in Table 2.3. Note that there is no rigorous and exact correlation, just a general trend, and we are considering only one contribution to the total shielding. Another trend is also revealed by this table, the shift to high frequency as the central atom gets heavier. Thus for CH_4, SiH_4 and GeH_4 the ^1H shifts are 0.1, 3.2 and

Table 2.3 Proton chemical shifts (ppm) of some methyl, silyl and germyl compounds

	M = C	M = Si	M = Ge
MH_4	0.1	3.2	3.1
MH_3I	2.0	3.4	3.5
MH_3Br	2.5	4.2	4.5
MH_3Cl	2.8	4.6	5.1
$(MH_3)_2O$	3.2	4.6	5.3
MH_3F	4.1	4.8	5.7

3.1 ppm, respectively, while for SnH_4 it is 3.9 ppm. PbH_4 is unknown, but shifts for alkyl lead hydrides are about 3 ppm higher than those for analogous tin hydrides. In Groups V to VII the changes are in the opposite direction.

A further major contribution to proton chemical shifts arises from neighboring groups or atoms having an anisotropic magnetic susceptibility. There is an excellent simple description of the origin of this effect in Ref. [5], but here we just note that this leads to deshielding of the protons in alkenes, but to shielding of alkyne protons. The effect is particularly pronounced for aromatic compounds, and causes marked deshielding for the aromatic protons themselves (the benzene ^1H resonance comes at 7.3 ppm), but has an opposite effect for any proton that may lie above or below the aromatic ring.

In transition metal complexes there are often low-lying excited electronic states, and the effect of an applied magnetic field is to mix these to some extent with the ground state. Consequently the paramagnetic term is important for the nuclei of transition metals themselves, which have large high-frequency shifts. The long range effect of this term, however, is shielding, and so protons bound to transition metals have exceptionally low chemical shifts, typically in the range 0 to -40 ppm. As few other shifts come in this region, these resonances are diagnostic and they can be studied with relatively slight problems caused by overlapping peaks from other resonances.

Most ^1H chemical shifts lie in a range of *ca.* 10 ppm, although this must be extended to *ca.* 70 ppm to include transition metal hydrides and other less common bonding situations, and to 1000 ppm to include all paramagnetic species (Section 2.16). The normal proton range is narrow compared to those for other elements, for which the normal ranges are hundreds of ppm (e.g. 400 for carbon, 900 for fluorine) or thousands for heavy metals (e.g. 13 000 for platinum). One disadvantage of this is that proton spectra are likely to involve overlapping groups of resonances, but an advantage is that solvent effects are likely to be small. However, there are solvent effects on chemical shifts, arising from the electric field of polar solvents such as acetonitrile, or, more significantly, from the anisotropic magnetic susceptibility of such solvents as benzene, and care must be taken when comparing data obtained in different solvents. The problem of spectral complexity due to small relative chemical shifts may be reduced by running spectra at higher magnetic fields, as shifts are constant in ppm, and therefore proportional to B_0 when measured in Hz. If a high field spectrometer is not available, addition of a small amount of a paramagnetic 'shift reagent' will induce large differential shifts for the various types of protons present, and so spread out the spectrum.

2.III   2.IV

Reagents commonly used are complexes of Eu[III] and Pr[III] with 2,2,6,6-tetramethyl-heptane-3,5-dionato (2.III) or 1,1,1,2,2,3,3-heptafluoro-7,7-dimethyloctane-4,6-dionato (2.IV) ligands.

Finally, we should emphasize again that much of the value of NMR spectroscopy lies in the application of empirical rules relating parameters to structure, and that experience is of great importance. To use the experience of others, it helps to have large compilations of relevant data. There is no single comprehensive collection of ^{1}H shift data (it would be enormous, and constantly growing) but there are some useful compilations and summaries, mainly for organic compounds [6] and each year the new literature is surveyed and summarized in the Royal Society of Chemistry Specialist Periodical Reports [7].

2.4.3   Chemical shifts of other elements

There is not room here to discuss chemical shifts of all of the many elements that have been studied by NMR, but some comments must be made on ^{13}C, ^{19}F and ^{31}P as they are so widely used. Then ^{77}Se and ^{125}Te will be mentioned as typical heavier non-metals, and ^{55}Mn and ^{195}Pt, as transition metals. For more details of these and other nuclei, see Ref. [1], and for up-to-date information the series of reports, Ref. [7].

There is a huge amount of information on ^{13}C chemical shifts. Interested readers should consult specialist books [8], reviews [9] and compilations of spectra (including a database) [10] devoted to ^{13}C. Typical ranges of ^{13}C shifts are illustrated in Fig. 2.4.

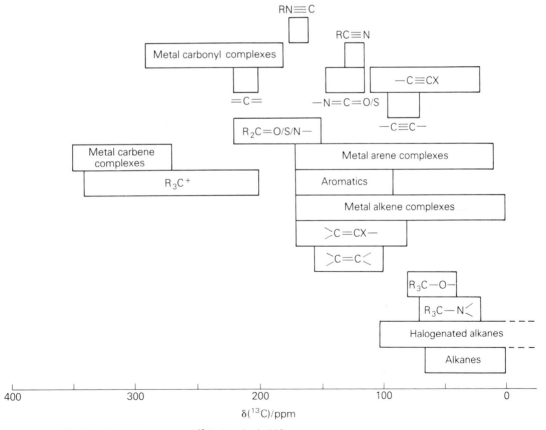

Fig 2.4   Typical ranges of ^{13}C chemical shifts.

For saturated carbon atoms the usual range of shifts is δ0-100 relative to SiMe$_4$, and there is a general tendency for electronegative substituents to increase the shift. The ^{13}C chemical shifts associated with the methyl groups in compounds CH$_3$X are directly related to the electronegativity of the group X and to the number of lone pairs of electrons on the central atom of the group. But the effects are not additive, and so shifts for multiply-substituted compounds CH$_2$XY *etc.* cannot be so easily predicted. Shifts for three-coordinate carbon atoms normally fall in the range δ80-240, while alkyne shifts are smaller again, between δ20 and 110 ppm. Shifts for aromatic compounds normally lie between δ110 and 170 ppm. However, in π-bonded metal alkene and arene complexes resonances may be shifted by up to 100 ppm to low frequency of their usual range, the shifts depending on the mode of coordination. Indeed the inorganic or organometallic chemist must often look for chemical shifts well outside the range normally considered by organic chemists – an important point to remember when recording spectra. At one extreme, CI$_4$ has a shift of -293 ppm. At the high frequency end, metal carbonyls are usually found between 170 and 290 ppm, while metal carbene complexes typically have resonances between 250 and 370 ppm. We should note here that it is often difficult to record ^{13}C spectra for metal carbonyl complexes, because the relaxation times for ^{13}C nuclei in these compounds can be very long. The problem can be alleviated by the use of paramagnetic relaxation agents, as described in Section 2.9.2.

Fluorine-19 chemical shifts cover a wider range, of some 900 ppm, and are not easy to interpret, as they are sensitive to electronegativity and oxidation state of the neighboring group, to stereochemistry, and to effects from more distant neighbors. Thus the shifts for the axial and equatorial fluorines in ClF$_3$ differ by 120 ppm, those for XeF$_2$ and XeF$_4$ by 180 ppm, and those for methyl and ethyl fluorides by 60 ppm. However, this great variability can be put to good use, and ^{19}F NMR is a very sensitive way of following reactions and changes of structure. A good example of this is the study of the reaction of WF$_6$ and WCl$_6$, in which all 9 isomers of WF$_n$Cl$_{6-n}$ (n = 1 to 6) (Fig. 2.5) were identified. There is one extensive compilation of ^{19}F chemical shifts, albeit somewhat dated [11]. Care should always be taken with literature data to check the reference standard (the accepted one is now CCl$_3$F) and sign convention used.

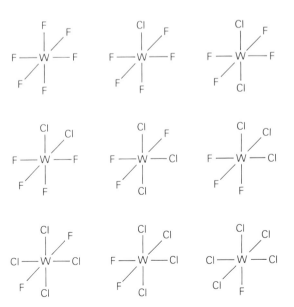

Fig 2.5 Structures of all isomers of WCl$_n$F$_{6-n}$ containing at least one fluorine atom.

Phosphorus-31 chemical shifts normally fall in the range ±250 ppm relative to 85% H_3PO_4. However, shifts can be as low as −460 (P_4), and the upper limit seems to change frequently, but certainly goes as far as +1362, in the phosphinidene complex [t-BuP{Cr(CO)$_5$}$_2$] [12]. So the extremes are reached with compounds having formal oxidation states of zero and one. The whole of the normal shift range is covered for P^{III} derivatives, and so the shifts are strongly substituent-dependent, but a much smaller range, $\delta - 50$ to $\delta 100$, covers most P^V compounds. The interpretation of these shifts is not at all easy, and there seem to be many contributing factors. However, we should note that it is often possible to predict unknown shifts if those for similar species are known, by simple extrapolation or interpolation, so that we can predict values for PX_2Y or PY_3, using those for PX_3 and PXY_2. This is illustrated for some phosphines with H, SiH_3 and PF_2 substituents in Fig. 2.6. But, even these simple additive relationships fail for some substituents, particularly electronegative ones such as Cl or F. So it is best to work by comparison with published data. There are some old compilations of data [13], but nothing modern. As with other nuclei, always check the reference standard and sign convention when using published data.

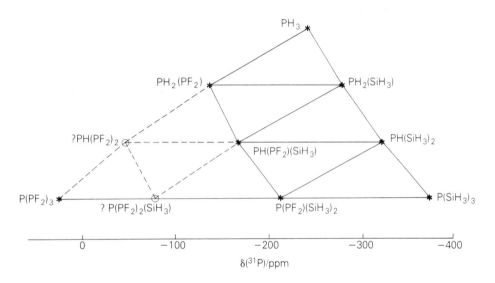

Fig 2.6 Representation of ^{31}P chemical shifts of phosphines with H, SiH_3 and PF_2 substituents. The regularity of the changes enables the chemical shifts of the unknown compounds $PH(PF_2)_2$ and $P(PF_2)_2$ (SiH_3), to be predicted.

Where there are analogies between phosphorus and selenium chemistry, there are also analogies between the chemical shifts of ^{31}P and ^{77}Se, but the effects are several times larger for selenium. Thus we find, for example, that $Se(SiH_3)_2$ and $P(SiH_3)_3$ have ^{77}Se and ^{31}P shifts close to the low frequency limits of their respective ranges, and that the shifts in the series SeR_2 and PR_3 increase in the order R = Me < Et < Pr^i < Bu^t. As with phosphorus, it is possible to extrapolate or interpolate shifts within groups of similar compounds. There is also a remarkably good correlation between ^{77}Se and ^{125}Te chemical shifts in analogous compounds, as illustrated in Fig. 2.7. This tells us that the same factors are important for both elements, although it does not help us to identify

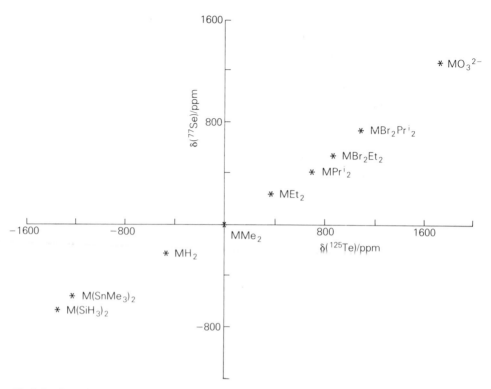

Fig 2.7 Correlation between chemical shifts of analogous selenium and tellurium compounds.

the factors. But as with a lot of NMR work, empirical observations are valuable, and the ability to extrapolate from one element to another one may be useful.

Manganese-55 is included here as an example of a nucleus easily observed by NMR, but having a fairly large quadrupole moment, and usually therefore giving broad lines (see Section 2.9.2). Line widths range from 10 Hz in symmetrical species such as the MnO_4^- ion, to 10 000 Hz in some carbonyl derivatives, and as the total chemical shift range is about 3000 ppm, this is not a good way to identify components of mixtures. However, the chemical shifts and linewidths give useful information and they have been correlated with isomer shifts and quadrupole splittings obtained by Mössbauer spectroscopy (Chapter 7). As with other metals, there is a relationship between oxidation state and shielding, with Mn^{VII} at $\delta 0$ (MnO_4^- is the reference standard), Mn^I at $\delta - 1000$ to $\delta - 1500$, and Mn^{-1} lower, down to $\delta - 3000$. A particular correlation has been noted between chemical shift and bond polarity for a series of complexes $Mn(CO)_5L$, from $\delta - 1000$ when L is negative (Cl) to $\delta - 2660$ ppm when L is positively polarized ($SnMe_3$). Both of these relationships are consistent with the idea that the ^{55}Mn chemical shifts reflect the total electron density on the metal atom.

Finally, we refer to platinum, as a heavy transition element, increasingly studied by NMR [14]. Many platinum complexes are studied by ^1H, ^{31}P and ^{13}C NMR, but when reactions are followed by these means, it is possible that products such as $PtCl_4^{2-}$ may be missed. The ^{195}Pt chemical shift range is wide (13 000 ppm), and it has been noted that shifts depend strongly on the donor atoms, but relatively little on longer range effects, so that complexes $[PtCl_2(PR_3)_2]$ have shifts that vary little if R is changed. Thus ^{195}Pt NMR is an extremely powerful tool for following the course of a reaction, but it is used surprisingly little. The major part of platinum chemistry studied so far has involved complexes with phosphine ligands, not because they are inherently more interesting than other ligands, but because the spin of ^{31}P makes the study easier. There is scope for great developments using, for example, sulfide or arsine ligands, and ^{195}Pt NMR as the primary means of investigation.

2.5 Information from NMR signal intensities

We have seen (Section 2.2) that the intensity of an NMR signal is proportional to the population difference for the two energy levels involved, and this in turn depends on the energy separation. For several nuclei of the same isotope, the resonance frequencies differ only by a few ppm, and the total intensities of the lines in a spectrum are proportional to the number of equivalent nuclei contributing to them. For continuous wave spectra, the intensities are given by the integrated areas of the peaks (or sets of peaks if there is coupling to other nuclei), but in Fourier transform spectra accurate integrals may not be obtained unless care is taken to allow for the effects of different relaxation times for different nuclei. However, it is possible to obtain reliable measurements of the relative numbers of nuclei in different environments, and this information may be of great chemical importance.

First, we can determine the relative numbers of nuclei in each distinct environment within a single compound, and this may help in identification. For example, treatment of $[Ta(C_5H_5)_2(CH_3)_2]^+$ with $P(CH_3)_3(CH_2)$ yields a neutral product. Its ^{1}H spectrum has three resonances, at $\delta - 0.2$, 4.8 and 9.9, with relative intensities 2:10:3 [15]. These intensities, taken with the chemical shift data, indicate that the product has CH_3 groups and C_5H_5 rings in the ratio 1:2, and another two protons, almost certainly in a $Ta=CH_2$ group. The reaction has involved deprotonation, and the identity of the product as monomeric $[Ta(C_5H_5)_2(CH_3)(CH_2)]$ was finally confirmed by X-ray crystallography.

The other application of intensity measurements is in the determination of the relative amounts present of the various components of a mixture. This can be invaluable in kinetic or thermodynamic studies of reactions and this aspect is discussed in Section 2.15. But the value to a synthetic chemist lies in the fact that he can study the progress of a reaction in solution, with control of conditions, particularly temperature, and so determine the feasibility of isolating particular components. For example, a base-induced coupling reaction can be used to prepare the unsymmetrical diphosphene (2.V), by reaction of $(2,4,6-Bu^t_3C_6H_2)PH_2$ with $(Me_3Si)_2CHPCl_2$ [16]. The choice of

2.V

base and its concentration are critical, and initially the reaction was studied under a variety of conditions by ^{31}P NMR. Once the optimum conditions had been determined, the way was open to isolation of the new compound.

2.6   Simple patterns due to coupling between spinning nuclei

2.6.1   General considerations

In dealing with chemical shifts (Section 2.4) we have assumed that the resonance spectrum of each spinning nucleus is characterized by just one parameter, its shielding constant. Of course this is not so. There are mutual interactions between spinning nuclei within a molecule, and these couplings give NMR spectroscopy a unique place in the study of chemistry, in that it is often possible to *count* the numbers of spinning nuclei present. We will discuss the applications of this by starting with the simplest systems, and adding complications as we go along.

The concept of magnetic equivalence must be grasped before the analysis of spectra is considered. This is most easily done by defining magnetic non-equivalence! A group of chemically equivalent nuclei are magnetically non-equivalent if they couple unequally to any other particular nucleus. This is best illustrated by an example. In both CF_2H_2 (2.VI) and in $CH_2=CF_2$ (2.VII) there are two groups of two spinning nuclei. In CF_2H_2 there is only one type of coupling between H and F, but in $CH_2=CF_2$ there are two distinct HF couplings, one between nuclei which are mutually *syn* with respect to each other, and one between *anti* nuclei. The two hydrogens are therefore magnetically non-equivalent, as are the two fluorines. In this system the ^1H and ^{19}F spectra are second order, and cannot be analyzed as simple multiplets, although some deceptively simple patterns may be observed. Methods of dealing with second order spectra are discussed briefly in Section 2.8.

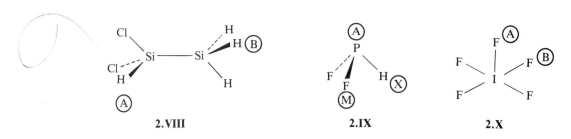

We can now describe a notation for classifying the spinning fragments of molecules (usually omitting nuclei with spin > 1/2). Each chemically distinct nucleus is assigned a letter, and a subscript indicates the number of such nuclei present. If the chemical shift difference for two sets of nuclei is much greater than the coupling constant between them, letters well apart in the alphabet are chosen. (Somewhat perversely, the order of choice is usually A, X, M, Q . . .). For smaller shift differences, consecutive letters are used, and for these systems second order spectra may be expected. Thus $SiCl_2HSiH_3$ (2.VIII) has an [AB_3] spin system. PF_2H (2.IX) has [AM_2X], and IF_5 (2.X) has

2.VIII   2.IX   2.X

[AB$_4$]. Subscripts outside square brackets are used to indicate sets of magnetically non-equivalent nuclei, so CH$_2$=CF$_2$ has a spin system [AX]$_2$, whereas CF$_2$H$_2$ has [A$_2$X$_2$]. Further refinements to this system are occasionally required, and they are fully described in all the standard NMR texts.

2.6.2 First order spectra of spin 1/2 isotopes of 100% abundance

If in a molecule there are two groups of equivalent spinning nuclei, there may be coupling between them. This results in the NMR spectra of each type of nucleus appearing as a multiplet, instead of a singlet resonance. If there is magnetic non-equivalence, or if the chemical shift difference for the nuclei is not much greater than the mutual coupling constant, the spectra will be second order. Such spectra will be briefly discussed in Section 2.8, but for now we consider only first order spectra arising from spin 1/2 nuclei. For the spin system [A$_p$X$_q$] the A spectrum appears as a $(q+1)$ line multiplet, with relative intensities given by Pascal's triangle (binomial coefficients) as shown in Table 2.4 and Fig. 2.8. The separations between lines are equal to the coupl-

Table 2.4 Relative intensities of resonances in NMR multiplets due to coupling to q equivalent spin 1/2 nuclei

q	Relative intensities
0	1
1	1:1
2	1:2:1
3	1:3:3:1
4	1:4:6:4:1
5	1:5:10:10:5:1
6	1:6:15:20:15:6:1

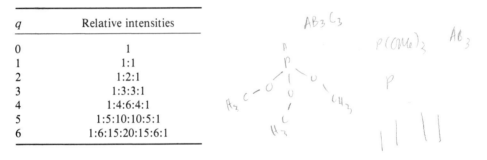

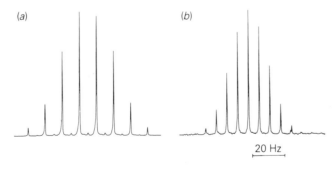

(a) (b)

20 Hz

Fig 2.8 Typical 'Pascal's triangle' multiplet patterns: (a) ^{31}P spectrum of P(OMe)$_3$ and (b) ^{29}Si spectrum of SiMe$_4$. In the second case the outermost lines are too weak to be seen. Weak lines between the strong ones arise from small amounts of ^{13}C (see Section 2.6.3).

ing constant. Thus observing A enables us to count how many X nuclei are present, but not the number of A nuclei. Observing the X spectrum, which will be a $(p+1)$ line multiplet, enables us to count the A nuclei. Note that the splittings in the A and X spectra are the same, and are usually written as J_{AX}. Sometimes the number of bonds between A and X is indicated, as in $^2J_{AX}$ for a two-bond coupling. The significance of coupling constant magnitudes is discussed in Section 2.7, but we should note now that

SECTION 2.6: COUPLING PATTERNS BETWEEN SPINNING NUCLEI   **45** one-bond couplings are typically hundreds or thousands of Hz, that they are an order of magnitude smaller per extra bond between the nuclei involved, and that one can normally resolve coupling over three or four bonds but not more. The usefulness of being able to count spinning nuclei is obvious and is illustrated by just one example. Reaction of methyl lithium with $P(SiH_3)_3$ gave a product whose ^{31}P NMR spectrum was a septet, showing coupling to six equivalent hydrogens. This suggests that it is $P(SiH_3)_2^-$, and this was subsequently proved using other data.

If there are more than two groups of equivalent nuclei present then several couplings may be observed in the spectrum of each group. For the spin system $[A_pM_qX_r]$ the A spectrum will have $(q+1)$ lines, separation J_{AM}, each of which is further split into $(r+1)$ lines with separation J_{AX}. This splitting can be repeated indefinitely. Fig. 2.9 shows the ^{31}P spectrum of $PF_2H(NH_2)_2$, labeled with ^{15}N, so that every nucleus has spin 1/2. There are two large splittings, by the unique ^1H nucleus into a doublet, and by the ^{19}F nuclei to give triplets. We have now accounted for the six main groups of lines. Each of these is split into a triplet by the two ^{15}N nuclei, and finally into quintets by the four amino protons. Altogether there are $2 \times 3 \times 3 \times 5 = 90$ lines, and the identification of the compound (made by adding NH_3 to PF_2NH_2) is effectively completed using a single spectrum.

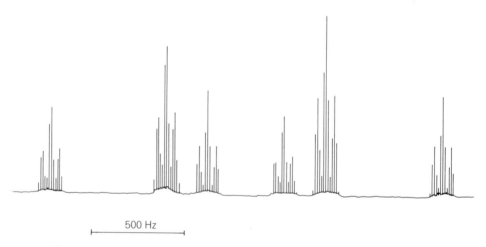

500 Hz

Fig. 2.9 ^{31}P NMR spectrum of $PF_2H(^{15}NH_2)_2$. It is a doublet (J_{PH}) of triplets (J_{PF}) of triplets (J_{PN}) of quintets (J_{PH}) – 90 lines in all.

2.6.3   **Effects of spin 1/2 isotopes of low abundance**

So far we have assumed that all the spinning nuclei present are of isotopes having 100% abundance, such as ^1H, ^{19}F, ^{31}P or ^{103}Rh. Samples enriched with ^{13}C or ^{15}N may also be encountered, but very many NMR studies involve isotopes of low natural abundance, such as ^{13}C, ^{29}Si and ^{183}W, known as 'dilute spins'. For example, the majority (86%) of the molecules of WF_6 in any sample have non-spinning isotopes of tungsten. Only 14% contribute to the ^{183}W spectrum, which shows a septet pattern due to coupling to six equivalent ^{19}F nuclei. Thus the spectra of the dilute spin nuclei themselves depend on the principles discussed earlier for abundant spins. However, all the WF_6 molecules contribute to the ^{19}F spectrum. For most of these the spectrum consists of

Example 2.1

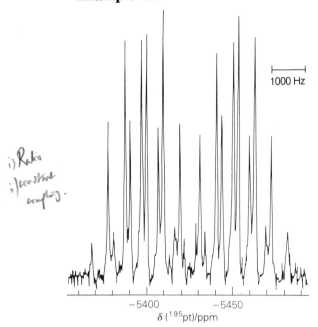

The figure alongside shows the ^{195}Pt NMR spectrum of a compound prepared by bubbling H_2 through a mixture of $[Pt(PPh_3)_3]$ and $[AuPPh_3][NO_3]$. The molecular composition of the product was shown by mass spectrometry to be $PtAu_7H(PPh_3)_8(NO_3)_2$. What can you deduce from the NMR spectrum?

1000 Hz

−5400   −5450

δ (^{195}pt)/ppm

The spectrum has one large doublet coupling (*ca.* 2300 Hz), and the rest of the pattern can be explained as a doublet (*ca.* 550 Hz) of multiplets (*ca.* 400 Hz). The multiplets clearly have even numbers of peaks, so the coupling must be to an odd number of nuclei. With five nuclei, the intensity ratio for the six lines would be 1:5:10:10:5:1, with seven it would be 1:7:21:35:35:21:7:1, and with nine it would be 1:9:36:84:126:126:84:36:9:1. The observed intensities fit best for the eight-line pattern, so we deduce that the platinum atom is probably bound directly to one PPh_3 ligand and to the unique hydrogen atom, and is surrounded by seven $AuPPh_3$ groups. This is confirmed by the crystal structure depicted below. All the metal and phosphorus atoms are shown, but the position of the hydrogen atom has to be inferred from the relatively long separations between Au(5), Au(6) and Au(9).

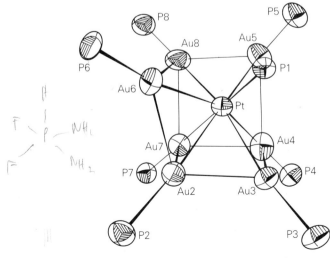

For more information, see R.P.F. Kanters, J.J. Bour, P.P.J. Schlebos, W.P. Bosman, H. Behmn, J.J. Skeggerda, L.N. Ito and L.H. Pignolet, *Inorg. Chem.*, **28**, 2591 (1989). The figures are taken, with permission, from this reference. Copyright (1989) American Chemical Society.

just one line, but for 14% it is a doublet, with a splitting of J_{WF}. The total spectrum therefore has a central line, plus satellites, with the intensity ratio being *ca.* 1:12:1.

For some symmetrical molecules, the satellite spectra can give us a great deal of structural information. About 95% of silicon atoms are ^{28}Si or ^{30}Si, with no spin, but 5% are ^{29}Si, with spin 1/2. Thus about 90% of Si_2H_6 molecules have non-spinning silicon atoms, and their ^1H spectrum is a single line. Nearly 10% of the molecules are $H_3{}^{28}Si^{29}SiH_3$, and these give rise to satellites in the ^1H spectrum. The resonance due to the three protons bound to the ^{29}Si atom is split into a wide doublet (splitting $^1J_{SiH}$), while that due to the other three is split into a narrow doublet (splitting $^2J_{SiH}$). But the two sets of protons are no longer equivalent, and so further splitting ($^3J_{HH}$) of the satellites into quartets is seen. In Fig. 2.10 the outer quartets are clearly visible, but the inner ones are only partially resolved from the intense central peak. So the satellites give us a count of the hydrogen atoms at each end, as well as indicating the presence of the silicon atoms, and the identity of the molecule is completely ascertained, using a single spectrum.

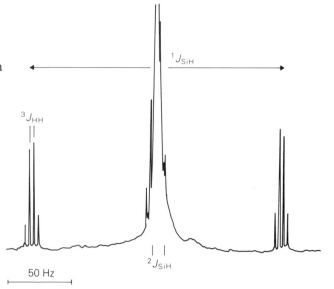

Fig. 2.10 ^1H NMR spectrum of Si_2H_6. The majority of molecules contribute only to the intense center line. Molecules of $^{29}SiH_3SiH_3$ give rise to two groups of resonances. The hydrogens attached to ^{29}Si are a wide doublet ($^1J_{SiH}$), split further into quartets by the other three hydrogens. The second group give a narrower doublet ($^2J_{SiH}$) of quartets ($^3J_{HH}$), but only four of these eight lines are resolved from the central resonance.

In Si_2H_6 there will also be some molecules of $H_3{}^{29}Si^{29}SiH_3$, but these are only 0.25% of the total, and can be ignored. As the majority of dilute spin elements have abundances of spin 1/2 isotopes of 15% or less (see Table 2.1) it is rarely necessary to worry about the probability of there being two or more of these spinning nuclei present. The exceptions are for xenon (not a problem for most chemists!) and platinum. In the latter case about one third of the nuclei have spin 1/2, and so the spectrum of a nucleus coupling to platinum will have three lines, with an intensity ratio *ca.* 1:4:1. This intensity pattern is diagnostic for platinum. For a nucleus coupling to two equivalent platinum nuclei the situation is more complicated, and three types of molecule can exist. In $2/3 \times 2/3$ of the molecules neither platinum nucleus has a spin, and the observed spectrum has a single resonance. In $1/3 \times 1/3$, *both* platinum nuclei have spin, and so there is a triplet resonance. The remaining molecules (4/9) have one spinning platinum nucleus, and the contribution to our spectrum is a doublet. Adding the various components together gives a quintet pattern, with intensity ratios *ca.* 1:8:18:8:1. This again is characteristic of a Pt_2X system, but the weak outer lines may easily be missed, leaving what appears to be a 1:2:1 triplet. This distinctive pattern occurs twice in the ^{13}C spectrum of the platinum complex 2.XI. The ^{13}C nuclei in the bridging isocyanide groups couple equally to their nearest-neighbor platinum atoms, while there is a two-bond coupling between ^{13}C nuclei in the terminal isocyanide groups and the further two platinum nuclei [17].

2.XI

Finally, we should note that the intensities of simple satellites may give some useful structural information. For example, each of the ^{31}P spectra of $P(SnPh_3)_3$, $PH(SnPh_3)_2$ and $PH_2(SnPh_3)$, with all couplings to hydrogen removed (see Section 2.10.4), has a main resonance with satellites due to ^{119}Sn (8.6% abundance) and ^{117}Sn (7.6%). However, these satellites are much bigger for the compounds with more tin substituents, as these molecules have much higher probability of containing one spinning tin nucleus, and so the three compounds can easily be distinguished.

2.6.4 Spectra of spin 1/2 isotopes of low abundance

We have seen, considering the example of WF_6, that the NMR spectrum of a single dilute spin is in principle no different from that of an abundant spin: the majority of molecules simply do not contribute to the spectrum. However, when a molecule has several atoms of elements having low-abundance spin 1/2 isotopes, coupling between these nuclei can usually be ignored as it is relatively unlikely that any one molecule will have more than one of these spin 1/2 nuclei. Splittings are therefore caused only by coupling to abundant spins such as hydrogen and fluorine. So if the coupling effects due to abundant spins are removed by decoupling (see Section 2.10.4), the resulting spectrum will consist of just one resonance for each chemically distinct environment occupied by the isotope being observed. The importance of this in ^{13}C NMR in organic chemistry is immense, and there are many books and reviews on the subject [8–10]. But carbon occurs in inorganic compounds too, and there are many other useful dilute spin elements (Table 2.1). The use of ^{13}C NMR in organometallic chemistry is of particular importance, and the subject has been reviewed [18]. For an example of the use of ^{13}C NMR in organometallic chemistry, consider the complex (2.XII) and the spectrum shown in Fig. 2.11. The tetramethylethylenediamine ligand has four non-equivalent types of methyl carbon and two distinct methylene carbons, while the allyl ligand has a methylene, two methinyl, and four chemically distinct aromatic sites. All thirteen resonances can be clearly seen, and with some knowledge of chemical shifts, it is not difficult to assign them, with some uncertainty about the order of the methyl

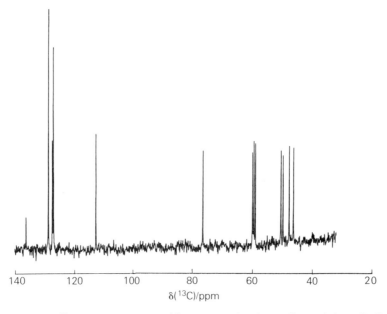

2.XII

group lines. Note that the spectrum does not prove the structure completely, but that it does provide extremely strong corroborative evidence.

There are similar applications for other dilute, spin 1/2 isotopes, illustrated by the use of ^{29}Si NMR in the study of polymeric siloxanes. Siloxanes are built up of terminal units, $R_3SiO—$, chain units $[R_2Si(O—)_2]$, branch units $[RSi(O—)_3]$, and quaternary units $[Si(O—)_4]$. The most common substituents are methyl and phenyl, although

Fig 2.11 ^{13}C NMR spectrum, with proton noise decoupling, of the palladium complex 2. XII.

many others may be used. It happens that the chemical shift range for ^{29}Si is such that the groups $Me_3SiO—$, $Me_2PhSiO—$, $MePh_2SiO—$, $Ph_3SiO—$, $Me_2Si(O—)_2$, $MePhSi(O—)_2$, $Ph_2Si(O—)_2$, $MeSi(O—)_3$, $PhSi(O—)_3$ and $Si(O—)_4$ all fall into distinct, non-overlapping regions, regardless of what neighboring groups they may have. It is therefore very easy to study the degree of polymerization or extent of cross-linking in a siloxane, or to investigate the distribution of substituents, by integrating the resonances in the appropriate sections of the spectrum. This provides invaluable information, but there is more to come! *Within* each region, the neighboring groups have an effect, so that it is possible to investigate the relationships between the various types of group.

Example 2.2

Account for the forms of the ^{1}H-decoupled ^{13}C NMR spectra of the *cis* and *trans* isomers of [RuCl(NO)(bpy)$_2$](PF$_6$) (bpy = 2,2'-bipyridine) illustrated below. Assume that each of the peaks marked with arrows includes two overlapping lines.

The spectra are adapted, with permission, from H. Nagao, H. Nishimura, H. Funato, Y. Ichikawa, F.S. Howell, M. Mukaida and H. Kakihana, *Inorg. Chem.*, **28**, 3955 (1989).

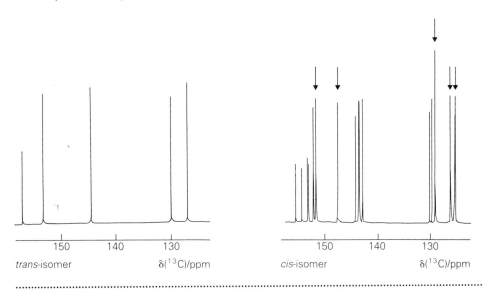

In *trans*-[RuCl(NO)(bpy)$_2$]$^+$ there are two planes of symmetry, so that each bpy ligand is bisected by one mirror plane, and the two ligands are related by the second plane (structure *a* below). There are just five chemically distinct carbon atoms, and so five resonances are seen in the ^{13}C NMR spectrum. The resonance at highest frequency is less intense than all the others. This is because all the other resonances arise from carbon bound to hydrogen, and their intensities are increased by the Nuclear Overhauser Effect (see Section 2.10.6).

In *cis*-[RuCl(NO)(bpy)$_2$]$^+$ (structure *b*) there are no planes of symmetry, and so the carbon nuclei are all chemically distinct. There are thus 20 resonances in all, of which four are due to quaternary carbons, and are less intense than the others. Each line in the spectrum of the *trans* isomer has become four lines, and the chemical shift changes are small, with one exception. One ring is now *trans* to chlorine ligand instead of to nitrogen, and so one of resonances associated with this ring has been shifted much more than the others.

Similar effects were shown by a study of a long chain polymer containing $Me_2Si(O-)_2$ [D] and $MeHSi(O-)_2$ [D'] chain units and Me_3SiO- end units [19]. The ^{29}Si NMR spectrum of the end [M] units showed two groups of peaks depending on the neighbors (MD— and MD'—) (Fig. 2.12). Then there were smaller chemical shift differences attributable to second neighbors (MDD— and MDD'—, MD'D— and MD'D'—) and finally even smaller ones due to third neighbors (MDDD—) etc. The repeated 1:1 doublet patterns showed that the groups were randomly mixed, with no systematic groupings recurring.

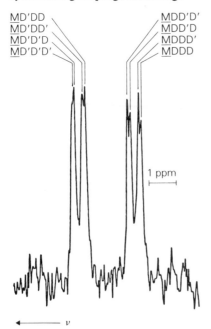

MD'DD
MD'DD'
MD'D'D
MD'D'D'

MDD'D'
MDD'D
MDDD'
MDDD

1 ppm

ν

Fig 2.12 ^{29}Si NMR spectrum, with proton decoupling, of the terminal Me_3SiO- (M) groups of a mixture of silicones of average composition $M_2D_8D_8'$. [Redrawn from R.K. Harris and B.J. Kimber, *J. Organomet. Chem.*, **70**, 43 (1974)].

It should by now be apparent that the spectra of abundant and dilute spin elements are to some extent complementary. The abundant spin (e.g. H,F) spectra are complicated by a lot of coupling, but contain a great deal of information, while the dilute spin (e.g. C,Si) spectra contain relatively little information, but in consequence are very much simpler. One suspects that when the universe was put together, the needs of NMR spectroscopists were taken into consideration! But of course, we are not satisfied, and occasionally we need the simplicity of dilute spins, but applied to proton spectra. This can be achieved in three ways. One is to observe ^2H spectra, either in the extremely low natural abundance, or in samples enriched with a few per cent of deuterium. The spectra must be recorded while decoupling ^1H, as described in Section 2.10.4. Another is to prepare a deuteriated sample, and observe the few residual protons, ideally with ^2H decoupling. The third method involves first obtaining a 2-dimensional spectrum, as described in Section 2.11.4. But whatever method is used, the spectrum consists of a single resonance for each chemically distinct hydrogen atom position. Fig. 2.13 shows such ^1H and ^2H spectra for toluene, obtained by the first two methods. Surprisingly, this technique has been applied very little to inorganic problems, although its potential is considerable. For example, reaction of trisilane, Si_3H_8, with various metal complexes should give a number of derivatives, which may contain $SiH_3SiH_2SiH_2-$, $(SiH_3)_2SiH-$ or other multiply-substituted groups. The ^1H spectra would be extremely complicated,

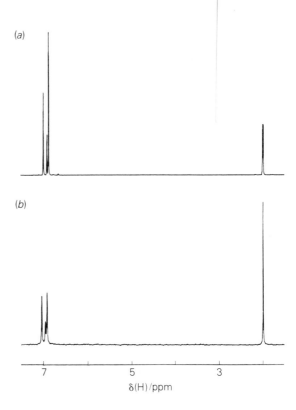

Fig 2.13 (a) ¹H NMR spectrum of the residual protons in a sample of deuteriated toluene. (b) Natural abundance ²H spectrum of normal toluene.

δ(H)/ppm

but by using Si_3D_8, and then observing residual protons, just three peaks would be expected for the "$SiD_3SiD_2SiD_2$—" group, two for "$(SiD_3)_2SiD$—" and so on, and the problem is thus readily solved.

2.6.5 **Coupling to quadrupolar nuclei**

So far we have only considered spectra in which multiplets arise from coupling to spin 1/2 nuclei. However, coupling to nuclei with spin $>1/2$ may sometimes be observed, and in this case a spin of $n/2$ causes a splitting into $n+1$ lines of *equal* intensity. Thus the ¹H spectrum of $^{14}NH_4^+$ has three equal intensity lines, and the spectrum of GeH_4 (Fig. 2.14) has a large central resonance, from molecules containing non-spinning germanium isotopes, with 10 roughly equal satellites, attributed to the 8% of the molecules containing ^{73}Ge, for which $I = 9/2$. Coupling to more than one quadrupolar nucleus is rarely observed. The most usual exception involves deuterium, for which $I = 1$. Coupling to two deuterium nuclei gives rise to a five-line pattern with intensities 1:2:3:2:1, while three give a seven-line multiplet, with relative intensities 1:3:6:7:6:3:1. These intensities are most easily worked out using a form of Pascal's triangle (Table 2.4).

In practice, sharp multiplets showing coupling to quadrupolar nuclei are not often observed. Such nuclei usually have very short relaxation times, and the efficiency of the quadrupolar relaxation may be such that the coupling gives broad lines, or is not seen at all (see Section 2.9.2). The relaxation is more efficient for nuclei with large quadrupole moments (see Table 2.2) and the other important factor is the electric field gradient, which is related to the local symmetry at the nucleus. The gradient is zero for regular tetrahedral, octahedral, cubic or spherical symmetry, and so narrow lines are

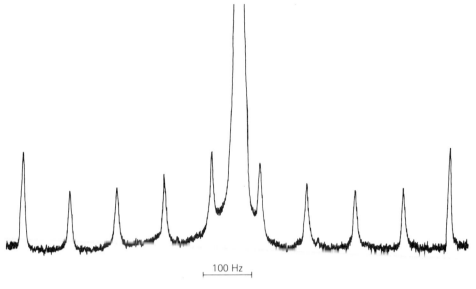

|———| 100 Hz |———|

Fig 2.14 ¹H NMR spectrum of GeH₄. The ten evenly spaced lines are due to the 8% of the molecules which contain ^{73}Ge ($I = 9/2$). The intense central line arises from all other isotopic species.

observed for ^{14}NH₄⁺ and ^{73}GeH₄, but the ¹H spectrum of ^{14}NH₃ shows a triplet of broad lines, while the spectrum of ^{73}GeH₃Cl shows no sign of coupling to ^{73}Ge at all. In by far the majority of cases couplings to quadrupolar nuclei are not observed, and it is usual to predict spectra assuming at first that all nuclei have spins of 1/2 or zero, and only later to consider the possible effects of coupling to quadrupolar nuclei. Fig. 2.15 shows

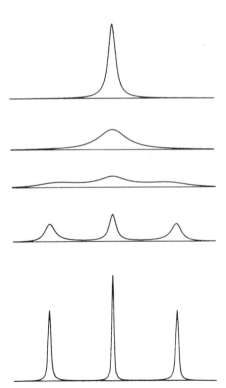

Fig 2.15 Calculated band shapes for a spin 1/2 nucleus (¹H) coupled to a spin 1 nucleus (^{14}N). The shape depends on the ratio of ^{14}N relaxation rate to the NH coupling. For very fast relaxation just a single line is observed, while for slow relaxation there are three lines of equal intensity.

theoretical line shapes for a spin 1/2 nucleus coupled to a spin 1 nucleus such as ^{14}N, for various values of the ^{14}N relaxation time.

The resonances of a quadrupolar nucleus itself are also broad if its relaxation time is very short. The linewidths themselves may give useful chemical information (see Section 2.9), but obviously very broad bands will not show evidence of small couplings to other spinning nuclei. However, couplings may been seen if the electric field gradient at the nucleus is very small or zero, and the patterns are exactly the same as for spin 1/2 nuclei. Thus the ^{27}Al spectrum of the tetrahedral AlH$_4^-$ ion is simply a 1:4:6:4:1 quintet.

2.7   Information from coupling constants

2.7.1   General principles

We now need to consider the chemical significance of the coupling constants observed in NMR spectra. The direct (dipole−dipole) coupling between nuclear spins is averaged to zero by molecular tumbling in isotropic fluids, and so does not concern us now. It is the indirect coupling, transmitted by the valence electrons of the molecule, that is observed. As with chemical shifts, the theoretical approach does not usually lead to accurate predictions of coupling constants, and a practising chemist normally uses empirical rules and correlations. Nevertheless we should consider briefly the factors contributing to coupling. This theory is well described in Ref. [20], and is covered in more or less detail in all the NMR textbooks. Details of all sorts of couplings are given and discussed in Ref. [1].

There are three components of a coupling constant arising from nucleus−electron interactions. First, the magnetic moment of one nucleus interacts with the field produced by orbital motion of the electrons, which in turn interacts with a second nuclear moment. Secondly, there is a dipole interaction involving the electron spin magnetic moments. The final contribution arises from spins of electrons in orbitals (derived from s atomic orbitals) which have a non-zero probability of being at the nucleus. This last term, known as the Fermi contact term, is by far the most important for proton−proton couplings, but for other nuclei the situation is not so simple.

As all these terms depend on the magnetogyric ratios of the two nuclei involved, it is often convenient to use the *reduced* coupling constant, K

$$K_{AB} = \frac{4\pi^2 J_{AB}}{h\gamma_A\gamma_B}. \tag{2.7}$$

This has units of NA^{-2}m^{-3}. It is particularly useful for comparing couplings for different elements, and also has the same value for different isotopes of an element. For an application of this, see Section 2.7.3

2.7.2   One-bond coupling

In situations where orbital and dipolar terms are not of great importance, one-bond couplings depend on the amount of s orbital character in the internuclear bond, *i.e.* on the hybridization of the atoms involved. This gives rise to empirical relationships which are very useful, provided care is taken. For example, typical values of $^1J_{CH}$ in hydrocarbons are 125, 160 and 250 Hz for carbon with nominal sp^3, sp^2 and sp hybridization, giving simple proportionality to percentage s character. But electronegative substituents increase the coupling greatly, to 209 Hz in CCl_3H for example. In other cases, the s contribution from each atom must be considered. For ^{13}C and ^{15}N, there is an empirical relationship:

$$\%s_N\%s_C = -80\,^1J_{NC}. \tag{2.8}$$

Other, less simply quantified, relationships exist for other couplings, such as $^1J_{PC}$, $^1J_{NH}$ (which has been related to bond angles at nitrogen) or $^1J_{PF}$. In the last case, the couplings can be used to determine the coordination number of the phosphorus atom, and in the case of five-coordinate species, to distinguish between axial and equatorial fluorine substituents. The coordination numbers of phosphorus atoms may also easily be distinguished by $^1J_{PH}$, typically 180 and 400 Hz for three- and four-coordinate atoms, respectively. Similarly, double and single phosphorus−selenium bonds may be recognized by the distinctive ranges of $^1J_{SeP}$, so enabling isomers of the types 2.XIII and 2.XIV to be distinguished. In the case of $^1J_{PN}$ care is needed, as the couplings may be positive or negative, and so their signs must be determined (see Section 2.10.3). Then derivatives of phosphorus(III) and phosphorus(V) may be distinguished.

2.XIII   2.XIV

Other useful applications have been made to transition metal complexes, where couplings between the metal and ligand atoms depend on the coordination number of the metal, and hence on its s orbital contribution to each bond. Thus values of $^1J_{PtP}$ for the four- and six-coordinate complexes 2.XV and 2.XVI are in the ratio 3:2, but care is needed to ensure that only complexes that are alike in all other respects are compared.

2.XV   2.XVI

2.7.3   Two-bond coupling

Couplings over two bonds seem to be ignored in many reference books. This may be because for many homonuclear systems the nuclei are equivalent and so the coupling is not observed. Nevertheless these couplings can give useful structural information. For example, the fact that *trans* nuclei in heavy-metal complexes normally couple much more strongly than *cis* nuclei in related complexes is of great value in assigning stereochemistry. So in structure 2.XVII $^2J_{PP}$ would be expected to be an order of

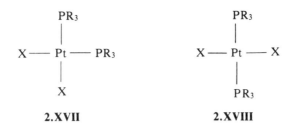

2.XVII   **2.XVIII**

magnitude less than in 2.XVIII. (This also has a bearing on the spectra of other nuclei in the PR_3 ligands, see Section 2.8.1.) In other groups of closely related compounds, there may be relationships between two-bond couplings and the angles between the bonds. Such relationships have been described, amongst others, for $^2J_{FF}$ and FCF or FPF bond angles, and for $^2J_{PH}$ and PCH angles. But in two of these three examples, and no doubt in others, the coupling constant range passes through zero, and it is therefore necessary to determine the sign as well as the magnitude of the coupling, if it is to be used in this way. Finally, note that two-bond couplings may depend on conformation. In structure 2.XIX $^2J_{PP}$ is typically 450 Hz, whereas in 2.XX it is much smaller, around 100 Hz.

2.XIX   **2.XX**

When two-bond HH couplings cannot be observed, because the hydrogen nuclei are chemically and magnetically equivalent, it may be helpful to use partial deuterium substitution. For example, the ^1H spectra of phosphines PH_2X consist of doublets, giving $^1J_{PH}$ but not $^2J_{HH}$. But the spectrum of PDHX shows an extra 1:1:1 triplet splitting arising from $^2J_{DH}$. Multiplication of this coupling constant by γ_H/γ_D, where γ is a magnetogyric ratio, gives $^2J_{HH}$.

2.7.4   **Coupling over three or more bonds**

For three-bond couplings we are back on better-trodden ground. In saturated XCCY units $^3J_{XY}$ depends primarily on the X—C—C—Y dihedral angle, ϕ. The coupling is described by:

$$^3J_{XY} = A \cos 2\phi + B \cos\phi + C \tag{2.9}$$

where A, B and C are empirical constants. This equation has been applied to $^3J_{HH}$, $^3J_{PH}$, $^3J_{PC}$ and $^3J_{FH}$ (with some complications due to other substitution effects) and undoubtedly is applicable in other situations including those with intervening atoms other than carbon. The difficulty lies in obtaining enough firm structural information to be able to determine the constants. The marked variability of three-bond couplings is illustrated by $PF_4{}^{15}NHP'F_2$ (2.XXI). In this compound, the coupling from P′ to one axial

$$2.XXI$$

fluorine is 15 Hz, while it is 209 Hz, and of opposite sign, to the other. The coupling from P′ to the equatorial fluorine atoms is too small to be resolved. Yet the fluorine nucleus coupling most strongly to P′ also couples most strongly to H − so there can be no simple explanation such as 'anti-couplings are the greatest'.

Small couplings over four or more bonds are sometimes observed. Very long-range couplings may be associated with extended π-electron systems, as in $CH_3C \equiv CC \equiv CC \equiv CCH_2OH$, for which $^9J_{HH}$ (0.4 Hz) has been observed. They may also be observed for atoms which are physically close in space although separated by many bonds. The observation of such a long-range coupling therefore provides structural information which may be important.

2.8   Not-so-simple spectra

2.8.1   Second order spectra

In Section 2.6 the first order splitting patterns arising from coupling between spinning nuclei are described, but in some circumstances second order spectra are observed, and these are more complicated. The first such circumstance occurs when the coupling between two nuclei is not several times greater than their chemical shift difference. The effect is to change the number, positions and intensities of the resonances, so that the resulting spectrum may appear incomprehensible at first sight. Often in these cases, by running the spectrum at higher field (thereby increasing the ratio of chemical shift to coupling constant), it can be made to approximate more closely to the first order form, as shown in Fig. 2.16.

For some spin systems, analysis of spectra can be achieved with the aid of equations for transition frequencies and intensities. It is not possible to go into spectral analysis in detail here, but the subject is usually covered fairly thoroughly in the major NMR textbooks, and there are some publications specifically devoted to it [21]. The other approach is to use one of the computer programs devised for the purpose referred to in Section 2.2. Suffice it to say that the analysis yields chemical shifts and coupling constants, as usual, but may also give the relative signs of some of the coupling constants, which is useful additional information.

Second order spectra also arise when there are several sets of chemically equivalent but magnetically non-equivalent nuclei. In this case the second order character is related to symmetry, and changing the magnetic field can never make the spectra become first order. Again, analysis of the spectra can be achieved by using equations for some line positions and intensities, or suitable computer programs. However, such systems often have large numbers of spinning nuclei, and cannot easily be handled by general purpose programs. It is therefore often useful to use the symmetry properties

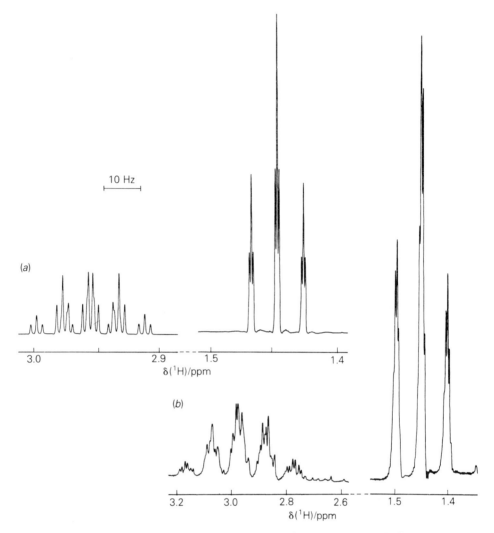

Fig 2.16 ^{1}H NMR spectra of $CH_3CH_2SPF_2$. The 360 MHz spectrum (*a*) is first order, but the 80 MHz spectrum (*b*) is second order, and not easy to interpret. The CH_2 resonances in spectrum (*a*) are a doublet ($^3J_{PH}$) of quartets ($^3J_{HH}$) of triplets ($^4J_{FH}$), and the CH_3 resonances are a triplet ($^3J_{HH}$) of triplets ($^5J_{FH}$).

of the spin system to factorize the spin Hamiltonian into sub-matrices, and so reduce the computational problems. This procedure has been fully described for $[A[X]_n]_2$ spin systems [22], which occur surprisingly often. The ^{19}F spectrum of $S(PF_2)_2$ is shown in Fig. 2.17. As with all spectra of this type it is centro-symmetric, and it is dominated by a pair of resonances, separated by the sum of couplings $|\,^1J_{PF} + \,^3J_{PF}\,|$. Each half of the spectrum has three other groups of weaker lines, and the positions of these groups enable $|\,^1J_{PF} - \,^3J_{PF}\,|$ and $|\,^2J_{PP}\,|$ to be determined. Thus the relative signs of the phosphorus–fluorine couplings can be obtained, and a coupling constant involving only ^{31}P is derived from the ^{19}F spectrum. The small splittings within the groups of resonances depend on the two different couplings, $^4J_{FF}$.

Spin systems of this type are also frequently encountered in transition metal complexes with phosphine ligands. For example, the *cis* and *trans* isomers of $[PtCl_2(PMe_3)_2]$ have $[A[X_3]_3]_2$ spin systems (ignoring the ^{195}Pt satellites), with small

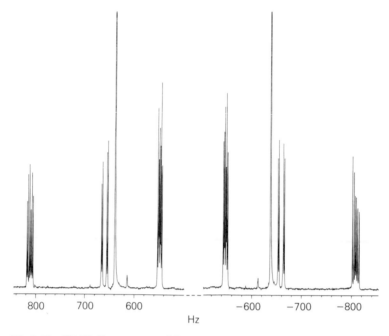

Fig 2.17 ^{19}F NMR spectrum of $S(PF_2)_2$. The spectrum is centro-symmetric, and half of the total intensity falls in the two most intense lines, which are truncated in the figure. (The two very weak lines are spinning sidebands.)

values for $^4J_{PH}$ and $^6J_{HH}$. In the *trans* isomer $^2J_{PP}$ is large, and the ^{1}H spectrum has half its intensity in a pair of lines, separated by $|\,^2J_{PH} + {}^4J_{PH}\,|$, and the other half in a series of lines, which all fall in the center of the group of resonances. The resulting pattern is thus apparently a simple triplet. For smaller values of $^2J_{PP}$ as, for example, in *cis*-$[PtCl_2(PMe_3)_2]$, the group of resonances accounting for the second half of the intensity no longer coincide, and the center line of the triplet may appear broadened, or split into a number of components, or, if $^2J_{PP}$ is zero, may simply add to the outer pair of lines, so that the resulting spectrum is a doublet. Thus the form of the spectrum indicates whether *cis* or *trans* isomers are present, for PMe_3, PMe_2R, $PMeR_2$, or other related ligands.

Finally, it should be noted that spectra appearing at first sight to be first order may in fact show some second order characteristics, and that this often causes confusion. If an unexpected small splitting is found, or the separations between lines in a multiplet are uneven, or intensities seem distorted, check whether the spectrum may in fact be second order, and if necessary, calculate a theoretical spectrum.

2.8.2   Chiral and prochiral non-equivalence

In this section we do not introduce any new physical principle, but merely attempt to clarify a matter which causes endless misunderstandings. Any molecule having no elements of rotation–reflection (S) symmetry (remembering that a reflection is S_1 and an inversion is S_2) is chiral. In such molecules some groups of nuclei normally expected to be chemically equivalent may be non-equivalent. For an —MX_3 or planar —MX_2 group, internal rotation usually makes the X nuclei equivalent on the NMR timescale, but a pyramidal —MX₂ group in an optically-active molecule must have non-equivalent X nuclei, no matter what rotation takes place. So the protons of —CH₂X or —SiH₂X groups, the fluorines of —PF₂ groups, or the methyl groups of —PMe₂R ligands will show non-equivalence in these circumstances. This can be put to good use. For example, to study $^2J_{FF}$ in difluorophosphines, compounds containing secondary butyl groups, such as PF_2NHBu^s, were synthesized. The two fluorine nuclei were thus non-equivalent and coupling between them could be observed.

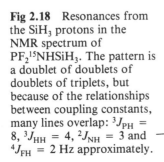

2.XXII

A similar situation arises in compounds such as 2.XXII, even though this is not chiral. Although the two —OPF₂ groups are equivalent, being related by symmetry, there can be no symmetry relationship between the two fluorine nuclei within one such group. The consequences are exactly the same as for chiral molecules, but the circumstances giving rise to the phenomenon (called prochirality) are not so easily spotted!

2.8.3 Coincidences

Another common cause of headaches is the failure to recognize that fortuitous relationships between NMR parameters may lead to spectra differing markedly from what may at first be expected. For example, we recently had a compound containing an ethyl group, but instead of the expected triplet and quartet in the ¹H spectrum, we observed just a single resonance. The explanation was straightforward – once it was recognized! The methyl and methylene protons were isochronous (having the same chemical shift), and so behaved as a group of chemically equivalent nuclei, showing no internal coupling. Changing the solvent led to a small change in the internal chemical shift, and the normal [A₃B₂] spectrum was then observed.

Fig 2.18 Resonances from the SiH₃ protons in the NMR spectrum of $PF_2{}^{15}NHSiH_3$. The pattern is a doublet of doublets of doublets of triplets, but because of the relationships between coupling constants, many lines overlap: $^3J_{PH}$ = 8, $^3J_{HH}$ = 4, $^2J_{NH}$ = 3 and $^4J_{FH}$ = 2 Hz approximately.

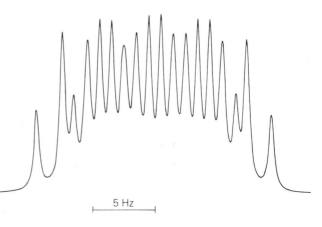

5 Hz

SECTION 2.8: NOT-SO-SIMPLE SPECTRA   61

Similarly, because coupling constants can be positive or negative, they can also be very close to zero, and so a coupling may apparently be absent. What appears to be a quartet may in fact be a doublet of triplets, with the two coupling constants having the same magnitude. Or unusual intensity patterns in multiplets may be seen when the magnitudes of coupling constants are in simple ratios to each other. The spectrum shown in Fig. 2.18, for the SiH_3 protons of $PF_2^{15}NHSiH_3$, indicates the difficulty. It has 18 lines, of which the middle 16 come at almost exactly 1 Hz intervals, and 14 of them have the same intensity. The pattern actually arises from couplings of almost exactly 8, 4, 3 and 2 Hz, to ^{31}P, ^{15}N, ^{1}H and ^{19}F, respectively. Knowing that, it is easy to account for the observed spectrum, but working the other way is not so easy.

2.8.4   The multi-nuclear approach

The emphasis in this book is on the broad approach to structural problems, using as many techniques as possible to elucidate the structural properties of a compound. In the same way, it is often valuable to apply a single spectroscopic method widely, and obtain more information than is given by a single spectrum. Too often we are content to run a proton NMR spectrum and puzzle over its interpretation, when spectra of other nuclei may resolve the problem quickly and easily. The analysis of the spectrum shown in Fig. 2.18 is not at all difficult when it is taken in conjunction with ^{19}F and ^{31}P spectra. Once the magnitudes of two of the coupling constants have been determined, the others follow quickly. Moreover, the various spectra, taken together, provide sufficient information to enable the compound to be completely identified.

The usefulness of the multinuclear approach is illustrated by a study of the unusual tetra-alkylmercurate, 2.XXIII [23]. The ^{1}H spectrum of this compound has three groups of resonances with the intensity ratio 12:4:1. The resonances due to the CH_3 and CH_2 groups show couplings to ^{31}P, and also ^{199}Hg satellites, while the third

2.XXIII

resonance is broad, and has no resolved couplings. This shows that the ligands are symmetrically bonded to mercury, and this is confirmed by observation of a single ^{31}P resonance, which also has ^{199}Hg satellites. But still there is no evidence that the mercury has two of these ligands. The proof is provided simply by observing the ^{199}Hg spectrum, with ^{1}H decoupling. This is a quintet, with coupling to four equivalent ^{31}P nuclei.

Of course, the temptation will be to run a spectrum for every conceivable nucleus in every sample. This can be wasteful of expensive instrument time. But many different nuclei can now be routinely observed, and it is worth remembering that in some circumstances, running a spectrum of one of the more unusual nuclei may save a lot of time or trouble.

2.9 Relaxation

2.9.1 General principles

Although chemical shifts and coupling constants are the NMR parameters most usually considered, useful chemical information may also sometimes be obtained by studying relaxation processes. Relaxation rates and mechanisms are harder to measure and to understand than the other parameters, but some discussion is necessary, as the information obtainable may be valuable, and an understanding of the phenomena may help in the planning of experiments involving Fourier transform spectra. We do not attempt to go into the theory of relaxation here because space is too limited, but there are accounts of various lengths in the more recent NMR textbooks, with Ref. [24] being more easily understood than some others! There are many examples of the applications of relaxation studies in Ref. [1].

There are two important relaxation times. The spin-lattice relaxation time, T_1, relates to the restoration of the Boltzmann equilibrium after excitation of a spin system by a radio-frequency pulse, by transfer of energy to its surroundings. It describes the changing of the component of the magnetization which is parallel to the applied magnetic field. Relaxation in the plane perpendicular to the applied field is described by the spin–spin relaxation time, T_2, which, as its name implies, relates to the redistribution of energy between spins. The two relaxation times can be measured independently, using techniques involving two or more pulses. For T_1, the inversion recovery method is commonly used, with the so-called Freeman–Hill modification to reduce errors, while spin-echo methods are used to measure T_2, with the pulse sequence known as the Carr–Purcell–Meiboom–Gill experiment giving the most accurate and precise results.

For non-viscous liquids and gases the 'extreme motional narrowing' condition applies, and T_1 and T_2 are essentially equal. For viscous liquids and solids T_2 is much smaller than T_1. As linewidths increase as T_2 decreases, there are extreme problems (discussed in Section 2.14) in obtaining useful spectra at all. In the following paragraphs we consider only non-viscous fluids, although the mechanisms discussed may also apply for other states.

2.9.2 Relaxation mechanisms

For a nucleus with spin 1/2, relaxation to the Boltzmann distribution can only occur by interactions of its magnetic moment with the random magnetic fields in the sample. These random fields originate in the rotations and translations (but not vibrations, which have too high frequencies) of the molecules. The overall relaxation rate therefore depends on the correlation times characterizing these motions, and on the coupling energies for the interactions between the spin system and the molecular motions. Nuclei in small molecules, which tumble faster than large ones, generally have relatively short relaxation times. Similarly, the ends of long chains are more mobile than the middles, and ^{29}Si T_1 measurements have been used to study the mobility of different parts of silicone chains, and the dependence of mobility on chain length, just as ^{13}C T_1 measurements yield information about organic polymers.

There are several mechanisms contributing to the exchange of energy between the spin system and its surroundings, and the rates for these processes are additive. Direct dipole–dipole interactions with other nuclei, both intra- and inter-molecular, are important, and it is when this mechanism is dominant that the Nuclear Overhauser Effect is most pronounced. This effect is discussed more fully in Section 2.10.6. Interaction between nuclear spin and molecular rotations is also important, and the energy transfer occurs in this case when the rotation is interrupted by collisions. This mechanism is particularly significant for low viscosity liquids at high temperatures, and for gases.

A third mechanism brings us to a practical problem. The shielding at a spinning nucleus is in general anisotropic. Normally this does not matter, as molecules tumble rapidly, but the tumbling can cause relaxation. The relaxation rate is proportional to the square of the applied magnetic field, and for some nuclei, particularly the heavier ones, this can cause difficulties. It is found in practice that for nuclei such as ^{195}Pt and ^{205}Tl, spectra obtained at high fields have broad lines, and the increase in sensitivity expected using higher field spectrometers is not achieved.

Further contributions to relaxation can come from scalar interactions with another spinning nucleus. These can arise if the coupling constant for the two nuclei is time-dependent as a result of chemical exchange (Section 2.15.2), or if the second nucleus is itself relaxing rapidly, and there is coupling between the two. The second situation arises with quadrupolar nuclei, and is discussed in Section 2.6.5. Of course, for the quadrupolar nucleus itself, the interaction of its nuclear quadrupole moment with an electric field gradient provides an extremely efficient process for nuclear relaxation via the molecular rotation. For example, for ^{14}N in MeCN, T_1 is only 22 ms, but for aqueous ammonium ion, which has no electric field gradient, it is more than 50 s. For quadrupolar nuclei in general quadrupolar relaxation is overwhelmingly dominant.

The final relaxation mechanisms involve unpaired electrons. Intramolecular nucleus–electron dipole–dipole interactions are normally very large, and NMR spectra of paramagnetic species usually have very broad lines, and may not even be observed at all. However, there are also smaller intermolecular effects, and these may be put to good use. One problem frequently encountered in Fourier transform NMR is that of observing nuclei that relax very slowly, such as ^{13}C in metal carbonyl complexes, as discussed in Section 2.4.3. For such nuclei, either pulse rates must be very slow, or only a small part of the full free induction decay can be monitored. Either way, the experiment takes a relatively long time, and good signal-to-noise ratios may be hard to obtain. Addition of a trace of a paramagnetic compound (such as tris-acetylacetonato-chromium(III)) can solve the problem: the relaxation times are reduced, and the pulse repetition rate can be increased accordingly.

2.10   Multiple resonance

2.10.1   Introduction

In the normal, single resonance, NMR experiments, an observing radiofrequency field, $B_1 \cos \omega_1 t$ (where ω_1 is the frequency), is applied at right angles to the static magnetic field, B_0. In double resonance experiments a second radiofrequency, $B_2 \cos \omega_2 t$, is also applied at right angles to B_0, and the spectrum observed may be perturbed. If we observe nucleus A while irradiating transitions associated with nucleus X (an experiment usually written as A–{X}) it may be possible to find out a great deal about the NMR spectrum of X without observing it directly. As the equipment needed for double resonance experiments is not very expensive (assuming one has a single resonance spectrometer to start with!) this is an easy, cheap and versatile way of extending the usefulness of a simple instrument. Moreover, there is much information that can *only* be obtained by multiple resonance methods.

The applied perturbing field can be controlled in several ways. The frequency is obviously variable, but it is also possible either to use a single frequency, or to irradiate with a band of frequencies. The power level is also important, and this can determine the form of the observed spectrum, in particular whether the line positions are changed, or just the intensities. Finally, controlling the timing of switching of B_2 on and off makes it possible to separate the intensity and transition frequency effects.

2.10.2 Low power irradiation – population transfer

Consider the energy level diagram for an [AX] spin system shown in Fig. 2.19. The terms $\alpha\alpha$, $\alpha\beta$ etc. refer to the spin quantum numbers of the A and X nuclei. At thermal equilibrium the population differences are proportional to the energy differences, as these are small compared with kT. If the transition X_1 is saturated by an applied radio-frequency field, the populations of the levels connected by this transition are equalized. As the intensities of lines A_1 and A_2 depend on the population differences for their associated energy levels, that of A_2 is enhanced by irradiating X_1, while that of A_1 is

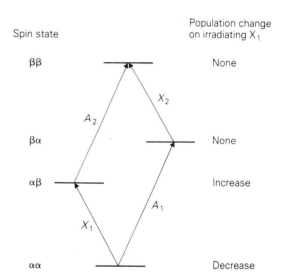

Spin state

Population change on irradiating X_1

$\beta\beta$ — None

$\beta\alpha$ — None

$\alpha\beta$ — Increase

$\alpha\alpha$ — Decrease

Fig 2.19 Energy level diagram for an [AX] spin system.

diminished. In the same way, irradiation of X_2 increases the intensity of A_1, but decreases that of A_2. One way to present this information is as a plot of intensity of A_1 (or A_2) against irradiation (X) frequency, as shown in Fig. 2.20. This can easily be done with continuous wave spectrometers simply by monitoring the chosen transition, and scanning the irradiation frequency. The experiment, known as internuclear double resonance (INDOR) thus provides information about the frequencies in the X spectrum, while observing only A. This sort of experiment is less important than formerly, since multinuclear spectrometers are now widely available, but it is still useful, particularly if X is an insensitive nucleus for NMR or, of course, if only a proton spectrometer is

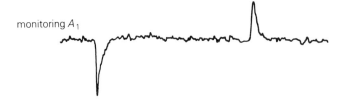

monitoring A_1

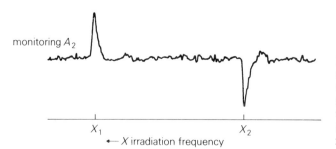

monitoring A_2

X_1 X_2

← X irradiation frequency

Fig 2.20 A—{X} INDOR spectra. The two traces are obtained by monitoring resonances A_1 and A_2, while varying the irradiation frequency in the vicinity of resonances X_1 and X_2.

available. For example, in a study of SiH_3 derivatives of mercury a compound was formed, having a proton spectrum showing ^{29}Si and ^{199}Hg satellites, but no other coupling. Spectra of these other nuclei could not at that time be observed, so a $^1H-\{^{199}Hg\}$ INDOR spectrum was recorded, and showed a seven-line multiplet. The compound therefore had six equivalent hydrogen atoms, and was identified as $Hg(SiH_3)_2$, and not $HgCl(SiH_3)$.

The perturbing field, B_2, does not have to be applied continuously. With Fourier transform spectrometers it is possible to achieve population equalization using a 90° pulse, and the effect on intensities is the same as for continuous wave irradiation. However, a 180° X pulse inverts the X spins and thus the populations of the associated levels. The effect of this is to change the intensities of the A transitions even more dramatically, and this forms the basis of the polarization transfer techniques described in Section 2.11.2.

Finally, we should note that intensity effects are not instantaneous, but take time, of the order of T_1, to build up and decay. There are other consequences of double resonance, involving changes in transition frequencies, and these depend on a modified spin Hamiltonian. They therefore occur immediately the irradiating field is switched on, and cease when it is switched off. Judicious switching enables the two types of effect to be separated.

2.10.3   Medium power irradiation – spin tickling

We have seen that weak irradiation of a particular transition of a two-spin system perturbs the populations of the two energy levels connected by the transition, and that the intensities of associated resonances are therefore altered. The same principles apply for any multi-spin system. As the strength of the irradiating field is increased, the population changes are still seen, but the energy levels themselves are also altered, and the *pattern* of the observed spectrum changes. The form of the new spectrum can be calculated using a Hamiltonian modified to take account of the extra radiofrequency field. Details of the theory are given in recent textbooks. All we need to note is that at moderate power levels, with continuous wave irradiation, transitions sharing an energy level with the irradiated transition split by $\gamma_X B_2/2\pi$, where γ_X is the magnetogyric ratio of the irradiated nucleus. This sort of experiment, known as spin tickling, may therefore help in assigning lines in complicated spectra, or in determining the relative signs of coupling constants. Note that INDOR spectra can be obtained under these conditions, but that as a line splits, the intensity at the original position falls to zero. The spectrum is therefore a series of negative peaks, and not some positive and some negative, as was shown in Fig. 2.20. Consider again the [AX] energy level diagram, Fig. 2.19. The transitions A_1 and A_2 do not have exactly the same frequency – they differ by J_{AX}, a few Hz in tens or hundreds of MHz. If J_{AX} is positive, the two middle spin states are reduced in energy, and A_1 is therefore of lower frequency than A_2, and X_1 is lower than X_2. For negative J_{AX} the reverse is true. But for first order spectra as considered so far it simply does not matter, because we have no way of knowing which line is which. However, for any spin system with three or more sets of spins coupled together it is possible to determine relative signs, but not absolute signs of coupling constants. The simplest three-spin system is [AMX], for which an energy level diagram is shown in

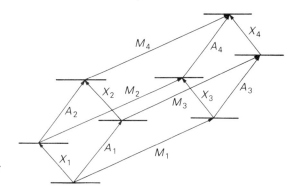

Fig 2.21 Energy level diagram for an [AMX] spin system.

Fig. 2.21. Irradiation of line X_1 affects only A_1 and A_2 in the A spectrum, and only M_1 and M_2 in the M spectrum. Irradiation of X_2 would have exactly the same effect on the A spectrum, but lines M_3 and M_4 would be affected in the M spectrum. Now consider the A and X spectra (Fig. 2.22), and the effect on A of irradiating X transitions. The figure shows that the pair of transitions A_1 and A_2 may lie to high or low frequency of A_3 and A_4, depending on the sign of J_{AM}. Similarly, J_{MX} determines the positions of X_1 and X_2 relative to X_3 and X_4. The experiment is therefore simple. Irradiate either of the high frequency lines in the X spectrum. If high frequency lines in the A spectrum are split, J_{AM} and J_{MX} have the same sign; if low frequency lines split, the signs are opposite. Note that the experiments A−{X} and X−{A} give the same information, but that A−{M} would relate the signs of J_{AX} and J_{MX}. Thus by observing only A and irradiating M and X we can now determine *all* the chemical shifts and coupling constants, including relative signs, for the spin system. Absolute signs can only be found if the absolute sign of one of the set is known. This is often possible as, for example, $^1J_{CH}$ and $^1J_{PtP}$ are always positive, and $^1J_{PF}$ is always negative.

It is in the cases where couplings may be either positive or negative that the information is valuable. For example, $^1J_{PN}$ normally has different signs in phosphorus(III) and

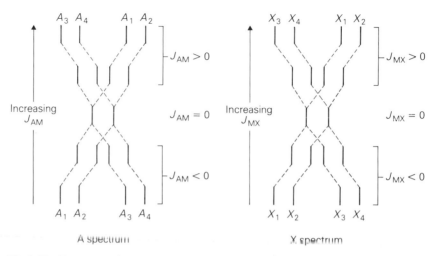

Fig 2.22 Representations of A and X spectra of an [AMX] spin system. The relative positions of transitions in the A spectrum depend on the magnitude and sign of J_{AM}. Similarly, positions in the X spectrum depend on J_{MX}.

phosphorus(v) compounds. Clearly the determination of the sign yields chemical information. Similarly, some coupling constants are dependent on geometrical parameters, particularly dihedral angles (Section 2.7.4). If the possible range includes zero, then the sign of the coupling must be known if unambiguous structural information is to be obtained.

<table>
<tr><td>2.10.4</td><td></td></tr>
</table>

High power irradiation – spin decoupling

We have seen for the [AX] spin system that irradiation of X transitions at moderate power levels can cause splittings of lines in the A spectrum. If the power is increased further the effects become more dramatic, and at high levels ($B_2 \gg 2\pi J_{AX}/\gamma_X$) the coupling between A and X is completely eliminated. The precise patterns observed depend on the frequency of irradiation (ideally it should be exactly in the center of the AX doublet in the X spectrum) and on the power level. In practice, spin decoupling is normally carried out using a broad band of frequencies (noise decoupling) so that the precise frequency chosen is not important. Details of decoupling theory are given in most NMR textbooks, but we will consider just two situations: full noise decoupling, and off-resonance decoupling.

Noise decoupling, completely removing the effects of the irradiated nuclei from observed spectra, is widely and routinely used. Without proton decoupling, ^{13}C spectra of organic compounds would normally be impossibly complicated, with each type of carbon coupling to several different sets of protons. With decoupling, as described in Section 2.6.4, the spectra normally consist of a single resonance for each chemically distinct carbon, and are easier to interpret. The same principles apply for many other spinning nuclei which couple to abundant nuclei (usually hydrogen). For example, Fig. 2.23 shows ^{31}P spectra, with and without ^{1}H decoupling, for a mixture of *trans*-[PtCl$_4$(PEt$_3$)$_2$] and *trans*-[PtBr$_4$(PEt$_3$)$_2$]. With proton coupling retained each ^{31}P nucleus couples to the thirty protons, and the spectrum shows a partially resolved series of broad resonances. With decoupling there is just one sharp line (with ^{195}Pt

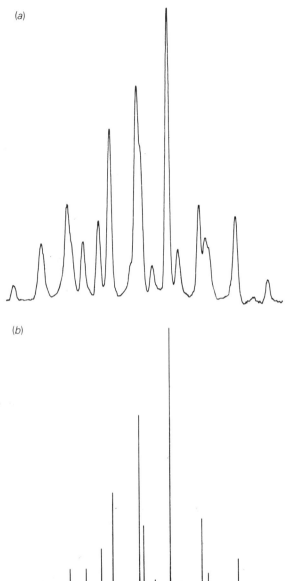

(a)

(b)

Fig 2.23 81.0 MHz ^{31}P NMR spectra of the mixture of products formed by the reaction of *trans*-[PtCl$_4$(PEt$_3$)$_2$] and *trans*-[PtBr$_4$(PEt$_3$)$_2$]. In spectrum (*a*) proton coupling is retained, and not all the resonances from the various products can be distinguished. Spectrum (*b*) was obtained with proton noise decoupling, and shows six resonances, each with ^{195}Pt satellites. The six complexes have from four chlorines and no bromines to no chlorines and four bromines: there are two isomers of the complex with two atoms of each halogen.

1000 Hz

satellites) for each compound, and it can be seen that six compounds are present. These are all species with *trans* PEt$_3$ groups, and four, three, two (two isomers), one and no chlorine substituents, with a more-or-less random distribution of the halide ligands.

As well as this general simplification, noise decoupling may help in assigning particular coupling constants. Fig. 2.24 shows the ^{1}H spectrum of Fig. 2.18, this time with ^{15}N decoupling. The change in the pattern shows unequivocally that the protons concerned are coupled to one nitrogen nucleus. Similarly, interpretation of the complex pattern shown in Fig. 2.18 could be simplified by recording ^{1}H$-\{^{19}$F$\}$, and ^{1}H$-\{^{31}$P$\}$ spectra, in each of which one particular coupling would be removed.

Example 2.3

Reaction between [Ir(CO)H(PPh$_3$)$_3$] and SiBrH$_3$ leads to the elimination of PPh$_3$ and the formation of the complex shown in the figure, whose ^{31}P-{^{1}H} NMR spectrum shows the presence of two equivalent ^{31}P nuclei. The proton spectrum (excluding the aromatic region) is shown below. How would you use double resonance to help to account for the splitting pattern observed in the spectrum for H$_t$, at δ-10.03?

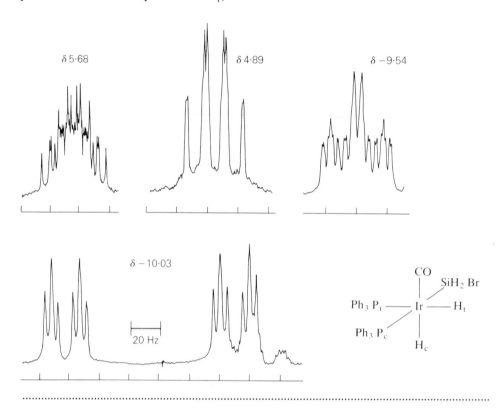

The pattern observed for H$_t$ is a wide doublet of narrower doublets of triplets. The wide doublet can be put down to coupling to the *trans*-^{31}P nucleus and the narrower to coupling to the *cis*-^{31}P nucleus; this may be confirmed by single-frequency ^{31}P-{^{1}H} double resonance. The triplet coupling is more puzzling. We would expect to observe coupling between H$_t$ and H$_c$, but this should lead to a doublet and not a triplet. There are three other proton resonances in the spectrum shown: one of these is at low frequency, and must be due to H$_c$, and the other two must be due to the SiH protons, which are made non-equivalent by the chirality. Irradiation at H$_c$ reduces the triplet splitting in H$_t$ to a doublet splitting; irradiation at δ4.89 has no effect, but irradiation at δ5.68 also reduces the triplets to doublets. This shows that the triplet splitting in H$_t$ is due to two doublet couplings that happen by chance to be equal; one is to H$_c$ and the other to one of the SiH$_2$ protons. The two SiH resonances look very complicated, but can be analyzed easily in terms of a series of doublet splittings; there are five nuclei with spin one-half that are involved in the spin system, and at high resolution all 32 lines expected in the group of resonances centered at δ5.86 can be resolved.

For more details, see E.A.V. Ebsworth, T.E. Fraser, S.G. Henderson, D.M. Leitch and D.W.H. Rankin, *J. Chem. Soc., Dalton Trans.*, 1010 (1981), from which the figure is taken, with permission.

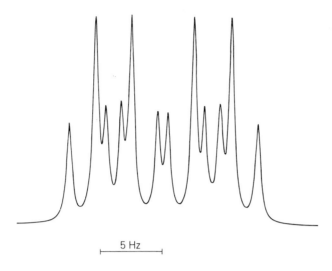

Fig 2.24 ^1H spectrum of $PF_2{}^{15}NHSiH_3$, with ^{15}N noise decoupling. This spectrum should be compared with that shown in Fig 2.18.

5 Hz

In off-resonance decoupling a single irradiation frequency is used, applied a few hundred Hz away from the resonances involved. The effect is to remove all long-range couplings, but not quite collapse the largest (normally one-bond) couplings, so that it is possible to see multiplet structure, and so determine the number of neighboring spinning nuclei. This method is very often used for ^{13}C spectra, which are extremely complicated with all the proton couplings retained. With full decoupling, the spectra become simple, with series of single lines, but assignment is difficult. With off-resonance decoupling, the single lines are replaced by narrow multiplets, and it is therefore easy to identify primary, secondary, tertiary and quaternary carbon atoms. The method is general, although mainly used for $^{13}C-\{^1H\}$, but to some extent it has been superseded by multi-pulse techniques (see Section 2.11).

One problem frequently encountered in decoupling experiments with inorganic systems arises from the size of the coupling constants involved. The power levels required when the constants are 1000 Hz or more are so high that it is difficult to avoid boiling the sample or even melting the probe. In these cases composite pulse decoupling [25] seems promising. This is a new technique, not fully developed at the time of writing, which in effect is similar to noise decoupling, but can deal with coupling constants of several kHz. It should be useful to inorganic chemists when it becomes widely available.

2.10.5 Triple resonance

The term triple resonance is used for any experiment in which two irradiation frequencies are applied while a third frequency is observed, but there are two distinct types of experiment involved. In the first type, one of the irradiation channels provides noise decoupling, usually of ^1H, while the other is used for a spin-tickling experiment. There is nothing conceptually new in this; it is simply a combination of techniques already discussed. It might be used, for example, to study the ^{195}Pt chemical shifts of the phosphine complexes involved in the spectrum shown in Fig. 2.23, using $^{31}P-\{^1H,$ noise$\}-\{^{195}Pt,$ tickle$\}$. The other sort of experiment is exemplified by a study of organophosphine selenides [26], in which no ^{77}Se coupling could be observed in the ^1H

spectrum. Nevertheless, because ^{77}Se is coupled to ^{31}P, and ^{31}P to ^{1}H, it was possible to determine the ^{77}Se chemical shift and coupling constant to ^{31}P, by ^{1}H $-\{^{31}$P, ^{77}Se$\}$ triple resonance. Such experiments are rarely required when multi-nuclear spectrometers are available but may still occasionally be needed for proton-only instruments.

2.10.6   The Nuclear Overhauser Effect

In all double resonance experiments there are changes in the intensities of observed resonances, whatever the irradiation power levels. In decoupling experiments the ratio of total integrated double resonance intensity to single resonance intensity is known as the Nuclear Overhauser Effect. (The name is applied both to the magnitude of the effect, and to the phenomenon itself.) Its magnitude depends on the balance of relaxation mechanisms (Section 2.9.2), and its maximum value for an A $-\{X\}$ experiment, occurring when dipole $-$ dipole mechanisms predominate, is given by

$$\text{NOE}_{\text{max}} = 1 + \gamma_X/2\gamma_A. \tag{2.10}$$

The use of the NOE to study relaxation mechanisms is discussed fully in Ref. [27], and in most NMR textbooks. Here we can only consider the most important practical points for chemists. First, the enhancement can give a useful gain in signal-to-noise ratio. Dipolar relaxation is particularly important for spin 1/2 nuclei, and as the rate is inversely proportional to the sixth power of the distance between the nuclei, it is most significant where the nucleus being irradiated is directly bonded to the one being observed. For ^{13}C $-\{^{1}$H$\}$ the maximum effect is nearly 3, and enhancements close to this are normally seen for all carbons bound to hydrogen, but not for quaternary carbons.

Secondly, we should note that Eqn (2.10) leads to *negative* enhancements if γ_X or γ_A is negative. For ^{29}Si $-\{^{1}$H$\}$ the maximum effect is *ca.* -1.5, while for ^{15}N $-\{^{1}$H$\}$ it is -4, but as the minimum effect is $+1$, the actual effect may in practice be zero, and expected resonances may be absent. Fig. 2.25 shows ^{29}Si spectra of SiHPh$_3$ showing the inversion of the signals and the improvement in signal-to-noise ratio obtained on proton decoupling.

Eqn (2.10) shows that the maximum enhancement is 50% for homonuclear decoupling. As dipole–dipole relaxation depends on distances between nuclei, molecular conformations can be studied by measuring Overhauser enhancements. Normally applied to ^{1}H–$\{^{1}$H$\}$ experiments, the procedure involves recording two spectra, one with no irradiation, and one while irradiating one particular group of protons. The

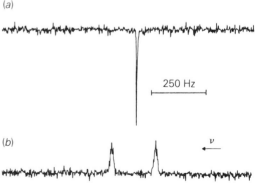

(a)

250 Hz

Fig 2.25   Proton-decoupled (a) and coupled (b) ^{29}Si spectra of triphenylsilane. [Reprinted from R.K. Harris & B.J. Kimber, *Appl. Spectrosc. Rev.*, **10**, 117 (1975) by courtesy of Marcel Dekker, Inc.].

(b)

ν

difference spectrum only has signals due to those protons which are physically close to those being irradiated. By a series of experiments, a map of the relative proton positions can be built up. Two-dimensional NOE spectroscopy is described in Section 2.11.4.

<table>
<tr><td>2.10.7</td><td>

Gated decoupling

The Nuclear Overhauser and decoupling effects that we have discussed are not always desirable. We may need accurate integration data which could be distorted by Overhauser enhancements. We might also want to avoid the Overhauser effect where it could lead to negative or zero intensities. Or we may want the intensity enhancement, while retaining a fully coupled spectrum. Separation of the two effects may be achieved by using gated decoupling in Fourier transform experiments. If the decoupler is switched on at the time of the observe pulse, decoupling occurs immediately, whereas the NOE builds up slowly, and so does not influence the FID. There must be a delay after acquisition of the FID to allow the NOE to decay, before the next pulse sequence starts. In this way decoupling is observed, without NOE. The reverse procedure, with the decoupler off during acquisition, but on for a period between pulses, gives an NOE without decoupling. Thus the intensity-change effect (*i.e.* inversion) and the decoupling effect, shown in Fig. 2.25, may be obtained separately.

The NOE can also be suppressed by adding a paramagnetic relaxation agent, such as tris-(acetylacetonato)chromium(III) (see Section 2.9.2), which changes the balance of relaxation mechanisms.
</td></tr>
</table>

2.11   Multi-pulse methods

<table>
<tr><td>2.11.1</td><td>

Introduction

We have already seen how it is possible, in Fourier transform NMR, to decouple nuclei or change populations of energy levels and thus alter intensities of transitions. In describing these phenomena, we have in fact referred to simple multi-pulse methods. Gated decoupling (Section 2.10.7), separating decoupling and Overhauser effects, is an example, and other pulse sequences are used to measure T_1 and T_2 (Section 2.9.1). Many other related methods have been described, and as this area of NMR is developing fast, others will no doubt appear. Most of them lead to simplification of spectra, which is good news for the chemist! A normal spectrum contains a lot of information, but it is presented as a plot of intensity against frequency which, for a complex system, may be difficult to analyze. In contrast, some special pulse methods can be used to select particular parts of a spectrum, or suppress particular peaks; some give just a gain in intensity, and others present the information in two dimensions, making analysis easier and providing new information. However, simplification of the spectra brings with it a marked increase in the complexity of the experimental theory. In this book, we have kept the theory to a minimum, and it is impossible to go very far into these methods here. But understanding their physical basis is essential, if they are to be applied in the best ways: good short descriptions are given in Refs [24], [28] and [29], and these give many other references. All the methods depend on control of the timing (start time and duration), frequencies, amplitudes and phases of series of pulses. Here we do not give the details of these sequences, but concentrate on the form of the spectra obtained, and the information that can be derived from them.
</td></tr>
</table>

Sensitivity enhancement

With multinuclear NMR spectrometers widely available, the range of nuclei commonly studied has become very great. However, there are still serious problems in recording spectra of nuclei of low sensitivity, particularly when they are also of low natural abundance. It is therefore important to utilize the techniques that are available to increase sensitivity. Using higher field magnets helps, although for heavier nuclei broadening of lines by chemical shift anisotropy (Section 2.9.2) may limit this benefit. The Nuclear Overhauser Effect (Section 2.10.6) can give signal enhancements of up to $1 + \gamma_X/2\gamma_A$ for $A-\{X\}$ experiments, but this is a maximum which is rarely achieved, and depends on the relaxation mechanisms effective for each nucleus. An alternative approach uses the inversion of populations obtainable for $A-\{X\}$ systems on applying a $180°$ pulse to one or more X transitions. The general principles of this method can be discussed with reference to the energy levels for an $[AX]$ spin system, shown in Fig. 2.19. Applying a $180°$ pulse to transition X_1 interconverts the $\alpha\alpha$ and $\alpha\beta$ spin states, so their populations are momentarily reversed. If the A spectrum is then observed, transition A_1 will have decreased in intensity, by a factor $1 - \gamma_X/\gamma_A$, while A_2 will have increased by $1 + \gamma_X/\gamma_A$. For $^{29}Si - \{^1H\}$ the two lines will have intensities of *ca.* -4 and 6, instead of 1 and 1, and so the net result is an average enhancement of γ_X/γ_A, with gross distortion of the normal multiplet pattern.

The same effect (called polarization or magnetization transfer) can be achieved without needing to irradiate a specific frequency, by using a pulse sequence called INEPT. But even that has been superseded by a sequence known as DEPT [30], which gives enhancement of γ_X/γ_A, without multiplet distortion. The only general requirement is that the AX couplings should be more or less the same for all sets of spins in the system, but then the intensity gain can be maximized, regardless of relaxation mechanism. The technique therefore has major advantages over the Nuclear Overhauser Effect for signal enhancement, being more generally applicable, never being zero, and being larger. The maximum effect is 5.03 for $^{29}Si - \{^1H\}$, compared with -1.52 for the Overhauser Effect.

2.11.3   **Selective observation**

In a normal Fourier transform NMR experiment, a short pulse is used to excite a broad band of frequencies. However, in some of the experiments described in this section it is necessary to pulse only at specific frequencies. This can be done using a long, low-power pulse, but in most spectrometers this is not possible for the observing channel. The alternative is to use the DANTE pulse sequence [31] to give 'tailored excitation', *i.e.* only at the one chosen frequency.

An elegant application of tailored excitation is widely used in ^{13}C NMR, but has great potential for other nuclei. If a single frequency in a $^{13}C - \{^1H\}$ spectrum is irradiated, only the carbon nucleus resonating at that frequency is observed. However, if the decoupler is turned off after excitation but before acquisition of the FID, all the couplings to that carbon are restored. Thus the coupling patterns of each carbon may be studied separately, avoiding the impossibly complicated overlap that would be seen in the spectrum without any decoupling. One application of this technique could be to the palladium complex 2.XII. The individual ^{13}C resonances could be observed, each with its coupling pattern retained, invaluable in assigning the allyl group resonances.

It is also occasionally necessary to excite all frequencies *except* a particular one, usually when one peak dominates all others, and may obscure important features. In this case the offending resonance is irradiated for a time at least comparable with T_1 immediately before each observing pulse (pre-saturation), and the experiment then proceeds normally. One useful extension of this technique is possible in cases in which the nucleus being irradiated is exchanging with others. The saturation is thus transferred by the exchange mechanism, and other peaks in the spectrum can be partly or totally saturated. The extent of saturation transfer depends on the exchange rate, which can thus be measured if it lies within certain limits. The uses of this are discussed further in Section 2.15.2.

Another extremely useful technique is multiplet selection, also known as spectrum editing. The best method uses the DEPT [30] pulse sequence, and is simple to perform, and quick. Again it is best illustrated by ^{13}C spectroscopy. The spectrum of 2.XII (Fig. 2.11) has thirteen lines, and it is not immediately obvious how they should be assigned. For more complicated systems, the problem is obviously greater. Using DEPT, it is possible to obtain the spectra shown in Fig. 2.26(a–c), in which the only carbon resonances seen are those having coupling to three, two or one protons, respectively. Fig. 2.26(d), obtained using a different pulse sequence, contains only resonances from quaternary carbons. Obviously this method has enormous potential, and it is not restricted to $^{13}C-\{^1H\}$, nor indeed to directly bonded pairs of nuclei. The only requirement is that all the couplings (in this case $^1J_{CH}$) should be similar, within 30% or so of an assumed average.

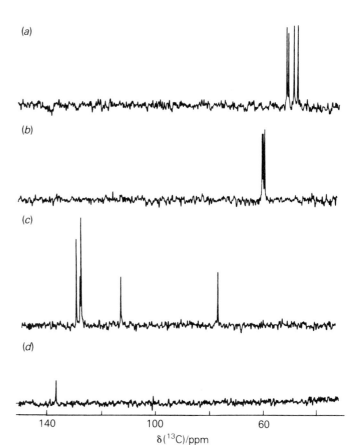

Fig 2.26 ^{13}C spectra of the palladium complex 2.XII, showing (*a*) CH_3 groups, (*b*) CH_2 groups, (*c*) CH groups and (*d*) quaternary carbons only. Spectra (*a*), (*b*), and (*c*) were obtained using the DEPT pulse sequence.

A variation of the multiplet selection method can be used to observe satellites showing coupling to a dilute spin nucleus, while rejecting the much more intense resonances due to molecules with non-spinning isotopes. For example, in a complicated proton spectrum, it may be helpful to record separate spectra for those protons bonded to ^{29}Si, ^{13}C, ^{15}N, ^{195}Pt etc. The best way to do this is to use the DEPT sequence, observing ^{1}H and irradiating the other relevant nucleus, and to select only doublets. The main resonances are basically singlets, and the satellites are doublets, so the desired effect is achieved. There are several other ways of obtaining similar results, but they generally give distorted spectra.

Finally, there is a method for the observation of homonuclear satellites in the spectrum of a dilute spin nucleus. So far it has been applied mainly to ^{13}C, but has potential for other nuclei. The technique, called INADEQUATE [32], involves excitation of double quantum transitions, which is only possible for the very small proportion of molecules (0.01% for a two carbon atom system) having two ^{13}C nuclei. The spectrum thus obtained contains only ^{13}C satellites of ^{13}C resonances, and so the $^{13}C-^{13}C$ coupling constants can be measured. These yield information about carbon–carbon connectivity, and thus about the chemical structure. A two-dimensional version of INADEQUATE, which provides the linkage information more clearly, is described in the next subsection.

2.11.4   Two-dimensional NMR spectroscopy

In Sections 2.11.2 and 2.11.3 we have described several pulse sequences, which have one or more intervals between the pulses. The duration of these intervals is of great importance, and the appearance of the spectrum obtained in each case changes as the intervals are varied. For any particular type of experiment a series of spectra with regularly incrementing delay intervals can be obtained, by Fourier transforming each FID with respect to the time after the observe pulse. Each of the spectra presents intensity as a function of frequency in the normal way. The series of intensities at any particular frequency can then be Fourier transformed again, this time with respect to the delay interval. If this is done for each spectral frequency a two-dimensional (2D) spectrum results, with each intensity point being a function of two distinct frequencies. What these two frequencies represent depends on the type of 2D spectrum, but they may be chemical shifts or coupling constants, or frequencies of double quantum transitions. Each type of experiment involves its own pattern of pulses and delays, details of which may be found in a book on the subject [33], while Refs [24], [28] and [29] provide easier theoretical introductions. Here we illustrate some of the methods which are proving to be important in the study of inorganic compounds, and as far as possible we have chosen spectra which are simple to interpret. Of course, one of the reasons for using 2D spectra is to simplify the analysis for systems which give complicated 1D spectra, so most 2D spectra that you will encounter in the chemical literature will be more confusing than those given here.

In Fig. 2.27 a 2D spectrum is shown as an intensity contour map. This spectrum was obtained from a solution of $B_{10}H_{14}$ (2.XXIV) by a technique known as correlated spectroscopy (COSY), which provides information about couplings between nuclei of a single isotope. Both axes relate to ^{11}B chemical shifts, and peaks appear on the diagonal at frequencies (the same frequency on each axis) corresponding to the resonances in the normal ^{11}B spectrum. The appearance of an off-diagonal peak at a frequency (f_1, f_2) implies that there is a coupling between the nuclei

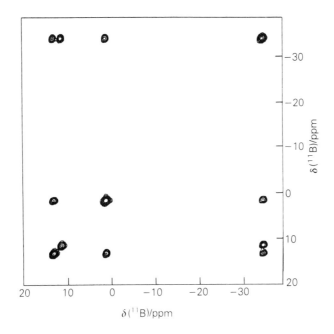

Fig. 2.27 COSY ^{11}B spectrum for $B_{10}H_{14}$, 2.XXIV.

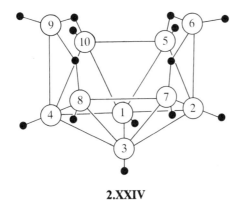

2.XXIV

resonating at f_1 and f_2. There is usually strong coupling between adjacent ^{11}B nuclei in boron hydride clusters, but not if the nuclei are linked by a hydrogen bridge. In the case of $B_{10}H_{14}$, B(1) and B(3) are chemically equivalent, as are B(2) and B(4), B(5), B(7), B(8) and B(10), and B(6) and B(9). Looking at structure 2.XXIV we can see that B(2) and B(4) are directly bound to atoms of all the other types, so they must have the lowest frequency chemical shift, as this resonance shows couplings to all the other three groups in Fig. 2.27. The third diagonal resonance, at about 11 ppm, shows coupling to only one other resonance, that from B(2) and B(4), so it must be assigned to B(6) and B(9). The other two resonances cannot be assigned using the COSY spectrum alone, but the one at 1 ppm is seen in the 1D ^{11}B spectrum to be twice as intense as that at 13 ppm, so it must be assigned to B(5), B(7), B(8) and B(10), thus completing the assignment [34].

A heteronuclear correlation (HETCOR or HCOR) 2D NMR spectrum of $B_{10}H_{14}$ is presented in Fig. 2.28(a). This differs from the COSY spectrum in having the two axes different from one another, one representing ^{11}B chemical shifts, and the other representing ^{1}H shifts. A peak in this spectrum at a frequency (f_1, f_2) indicates that a

Example 2.4

The figure below shows 1D (proton decoupled) and 2D COSY ^{11}B NMR spectra of $B_9H_{11}NH$. Assuming that coupling is only observed between adjacent boron nuclei, deduce the structure of the boron cage.

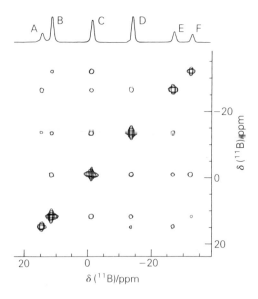

Three of the resonances in the 1D spectrum are twice as intense as the others, so they represent two boron nuclei each. The pairs of nuclei B, C and D couple with one another, so they must be arranged as in structure *a* below. Nucleus F couples to the pairs B and C, and must therefore cap that face of the prism, while E couples to the C and D pairs, and thus lies over that face (structure *b*). The remaining nucleus, A, shows coupling to the unique E and the pair of D nuclei, and so should be placed over the triangular face DDE. The NH group is in fact linked to the boron atoms CCF, and if we add a bond between the two B atoms we complete the cage (structure *c*). Each boron atom carries a terminal hydrogen atom, and there are also hydrogens bridging the two A—D bonds.

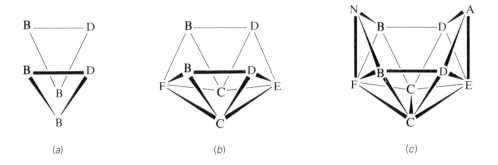

The spectra are adapted, with permission, from J.G. Kester, J.C. Huffman and L.J. Todd, *Inorg. Chem.*, **27**, 4528 (1988). Copyright (1988) American Chemical Society.

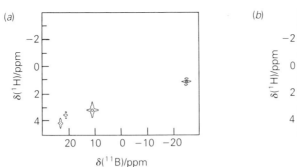
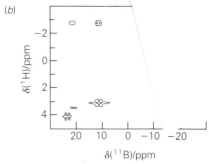

Fig. 2.28 Heteronuclear 2D ^{11}B/^{1}H correlation spectra for $B_{10}H_{14}$, 2.XXIV. In spectrum (a) proton decoupling was used during data acquisition, but it was not used while spectrum (b) was being recorded. The peaks relate the ^{1}H chemical shifts of protons to the ^{11}B chemical shifts of neighboring boron atoms. Adapted, with permission, from D.P. Burum, *J. Magn. Reson.*, **59**, 430 (1984).

proton at f_2 couples strongly with a boron nucleus resonating at f_1. Thus the boron with chemical shift – 25 ppm is coupled to a proton at just over 1 ppm, and in this way the proton resonances can also be assigned, using the knowledge that a strong ^{11}B-^{1}H coupling implies the presence of a bond to a terminal hydrogen atom. The rather curious contours in this spectrum · arise from inhomogeneity in the ^{11}B radiofrequency pulses, but that turns out to be useful when proton decoupling is not used during the data acquisition. The resulting spectrum, Fig. 2.28(b), now also includes two weak peaks, corresponding to bridging protons which couple weakly to their boron neighbors. There is just one such set of hydrogen atoms, which must therefore bridge between B(6) and B(9) and B(5), B(7), B(8) and B(10). Projection of this 2D spectrum onto the ^{11}B axis yields a normal ^{11}B spectrum with proton coupling retained, while projection onto the ^{1}H axis gives a ^{1}H spectrum with both ^{11}B and ^{1}H decoupling.

A 2D NOESY (Nuclear Overhauser Effect Spectroscopy) spectrum looks rather like the COSY spectrum shown in Fig. 2.27, but the origin of the off-diagonal peaks is different. In the COSY case, a peak indicates coupling between the two nuclei concerned, whereas the NOESY peak identifies a Nuclear Overhauser Effect, which arises from the proximity of the two nuclei in space. Figure 2.29(a) shows a hetero-nuclear NOESY (sometimes called HOESY) spectrum for the tetramethylethylene-diamine (tmeda) adduct of 2-lithio-1-phenylpyrrole, whose dimeric structure is shown in Fig. 2.29(b). The normal ^{1}H and ^{6}Li spectra are shown along the axes of the 2D contour plot, which contains just four peaks, one of which does not correspond to any ^{1}H peak and is an artefact of the experiment. The lithium atom is therefore close (i.e. less than about 3.5 Å) to three different sets of protons, which can be readily identified as H(7) and H(11), equivalent by virtue of fast rotation about the N(1)– C(6) bond in solution, H(3) and methyl protons of the tmeda ligand. The close contact between Li and H(11) seen in the crystal structure is thus maintained in solution, and it is of chemical significance, as it leads to activation of this hydrogen atom, so that introduction of a second lithium atom with *n*-BuLi leads to substitution at this site.

In Fig. 2.30 the ^{13}C COSY and NOESY spectra for $[Os_3H_2(CO)_{10}]$ (2.XXV) 75% labeled with ^{13}C are shown. As both spectra are symmetrical about the diagonal line for homonuclear experiments it is possible to show the two sets of results in a single

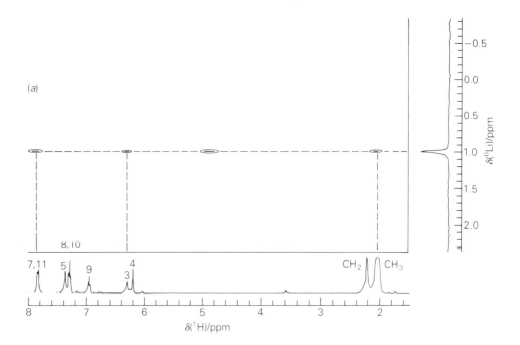

(a)

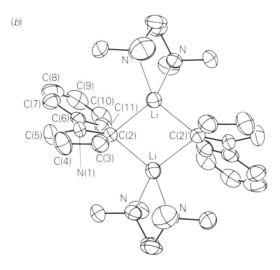

(b)

Fig. 2.29 (*a*) Heteronuclear 2D ^{1}H/^{6}Li Nuclear Overhauser Effect (HOESY) spectrum for the tetramethylethylenediamine adduct of 2-lithio-1-phenylpyrrole (*b*). Taken, with permission, from W. Bauer, G. Miller, R. Pi and P.v.R. Schleyer, *Angew. Chem. Int. Ed. Engl.*, **25**, 1103 (1986).

2D spectrum. The intensity of the peak at 175.2 ppm in the 1D spectrum identifies it as being due to CO(7), while a ^{13}C spectrum with ^{1}H coupling retained indicates that the CO(6) resonance is at 176.3 ppm. The COSY spectrum shows that these two couple with one another, while the peak in the NOESY spectrum is in this case indicative of chemical exchange between the CO(6) and CO(7) sites (see Section 2.15.2). The COSY spectrum also shows coupling between ^{13}C nuclei of CO(4) and

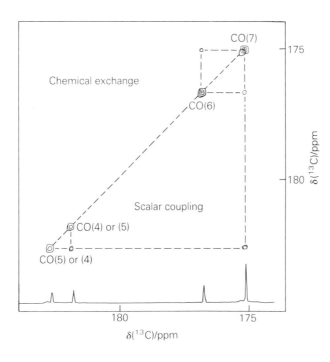

Fig. 2.30 Homonuclear 2D ^{13}C scalar coupling (COSY) and chemical exchange (NOESY) spectra for $[Os_3H_2(CO)_{10}]$, 2.XXV, obtained in two separate experiments. The one-dimensional ^{13}C spectrum is shown at the bottom. Taken, with permission, from G.E. Hawkes, L.Y. Lian, E.W. Randall, K.D. Sales and S. Aime, *J. Chem. Soc., Dalton Trans.*, 225 (1985).

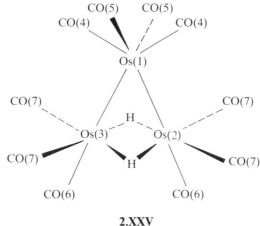

2.XXV

CO(5) groups as expected, although the assignment of these two resonances is uncertain. However, the spectrum indicates one further coupling, between C(7) and either C(4) or C(5). As the couplings C(7)—C(4) and C(6)—C(5) both involve a *trans* relationship at one Os atom and a *cis* relationship at another, we might expect that if we sec one such coupling we should also see the other. So the most likely explanation is that the observed coupling is C(7)—C(5), with *cis* relationships at both Os atoms.

Two-dimensional NMR methods, making use of ^{183}W, which is 14% abundant and has a spin of $\frac{1}{2}$, have been useful in studies of heteropolytungstates [35, 36]. These compounds contain tungsten atoms, each of which is octahedrally coordinated to six oxygen atoms, with the octahedra linked across corners or edges. Fig. 2.31 shows how a set of eleven such octahedra are arranged. In the heteropolytungstates the twelfth octahedral site, linking at corners to 1, 1′, 4 and 4′, is occupied by a different metal ion, and there is a silicon or phosphorus atom at the center of the ion. Assigning the

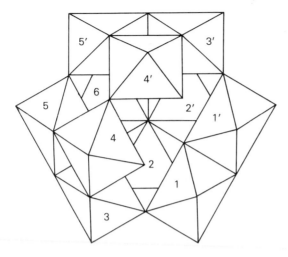

Fig. 2.31 Relationship between the WO_6 octahedra in the structure of $Na_5PPbW_{11}O_{39}$. Redrawn from Ref. [35].

six observed ^{183}W resonances for each compound is difficult, even making use of the observation that $^{183}W-^{183}W$ coupling is only significant for nuclei in adjacent octahedra. In principle, COSY spectra should suffice, but because tungsten is a dilute spin element, relatively few spinning nuclei also have spinning neighbors, so the off-diagonal peaks in the spectra are very weak. The solution to the problem is to use 2D INADEQUATE spectra, based on a two-dimensional version of the technique described in Section 2.11.3. Spectra for $[PPbW_{11}O_{39}]^{5-}$ and $[SiZnW_{11}O_{39}]^{6-}$ are shown in Fig. 2.32, with the double quantum frequencies shown in one dimension,

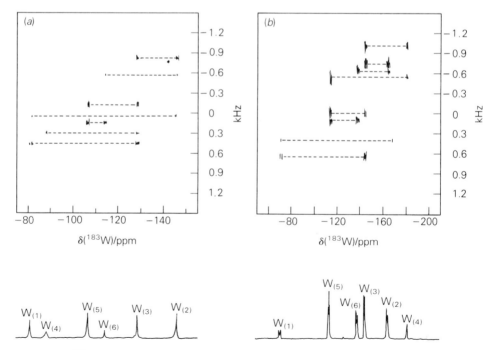

Fig. 2.32 ^{183}W 2D INADEQUATE spectra for (*a*) $[PPbW_{11}O_{39}]^{5-}$ and (*b*) $[SiZnW_{11}O_{39}]^{6-}$. The pairs of off-diagonal peaks indicate connections between the associated WO_6 octahedra. Normal ^{183}W spectra are shown underneath the 2D spectra. Redrawn from Refs [35] and [36].

and the normal ^{183}W chemical shifts in the other. Double quantum transitions are only possible for molecules which have two nuclei coupling to each other, which in this case means in adjacent octahedra. The spectrum consists of pairs of signals, equidistant from the diagonal line (any signals lying on the line arise from imperfections in the experiment), and each pair, which lies on one horizontal line, indicates that the nuclei with those chemical shifts are coupled. For example, in each of the spectra of Fig. 2.32 there is one tungsten coupling to four others, and this must be W(3). The unique tungsten atom, W(6), can be recognized in the 1D ^{183}W NMR spectra as its resonances are half the intensity of the others, and it can be seen to be coupled to W(2) and W(5), and so on. Thus, the whole of the skeleton of each ion can be deduced from a single spectrum, with the exception of the connection between W(4) and W(5) in $[PPbW_{11}O_{39}]^{5-}$. The full analysis shows that the two ions have similar structures, and reveals a large difference between the W(4) chemical shifts, which probably arises from different placements of the Pb and Zn atoms in the ions.

Exactly the same procedures are applied to the determination of the carbon skeletons of unknown organic and organometallic compounds, and there is potential for studying metal clusters in the same way.

2.12 Gases

In the preceding part of this chapter we have assumed that samples are in the liquid phase, whether in solution or as pure liquids. In studies of gases, the principles and techniques that we have described still apply, but there are two major extra problems to be considered. In the first place, the amount of gaseous material that can be observed is limited by its vapor pressure and by the volume of the sample tube. By using large diameter tubes and high field magnets it is possible to record good Fourier transform spectra for the more sensitive nuclei with vapor pressures down to 1 torr or less, but at such low pressures it is necessary to use isotopically enriched samples if nuclei such as ^{13}C are to be observed. The second problem, particularly severe at low pressures, is that relaxation is not very efficient, so that T_2 becomes long and linewidths unacceptably large. This is dealt with by the addition of 100–1000 torr of an inert gas such as argon or SF_6, and allowing a relaxation interval of several seconds between pulses.

Gas phase spectra are often used in studies of conformational and other equilibria. NMR spectroscopy can give accurate information about the relative proportions of species present, whereas other gas-phase structural methods such as microwave spectroscopy (Chapter 4) or electron diffraction (Section 8.3) are poor in this respect. Positions of equilibria may be quite different in solution and gas phases, and rates of exchange between conformations are often as much as two orders of magnitude faster in the gas phase than in solution. The effects of solvent are therefore significant, and so experimental gas phase data must be used for comparison with molecular energies calculated theoretically. Thus it is possible, for example, to compare theoretical and experimental results for the relative energies of *syn* and *anti* conformers of alkyl nitrites and for their rate of interconversion, or to study the barriers to pseudorotation of PF_5 or to internal rotation in compounds such as dimethylformamide (see Section 2.15.2)

2.13   Liquid crystals

When NMR spectra of compounds dissolved in liquid crystal solvents are studied, it is possible to obtain information which relates directly to the molecular structure (i.e. bond lengths and angles) of the compounds. This is something quite new, as up to this point we have considered only information relating to the number of spinning nuclei present, and their interconnections, with at best very indirect information about geometry. The couplings observed in spectra obtained in fluid phases are the *indirect* couplings (*J*), transmitted via the electrons. But there are also *direct* dipolar couplings (*D*), which depend only on the geometrical relationships between pairs of nuclei, and are totally independent of the bonding relationships. However, these dipolar couplings are normally completely lost by the tumbling of the molecules, which is usually fast on the NMR timescale.

Liquid crystals are compounds or mixtures which have one or more phases between liquid and solid, in which there is local alignment of molecules, but no overall alignment. The material in these phases is therefore fluid, and isotropic in bulk, but locally anisotropic. In the presence of an electric or magnetic field, the fluid becomes anisotropic in bulk, as all the molecules tend to align in one particular direction, although they remain mobile. At the same time, any solute molecules also become oriented with respect to the field, and dipolar couplings are no longer lost by isotropic tumbling.

Spectra of the solute in a liquid crystal solvent may be dramatically different from those of the same compound in an isotropic solvent. In the first place, coupling can now be observed between nuclei which are chemically and magnetically equivalent. Thus in Fig. 2.33(a) there is a triplet splitting, corresponding to coupling between the three equivalent protons. This splitting, which is in fact $3D_{HH}$, is much larger than we would expect for a two-bond indirect coupling, and this is generally true for direct couplings, which are normally hundreds or even thousands of Hz. Where a first order splitting J_{AX} is observed for an isotropic solution, it is replaced by $J_{AX} + 2D_{AX}$, and so the direct coupling may be determined. Because the couplings in liquid crystal solutions are large, second order spectra are frequently encountered, as shown in Fig. 2.33(b), and such spectra must be analyzed in the normal way, using variants of spectral analysis programs whch take account of dipolar coupling. Second order spectra are also observed when a group of magnetically equivalent nuclei are not fully equivalent. This happens when the couplings within the group (which are now observable) are not all equal, as for example in benzene, which has three different proton-proton dipolar couplings.

Assuming that the dipolar couplings can be measured or derived, they can be related to molecular geometry by the equation:

$$D_{ij} = \frac{-\gamma_i\gamma_j h}{4\pi^2 r_{ij}^3}(S_{zz}\cos^2\theta_{ijz} + S_{yy}\cos^2\theta_{ijy} + S_{xx}\cos^2\theta_{ijx} + 2S_{xz}\cos^2\theta_{ijx}\cos\theta_{ijz}$$

$$+ 2S_{yz}\cos\theta_{ijy}\cos\theta_{ijz} + 2S_{xy}\cos\theta_{ijx}\cos\theta_{ijy}) \tag{2.11}$$

In this equation, θ_{ijx} refers to the angle between the x axis and the vector joining atoms i and j, which is simply a function of the molecular geometry, as is r_{ij}, the distance between the atoms. Thus the coupling depends on some fundamental constants, the molecular geometry, and the six terms S_{zz}, S_{xy}, etc., which reduce to

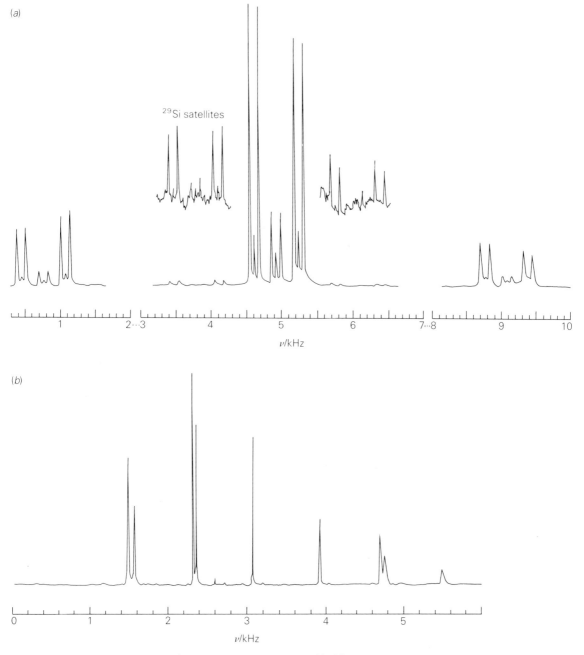

Fig. 2.33 (a) ^1H NMR spectrum of $SiH_3{}^{13}C^{15}N$ in a liquid crystal solvent. The spectrum is a wide triplet (HH coupling) of doublets (CH) of doublets (NH), with satellites due to 5% ^{29}Si present in natural abundance. Weaker peaks between the main ones are due to molecules containing ^{12}C and/or ^{14}N. (b) ^1H NMR spectrum of $SiH_3C\equiv CH$ in a liquid crystal solvent. The spectrum is second order because the HH couplings are large.

five because $S_{xx} + S_{yy} + S_{zz} = 0$. These are called orientation parameters, and together they describe the average orientation of the molecule with respect to the applied field. Initially their values are unknown. Fortunately, some of these parameters are zero for molecules with symmetry, as shown in Table 2.5. For symmetric tops, there is only one unknown orientation parameter, and so it is possible to determine the ratios of

Table 2.5 Numbers of independent orientation parameters

Point group	No. of orientation parameters
C_1, C_i	5
C_2, C_{2h}, C_s	3
C_{2v}, D_2, D_{2h}	2
All symmetric tops	1
All spherical tops	0

distances between spinning nuclei, but never the absolute distances. Thus, using the dipolar couplings determined from the spectrum shown in Fig. 2.33(a), it is possible to determine the ratios of the SiH, CH and NH distances to the HH distance, and so derive the whole of the geometry, including the angle at silicon, apart from the overall size. Absolute distances can never be determined unless the orientation parameter is known. The usual procedure is to assume one geometrical parameter (often using a rotational constant measured by microwave spectroscopy) or to use electron diffraction data, which can also provide the required scaling factor.

For more complicated systems it is useful to consider how many pieces of structural information can be obtained by this method. For example, PF_2CN has C_s symmetry, and so has three orientation parameters. Assuming that couplings to ^{13}C and ^{15}N can be observed, perhaps using labeled samples, a total of seven couplings can be measured. These are D_{PF}, D_{PC}, D_{PN}, D_{FF}, D_{FC}, D_{FN}, and D_{CN}. Allowing for the three orientation parameters, this means that just four useful pieces of geometrical information are available, and this is not enough to determine the full structure, which requires three bond lengths and three angles. In this case data from some other structural technique are also required.

Although this method gives extremely precise structural data, from experiments which are easy to perform, there are severe limitations to its usefulness. First, dipolar couplings can normally only be measured for nuclei with spin 1/2, so no information can be obtained about the positions of other nuclei. Secondly, there is the problem of the unknown orientation parameters, with the consequent need to make some structural assumptions. That introduces a further problem, as there can be no certainty that the structure in solution is identical to that in the gaseous or crystalline phase. There is also a need to make allowance for the effects of vibrations on the observed couplings, and this involves the determination of a complete force field analysis for the molecule. There may also be problems arising from anisotropy in the indirect (J) couplings, particularly when heavy atoms are involved. As it is not possible to make proper allowance for this at present, it is best to ignore couplings to heavy atoms (i.e. beyond the second main group of the periodic table). Nevertheless, despite this catalog of woes, the technique is a powerful one, providing useful structural data which are otherwise very difficult to obtain for molecules in solution. More information can be obtained from a book [37] and a review [38] on the subject.

2.14   Solids

The development of NMR spectroscopy of solids has been much slower than that of fluids, because of three major difficulties, all associated with the immobility of the nuclei in solids. However, advances have been made which make it possible to obtain high resolution spectra, more or less routinely, and so the potential of NMR for solving structural problems can be applied to solid materials, including polymers, minerals, insoluble compounds, *etc.* For a good short account of the methods see Ref. [24], which also refers to other reports and reviews.

The first major difficulty with solids is that dipolar couplings are not averaged to zero by molecular tumbling, and that long range couplings are also important (unlike in liquid crystals) because the material is not fluid. The combined effect of the many couplings is to give very broad resonances, often tens of kHz wide. (Changes in line widths can sometimes be used to give information about internal motions in solids.) Secondly, the chemical shift of a nucleus depends on the orientation of the molecule with respect to the magnetic field, and in solids the effect of this chemical shift anisotropy is not averaged out by molecular tumbling, and so leads to line broadening, again of thousands of Hz. Study of this anisotropy may be useful, but for our purposes now we regard it only as something to be overcome. Finally, again because of the immobility of the nuclei, the relaxation time T_1 is very long, and this means that multipulse methods are not very efficient. It is therefore very difficult to get spectra with good signal-to-noise ratios. The combination of these problems meant that solid spectroscopy was neglected; after all, it is not easy to obtain useful information from spectra in which each resonance is as wide as the whole normal range of chemical shifts.

The very large linewidths associated with solids can be markedly reduced. First, the effects of chemical shift anisotropy can be averaged out by a technique known as magic angle sample spinning (MASS, MAS or MAR). If a sample is rotated rapidly about an axis inclined at an angle θ to the magnetic field, the expression describing the line broadening due to chemical shift anisotropy includes a term $(3 \cos^2\theta - 1)$. When θ is 54.7° this term vanishes, and so the anisotropy effects can be eliminated, providing the rotation is rapid enough.

In practice the rotation rate is limited by the mechanical strength of sample containers. Modern materials enable rotation rates sufficient for the removal of line broadening due to chemical shift anisotropy for nuclei such as ^{13}C or ^{29}Si to be achieved, but for some other nuclei, particularly those of heavy elements, the rotation is not fast enough. Consequently each single resonance in a spectrum is replaced by a central line and a series of uniformly spaced spinning sidebands. The central line is not necessarily the most intense, so it may be necessary to vary the spinning speed to distinguish side and center bands.

Provided the rotation rate is greater than the linewidths, dipolar couplings are also averaged to zero by magic angle spinning, just as they are for molecules tumbling in solution. But dipolar couplings are often 50 kHz or more, and so the rotation cannot be fast enough to remove them completely. These dipolar interaction effects can be eliminated for heteronuclear systems by decoupling in the normal way (Section 2.10.4), but very high decoupling powers are needed for the large couplings. So for nuclei such as ^{13}C or ^{29}Si, for which only heteronuclear couplings are important, high resolution spectra, comparable with those obtained for solutions, can be obtained.

The problem of long relaxation times remains. This is particularly important for dilute spin, insensitive nuclei, but it can be overcome for dilute nuclei with spin 1/2 in the presence of abundant spins, by a technique known as cross-polarization. The theory is too complicated to explain here (see Ref. [24] for details) but the effect is to transfer magnetization from the abundant spin (usually ^{1}H) to the dilute spin,

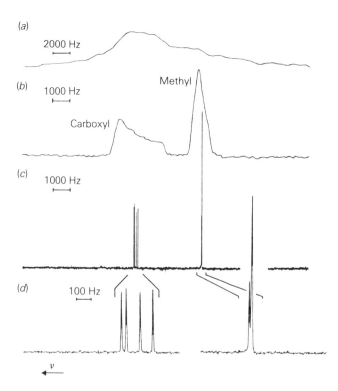

Fig. 2.34 ^{13}C NMR spectra of $2Ca(CH_3CO_2)_2.H_2O$ at 22.6 MHz. (*a*) static, (*b*) static, with high-power proton decoupling, (*c*) with decoupling and magic angle rotation and (*d*) expansion of (*c*) showing fine structure due to the existence of different molecules in the unit cell, unrelated by symmetry. Taken, with permission, from Ref. [24].

commonly ^{13}C, giving an intensity gain of γ_H/γ_C. Moreover, the cross-polarization can be performed many (typically 10 to 20) times in one pulse sequence, giving a corresponding saving of time, or gain of signal-to-noise ratio, in the whole experiment. For these purposes, the dilute spin can include 100% natural abundance isotopes such as ^{31}P or ^{19}F, provided they comprise only a small proportion of the total sample. This may occur naturally, for example for ^{31}P in a transition metal phosphine complex, or be achieved by matrix isolation.

Using one or more of these methods (the combination of magic angle spinning with cross-polarization is often referred to as the CP–MAS technique), high resolution spectra of solids, including crystalline powders, rubbers, gels and composite materials, may be obtained. Fig. 2.34 illustrates the dramatic narrowing effects of high power decoupling and magic angle rotation, which allow us to resolve fairly small splittings in the two resonances. This fine structure arises from four molecules which are unrelated by symmetry in the crystal, and there is therefore a useful link between crystallography and solid phase NMR work. In fact, in the particular case of calcium acetate, it has not been possible to grow crystals for X-ray analysis, and so our knowledge of the crystal structure is derived from NMR data alone. But we can be confident that the explanation of the observed splittings is correct, because they are also seen in the spectrum of the analogous crystalline barium acetate, for which the crystal structure has been determined.

Fig. 2.35 shows a ^{77}Se spectrum of $[(CH_3)_2SnSe]_3$ (2.XXVI). There are two main resonances, one twice as intense as the other, and this indicates that there is probably half a molecule in the asymmetric unit of the unit cell, with a mirror plane or two-fold axis passing though the unique selenium atom [39]. There must therefore be two different sites for tin atoms, and this is confirmed by the ^{119}Sn spectrum, which also has two main resonances, as well as a series of satellites due to coupling to ^{77}Se and

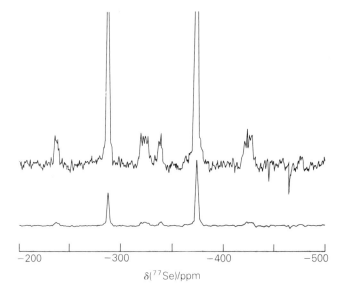

Fig. 2.35 Solid-state ^{77}Se NMR spectrum of [(CH$_3$)$_2$SnSe]$_3$, 2.XXVI. Taken, with permission, from Ref. [39].

$\delta(^{77}$Se)/ppm

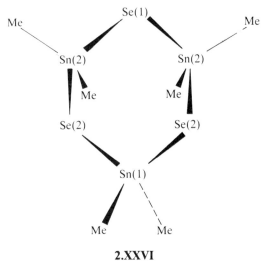

2.XXVI

^{117}Sn, as well as between non-equivalent ^{119}Sn nuclei. The satellites in Fig. 2.35 arise from coupling to ^{119}Sn and ^{117}Sn, which have similar natural abundances and only slightly differing magnetogyric ratios. The weaker, higher frequency resonance, assigned to Se(1), has one pair of satellites for each tin isotope due to coupling with Sn(2). However, Se(2) shows non-equal couplings to Sn(1) and Sn(2), and so the stronger resonance has a total of four pairs of satellites. This is again consistent with the molecule having less than three-fold symmetry, which is confirmed by an X-ray study of the crystal. The molecule in fact has a twisted boat conformation of C_2 symmetry, as shown in structure 2.XXVI.

The fact that we can observe site splittings indicates that chemical shifts are very sensitive to environment. Thus solid phase NMR spectroscopy provides us with a powerful tool for investigating interactions in solids. For example, the ^{119}Sn chemical shift of Ph$_3$SnOH in solution is -80 ppm, but in the solid phase this changes to -298 ppm. Clearly there is some dramatic difference between the structures in the two phases. In solution the tin atoms are tetrahedral, four-coordinated, but in the

Example 2.5

The compound $[(CH_3)_2SnS]_3$, which has a structure similar to 2.XXVI, exists in two crystalline forms. In the monoclinic form there is no molecular symmetry, while there is a two-fold axis through the molecules in the tetragonal form. Describe the ^{13}C and ^{119}Sn NMR spectra expected for each form.

....

The ^{13}C spectrum of the monoclinic form will have six resonances, each with ^{117}Sn and ^{119}Sn satellites, as the methyl groups are all inequivalent. In fact, only five resonances are observed, as two of them overlap. The three tin nuclei are also inequivalent, so there will be three resonances. Each of these will have eight satellite lines, due to coupling with ^{117}Sn and ^{119}Sn at each of the other tin sites.

In the tetragonal form none of the methyl groups lies on the two-fold axis, so there are three pairs of equivalent carbon atoms, and three ^{13}C resonances, with the usual satellites due to coupling with the adjacent tin atoms. Two of the tin atoms are symmetry-related, while the third one is unique, so the ^{119}Sn spectrum has two resonances, one twice as intense as the other. The satellite pattern, shown in the spectrum below, is quite complex. The resonance due to the unique tin, which is at the high-frequency end of the spectrum, has satellites due to ^{119}Sn and ^{117}Sn in the other position. However, the ^{119}Sn satellites are part of a second-order [AB] pattern, and are therefore not of equal intensity, nor centered on the main resonance. The same applies to their counterpart satellites in the low-frequency half of spectrum. In this part can also be seen *two* pairs of ^{117}Sn satellites, one arising from coupling to ^{117}Sn at the unique position, and the other from coupling between ^{117}Sn and ^{119}Sn in the two symmetry-related sites.

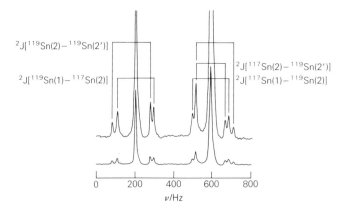

The spectrum is taken, with permission, from Ref. 39.

crystal there are strong intermolecular interactions, and the tin atoms are five-coordinate, with a trigonal bipyramidal arrangement of the bonds. The same sort of interaction is found in many tin compounds, and the ^{119}Sn spectrum of solid Me$_3$SnF provides additional evidence. The spectrum is complicated by the presence of many spinning sidebands, but there are just three series of lines, with integrated intensities in the ratio 1:2:1. The triplet arises from coupling to two equivalent fluorine nuclei, and the tin chemical shift confirms that the metal atom is five-coordinated, consistent with the polymeric structure 2.XXVII.

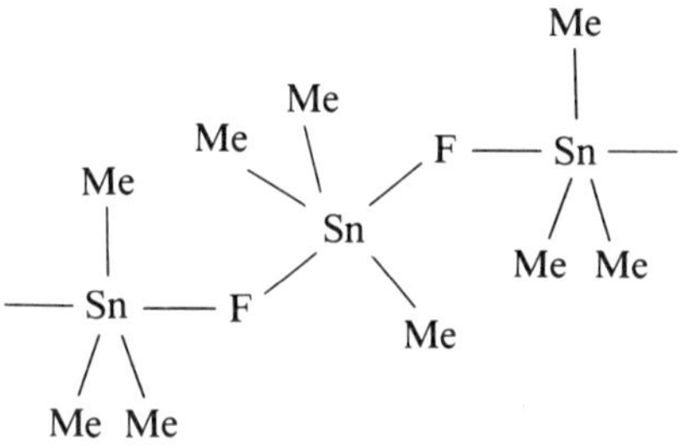

2.XXVII

These examples have shown how we can use solid phase NMR studies to investigate interactions between molecules. The technique will also show up any major change in structure within a molecule on going from solution to solid phase. The solution ^{31}P NMR spectrum of the phospha-alkene complex 2.XXVIIIa contained resonances attributable to the triphenylphosphine ligands, and another set with δP – 31 ppm and J_{PtP} 498 Hz, characteristic of a π-bonded ligand [40]. But the crystal structure showed unequivocally that the structure was 2.XXVIIIb, with a σ-bonded ligand. Was the interpretation of the solution spectrum wrong, or did the structure change on going from solution to solid? The solid phase ^{31}P NMR spectrum (Fig. 2.36) shows that a structural change does indeed occur. The chemical shift is now 247 ppm, and the PtP coupling constant has increased nearly ten-fold, to 4720 Hz, showing that there is direct bonding between these two atoms.

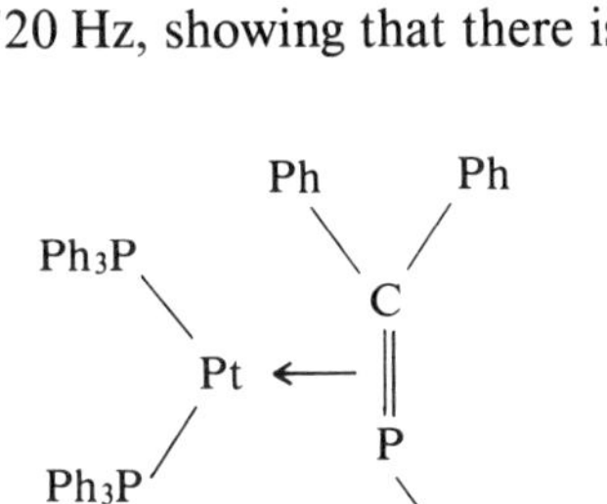

2.XXVIIIa 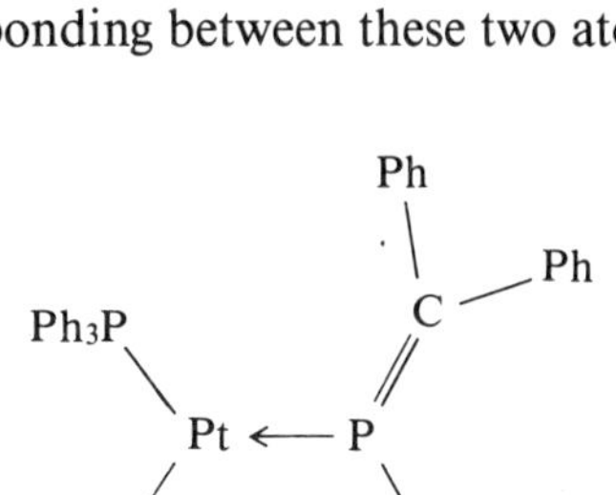   **2.XXVIIIb**

Although routine solid state NMR is a relatively recent development, it has many and wide-ranging applications. One area where it is proving to be important is in studies of silicates and zeolites, using ^{29}Si and ^{27}Al nuclei. The future potential of the method for studying structures of zeolites is illustrated by Fig. 2.37, which shows a ^{29}Si COSY spectrum for ZSM-39, which has the structure shown in the figure [41]. There are tetrahedral silicon atoms at all the junctions of polyhedra, with oxygen

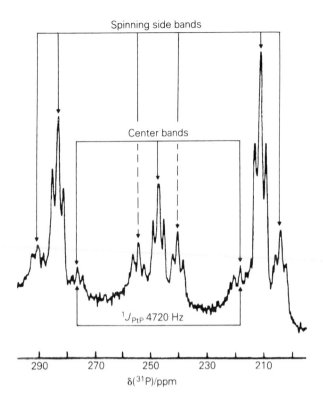

Fig. 2.36 Solid phase ^{31}P NMR spectrum of the platinum complex 2.XXVIII. The spectrum contains many spinning sidebands, but the central resonances attributed to the phospha-ethene ligand show that δP is 247 ppm and J_{PtP} is 4720 Hz. These parameters are characteristic of a σ-bonded ligand. Adapted, with permission, from Ref. [40].

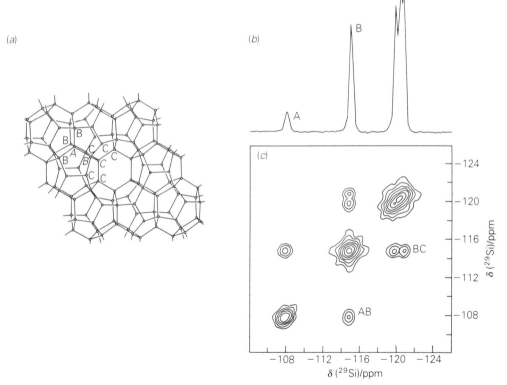

Fig. 2.37 (*a*) Schematic representation of the lattice framework of the zeolith ZSM-39. Three inequivalent silicon sites are labeled *A*, *B* and *C*. (*b*) ^{29}Si CP-MAS NMR spectrum of a sample of ZSM-39 enriched in ^{29}Si. (*c*) COSY 2D ^{29}Si NMR spectrum of the enriched sample. Adapted, with permission, from Ref. [41]. Copyright (1989) American Chemical Society.

atoms linking each pair of adjacent silicons. The silicon atoms are in three different sites, labeled A, B and C in the figure, with relative proportions 1:4:12, as is clearly shown by the CP-MAS 1D spectrum. Splitting of the C signal into three equal components arises from deviation of the material from cubic symmetry, as the tetragonal form studied lacks the three-fold symmetry axes. The COSY spectrum indicates that there is coupling between Si(A) and Si(B), and between Si(B) and Si(C), but not between Si(A) and Si(C). Coupling is only expected between nuclei at adjacent vertices, so the pattern is consistent with the structure shown in the figure, where all the neighbors of particular atoms of types A, B and C are indicated. The results are impressive, but we should not get the impression that such work is easy – the spectrum depicted took nearly 24 hours to obtain, even using a sample 80% enriched in ^{29}Si!

Solid state NMR is also important in following solid phase reactions, and in studies of phase changes and polymorphism. It is not restricted to cystalline materials. In amorphous materials such as glasses and gels, a particular type of nucleus may be in a wide range of environments, rather than in just a few specific ones. It will therefore give rise to a broad NMR signal, rather than a few sharp lines. For example, the mixed calcium barium metasilicate gives a broad ^{29}Si resonance, thus showing that the material is glassy, with variable composition, rather than consisting of separate regions of crystalline calcium metasilicate and barium metasilicate: the latter would give a spectrum with just two sharp resonances. NMR spectroscopy is one of the very few structural methods to give useful information about glasses, and it has particular promise in this area.

For abundant spin nuclei, particularly ^{1}H and ^{19}F, it is not possible to use high power decoupling to remove homonuclear couplings, nor can rotation speeds be high enough to suppress the dipole-dipole interactions. In these circumstances, line-narrowing can be achieved by using a complicated series of pulses, the most frequently encountered of which is called MREV-8. For our purposes it is not necessary to understand how these methods work, but it is worth realizing that they make it possible to obtain proton or fluorine spectra with linewidths down to a few tens of Hz. In Fig. 2.38 a magic angle spinning ^{19}F spectrum of a mixture of fluoroapatite and calcium fluoride is shown. The fluoroapatite appears as a peak with long series of sidebands, while the calcium fluoride gives an unresolved resonance more than 50 kHz wide. In contrast, the CRAMPS spectrum (Combined Rotation And Multi-Pulse Spectroscopy) contains two clearly resolved peaks with only weak spinning sidebands, enabling the relative amounts of the two components to be estimated.

Quadrupolar nuclei present particularly severe problems for solid phase NMR, and linewidths may even be measured in MHz. However, these effects can be avoided for nuclei with half-integral spins by looking only at transitions between the $m_I = +\frac{1}{2}$ and $m_I = -\frac{1}{2}$ levels, and some progress has been made in obtaining useful spectra for such nuclei.

One other solid phase technique which may become very important is single crystal NMR. This is very difficult experimentally, but in principle and sometimes in practice, it is possible to measure dipole-dipole couplings, just as in liquid crystal solution experiments. A single crystal can be held stationary, and by orientating it in different directions the problems of unknown orientation parameters, which limit the usefulness of liquid crystal NMR, are eliminated, and internuclear distances can be determined directly. Inorganic examples are still hard to find, but a study of glycine [42] illustrates the potential of the method. The C—C and C—N distances were found

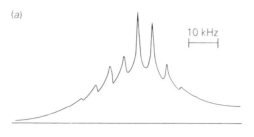

(a)

10 kHz

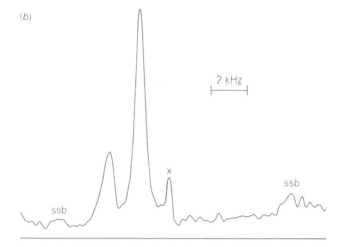

(b)

2 kHz

x

ssb

ssb

Fig. 2.38 ^{19}F NMR spectra of a mixture of fluoroapatite and calcium fluoride. (*a*) Spectrum obtained with magic angle spinning and (*b*) a spectrum of the same sample obtained using the CRAMPS technique. Note the greatly expanded scale in spectrum (*b*). Taken, with permission, from K.A. Smith and D.P. Burum, *J. Magn. Reson.*, **84**, 85 (1989).

to be 1.543 ± 0.008 and 1.509 ± 0.009 Å, respectively, and the CCN angle was $111.1 \pm 1.0°$. These values can be compared with those found in a neutron diffraction study: 1.539 ± 0.001 Å, 1.490 ± 0.001 Å, and $111.85°$.

2.15   Monitoring reactions

2.15.1   Concentration measurements

The intensity of an NMR signal for a particular nuclear isotope is proportional to the number of nuclei contributing to the signal. Integrated intensities can be measured to an accuracy of about 1%, and as resonances from different compounds may well not overlap at all, NMR spectroscopy can be an excellent means of monitoring the course of a reaction, or of measuring the relative concentrations of components of a mixture. In this respect it is greatly superior to most other spectroscopic techniques, including infrared and ultraviolet spectroscopy, where there is no direct relationship between intensity and concentration, and absorption bands are much more likely to overlap. However, care is needed for NMR intensity measurements, for it is quite easy to introduce errors of 10% or more, and there can be intensity variations dependent on relaxation rates, Overhauser effects, etc. which can have profound influences on apparent intensities in Fourier transform spectra. We should also note that absolute intensity measurements are not practicable, and intensities should ideally be measured relative to some unchanging standard.

If we can identify and measure the relative quantities of the components of reaction mixtures, then we may be able to deduce the mechanisms of reactions, determine equilibrium constants, and acquire kinetic data as well. The feasibility of following the course of a reaction by NMR, so that intermediates can be identified or the reaction rate measured, depends on that rate. Other techniques are usually more suitable for extremely slow reactions, or for very fast ones going to completion in a few seconds or less, although specialized flow systems can be used to study rapid reactions. But for reactions taking between a few minutes and a few weeks to reach completion, NMR can be ideal. Spectrometers can be programmed to collect sample spectra at pre-determined intervals, and to raise or lower the temperature as desired. Thus by plotting concentrations of reagents, products and intermediates much information can be obtained about the mechanism of a reaction and the conditions under which it occurs, and varying the temperature may indicate whether it is reversible. Often more information can be obtained by using labeled reagents, perhaps enriched in ^{13}C or ^{15}N, or containing deuterium instead of hydrogen.

The identification of significant intermediates in a complicated reaction is illustrated by Fig. 2.39. Cluster complexes of the type $[Rh_4(CO)_{12-x}L_x]$, where L is a phosphine, are catalysts for the hydroformylation reaction, in which hydrogen and carbon monoxide are added to an alkene to give an aldehyde. The ^{31}P spectrum, recorded for a sample of $[Rh_4(CO)_9\{P(OPh)_3\}_3]$ under 400 atmospheres pressure of CO at 300 K, shows that the Rh_4 cluster has broken down, and there are resonances from two Rh_2 complexes. One is a second order pattern from an $[AX]_2$ spin system, and is assigned to $[Rh_2(CO)_6\{P(OPh)_3\}_2]$, 2.XXIX, which has two phosphorus nuclei and two rhodium nuclei (^{103}Rh, 100% abundant, $I = 1/2$) coupling with one another. The second group of resonances is a first order doublet of doublets, attributed to a single phosphorus coupling to two ^{103}Rh nuclei which are in different environments. This arises from the complex 2.XXX. These dirhodium complexes are probably the reactive catalytic species, and the next stage of the reaction may involve addition of hydrogen, possibly involving a further breakdown to give compounds containing only single rhodium atoms.

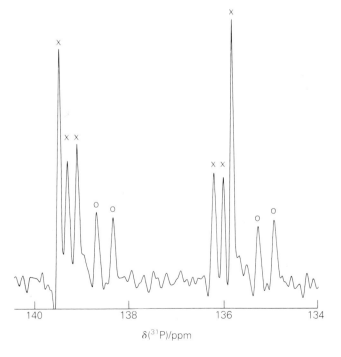

$\delta(^{31}P)/ppm$

Fig. 2.39 ^{1}H-decoupled ^{31}P NMR spectrum of products derived from $[Rh_4(CO)_9\{P(OPh)_3\}_3]$ under 400 atmospheres of CO at 300 K. Peaks marked x are due to $[Rh_2(CO)_6\{P(OPh)_3\}_2]$ and those marked ○ to $[Rh_2(CO)_7\{P(OPh)_3\}]$.

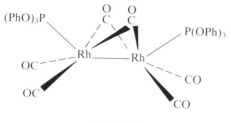

2.XXIX

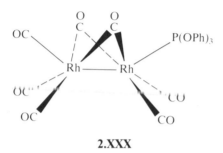

2.XXX

It is possible to obtain kinetic data by following reagent concentrations by NMR, but there are several problems which are not easy to overcome. In the first place, it is difficult to obtain highly accurate intensity integrals, particularly when, in the nature of the experiment, it is not possible to spend a lot of time acquiring each spectrum. Secondly, temperature stability and uniformity in NMR tubes are hard to achieve, nor is it easy to measure the temperature inside a spinning NMR tube. In addition, there is a problem in defining the starting time of a reaction, as it takes some time to get a sample prepared, into the spectrometer, and start collecting data.

On the other hand, measuring equilibrium constants, and so obtaining thermodynamic data, is somewhat easier. There are still problems in getting reliable intensities and temperature measurements, but at least the target is stationary! There may be other problems: not all the components may have distinguishable resonances in one spectrum, or absolute concentrations may be needed. Moreover, we must have all components present in solution (no solids or gases) and the exchange rate for the equilibrium reaction must be sufficiently slow that the NMR signals are not averaged (see next section). We must also ensure that true equilibrium is reached, ideally by approaching it from both sides. Finally, concentrations of all components must be comparable if they are to be measured accurately. NMR is not much use if there is 1000 times as much of one compound as of another. But when all these conditions are satisfied, as they often are, then it is a simple matter to determine equilibrium constants. If the experiments can be repeated at several temperatures, then enthalpy and entropy changes for the reaction can be measured.

2.15.2   **Exchange reactions**

Consider a simple process of exchange between two sites. It could be a chemical process, such as exchange of protons between the HCl and HBr, or it could be positional, such as the switching of two methyl groups by rotation about a carbon–nitrogen bond in dimethylformamide, 2.XXXI. The two sites have resonant frequencies ν_A and ν_B, and the mean lifetimes in the two states are τ_A and τ_B. The

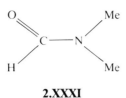

2.XXXI

lifetimes are, of course, equal if the two concentrations are equal. For very slow exchange, the NMR spectrum of A and B would just have their two distinct resonances, the situation we normally presume to hold for NMR spectroscopy. However, if the exchange rate increases, then by the Uncertainty Principle the width of the A and B resonances must increase, with half-width inversely proportional to the lifetimes, τ_A and τ_B. At still faster rates, the lines broaden further, then coalesce, and then the line sharpens to give a single resonance, at a frequency midway between ν_A and ν_B, as shown in Fig. 2.40. In practice, this last situation is very common, and we have tacitly assumed that it is the case in many examples in this chapter. For example, on symmetry grounds the three B protons in $SiHCl_2SiH_3$ (2.VIII) are *not* equivalent, but they become so when there is fast rotation about the Si—Si bond.

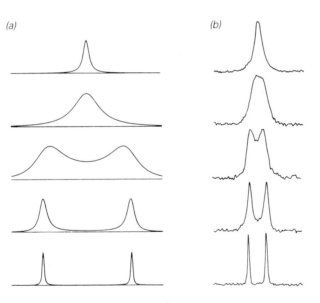

Fig. 2.40 (*a*) Calculated band shapes for various values of exchange rate relative to frequency difference for two equally populated sites. (*b*) Gas-phase ^{13}C NMR spectra of N,N-dimethyl-$^{13}C_2$-formamide, 2.XXXI. The spectra, recorded at 427, 415, 410, 397 and 373 K, show two distinct resonances at low temperatures, when rotation about the central C—N bond is rapid. Spectra in (*b*) are taken, with permission, from B.D. Ross and N.S. True, *J. Am. Chem. Soc.*, **106**, 2451 (1984). Copyright (1984) American Chemical Society.

In principle any of the line shapes can be analyzed, but a particularly useful point is the coalescence point, at which the two peaks combine so that there is just one maximum. At this temperature the lifetime, τ, is equal to $\sqrt{2}/\pi(\nu_A - \nu_B)$. Thus the rate constant may be determined, and hence the barrier or activation energy for the process may be derived.

For exchange between more than two sites, or involving non-equal concentrations, the situation is a lot more complicated, but the principles are the same. Lineshape analysis, usually by a least-squares fitting procedure, can yield the activation energies of processes, and may also give information about the mechanisms involved. Details of the methods used are given in books devoted to dynamic NMR spectroscopy [43].

SECTION 2.15: MONITORING REACTIONS   97

The spectra presented in Fig. 2.41 indicate that two separate exchange processes can occur as the ruthenium–silver complex 2.XXXII is warmed from 180 K to room temperature. The Ag(PEtPh$_2$) group is bound to one of two equivalent Cl$_3$ faces of the octahedron associated with one of the ruthenium atoms, and as the molecule therefore has no plane of symmetry, all five phosphorus nuclei are inequivalent. The ^{31}P spectrum shows two [AB] patterns arising from the pairs of phosphines bound to ruthenium, and the phosphorus bound to silver appears as two doublets, the splitting being caused by coupling to the two spin-1/2 isotopes of silver. On warming to 220 K the lines in the [AB] patterns collapse to a single resonance, while the silver phosphine pair of doublets is unchanged. The coalescence of the ruthenium phosphine resonances can be attributed to the Ag(PEtPh$_2$) group switching between four equivalent sites, two on each ruthenium atom, so that all four phosphines become equivalent. On

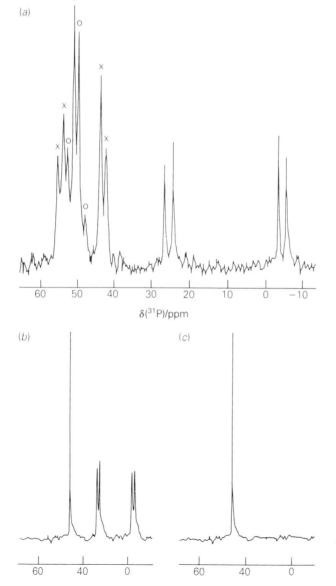

Fig. 2.41 ^{1}H-decoupled ^{31}P NMR spectra of [RuCl$_5$(PEtPh$_2$)$_4$.Ag(PEtPh$_2$)] (*a*) at 180 K, (*b*) at 220 K and (*c*) at 300 K. The lines marked ○ and x in spectrum (*a*) indicate the two separate [AB] spin systems associated with phosphines bound to ruthenium.

CHAPTER 2: NUCLEAR MAGNETIC RESONANCE SPECTROSCOPY

EtPh$_2$P,   Cl   PEtPh$_2$

EtPh$_2$P --- Ru ◄— Cl ►— Ru --- PEtPh$_2$

Cl   Cl   Cl

Ag

PEtPh$_2$

2.XXXII

further warming, all the resonances coalesce, so we can deduce that there is exchange of phosphines between coordination sites on ruthenium and silver.

Important information about exchange mechanisms can be obtained by considering coupling constants. For an intramolecular process, the couplings in the fast-exchange limiting spectra are simply the weighted means of the couplings in the individual exchanging forms. (Note that if the signs of couplings in two forms are opposite, their mean value may be much less than is at first expected.) For intermolecular processes, couplings between parts of the molecule which dissociate from each other are averaged to zero. In the highest temperature spectrum in Fig. 2.41 there is no coupling between silver and phosphorus, and so the exchange of phosphines between silver and ruthenium sites must take place by an *inter*molecular process. If it were *intra*molecular, all the phosphorus nuclei would remain associated with the same silver nucleus, and coupling would be retained, with the mean value of J_{AgP} being one-fifth of the value for the unique coupling in the low-temperature spectrum.

An *intra*molecular exchange process between two isomers of a ruthenium–tungsten cluster is implied by the spectra shown in Fig. 2.42. At the lowest temperature the major isomer (2.XXXIIIa, 87%) has two hydride resonances, with intensity ratio 2:1, while the minor isomer (2.XXXIIIb) has three equal resonances. The strongest line in the spectrum shows ^{183}W satellites. On warming all the lines broaden, coalescing at 220 K and finally giving one sharp resonance with ^{183}W satellites by the time room temperature has been reached. As only one signal remains, the hydrogen atoms must be exchanging rapidly between all five sites, so the two isomers are interconverting, and this provides a mechanism for all the hydrogen atoms to move between all the sites. The retention of coupling to ^{183}W demonstrates that the process must be intramolecular.

Thus NMR spectroscopy provides a convenient and extremely important way of studying both mechanisms and rates of exchange reactions, provided the lifetimes of the participating species are comparable with the NMR timescale, within two or three orders of magnitude of 10^{-5} s. Reactions which are somewhat slower than this can also be studied, by using the technique of saturation transfer. Suppose nuclei are exchanging between sites A and B, with lifetimes long compared with the NMR timescale, but short compared with their relaxation times, say of the order of one second. They will then give distinct resonances, but if one frequency is strongly radiated so that saturation occurs, the exchange process will then transfer saturation to the other signal, which will diminish in intensity. From the intensity change and the mean relaxation time of the nuclei in sites A and B the exchange rate can be calculated. Thus, the method fills a gap in determining exchange rates, between those

Example 2.6

Reaction between *trans* [Ir(CO)Cl(PMe$_3$)$_2$] and SF$_4$ in CCl$_2$H$_2$ gives a single product whose ^{19}F spectra at 180 K are shown in the figure below. Three groups of resonances are observed: a doublet of triplets of doublets at 68 ppm (integrated intensity 2), a triplet of doublets of narrow triplets at – 61 ppm (intensity 1), and a triplet of quartets at – 383 ppm (intensity 1). At 230 K the two resonances at higher frequencies broaden and lose detail, and at 300 K they have coalesced to a single broad peak; the lowest frequency set of resonances is unchanged. Account for these observations, and predict the ^{31}P spectrum of the product.

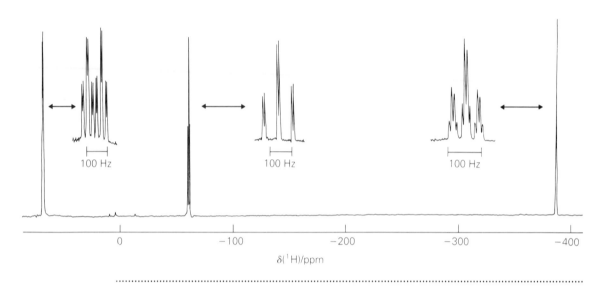

There is a resonance of intensity 1 to very low frequency, where peaks due to fluorine bound to a transition metal appear. The peaks in the high frequency region, where we would expect to see SF resonances, are of total relative intensity 3. Moreover, the low frequency resonance is split into a triplet of quartets. All this suggests that SF$_4$ has reacted with *trans*-[Ir(CO)Cl(PMe$_3$)$_2$] by oxidative addition of S—F to Ir, giving the complex [Ir(Co)ClF(PMe$_3$)$_2$(SF$_3$)] with F *trans* to CO and *trans* to SF$_3$.

Sulfur has a lone pair of electrons, and so coordination is likely to be based on a trigonal bipyramid, with the lone pair and the metal atom in the equatorial plane. With easy rotation about the Ir—S bond the axial fluorine atoms bound to sulfur become magnetically equivalent. The spin system then becomes [AM$_2$QX$_2$], with the high frequency resonance assigned to the two axial fluorines, the middle-frequency resonance assigned to the equatorial SF nucleus, and the low frequency resonance to the IrF nucleus. All three SF nuclei couple more or less equally to the IrF nucleus.

As the system is warmed, pseudorotation becomes faster on the NMR timescale. This leads to blurring of the distinction between axial and equatorial fluorine sites, and at 300 K the spectrum shows that exchange of position is beginning to lead to effective equivalence. The interchange of SF sites is intra- and not inter-molecular. The P—{H} spectrum at low temperature appears as a doublet of triplets of narrow doublets, but on warming this changes to a doublet of quartets as the SF nuclei become equivalent.

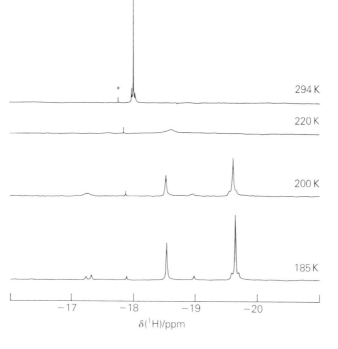

294 K

220 K

200 K

185 K

-17   -18   -19   -20

$\delta(^1H)$/ppm

Fig. 2.42 ^{1}H NMR spectra of [Ru$_3$W(C$_5$H$_5$)(CO)$_{11}$H$_3$], 2.XXXIII, at temperatures between 185 K and 294 K. The peak marked with an asterisk is due to an impurity. Taken, with permission, from Y. Chi, C.-Y. Cheng and S.-L. Wang, *J. Organometal. Chem.*, **378**, 45 (1989).

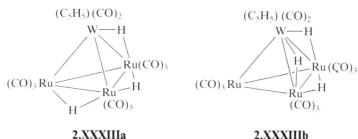

2.XXXIIIa   **2.XXXIIIb**

measurable by conventional NMR methods, and slower ones which can be measured by following the course of a reaction toward equilibrium, over a period of minutes or longer.

A two-dimensional NMR technique called EXSY (for exchange spectroscopy) also provides rate information for slow exchange reactions. The method is the same as that used to evaluate Nuclear Overhauser Effects, and the proximity of spinning nuclei to one another (NOESY, Section 2.11.4). Fig. 2.43 shows an EXSY spectrum for the methyl group protons of the platinum complex 2.XXXIV. Here there are four resonances due to methyl groups, and the off-diagonal peaks clearly indicate that group B is exchanging with C, and A with D. Analysis of the cross-peak intensities yields the rate constants for the exchange processes, and from the constants measured at a series of temperatures it is a simple matter to derive the activation energy and entropy for the exchange process, which evolves the free alkene group displacing the coordinated one [44].

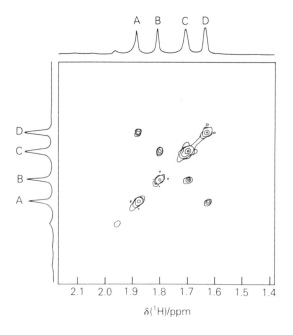

Fig. 2.43 EXSY 2D ^{1}H NMR spectrum of the platinum complex 2.XXXIV. The off-diagonal peaks show that the methyl group giving the resonance B is exchanging with that giving resonance C, and groups A and D are also exchanging. Taken, with permission, from Ref. [44].

2.XXXIV

2.16   Paramagnetic compounds

There have been few mentions of paramagnetic molecules in this chapter, for the simple reason that their nuclear magnetic resonances are normally so broad as to be useless. Nevertheless, some paramagnetic compounds are useful reagents in high resolution NMR studies, and there are also occasions when the study of paramagnetic species may be profitable, and the consequences of the paramagnetism may be positively helpful. There are some new concepts for us to consider, and in this brief account all we can do is indicate what can be achieved. For fuller information, specialist literature should be studied [45].

The two most obvious characteristics of NMR spectra of paramagnetic materials are the breadth of the resonances, and a great expansion of the normal chemical shift scale (for example, to more than 200 ppm for protons). The signals from diamagnetic materials may also show these effects in the presence of paramagnetic species, even in low concentrations. Both effects are apparent in Fig. 2.44. The ^{31}P spectrum *a* shows two [AB] patterns arising from the four chemically non-equivalent triphenylphos-

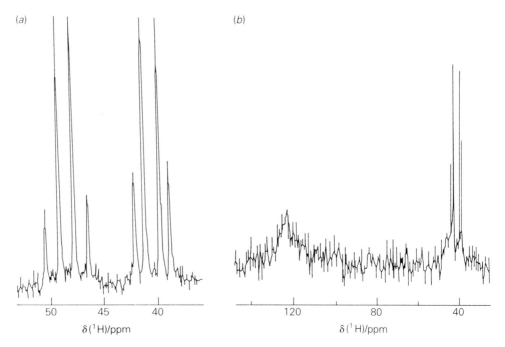

Fig. 2.44 (a) ^{31}P-{^{1}H} noise-decoupled NMR spectrum of the ruthenium complex 2.XXXV, showing two [AB] sub-spectra. (b) ^{31}P-{^{1}H} spectrum of the same complex after undergoing a one-electron oxidation.

phine ligands in the diamagnetic ruthenium complex, 2.XXXV. This compound undergoes a one-electron oxidation to give a cation in which the unpaired electron is localized on one ruthenium atom. This manifests itself in the ^{31}P NMR spectrum in a shift of nearly 80 ppm in two of the resonances, which become so broad that their coupling pattern is obscured, while the other [AB] pattern is virtually unchanged (Spectrum (b)).

2.XXXV

The line broadening in spectra of paramagnetic compounds is caused by short electronic spin-lattice relaxation times and/or hyperfine electron-nuclear coupling. Consequently it is usually the case that materials giving useful ESR spectra (see Chapter 3) have nuclear magnetic resonances so broad as to be unobservable. The two methods are therefore to a large extent complementary.

The paramagnetic shifts which are observed can be used to derive important information, and are made up of two components. First, there is a through-space dipolar interaction between the magnetic moments of the electron and the resonating nucleus, which gives a dipolar shift, also sometimes known as a pseudo-contact shift. This contribution may be small, and much effect has been expended in calculating it, so that the other component, known as the contact shift, may be identified. The origin of the contact shift may be understood if one considers coupling between electron and nucleus, which would give a doublet in the NMR spectrum under conditions of slow electronic spin-lattice relaxation, but with a coupling constant of millions of Hz! With such a large coupling, the intensities of the two resonances are not equal, and so their weighted mean position is not midway between them. Under fast relaxation conditions the doublet collapses back to a singlet at the weighted mean position, and so may fall some thousands of Hz away from the expected position – the contact shift.

Contact shifts give a measure of unpaired spin density at the resonating nucleus, and so are particularly useful in studying the spin distribution in organic radicals, or in ligands in organometallic compounds. For example, the bis(cyclopentadienyl) complexes of vanadium, chromium, cobalt and nickel (but not iron, as ferrocene is diamagnetic) have been studied, both ^1H and ^{13}C contact shifts being determined. Various theoretical bonding models have thus been compared, and it has been possible to gain information about the bonding orbitals involved in metal-ring interactions, and the mechanisms of electron delocalization.

When large paramagnetic shifts are obtained without undue broadening of lines, the resulting spectra can be particularly useful. Fig. 2.45 shows a spectrum of an oxidized iron(III) cytochrome b_2, which has an iron protoporphyrin IX center bound axially to two histidines (2.XXXVI), which are part of a protein containing 95 amino acids. The ^1H NMR spectrum of this compound, which has a molecular weight of 10500, is extremely complex. The aliphatic protons with chemical shifts between 0 and 5 ppm give an enormous peak, which is truncated in the figure, and aromatic protons give a series of peaks between 6 and 8 ppm. Many individual peaks can be resolved, but assignment is difficult. However, protons which are part of the porphyrin are close to the paramagnetic centre, and so large contact shifts move their resonances into more easily observed parts of the spectrum. With the aid of 2D COSY spectra it has been possible to assign many of the resonances, including, for example, three associated with one of the vinyl groups (marked v), a methyl group (m), and the protons of one CH_2 group of a propionate substituent (p).

The unusual properies of paramagnetic compounds which we have just discussed can be put to good use in a variety of ways. In particular, the relaxation and chemical shift effects can be so large that even small concentrations of paramagnetic materials

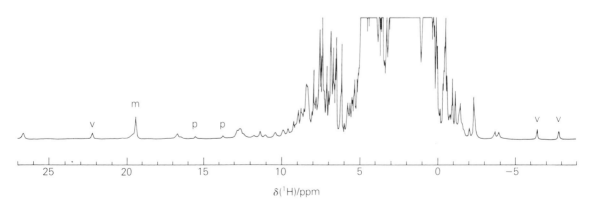

Fig. 2.45 ^1H NMR spectrum of the oxidized iron (III) cytochrome b_2, the central part of which is shown as structure 2.XXXVI. Intense peaks marked v, m and p have been assigned to protons which are parts of vinyl, methyl and propionate substituents on the porphyrin.

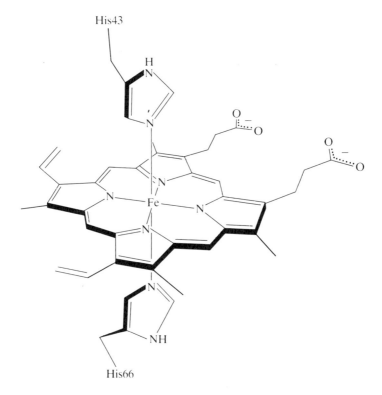

His43

His66

2.XXXVI

can have dramatic effects on the NMR spectra of diamagnetic compounds. One such application involves 'shift reagents', paramagnetic complexes of europium or praseodymium (see Section 2.4.2), which can be used to spread out complex spectra, usually of protons, and so simplify spectral analysis. In other situations paramagnetic compounds have been used to induce decoupling, by providing a rapid relaxation mechanism, and in others to eliminate the Nuclear Overhauser Effect (Section 2.10.6). All of these effects are now more often obtained by other means (use of high field magnets, double resonance or multi-pulse methods) but one application still of some importance is the use of traces of paramagnetic compounds to reduce the relaxation times of slowly-relaxing nuclei (Section 2.9.2), and so facilitate their study by Fourier transform methods.

Finally, we should note that the shifts in resonance frequencies caused by paramagnetic substances can be used to measure the magnetic susceptibilities of those materials [46]. The method, commonly known as Evans' method, involves the use of two solutions of equal concentration of an inert substance, typically 2% of t-butanol, one of which also contains the paramagnetic material. One solution is placed in a capillary, inside the other. Two resonances for the butyl groups are then observed, and the shift between them is attributed to the magnetic material, and is in fact directly proportional to the difference in volume susceptibilities for the two solutions. The mass susceptibility χ of the dissolved substance is then given by Eqn (2.12), in which Δv is the shift in the resonance frequency v, m is the mass of material per ml of solution, χ_0 is the mass susceptibility of the solvent (-0.72×10^{-6} for dilute aqueous t-butanol) and d_0 and d_s are the densities of the solvent and solution, respectively.

$$\chi = \frac{3\Delta\nu}{2\pi\nu m} + \chi_0 + \frac{\chi_0(d_0 - d_s)}{m} \qquad (2.12)$$

For strongly paramagnetic materials such as many transition metal complexes the last term is small, and may be neglected.

Problems

2.1 Describe the ^{1}H NMR spectra you would expect to observe for (a) GeClH$_2$GeClH$_2$, (b) SiH$_3$SiCl$_2$H, (c) PF$_2$H, (d) PF$_4$H and (e) SiH$_3$PH$_2$. Ignore ^{13}C and ^{29}Si satellites, and make reasonable assumptions about the magnitudes of coupling constants.

2.2 Describe the ^{1}H NMR spectra you would expect to observe for (a) SiF$_2$H$_2$, (b) P(OCH$_3$)$_3$, (c) SiH$_3$SiCl$_2$H, (d) CH$_3$SiH$_3$ and (e) P(SiH$_3$)$_3$, including ^{13}C and ^{29}Si satellites. Make reasonable assumptions about the magnitudes of coupling constants.

2.3 A silicone fluid consists of a mixture of straight-chain siloxanes, R$_3$SiO(SiR$_2$O)$_n$SiR$_3$ with n = 2,3,4,5 and 6. Describe the ^{1}H-decoupled ^{29}Si NMR spectrum of the fluid.

2.4 The ^{13}C NMR spectrum of the carbonyl groups in one isomer of [W(CO)$_4$P(OMe)$_3$(SPh)]$^-$ is shown in Fig. 2.46. Identify the isomer and account for the form of the spectrum. What would you expect to see in the ^{13}C spectrum of the other isomer? [The figure is adapted, with permission, from D.J. Darensbourg, K.M. Sanchez and J. Reibenspies, *Inorg. Chem.*, **27**, 3636 (1988). Copyright (1988) American Chemical Society.

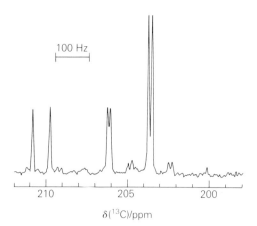

100 Hz

210   205   200

$\delta(^{13}$C)/ppm

Fig. 2.46

2.5 A notation for classifying spin systems is described in Section 2.6.1. What are the spin systems for the following molecules? Ignore ^{13}C and ^{29}Si satellite spectra, except where indicated.
Si$_2$H$_5$Cl N(PF$_2$)$_3$ SF$_5$C≡CH PF$_5$ *fac*-[Mo(CO)$_3$(PMe$_3$)$_3$] *mer*-[Mo(CO)$_3$(PMe$_3$)$_3$] Si$_3$H$_8$, including forms containing one ^{29}Si.

2.6 Which of the following species contains sets of nuclei which are chemically but not magnetically equivalent?
PF$_2$OCH$_3$ PF$_2$OPF$_2$ ^{12}CH$_3$O^{12}CH$_3$ ^{12}CH$_3$O^{13}CH$_3$ ^{13}CH$_3$O^{13}CH$_3$ TeF$_5$Br   S$_2$F$_{10}$
Ge$_2$H$_6$ Ge$_3$H$_8$ P$_2$H$_4$

F$_3$P, Cl, Cl
Pt, Pt
Cl, Cl, PF$_3$

2.7 Sketch the $^{13}C\text{-}\{^1H\}$ noise-decoupled NMR spectra of the following. For the purposes of this exercise the values of chemical shifts and coupling constants are not important, and species containing more than one ^{13}C nucleus should be ignored.

$$Fe(C_5H_5)_2 \quad (n\text{-}C_8H_{17})_3SiCl \quad PEt_3 \quad CD_2Cl_2 \quad Ru_3(CO)_{12}$$

2.8 Describe qualitatively what you would expect to see in the ^{13}C NMR spectrum of the diphosphene 2.V.

2.9 Describe the ^{19}F NMR spectra you would expect to observe for (a) ClF_6^+, (b) ClF_5 and (c) ClF_3.

2.10 Explain how it would be possible to distinguish the four isomers of hexasilane, Si_6H_{14}, using ^{29}Si NMR spectra obtained with ^1H noise decoupling.

2.11 Reaction between IF_7 and SbF_5 gives a 1:1 product. The ^{19}F NMR spectrum of this material contains two sets of resonances: one consists of 6 lines of equal intensity; the other consists of overlapping patterns of 6 lines of equal intensity and 8 lines of equal intensity, the former being somewhat stronger. Account for these observations.

2.12 Reaction of phosphorus trifluoride with hydrogen fluoride gives a number of products, including an anionic species, the ^1H NMR spectrum of which consists of a doublet (1000 Hz) of doublets (125 Hz) of quintets (5 Hz). Its ^{31}P spectrum is a doublet (1000 Hz) of quintets (820 Hz) of doublets (730 Hz). Identify this ion, and describe as fully as possible its ^{19}F NMR spectrum, assuming this to be of first order.

2.13 The 75.4 MHz ^{19}F spectrum of TeF_5^- includes lines with the following frequencies (relative to CCl_3F) and relative intensities.

Frequency/Hz	Intensity
− 2230.6	1.1
− 2282.4	6.2
− 2332.9	5.7
− 2384.2	5.3
− 2433.6	1.9
− 3131.4	34.1
− 3182.4	30.3

There are groups of satellite lines centered at *ca.* − 890, − 2469 (a doublet) and − 3840 Hz. The last group clearly has a second order pattern. Account for these observations, and calculate as many NMR parameters as you can. (Note that the frequencies and intensities are subject to random errors.)

2.14 The ^1H NMR spectrum of $GeFH_3$ consists of two lines separated by 42 Hz. What are the relative positions and intensities of all the lines in the ^{19}F spectrum of (a) $GeDFH_2$, (b) GeD_2FH and (c) GeD_3F?

2.15 Describe the ^1H NMR spectra of BH_4^-, AlH_4^- and GaH_4^-.

2.16 Using information given in Tables 2.1 and 2.2 describe the ^1H NMR spectra of (a) SnH_4, (b) TeH_2 and (c) NH_4^+, giving the relative positions and intensities of all lines. State what assumptions you make about coupling constants.

2.17 What would you expect to observe in the ^{29}Si NMR spectrum of a mixture of SiF_4 and $SiCl_4$, after halogen exchange has occurred to give a random mixture of products?

2.18 The intensities and frequencies of resonances in the 121.4 MHz $^{31}P\text{-}\{^1H\}$ noise-decoupled NMR spectrum of $[Pt(PPh_3)\{N(CH_2CH_2PPh_2)_3\}]$ are listed below. Calculate as many NMR parameters as you can. What can you deduce about the structure of the complex?

Frequency	Intensity	Frequency	Intensity	Frequency	Intensity
− 3393.4	7.8	− 673.4	2.0	1545.8	7.6
− 3313.5	7.7	− 593.6	0.6	1625.6	2.7
− 1484.8	31.1	263.7	7.9	3604.5	0.7
− 1404.9	31.3	343.7	7.7	3684.6	1.9
− 833.3	0.7	1385.5	2.6	3764.4	1.9
− 753.5	1.9	1465.6	7.8	3844.7	0.6

2.19 Explain qualitatively how the linewidths would differ in the ^{53}Cr NMR spectrum of CrO_4^{2-}, $CrCl_2O_2$, CrF_4 and $CrCl_6^{3-}$.

2.20 Describe the ^{1}H, ^{19}F, ^{31}P and ^{29}Si NMR spectra of SiF_3PH_2, making reasonable assumptions about the magnitudes of coupling constants.
  Describe qualitatively how the spectra would be modified if
(a) rotation about the Si—P bond was stopped,
(b) there was rapid intermolecular exchange of protons,
(c) the compound was dissolved in a liquid crystalline solvent.

2.21 Account for the following observations.
(a) The ^{1}H NMR spectrum of GeH_4 shows ^{73}Ge satellites but the spectra of $GeFH_3$ and $GeClH_3$ do not.
(b) The ^{13}C-{^{1}H} NMR spectrum of bis(allyl)nickel at 200 K shows three resonances, of equal intensity; at 250 K, only two resonances are observed, in intensity ratio 2:1.
(c) The ^{1}H NMR spectrum of rigorously dried ^{14}N-ammonia is a broad 1:1:1 triplet that becomes broader as the temperature is lowered; the ^{1}H spectrum of ^{15}N-ammonia containing a little $Na^{15}NH_2$ is a broadened doublet that becomes sharper as the temperature is lowered.
(d) The ^{19}F NMR spectrum of PF_3 in an isotropic solvent is a doublet; in a nematic-phase liquid crystal solvent, the spectrum is a doublet of triplets.

2.22 Describe the splitting pattern you would expect to see in the ^{1}H NMR spectrum of SiF_2H_2 dissolved in a liquid crystal solvent. How many dipolar coupling constants could you measure? What other NMR experiments must also be performed to gain the maximum possible amount of structural information?

2.23 For each of the following compounds, work out how many different dipolar couplings between spin 1/2 nuclei could be determined by liquid crystal NMR experiments. In each case state how many parameters are needed to describe the molecular orientation, and so deduce the maximum number of independent structural parameters that can be determined by this method.

$SiFH_3$ PF_2HSe PF_2HS $CH_3{\equiv}C{-}C{\equiv}N$ $[W(CO)_5(PH_3)]$ $[Fe(CO)_4H_2]$

2.24 How would the highest temperature spectrum of Fig. 2.42 appear (a) if the exchange process was intermolecular and (b) if there was scrambling of ligands on each isomer, but no interconversion of the two forms?

2.25 The structure of $PF_3(C_6H_5)_2$ is trigonal bipyramidal, with one equatorial and two axial fluorine atoms, which interchange positions when the compound is heated. Describe the low and high temperature ^{31}P and ^{19}F NMR spectra.

2.26 Using data given in Table 2.1, calculate the maximum possible Nuclear Overhauser Effect and maximum intensity gain possible by polarization transfer for (a) ^{15}N-{^{1}H}, (b) ^{77}Se-{^{19}F} and (c) ^{195}Pt-{^{1}H} experiments.

2.27 How would the spectrum shown in Fig. 2.9 be changed by (a) ^{1}H noise decoupling and (b) ^{15}N noise decoupling?

2.28 If the lowest frequency quartet of satellite lines in the ^{1}H NMR spectrum of Si_2H_6 (Fig. 2.10) is irradiated with moderate power, the satellites just to high frequency of the center collapse into the main resonance. What can you deduce from this?

2.29 The ^{11}B proton-decoupled and 2D COSY spectra shown in Fig. 2.47 were recorded for a metalloborane cluster containing a B_9Rh cage. Assuming that couplings are observed only for directly bonded (i.e. not hydrogen-bridged) pairs of boron atoms, deduce the structure of the boron part of the cage. The figure is taken, with permission, from X.L.R. Fontaine, N.N. Greenwood, J.D. Kennedy, P. MacKinnon and M. Thornton-Pett, *J. Chem. Soc., Dalton Trans.*, 2809 (1988).

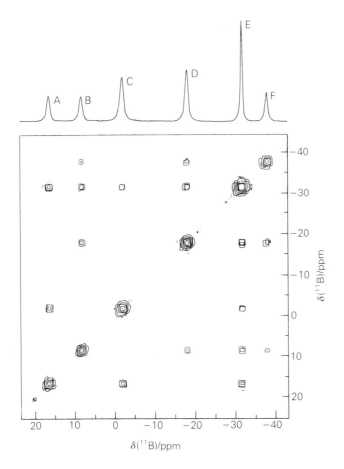

Fig.2.47

2.30 Sketch the ^{31}P-$\{^1$H$\}$ (decoupled) spectra of the following molecules. Make reasonable assumptions about the magnitudes of coupling constants. Ignore ^{13}C satellites.

    trans-[PtCl$_2$(PEt$_3$)$_2$]   PF$_4$(C$_6$H$_5$)   PF$_2$PH$_2$ P$_3$H$_5$

2.31 Sketch the ^{31}P NMR spectra of the following molecules and ions. Make reasonable assumptions about the magnitudes of coupling constants. Include ^{13}C and ^{77}Se satellites where appropriate.

    PF$_6$$^-$ PF$_2$H   P(OMe)$_3$ P(CH$_3$)H$_2$Se   P(CH$_3$)$_2$$^-$

2.32 Using the information provided by the ^{183}W 2D INADEQUATE spectrum of [SiZnW$_{11}$O$_{39}$]$^{6-}$ in Fig. 2.32(b) construct the ^{183}W COSY spectrum for this anion.

2.33 Assuming that J_{NH}, J_{PH} and J_{FH} are 80, 16 and 8 Hz respectively, sketch the ^{1}H NMR spectra of $PF_2{}^{15}NH_2$ (a) fully coupled, (b) with ^{19}F decoupling and (c) with ^{31}P decoupling.

2.34 The ^{1}H NMR spectrum of $H^{13}C^{15}N$ consists of four lines, A, B, C, and D, which are 535, 525, 275 and 265 Hz respectively to high frequency of the resonance of tetramethylsilane. Irradiation in the ^{13}C resonance region (*ca.* 25 MHz) or in the ^{15}N region (*ca.* 10 MHz) leads to splitting of lines in the ^{1}H spectrum, as follows:

Irradiation frequency/Hz		Lines affected		
25 147 839	*or* 25 147 579	B	*and*	D
25 147 821	*or* 25 147 561	A	*and*	C
10 135 866	*or* 10 135 856	A	*and*	B
10 135 848	*or* 10 135 838	C	*and*	D

If $^1J(^{13}C^1H)$ is assumed to be positive, what are the signs and magnitudes of the three coupling constants for the molecule?

2.35 Describe qualitatively the ^{13}C NMR spectrum you would expect to observe for crystalline samples of the rhodium complex shown in Fig. 8.16. How would the spectrum depend on the symmetry of the molecules in the crystal?

2.36 Describe the ^{13}C NMR spectrum of the complex $[Rh(\eta^5\text{-}C_5H_5)(\eta^2\text{-}C_2H_4)_2]$ in the crystalline phase under conditions of optimum resolution, assuming that there is one molecule per asymmetric unit in the unit cell. How would the appearance of the spectrum be altered (a) if there was a plane of symmetry through each molecule in the crystal and (b) if one of the C_2H_4 ligands was replaced by C_2F_4?

2.37 $W_2(OPr^i)_6$ reacts with PH_3 to give a product A. The ^{31}P NMR spectrum of A shows a wide triplet ($^1J_{PH}$, 400 Hz) of narrower doublets (20 Hz) with a satellite pattern whose total intensity is about 26% of that of the whole group of resonances. The ^{1}H NMR spectrum shows peaks due to coordinated OPr^i groups, and two other resonances: a doublet (20 Hz) of triplets (5 Hz) at -10 ppm and a doublet (400 Hz) of doublets (5 Hz) at 2.1 ppm. Treatment of A with strong base affords B, the ^{31}P NMR spectrum of which consists of a doublet ($^1J_{PH}$, 160 Hz) of doublets (17 Hz) with a satellite pattern whose total intensity is about 26% of that of the whole group of resonances. Assign structures to A and B and account for the spectroscopic data. (N.B. $^1J_{PH}$ is 180 Hz in PH_3 and 545 Hz in $PH_3{}^+$).

2.38 Reaction of disilane with tin (IV) chloride gives a mixture of products, the most volatile of which has a triplet and a less intense quartet in its ^{1}H NMR spectrum. Both groups of resonances have two sets of satellite lines, with coupling constants *ca.* 220 and 15 Hz. There are two rather less volatile compounds. One has a single resonance, with two sets of satellites, couplings *ca.* 220 and 15 Hz, but with each satellite further split into a triplet, splitting 5 Hz. The other compound has doublet and quartet resonances, again each with satellites corresponding to couplings of 220 and 15 Hz. Account for these observations.

2.39 The ^{1}H NMR spectrum of $Hg(C_5H_5)_2$ consists of three closely-spaced lines with intensity ratio roughly 1:10:1; the pattern has ^{13}C satellites. The ^{13}C-{^{1}H} spectrum also consists of three lines in approximate intensity ratio 1:10:1. The ^{1}H spectrum of $Tl(C_5H_5)$ consists of four lines, with intensity ratio 7:3:3:7, with the gap between the middle two very much wider than the other splittings. There are also ^{13}C satellites. These patterns are not affected by temperatures in the range 170 to 320 K. What can you deduce from these observations about the structures of $Hg(C_5H_5)_2$ and $Tl(C_5H_5)$?

2.40 At 170 K the ^{13}C NMR spectrum of a complex $[CuW(C_5H_5)(CO)_3(PPh_3)_2]$ has three resonances, intensity ratio 5:2:1, in addition to those associated with the PPh_3 ligands. The ^{31}P spectrum with ^{1}H decoupling consists of an [AB] pattern, with no ^{183}W satellites. At room temperature the corresponding region of the ^{13}C spectrum has one resonance of intensity 5, and one of intensity 3 with ^{183}W satellites. The ^{31}P spectrum contains just one resonance. What can you deduce from these observations?

CHAPTER 2: NUCLEAR MAGNETIC RESONANCE SPECTROSCOPY

2.41 The dipolar couplings $D_{SiH'}$, $D_{CH'}$ and $D_{C'H'}$, measured for a sample of $SiH_3C{\equiv}C'H'$ in a liquid crystal solvent are 64.4, -487.7, and -4988.0 Hz respectively. Assuming that the length of the Si—C bond is 1.8251 Å, calculate the C≡C and C—H bond lengths.

2.42 Tungsten hexacarbonyl reacts with sodium tetrahydroborate to give an anionic product. The tetraethylammonium salt has empirical composition $C_{18}H_{21}NO_{10}W_2$. Its ${}^1H$ NMR spectrum includes a triplet, intensity ratio 1:6:1, at $\delta -22.5$ ppm, with a separation between the outer peaks of 42 Hz. What can you deduce about the anion?

2.43 A platinum compound is thought to be $[Pt(PF_3)(PMe_3)(NCS)_2]$. What information about the composition and structure of the compound could you obtain by an NMR study? What spectra would you record?

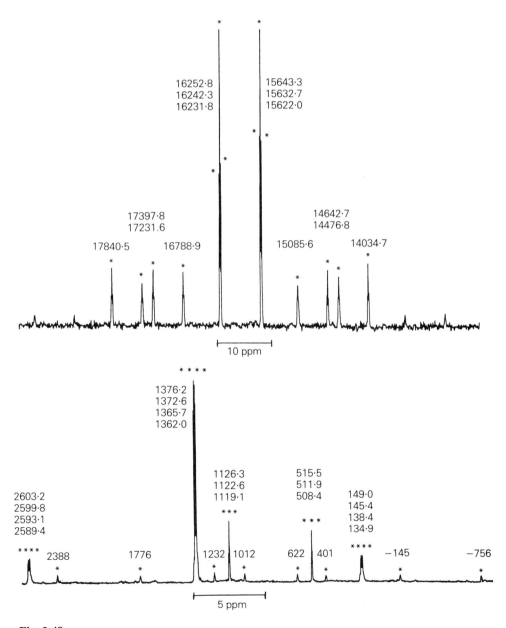

Fig. 2.48

2.44 How would a ^{19}F NMR spectrum enable you to tell whether a solution of SF_4 and BF_3 consisted of the unreacted compounds or the ionic product $SF_3^+ BF_4^-$?

2.45 At very low temperatures the ^1H NMR spectrum of a metal tetrahydroborate, $M(BH_4)_n$, has two broad groups of resonances, of equal intensity, centered 500 Hz apart. On decoupling ^{11}B the resonances sharpen, and each shows a small triplet coupling. On warming these resonances broaden, coalesce at 180 K, and eventually become a single sharp line, which becomes a 1:1:1:1 quartet when ^{11}B is no longer decoupled. Account for these observations, and calculate the mean lifetimes in the protons' two distinct positions at 180 K.

2.46 A solution was prepared containing 7.4 mg of a nickel complex with formula $[Ni(SOCPh)_2(NC_6H_7)_2]$ in 1 ml CCl_2H_2, with 3% $SiMe_4$. This solution was put into an NMR tube containing a capillary full of 3% $SiMe_4$ solution in CCl_2H_2. The ^1H NMR spectrum (100 MHz) included two lines attributable to $SiMe_4$, separated by 14.9 Hz at 301 K, 16.2 Hz at 273 K, 18.6 Hz at 249 K and 20.7 Hz at 223 K. The mass susceptibility of the solvent is -0.549×10^{-6}. Assuming that the densities of the two solutions are not significantly different, calculate the mass susceptibility and molar susceptibility at each temperature and show that $\chi_{molar} T$ is approximately independent of temperature.

2.47 The platinum complex $[Pt_2Cl_4(PEt_3)_2]$, which has two chlorine bridges between the platinum atoms, reacts with $trans$-$[PtCl(P'Cl_2)(PEt_3)_2]$ to give a product whose 81.02 MHz ^{31}P-$\{^1H\}$ spectrum has two groups of resonances, as shown in Fig. 2.48. Positions of lines (relative to 85% H_3PO_4) marked with asterisks are listed on the spectra. Deduce what you can about the nature of the product, and calculate as many NMR parameters for the product as you can.

2.48 SF_4 forms an equimolar adduct with trimethylamine. Describe the ^{19}F NMR spectrum you would expect to observe for this species (a) at very low temperatures, (b) if there was fast exchange between coordinated NMe_3 and excess free amine at low temperature and (c) as in (b), but at room temperature.

Further reading

H. Günther, *NMR Spectroscopy: An Introduction*, Wiley, New York (1980).

E.D. Becker, *High Resolution NMR Theory and Chemical Applications*, Academic Press, London and New York (1980).

J.K.M. Saunders and B.K. Hunter, *Modern NMR Spectroscopy: A Guide for Chemists*, Oxford University Press, Oxford (1987).

Also Refs [1], [3] and [24] below.

P. Laszlo (ed.), *NMR of Newly Accessible Nuclei*, Academic Press, London and New York (1983): applications of NMR to a wide range of nuclei rarely studied.

C. Brévard and P. Granger, *Handbook of High Resolution Multinuclear NMR*, Wiley, New York (1981): a sourcebook for technical data.

R. Freeman, *A Handbook of Nuclear Magnetic Resonance*, Longman, Harlow (1987).

M.L. Martin, G.J. Martin and J.-J. Delpuech, *Practical NMR Spectroscopy*, Heyden, London (1980): very mathematical, lots of practical details.

D. Neuhaus and M.P. Williamson, *The Nuclear Overhauser Effect in Structural and Conformational Analysis*, VCH Publishers, New York (1989).

The *Specialist Periodical Reports on NMR* (Ref. [7] below, published annually) include lists of recent books and reviews on the subject.

References

1 R.K. Harris and B.E. Mann, *NMR and the Periodic Table*, Academic Press, London and New York (1978).

2 J.C. Lindon and A.G. Ferrige, *Prog. NMR Spectrosc.*, **14**, 27 (1980).

3 T.C. Farrar and E.D. Becker, *Pulse and Fourier Transform NMR*, Academic Press, London and New York (1971).

4 D. Shaw, *Fourier Transform NMR Spectroscopy*, Elsevier, Amsterdam (1976).

5 B.P. Straughan and S. Walker, *Spectroscopy*, Chapman and Hall, London (1976), Vol. 1.

6 N.F. Chamberlain, *The Practice of Nuclear Magnetic Resonance Spectroscopy with Spectra–Structure Correlations for Hydrogen-1*, Plenum, London and New York (1974).

7 Specialist Periodical Reports, *NMR* and *Spectroscopic Properties of Inorganic and Organometallic Compounds*, Royal Society of Chemistry, London, annually.

8 G.C. Levy, R.L. Lichter and G.L. Nelson, *Carbon-13 Nuclear Magnetic Resonance Spectroscopy*, 2nd edn, Wiley-Interscience, New York (1980).

9 G.C. Levy (ed.), *Topics in ^{13}C NMR Spectroscopy*, Wiley-Interscience, New York, Volumes from 1974 on.

10 J.F. Johnson and W.C. Jankowski, *^{13}C NMR Spectra: A Collection of Assigned, Coded and Indexed Spectra*, Wiley-Interscience, New York (1972); *^{13}C Nuclear Magnetic Resonance Data Base*, (producer) W. Bremser, Fachinformationszentrum Energie.Physik.Mathematik GmbH, Karlsruhe (1983).

11 C.H. Dungan and J.R. Van Wazer, *Compilation of Reported ^{19}F Chemical Shifts*, Wiley Interscience, New York (1970).

12 G. Huttner, J. Borm and L. Zsolnai, *J. Organomet. Chem.*, **263**, C33 (1984).

13 M.M. Crutchfield, C.H. Dungan, J.H. Letcher, V. Mark and J.R. Van Wazer, *Topics in Phosphorus Chemistry*, (eds M. Grayson and E.J. Griffiths) Interscience, New York (1967) Vol. 5; G. Mavel, *Ann. Rep. NMR Spectrosc.*, **5B**, 1 (1973).

14 P.S. Pregosin, *Coord.Chem.Rev.*, **44**, 247 (1982).

15 R.R. Schrock, *J.Am.Chem.Soc.*, **97**, 6577 (1975).

16 A.H. Cowley, J.E. Kilduff, S.K. Mehrotra, N.C. Norman and M. Pakulski, *J.Chem.Soc., Chem.Commun.*, 520 (1983).

17 M. Green, J.A.K. Howard, M. Murray, J.L. Spencer and F.G.A. Stone, *J.Chem.Soc., Dalton Trans.*, 1509 (1977).

18 B.E. Mann and B.F. Taylor, *^{13}C NMR Data for Organometallic Compounds*, Academic Press, London and New York (1981).

19 R.K. Harris and B.J. Kimber, *J. Organomet.Chem.*, **70**, 43 (1974).

20 J.N. Murrell, *Prog. NMR Spectrosc.*, **6**, 1 (1971).

21 R.J. Abraham, *The Analysis of High-resolution NMR Spectra*, Elsevier, Amsterdam (1971); R.A. Hoffmann, S. Forsen and B. Gestblom, *NMR Basic Principles and Prog.*, **5**, 1 (1971).

22 G. Hägele and R.K. Harris, *Ber. Bunsenges. phys.Chem.*, **76**, 910 (1972).

23 H. Schmidbaur, O. Gasser, T.E. Fraser and E.A.V. Ebsworth, *J.Chem.Soc., Chem.Commun.*, 334 (1977).

24 R.K. Harris, *Nuclear Magnetic Resonance Spectroscopy*, Pitman, London (1983).

25 A.J. Shaka, J. Keeler, T. Frenkiel and R. Freeman, *J.Magn.Reson.*, **52**, 335 (1982).

26 W. McFarlane and D.S. Rycroft, *J.Chem.Soc., Dalton Trans.*, 2162 (1973).

27 J.H. Noggle and R.E. Schirmer, *The Nuclear Overhauser Effect: Chemical Applications*, Academic Press, London and New York (1971).

28 R. Benn and H. Gunther, *Angew.Chem.Int.Ed.Engl.*, **22**, 350 (1983).

29 G.A. Morris, *Magn.Reson.Chem.*, **24**, 371 (1986).

30 D.M. Dodrell, D.T. Pegg and M.R. Bendall, *J.Magn.Reson.*, **48**, 323 (1982).

31 G.A. Morris and R. Freeman, *J.Magn.Reson.*, **23**, 171 (1976).

32 A. Bax, R. Freeman and S.P. Kempsell, *J.Am.Chem.Soc.*, **102**, 4849 (1980).

33 A. Bax, *Two-dimensional NMR in Liquids*, D. Reidel, Dordrecht, Netherlands (1982).

34 D. Reed, *J.Chem.Res.(S)*, 198, (1984).

35 C. Brévard, R. Schrumpf, G. Tourne and G.M. Tourne, *J.Am.Chem.Soc.*, **105**, 7059 (1983).

36 T.L. Jorris, M. Kozik, N. Casan-Pastor, P.J. Domaille, R.G. Finke, W.K. Miller and L.C.W. Baker, *J.Am.Chem.Soc.*, **109**, 7402 (1987).

37 J.W. Emsley and J.C. Lindon, *NMR Spectroscopy Using Liquid Crystal Solvents*, Pergamon, New York (1975).

38 P. Diehl and C.L. Khetrapal, *NMR Basic Principles and Prog.*, **1**, 1 (1969).

39 I.D. Gay, C.H.W. Jones and R.D. Sharma, *J.Magn.Reson.*, **84**, 501, (1989).

40 H.W. Kroto, S.I. Klein, M.F. Meidine, J.F. Nixon, R.K. Harris, K.J. Packer and P. Reams, *J. Organomet. Chem.*, **280**, 281 (1985).

41 C.A. Fyfe, H. Gies and Y. Feng, *J.Am.Chem.Soc.*, **111**, 7703 (1989).

42 R.A. Haberkorn, R.E. Stark, H. van Willigen and R.G. Griffin, *J.Am.Chem.Soc.*, **103**, 2534 (1981).

43 L.M. Jackman and F.A. Cotton, *Dynamic Nuclear Magnetic Resonance Spectroscopy*, Academic Press, London and New York (1975); J. Sandström, *Dynamic NMR Spectroscopy*, Academic Press, London and New York (1982).

44 E.W. Abel, D.G. Evans, J.R. Koe, V. Sik, M.B. Hursthouse and P.A. Bates, *J.Chem.Soc., Dalton Trans.*, 2315 (1989).

45 G.N. La Mar, W. De W. Horrocks and R.H. Holm, *NMR of Paramagnetic Molecules, Principles and Applications*, Academic Press, London and New York (1973).

46 D.F. Evans, *J.Chem.Soc.*, 2003 (1959).

3 Electron Spin and Nuclear Quadrupole Resonance Spectroscopy

3.1

The electron spin resonance experiment

The technique of electron spin resonance spectroscopy can only be applied to species having one or more unpaired electrons. This covers a fairly wide range of substances, including free radicals, biradicals and other triplet states, and many transition metal compounds, but the breadth of application is nothing like that achieved by NMR spectroscopy (Chapter 2). The simple fact that an ESR spectrum can be observed for a sample is an indication that at least some proportion of the sample has unpaired electrons, and the technique is particularly valuable for the study of unstable para-magnetic species, which may be generated *in situ* by such means as electrochemical oxidation or reduction. For pure paramagnetic compounds, bulk magnetic suscep-tibility measurements provide information about numbers of unpaired electrons, which may help in interpreting ESR spectra and in understanding the electronic struc-ture of the magnetic species. With the exception of Evans' method, using NMR (Sec-tion 2.16), these bulk methods are beyond the scope of this book.

The spin of an electron and its associated magnetic moment (possibly with some contribution from orbital angular momentum) are the basis of ESR spectroscopy, also sometimes known as electron paramagnetic resonance (EPR) spectroscopy. In the presence of a magnetic field, B_0, a molecule or ion having one unpaired electron has two electron spin energy levels, given by:

$$E = g\mu_B B_0 M_S. \tag{3.1}$$

Here μ_B is the Bohr magneton, M_S is the electron spin quantum number ($+1/2$ or $-1/2$) and g is a proportionality factor, equal to 2.002 32 for a free electron. For radicals this factor is normally very close to the free-electron value, between 1.99 and 2.01, but for transition metal compounds spin–orbit coupling and zero-field splitting, discussed in Section 3.3, can lead to major variations in g values. Typical values are in the range 1.4 to 3.0. It is the transitions between such energy levels, induced by the application of an appropriate frequency radiation, that are studied by ESR. The frequencies required depend on the strength of the magnetic field: most commonly fields of 0.34 and 1.24 T are used, with corresponding frequencies of 9.5 and 35 GHz, which are in the microwave region. Note that g factors are indepen-dent of field direction only in isotropic systems; gases, liquids or solutions of low viscosity, and solid sites with cubic symmetry.

The simple picture of two energy levels must be modified if the electron can interact with a neighboring nuclear magnetic dipole. The energy levels then become

$$E = g\mu_B B_0 M_S + aM_S m_I \tag{3.2}$$

where m_I is the nuclear spin quantum number for the neighboring nucleus. The constant a is called the hyperfine coupling constant. A single nucleus with spin 1/2 will split

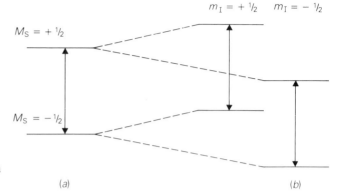

Fig. 3.1 Energy levels and transitions for a single unpaired electron in an external magnetic field (a) with no coupling and (b) coupling to one nucleus with spin ½.

each level into two as shown in Fig. 3.1, and the selection rule is such that two transitions can then be observed. The energy difference between them is equal to the hyperfine coupling constant: analysis and interpretation of hyperfine splittings is discussed in Section 3.2.

During a normal ESR experiment the population difference between the energy states involved may change. Energy is supplied by the applied microwave radiation, and this is transferred from the spin system to its surroundings, a process known as spin−lattice relaxation. Relaxation can also occur by spin−spin transfer mechanisms. If the relaxation time is long, the population of the upper state will increase during observation, and the signal intensity will thus saturate or decrease in intensity. If the relaxation time is short, then by the Uncertainty Principle the resonance lines must be wide. This is often the case for transition metal complexes, and it is often necessary to cool samples to liquid nitrogen or even liquid helium temperatures in order to observe a spectrum at all. Line broadening can also be caused by exchange processes. The effects of such processes in NMR spectroscopy are discussed in Section 2.15, and the same principles apply in ESR spectroscopy. However, we should note that relevant processes now include electron exchange, and not just exchange of spinning nuclei.

Although absolute intensities cannot be measured reliably, relative intensities of signals are in principle proportional to the relative total numbers of unpaired electrons in the systems. Integrated ESR absorption intensities (obtained by double integration of the normal derivative spectra, or less accurately by measuring peak-to-peak amplitudes) can therefore be used for any purpose requiring estimation of concentrations, such as kinetic studies of the decay of unstable radicals.

A typical ESR spectrometer has a radiation source (usually a klystron), a sample chamber situated between the poles of a magnet, and a detection and recorder system. The sample chamber is situated in a resonant cavity, which works on the same principle as an organ pipe. The cavity size is chosen so that a standing wave is set up, and the sample is then placed in the region of highest energy density. The cavity is placed so that this point coincides with a region of uniform magnetic field. The detection system normally employed utilizes a small-amplitude magnetic field modulation and a phase-sensitive detector as a means of reducing noise. A consequence of this is that the output to the recorder is in the form of the first derivative of the absorption curve (Fig. 3.2).

For a clear and readable account of the theory and applications of ESR, and of the principles of spectrometer design, see Ref. [1]. This book gives references to further sources of any required degree of complexity.

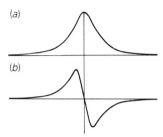

Fig. 3.2 Absorption spectrum (*a*) and first-derivative spectrum (*b*) for a Lorentzian line-shape.

3.2 Hyperfine coupling in isotropic systems

Interactions between electron and nuclear spin magnetic moments lead to the appearance of fine structure in ESR spectra, and this can provide much information about the species involved. The couplings arise in two distinct ways. The first of these is by direct dipole–dipole interaction, which depends on the angle between the vector joining the two dipoles and the applied magnetic field. In many cases this coupling is averaged to zero by rapid tumbling: the situations in which this is not the case are discussed in Section 3.3. The second mechanism by which coupling can occur is the Fermi contact interaction, which results from unpaired electron density at the nucleus in question. This is only present when the electron is in an orbital having some *s* character. Observation of coupling by this mechanism clearly gives direct indication of the extent of electron delocalization.

The coupling patterns observed in ESR spectra are determined by the same rules that apply to NMR spectra (Section 2.6) except that couplings to nuclei with spin $> 1/2$ are more frequently observed. Coupling to a single nucleus of spin $n/2$ gives $(n+1)$ lines, which are of equal intensity and equally spaced. Coupling to n equivalent nuclei of spin $1/2$ again gives $(n+1)$ lines, but the intensities follow the binomial distribution (Table 2.4). As a simple example, the radical anion formed by reaction of benzene with an alkali metal, $[C_6H_6 \cdot]^-$, has an ESR spectrum showing seven lines (Fig. 3.3). This implies equivalent coupling to all six hydrogen atoms, as expected if the electron is delocalized over all six carbon atoms.

If an electron couples with several sets of nuclei, then the overall pattern must be determined by first applying one of the couplings, then splitting each of these lines by

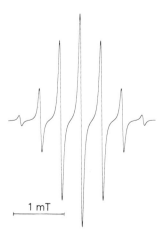

1 mT

Fig. 3.3 ESR spectrum of the benzene radical anion, $[C_6H_6 \cdot]^-$.

applying the second coupling, and so on. Finally, as with NMR spectra, the total spectrum may be the superposition of spectra due to several different isotopic species. Lists of important spinning nuclei are given in Tables 2.1 and 2.2.

The spectrum of the radical anion of pyrazine, 3.I, (Fig. 3.4) illustrates several of these points. The potassium salt shows couplings to two equivalent ^{14}N nuclei, giving a 1:2:3:2:1 quintet (see Section 2.6.5 for an explanation of the origin of this pattern), and these lines are split further into 1:4:6:4:1 quintets, showing that the electron is delocalized round the ring, coupling to the four hydrogens as well as to the nitrogen nuclei. In the spectrum of the sodium salt, a further splitting into 1:1:1:1 quartets is observed. The observation of this coupling, to ^{23}Na, is at first surprising, and it indicates the existence of ion pairs in solution. Equivalent couplings are also observed in the lithium and cesium derivatives, so it appears that it is the potassium salt which is the odd man out.

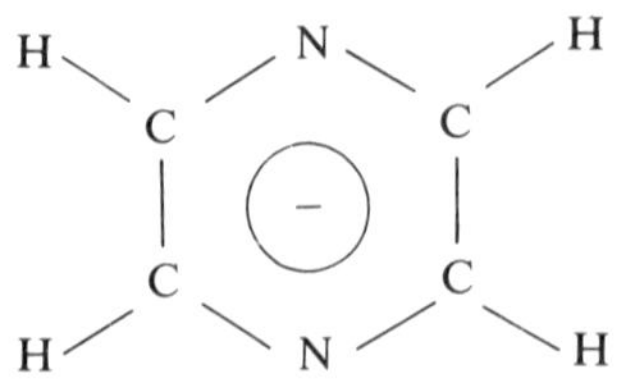

3.I

The separation between the lines in a multiplet pattern is called the hyperfine splitting constant, and is measured in terms of the change in the applied field necessary to give the lines the same frequency. Units usually given are gauss or millitesla (10^4 gauss = 1 tesla). The interaction energy, or hyperfine coupling constant, obtained on multiplication by $g\mu_B$, is usually quoted in MHz or cm^{-1}, neither of which, strictly speaking, is a unit of energy.

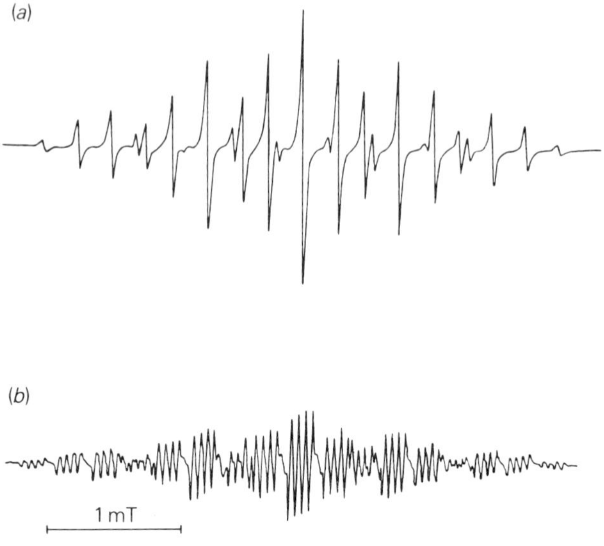

Fig. 3.4 ESR spectra of the pyrazine radical anion, (a) with K$^+$ counterion, and (b) with Na$^+$ counterion. Both spectra show hyperfine coupling to two ^{14}N and four ^{1}H nuclei, and in (b) further coupling to ^{23}Na is observed. Redrawn from A. Carrington and J. dos Santos Veiga, *Mol. Phys.*, **5**, 21 (1962); and J. dos Santos Veiga and A.F. Neiva-Correia, *Mol. Phys.*, **9**, 395 (1965).

The magnitude of the hyperfine splitting constant due to delocalization of an unpaired electron depends on the spin density at the nucleus in question. This is not the same thing as the unpaired electron density, which relates to the contribution to the atom of the molecular orbital containing the unpaired electron. An unpaired electron in an orbital associated with an atom can polarize the paired spins in an adjacent σ bond, so that one electron is associated more with one atom than with the other. The consequence of this is that there is then unpaired spin density at both nuclei involved in the bond, even though one of them has no unpaired electron density. Thus in the pyrazine anion, 3.I, the odd electron is delocalized around the ring carbon and nitrogen atoms. The direct interaction with nitrogen gives a large coupling, but hyperfine coupling to hydrogen is also observed as a result of spin polarization (Fig. 3.4).

As hyperfine splitting originates in electron delocalization, the experimental values can be used to estimate spin densities at the various atoms in a radical. This may be done in a qualitative way, but often results can be compared with those predicted using molecular orbital calculations, of various degrees of sophistication. Fig. 3.5 shows an ESR spectrum of $[BH_3\cdot]^-$ radicals, generated by hydrogen-abstraction from BH_4^- ions using t-butoxyl radicals. The spectrum shows coupling to ^{11}B and ^1H, with weaker resonances from the ^{10}B species, and from the hyperfine splittings the spin densities on boron and hydrogen have been calculated. These show that the radical is planar.

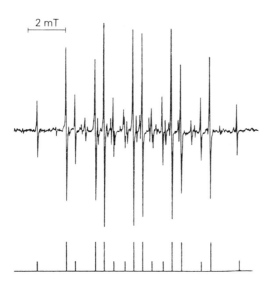

Fig. 3.5 ESR spectrum of $[BH_3\cdot]^-$ in solution. The stick diagram marks the resonances for the ^{11}B species, which shows couplings to ^{11}B ($I = 3/2$) and to three protons. The remaining weak resonances are due to the radicals containing ^{10}B ($I = 3$). Redrawn from J.R.M. Giles and B.P. Roberts, *J. Chem. Soc., Chem. Commun*, 360(1981).

The spectrum shown in Fig. 3.6 is of a paramagnetic species formed by oxidation of the chromium(III) porphyrin derivative, 3.II. The central multiplet pattern is due to coupling to four equivalent ^{14}N nuclei ($I = 1$), and there are weak satellite peaks which show additional coupling to ^{53}Cr (9.6% abundant, $I = 3/2$). The strength of this coupling shows that the product is a metal-centered radical, and further experiments using an ^{17}O-labelled oxidizing agent showed that an oxygen atom was bound to chromium [2]. On this basis, the product was formulated as a d^1 oxochromium(V) species, formed by a two-electron oxidation reaction. The alternative chromium(IV)−porphyrin π cation−radical (with the odd electron ligand-based rather than metal-based) would be expected to show some proton hyperfine interactions, and to have a g

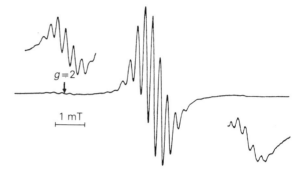

Fig. 3.6 ESR spectrum of the oxidation product of a chromium (III) porphyrin complex (3.II), showing hyperfine coupling to four equivalent ^{14}N nuclei, and satellites due to coupling to ^{53}Cr. Redrawn from Ref. [2].

3.II

value very close to 2.002. The observed value of 1.982 is consistent with the proposed structure, in which spin−orbit coupling is significant.

Thus the patterns of hyperfine splittings observed in ESR spectra provide direct information about the numbers and types of spinning nuclei coupled to the electrons: this information is exactly analogous to that obtained from coupling patterns in NMR spectra. The magnitudes of the hyperfine couplings can indicate the extent to which the unpaired electrons are delocalized, while *g* values may also show whether unpaired electrons are based on transition metal atoms or on adjacent ligands. Paramagnetic compounds do not in general give useful NMR spectra, as the resonances are normally extremely broad (see Section 2.16), and so ESR spectroscopy is in many ways complementary to NMR spectroscopy. It is a necessary and powerful tool for the study of paramagnetic systems. Many other examples of the application of ESR to chemical problems have been reviewed [3], and recent developments are reviewed annually [4]. There is one extensive compilation of relevant data [5].

3.3 Anisotropic systems

3.3.1 Hyperfine splittings and *g* factors

So far we have only considered ESR spectra obtained for isotropic systems, such as solutions. In practice, very many spectra are recorded for anisotropic systems, which can include unstable species prepared by irradiation of crystalline materials or of substrates trapped in other host matrices; solids, including frozen solutions; and paramagnetic point defects in single crystals. In such circumstances the **g** factors, which we have hitherto regarded as simple scalar quantities, must be considered as symmetric tensors. Such a tensor can always be diagonalized, to give the three principal **g**-factors, g_{xx}, g_{yy} and g_{zz}. In solution these are averaged by rapid tumbling to give the single isotropic factor, and for systems with spherical or cubic symmetry they are equal. However, for a system with lower, but at least axial, symmetry one of these terms $(g_{\parallel})$ is different from the other two $(g_{\perp})$, and for lower symmetries the three terms are all different.

In exactly the same way, the hyperfine splitting constants of isotropic systems must be replaced by tensors for anisotropic systems: again there are two independent terms for symmetric tops and three terms for species with lower symmetry. It is useful to separate the coupling constant tensor into an isotropic and an anisotropic part, as these components have different physical origins (Section 3.2): the new anisotropic term which we must now consider is due to direct dipole–dipole nucleus–electron interactions. These depend on the angle of the nucleus–electron vector to the magnetic field (which is why they are averaged to zero by tumbling), and are inversely proportional to the cube of the distance between the dipoles. Thus the effects of anisotropy on **g** values and hyperfine couplings in ESR spectroscopy are exactly analogous to the effects on chemical shifts and dipolar couplings in the NMR spectra of solids (Section 2.14).

If a single crystal is studied by ESR, it is found that the spectra depend on the orientation of the sample. By taking measurements at various angles it is possible in principle, and in some cases in practice, to determine all the elements of the **g** factor and hyperfine coupling tensors. This can yield valuable information about electron distribution, but details of such work are beyond the scope of this book. Much more frequently spectra are obtained for anisotropic systems such as powders or frozen solutions, which contain random mixtures aligned at all possible angles to the magnetic field. In the absence of hyperfine splitting, the resonance absorption has a distinctive envelope,

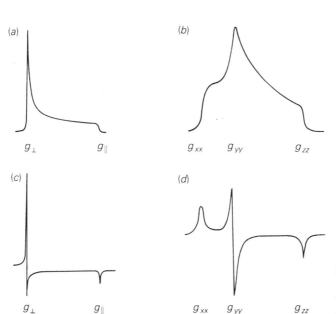

Fig. 3.7 Absorption line shapes for randomly orientated systems having (a) an axis of symmetry and (b) no symmetry. (c) and (d) are first derivatives of the curves (a) and (b), respectively.

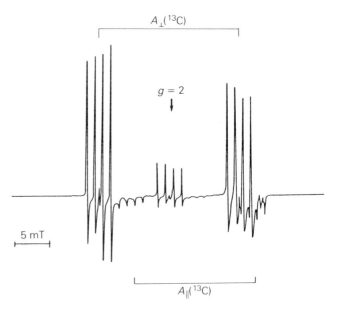

Fig. 3.8 ESR spectrum of $Li^+ - {}^{13}CO_2^-$ ion pairs in a CO_2 matrix. Parallel and perpendicular components can be seen, each showing splittings due to ${}^{13}C$ and ${}^7Li$ ($I = 3/2$). The weak central lines are due to $Li^+ - {}^{12}CO_2^-$. Redrawn from J.-P. Borel, F. Faes and A. Pittet, *J. Chem. Phys.*, **74**, 2120 (1980).

as shown in Fig. 3.7. These envelopes have clear upper and lower cut-offs, and that for a system having no symmetry also has a well-defined maximum. In consequence, the derivative curves, which are also shown, have two or three peaks, from which the anisotropic *g* values can easily be determined.

When hyperfine splittings, which are also anisotropic, are then introduced, the spectra rapidly become extremely complex. Fig. 3.8 shows a spectrum for the $Li^+-CO_2^-$ ion pair, labelled with ${}^{13}C$, in a CO_2 matrix. The patterns are basically those of Fig. 3.7(*c*), with large ${}^{13}C$ and small ${}^7Li$ ($I = 3/2$) hyperfine splittings. In this case, the spectrum is easy to analyze, but often computer fitting is necessary. There are many programs designed for this purpose, few of which can be applied to a wide range of problems. The annual series of Specialist Periodical Reports [4] has a section on spectral analysis, and provides a convenient way of tracking down suitable information in a rapidly changing field. In any spectral simulation, the spin Hamiltonian is set up, and we should note that this includes some terms which we have not discussed here, such as quadrupolar nucleus–electron interactions. Full details are given in the major books devoted to ESR [1, 3].

3.3.2   Electron–electron interactions

In discussion of ESR spectra up to this point we have only considered radicals which have just one unpaired electron. However, there are many systems which have several unpaired electrons, and these may also be studied by ESR. For example, molecules having diamagnetic ground states may have excited triplet states which have lifetimes long enough for their ESR spectra to be recorded. In a magnetic field a triplet state splits into its three components, as shown in Fig. 3.9, giving two possible transitions, $M_S = -1 \rightarrow 0$ and $M_S = 0 \rightarrow +1$, whose energies are identical. In a crystal or frozen sample, dipole–dipole interactions are anisotropic, and the levels with $M_S = \pm 1$ are shifted relative to that with $M_S = 0$ even in the absence of a magnetic field. Consequently, when a magnetic field is applied, the two transitions no longer have the same energy, or are observed at different applied magnetic fields, as shown in Fig. 3.10.

Example 3.1

When Mn(CO)$_5$H is held in a matrix of solid krypton at 77 K and radiolyzed, a paramagnetic product is formed, giving the ESR spectrum (a) shown in the figure below. If the krypton used is enriched to 42% with ^{83}Kr ($I = 9/2$) spectrum (b) is obtained. This corresponds to an expansion of the left-hand half of spectrum (a). What can be deduced from these spectra? How would spectrum (a) be changed if Mn(^{13}CO)$_5$H was photolyzed?

(a)

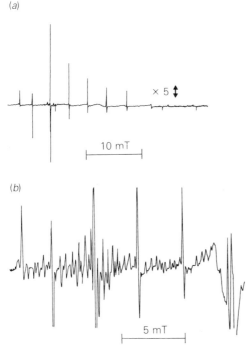

× 5

10 mT

(b)

5 mT

The spectra are taken, with permission, from S.A. Fairhurst, J.R. Morton, R.N. Perutz and K.F. Preston, *Organometallics*, **3**, 1389 (1984). Copyright (1984) American Chemical Society.

....

In spectrum (a) six of the seven most intense resonances form a regular progression, indicating that there is hyperfine coupling to ^{55}Mn, 100% abundant, $I = 5/2$. There is a second series of six resonances, rather difficult to pick out, extending from the strong line on the left-hand side of the spectrum, to the two weak lines on the right-hand side. One of these lines is obscured by a member of the first series. Measurement gives $A_\parallel$ (^{55}Mn) 6.5 mT (the larger hyperfine splitting) and $A_\perp$ (^{55}Mn) 3.5 mT. The spectrum is therefore characteristic of a species having axial symmetry, and one would expect it to be ·Mn(CO)$_5$. However, spectrum (b) clearly shows additional hyperfine coupling to ^{83}Kr, as each intense line is accompanied by a set of ten satellites. These groups overlap so that the overall pattern is quite complex, but most of the lines can be distinguished, and there can be no doubt about the assignment. The implication is that there is one krypton atom bound to the manganese atom, so that the product is KrMn(CO)$_5$. The hyperfine splitting, $A_\perp$(^{83}Kr), is 0.40 mT.

If ^{13}C-labeled CO was used to make the starting material for this experiment each resonance in spectrum (a) would be split into a doublet of quintets, by the axial and four equatorial CO ligands The experiment was in fact done using only 30% enrichment, so the spectrum is further complicated by species containing from zero to five ^{13}C nuclei. The spectrum is given in the original paper, which should be consulted for further information.

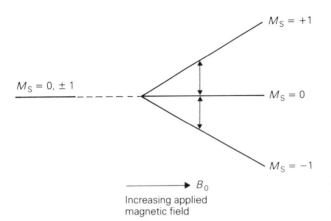

$M_S = +1$

$M_S = 0, \pm 1$

$M_S = 0$

$M_S = -1$

B_0

Increasing applied
magnetic field

Fig. 3.9 Splitting of a
triplet state in an applied
magnetic field.

Kramer's rule states that for a system with an *even* number of unpaired electron spins
the lowest energy state will be that with $M_S = 0$, as illustrated here for a triplet. All
higher energy states in this case, and *all* states for systems with an odd number of unpaired spins, will be doubly degenerate, corresponding to the two signs possible for M_S
$\neq 0$. This degeneracy is known as Kramer's degeneracy; it is, of course, removed by
the applied magnetic field, which shifts the two levels $M_S = -1$ and $M_S = +1$ in opposite directions (see Fig. 3.10).

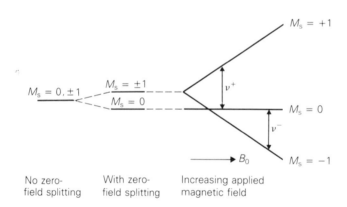

$M_S = +1$

$M_s = 0, \pm 1$
$M_s = \pm 1$
$M_s = 0$
ν^+
$M_s = 0$

ν^-

B_0
$M_s = -1$

No zero-
field splitting

With zero-
field splitting

Increasing applied
magnetic field

Fig. 3.10 Effect of a
magnetic field on the
energy levels of a triplet
state which exhibits
zero-field splitting.

The important consequence is that the two allowed transitions ($M_S = -1 \rightarrow 0$ and
$M_S = 0 \rightarrow +1$), which are of equal energy at any particular applied magnetic field in
the absence of zero-field splitting, are now of quite different energy, and *two* signals
are observed. If the zero-field splitting is small compared with $g\mu B_0$ this results in a
pair of signals near $g = 2.0$, but if the splitting is large, as is often the case for transition
metal ions, the apparent g-values may be very different from 2, and it may be
impossible to observe the expected transitions. For systems with very large splittings,
the energy gap between the lowest energy level and higher ones may be so large that
no crossings at all occur at magnetic fields normally used (see Fig. 3.11). In the case
which is illustrated the only transitions observed are those from $M_S = -1/2$ to
$M_S = +1/2$. In general there is an *effective spin quantum number*, which is equal to

CHAPTER 3: ESR AND NQR SPECTROSCOPY

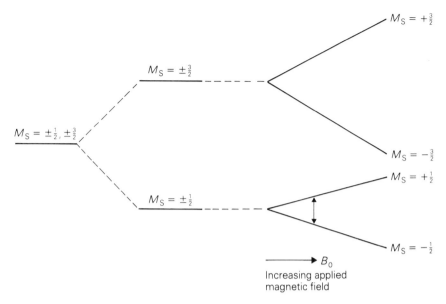

$M_S = +\frac{3}{2}$

$M_S = \pm\frac{3}{2}$

$M_S = \pm\frac{1}{2}, \pm\frac{3}{2}$

$M_S = -\frac{3}{2}$

$M_S = +\frac{1}{2}$

$M_S = \pm\frac{1}{2}$

$M_S = -\frac{1}{2}$

B_0

Increasing applied
magnetic field

Fig. 3.11 Effect of a large zero-field splitting on a quartet system ($S = 3/2$). The separation induced between the $M_S = \pm 1/2$ and $M_S = \pm 3/2$ states is so large that only a single ESR signal, due to the $M_S = -1/2 \rightarrow +1/2$ transition, is observed near $g = 2.0$. The energies of the $M_S = -1/2 \rightarrow -3/2$ and $+1/2 \rightarrow +3/2$ transitions are outside the observable frequency range.

$|M_S|$ for the lowest state. This will be zero for an even number of unpaired electrons, giving no signal at all, and is often 1/2 for systems with an odd number of unpaired electrons, giving a simple spectrum in the normal frequency range.

Another possible consequence of zero-field splitting is that the $\Delta M_S = \pm 1$ selection rule may break down, giving more observed transitions. So the spectra, particularly for transition metals with several unpaired electrons, are complicated. Also, the zero-field splitting is anisotropic, which further complicates matters for solids of less than cubic symmetry. Of course, tumbling averages it to zero, but in solution the spin–spin interactions cause rapid relaxation, and in practice spectra are usually too broad to be observed. Intramolecular spin–spin interactions may similarly make spectra impossible to observe, and it is often necessary to isolate spin centers from each other by freezing solutions to give glassy solids.

3.4 Transition metal complexes

Consideration of the ESR spectra of transition metal complexes is complicated by the fact that they have several approximately degenerate orbitals, and several unpaired electrons. These manifest themselves first in an orbital contribution to the magnetic moment, which leads to anisotropy in g factors, and secondly in zero-field effects, like those described in the last section. Isotropic spectra of transition metal complexes are straightforward, and are described in Section 3.2.

In a free transition metal ion, the five d orbitals are degenerate, but in a complex the orbitals do not interact equally with the ligands, and so the levels are split into two or more groups. For an octahedral complex there are three degenerate lower levels and an upper pair, but there are other arrangements of orbital energies for other types of complex. We will restrict our discussion to octahedral first row transition metal complexes, for which the crystal field energy is greater than the spin−orbit interaction energy. Within this limited group of complexes there is still enormous variety, as there can be up to ten d electrons, and from one to five unpaired electrons in high spin complexes, or from none to three in low spin complexes. For fuller descriptions of a wider range of situations see Refs [1] and [6], and Ref. [7] for a comprehensive account.

In analyzing any one system, we must ascertain the number of d electrons, and whether the complex is high or low spin. The consequences of Jahn−Teller distortion, which occurs to remove the degeneracy of any orbitally degenerate ground state, and of zero-field splitting and Kramer's degeneracy, which we have mentioned earlier, must also be considered. It is worth pointing out that the major reason for zero-field splitting is spin−orbit coupling, so it is much larger for heavier atoms than for lighter ones. Partly for this reason, ESR spectra of second- and third-row transition metal complexes are often harder to observe and interpret than those of first-row species. On the other hand, rare-earth metal complexes, for which spin−orbit coupling is small, often give clear, useful spectra. Finally we must note that if excited states lie close to the ground state, spin−lattice relaxation times will be short and spectral lines broad, and low temperatures will be needed to observe spectra. This is also true for systems with orbitally degenerate ground states, even for isotropic spectra.

We illustrate these principles with three examples. In a d^3 system, such as chromium(III), the free ion ground state is ^4F, and in an octahedral field the electrons occupy the three lower (t_{2g}) orbitals, giving a $^4A_2$ ground state (Fig. 3.12). There is no Jahn−Teller distortion in this case, but if there is tetragonal distortion, as in a complex

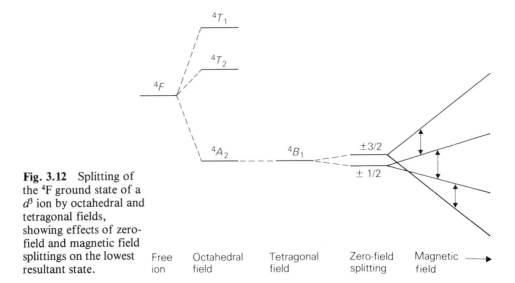

Fig. 3.12 Splitting of the ^4F ground state of a d^3 ion by octahedral and tetragonal fields, showing effects of zero-field and magnetic field splittings on the lowest resultant state.

with four ligands of one type and two (mutually *trans*) of another, the ground state becomes $^4B_1$. There is then zero-field splitting, and by Kramer's rule the lowest state is degenerate, and must give observable resonances. Other transitions, as shown in Fig. 3.12, will also be observed so long as the zero-field splitting is not too large. Also, ground and excited states are well separated in energy, so spin−lattice relaxation times are long, and spectral lines are narrow. A typical spectrum for a d^3 system is shown in Fig. 3.13.

CHAPTER 3: ESR AND NQR SPECTROSCOPY

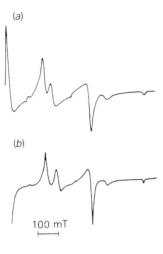

(a)

(b)

100 mT

Fig. 3.13 ESR spectra of *trans*-[Cr(pyridine)$_4$Cl$_2$]$^+$ at *ca.* 9 GHz: (*a*) frozen solution in DMF/H$_2$O/MeOH and (*b*) in *trans*-[Rh(pyridine)$_4$Cl$_2$]Cl.6H$_2$O powder. Computer fitting of the complicated patterns is necessary to obtain information about *g* tensors and zero-field splitting, but the spectra of powder and glass samples are very similar. Redrawn from E. Pedersen and H. Toftlund, *Inorg. Chem.*, **13**, 1603 (1974).

Low-spin d^6 systems are diamagnetic, and high-spin systems have ^5D ground states which split to give $^5T_2$ ground states in octahedral fields, and $^5B_2$ with tetragonal distortion. The ground states are connected to excited states by spin–orbit coupling, giving short relaxation times and broad resonances, and the coupling also gives large zero-field splittings. The lowest level is not degenerate (Kramer's rule), and so no resonances are observable.

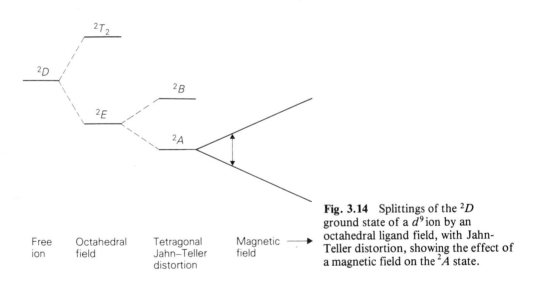

Free ion   Octahedral field   Tetragonal Jahn–Teller distortion   Magnetic field

Fig. 3.14 Splittings of the ^2D ground state of a d^9 ion by an octahedral ligand field, with Jahn-Teller distortion, showing the effect of a magnetic field on the ^2A state.

For a d^9 system the ^2D ground state of the free ion becomes ^2E in an octahedral field, and Jahn–Teller distortion yields a tetragonal structure whose ground state is ^2A (Fig. 3.14). A doublet ground state gives ESR without any further problem. Provided the Jahn–Teller distortion is large, which it usually is, spectral lines are narrow. A typical spectrum for a d^9 complex is shown in Fig. 3.15. This spectrum is of the copper(II) tetraphenylporphine complex, with the same ligand as in the chromium complex 3.II. The complex is square planar, and has the same symmetry as a tetragonally distorted octahedron, so there are two sets of resonances, with $g_\parallel$ less than $g_\perp$. Copper has two isotopes of about 70 and 30% abundance, both with spin 3/2, and so there are four

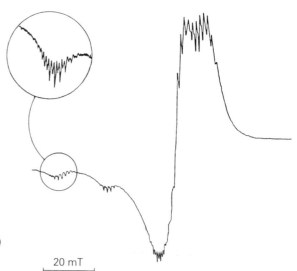

Fig. 3.15 ESR spectrum of a
copper(II) tetraphenylporphine
complex (frozen solution in CCl_3H)
Redrawn from J.M. Assour, *J.
Chem. Phys.*, **43**, 2477 (1965).

20 mT

groups of parallel resonances, three of which are clearly seen at low field. The lowest
frequency group is presented in more detail and shows the non-coincident ^{63}Cu and
^{65}Cu transitions, and hyperfine coupling to four ^{14}N nuclei. The perpendicular
resonances are more intense, but not so clearly resolved.

Although the ESR spectra of transition metal species can be very difficult to interpret, they can be very important, first simply as aids to identification, particularly
using hyperfine splitting patterns, and secondly as a means of studying the electron
distribution. Values of $g_\parallel$ and $g_\perp$ can show which d orbitals are occupied, and thus
indicate the direction in which Jahn−Teller distortion has occurred, while hyperfine
coupling constants can be used to determine the spin densities on various nuclei, and to
distinguish the components in s orbitals (isotropic) from those in p and d orbitals
(anisotropic).

3.5 Multiple resonance

There are many multiple resonance techniques that have been applied to ESR spectroscopy, but we can only refer briefly to two of them here. For more information,
consult specialist books on the subject [8] and the periodical reports on developments
in ESR [4].

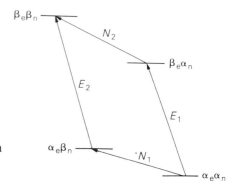

Fig. 3.16 Energy level diagram for a system
with one unpaired electron and one nucleus with
spin 1/2. The terms $\alpha_e\beta_n$ *etc.* refer to the spin
quantum numbers of the electron and nucleus.

CHAPTER 3: ESR AND NQR SPECTROSCOPY

Fig. 3.16 shows an energy level diagram for a system which has a single electron spin and a nucleus with spin 1/2. There are two allowed electron spin transitions, labelled E_1 and E_2, and the two transitions involving the nuclear spin, N_1 and N_2. These latter transitions would not normally be observed in an NMR spectrum, as the resonances would be extremely broad. We should also notice that the electron−nucleus coupling constant may be comparable with or even greater than the resonance frequency of the nucleus in the absence of coupling, and that the $\alpha_e\beta_n$ level may be lower in energy than $\alpha_e\alpha_n$.

In an ENDOR experiment (electron−nuclear double resonance) the ESR spectrum is observed while nuclear transitions are irradiated. The situation is similar to that in an all-nuclear INDOR experiment (see Section 2.10.2) in that the populations of the various energy levels are perturbed, and the ESR transitions change in intensity. Thus, for the system shown in Fig. 3.16, transition E_1 or E_2 would be monitored, and the frequencies N_1 and N_2 irradiated. Their frequencies depend on the magnetogyric ratio of the nucleus, and one obvious application of ENDOR is in the identification of nuclei responsible for particular hyperfine splittings. Another major application is to systems for which the ESR spectrum is complex. A series of ENDOR spectra, each with as few as two lines, can greatly simplify the task of spectral assignment, and this is particularly important when small couplings have to be identified and measured, so that spin densities at a number of different nuclei can be determined. Finally, all double resonance methods are used in studies of relaxation phenomena.

(a)

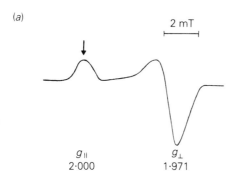

2 mT

$g_{\parallel}$
2·000

$g_{\perp}$
1·971

(b)

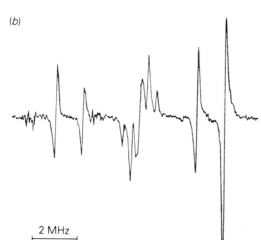

2 MHz

Fig. 3.17 Spectra of a frozen solution of $[\mathrm{Ti(C_8H_8)}\,\mathrm{(C_5H_5)}]$ in toluene: (a) ESR spectrum and (b) ^{1}H ENDOR spectrum measured at the arrowed point in spectrum (a). Redrawn from G. Labrauze, J.B. Raynor and E. Samuel, *J. Chem. Soc., Dalton Trans.*, 2425 (1980).

For example, Fig. 3.17(a) shows an ESR spectrum of a frozen solution of the titanium complex 3.III. This is a simple d^1 system with an axis of symmetry (assuming free rotation of the rings), but no hyperfine splittings are resolved. By monitoring the arrowed point and scanning across the ^{1}H NMR frequency region appropriate to the magnetic field, the ^{1}H ENDOR spectrum (b) was obtained. This clearly shows two wide doublets, which give the hyperfine couplings to protons in the C_8H_8 and C_5H_5 rings directly. The central group of lines arises from protons in the toluene solvent. A similar experiment monitoring the parallel line gave the parallel components of the couplings, and spin densities were then calculated.

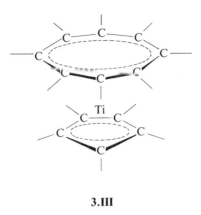

3.III

An ELDOR (electron−electron double resonance) experiment is somewhat different. In terms of the simple system illustrated in Fig. 3.16, one electron spin transition is observed while the other is irradiated. As these transitions do not share energy levels, the effects of irradiation (also called pumping) are only seen as a result of spontaneous nuclear spin transitions. In an ELDOR spectrum the intensity change in the monitored resonance is plotted against frequency difference, and in our example a peak will be seen at $|E_2 - E_1|$, which equals the hyperfine splitting constant. For a more complicated system, the spectrum may have several lines, each single line giving a direct measurement of a splitting constant. This is particularly useful for measuring very large splittings, or on occasions when a spectrum is complicated by numerous hyperfine splittings. ELDOR spectra also contain lines corresponding to pumping of the forbidden transitions in Fig. 3.16, that is $\alpha_e\alpha_n$ to $\beta_e\beta_n$, and $\alpha_e\beta_n$ to $\beta_e\alpha_n$, and these contain information about the mean nuclear resonance frequency, as well as the hyperfine coupling constant.

3.6 The nuclear quadrupole resonance experiment

Nuclear quadrupole resonance (NQR) spectroscopy is a technique which is infrequently used, except by specialists, partly because the range of problems to which it can be applied is narrow, but also because the necessary equipment is not widely available. Yet it is an analytical method which can be of great importance, particularly in dealing with solid materials which do not give large single crystals.

Any nucleus having a nuclear spin of 1 or more has, in addition to a nuclear magnetic dipole moment, a nuclear electric quadrupole moment. If there is also an electric field gradient at the nucleus, produced by an asymmetric charge distribution in its environment, then interaction between the field and the quadrupole gives rise to a series of energy levels. These levels are fixed molecular properties, unlike those involved in NMR, which depend on an external applied magnetic field. Transitions between these levels, at frequencies which may be from 100 kHz up to 1 GHz, are observed in NQR spectroscopy.

In order to observe an NQR spectrum we therefore need a nucleus with spin greater than 1/2, in an environment which is of less than cubic or tetrahedral symmetry. The sample must be quite large, typically about 2 g for an isotope of high abundance, and solid, because the electric field gradient at the nucleus is averaged to zero by tumbling in liquids or gases. The spectrometer is essentially a fairly simple device, in which a sample is placed in a coil, which is part of a radio-frequency circuit. The frequency is varied, and at resonance energy is absorbed. This sounds easy, but in practice the resonances are very weak and broad (typically 10 kHz), and may be found over a wide frequency range, so choice of detection system is important. Two types of oscillator have been widely used, the super-regenerative oscillator allowing broad band scanning, but distorting the peaks into potentially confusing multiplets, and the marginal oscillator, which is particularly useful at low frequencies. Spectra can also be obtained by observation of the free induction decay following pulse excitation, just as in Fourier transform NMR (Section 2.3.3). Another way of detecting quadrupole resonances is known as double resonance level crossing. This uses an easily-observed NMR nucleus which is bonded to or close to the quadrupolar nucleus, and a sequence of operations which allows energy to transfer from the NMR system to the NQR system. The outcome is that NQR frequencies can be measured by observing changes in the NMR spectrum, thus giving an enormous gain in sensitivity.

An NQR spectrum consists of one set of resonances for each chemically-distinct quadrupolar nucleus which has an asymmetric environment in the sample. A nucleus with nuclear quadrupole moment eQ (values are listed in Table 2.2) in an axially-symmetric electric field with gradient eq has energy levels

$$E = \frac{e^2Qq}{4I(2I-1)} [3m^2 - I(I+1)] \tag{3.3}$$

where I is the nuclear spin quantum number, and m is the nuclear magnetic quantum number, which takes values of I, $I - 1$, ... $-I$. Here e^2Qq is called the nuclear quadrupole coupling constant. This equation gives rise to energy levels and transitions as shown in Fig. 3.18, with the selection rule $\Delta m = \pm 1$. However, for a general system, the electric field gradient must be described by a tensor. This can be defined in diagonal form, and as the sum of the diagonal terms is zero, only two parameters are required. The largest field gradient term (which is a second derivative of the electrostatic potential) is defined as eq, and an asymmetry parameter, η, is defined as the difference between the remaining two terms, divided by eq. Thus Fig. 3.18 represents the levels for $\eta = 0$. If η is not zero, the upper pair of levels for the $I = 1$ case split, and two transitions are normally seen, allowing both eq and η to be determined. For spins 3/2, 5/2 etc., the degeneracy of the pairs of levels is not lifted, but the energy levels change, and for spins of 5/2 or more, eq and η may again be determined separately, provided two or more transitions can be observed. But for $I = 3/2$ there is still just one transition, and in this case the two terms can only be evaluated if the degeneracy of the levels is lifted by applying a weak magnetic field. Fig. 3.19 shows a typical NQR spec-

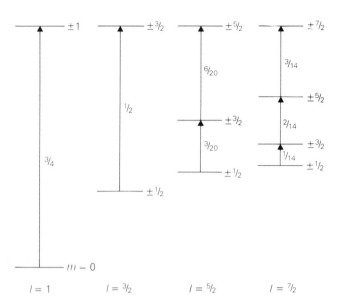

Fig. 3.18 Energy levels and transitions for quadrupolar nuclei in axially-symmetric electric fields. The transition energies are given as functions of e^2Qq.

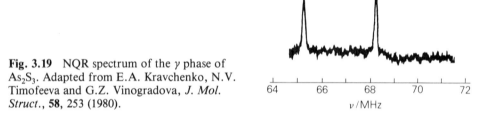

Fig. 3.19 NQR spectrum of the γ phase of As_2S_3. Adapted from E.A. Kravchenko, N.V. Timofeeva and G.Z. Vinogradova, *J. Mol. Struct.*, **58**, 253 (1980).

trum, in this case showing two resonances from two non-equivalent spin 3/2 ^{75}As nuclei.

Thus the main information obtainable by NQR relates first to the number of chemically distinct sites occupied by the quadrupolar nuclei in the sample, and secondly to the local electric environment of the nuclei in each of these sites. In this chapter we will concentrate on applications of NQR to chemical and structural problems. For more information about the technique itself, see the standard textbooks [9, 10], or shorter review articles [11–13].

3.7   Structural information from NQR spectra

In the simplest application of NQR spectroscopy, merely counting the number of resonances observed can provide useful information about species present in a crystal. For example, phosphorus pentachloride is known to crystallize as $PCl_4^+ PCl_6^-$. The NQR spectrum of a low temperature phase has ten ^{35}Cl resonances, a group of four assigned to the chlorine nuclei in the cation, and a group of six at lower frequencies assigned to the anion. But it often happens that several of the nuclei are in equivalent positions in the lattice, and so fewer resonances will be seen. If phosphorus pentachloride vapor is rapidly cooled, a metastable phase is formed, and this only gives three resonances. Two of these have been assigned to non-equivalent chlorine nuclei in axial positions of trigonal-pyramidal PCl_5, but the remaining resonance arises from the three equatorial chlorines, which are apparently equivalent. There are, of course, equivalent sets of ^{37}Cl resonances, whenever ^{35}Cl resonances are observed.

If one of the chlorine atoms of PCl_5 is replaced, there are two possible structures, depending on whether the substituent occupies an axial or an equatorial position. The NQR spectrum of $PFCl_4$ has two resonances, one of three times the intensity of the other, and this is consistent with the presence of the axial isomer, 3.IV, but the spectrum of PCl_4Ph, for which the equatorial isomer, 3.V, is stable, has two groups of two

resonances. In the first of these two cases, the molecular symmetry is apparently preserved in the crystal, and so the total number of resonances is less than the number of chlorine atoms. For $PCl_4^+ PCl_6^-$ each chlorine gives a separate resonance, and in other cases there may be two or more molecules in the asymmetric unit of the unit cell, thus multiplying the number of resonances. For example, $Na^+ GaCl_4^-$ has four chlorine resonances, whereas $Cs^+ GaCl_4^-$ has eight, and must therefore have a different arrangement in the crystal. The related compound, $Ga^+ GaCl_4^-$, has only two chlorine resonances at low temperatures, but four at higher temperatures. Fig. 3.20 shows how the resonance frequencies change with temperature, and the phase change at 320 K is easily observed. NQR spectroscopy is particularly useful for detecting and studying phase changes in crystalline solids.

Reactions between alkali metal halides and aluminium trihalides give a number of ionic products which are not easy to characterize. Compounds such as $Na^+ [Ga_2Cl_7]^-$ have been identified by NQR [14]. The spectrum has two ^{69}Ga resonances and seven ^{35}Cl resonances, one of which is of substantially lower frequency than the other six. Comparison of frequencies with those recorded for other gallium compounds show the $[Ga_2Cl_7]^-$ ion must have two gallium tetrahedra, bridged by one corner chlorine atom (3.VI). The spectrum also has equivalent sets of resonances due to ^{71}Ga and ^{37}Cl isotopes.

Tellurium tetrachloride is another interesting example. Its NQR spectrum has six ^{35}Cl resonances, of equal intensity [15], and the explanation for this is not immediately obvious. In this case, the crystal structure is known, and it has been shown that there are Te_4Cl_{16} units, as shown by structure 3.VII, with $TeCl_3^+$ and Cl^- ions. The Cl^- ions do not give observable resonances, and within each Te_4Cl_6 unit two pairs of $TeCl_3^+$ ions are crystallographically inequivalent — hence the six resonances. $TeBr_4$ also has six resonances due to ^{79}Br and six due to ^{81}Br, and it is reasonable to assume that it has a similar structure, with $TeBr_3^+$ ions. Of course, once a species such as

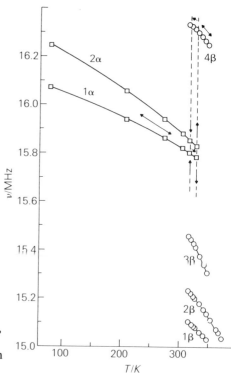

Fig. 3.20 ^{35}Cl NQR frequencies of $Ga^+GaCl_4^-$ as a function of temperature. At low temperatures there are just two resonances, but there is a phase change at *ca.* 320 K, above which there are four resonances. Redrawn from Ref. [14].

3.VI

3.VII

$TeCl_3^+$ has been identified, it is relatively easy to recognize it when it occurs elsewhere. Thus $AlCl_3$ and $TeCl_4$ give a solid product which has seven ^{35}Cl resonances. Four of these are at the low frequencies associated with aluminum—chlorine bonds, and three are at higher frequencies. The product is clearly $TeCl_3^+ AlCl_4^-$.

3.8 Interpretation of nuclear quadrupole coupling constants

Analysis of an NQR spectrum yields nuclear quadrupole coupling constants and asymmetry parameters for each of the contributing nuclei, and from these the electric field gradient (or gradients, for an asymmetric center) at the nuclei may be determined.

Example 3.2

When antimony trichloride and methoxybenzene are melted and mixed together in molar ratio 2:1, a molecular complex with formula $2SbCl_3.C_6H_5OCH_3$ is formed. The NQR spectrum of this product contains many lines, the number and positions of which vary with temperature. The figure below shows the temperature dependence of the lowest frequency ^{121}Sb resonances. There are six ^{35}Cl resonances of equal intensity at temperatures below 141 K, and above this temperature there are three equal resonances. Explain these observations, and work out how many other lines should be observed in the low temperature spectrum.

The figure is taken, with permission, from H. Ishihara, T. Okuda, K. Yamada and H. Negita, *Bull. Chem. Soc. Jpn*, **58**, 2614 (1985).

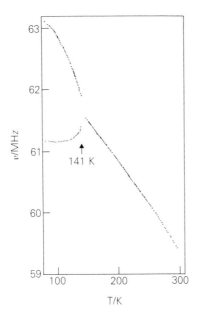

The six ^{35}Cl resonances at low temperature indicate that there are six chemically distinct chlorine atoms in the crystal, i.e. that there is one $2SbCl_3.C_6H_5OCH_3$ complex per asymmetric unit. This must have two distinct antimony atoms. The ^{123}Sb isotope (57% abundant, $I = 5/2$) will give two resonances for each of these, with the higher frequency being twice the lower one for an axially symmetric site. The two resonances observed near 60 MHz are clearly consistent with there being two antimony sites. At 141 K the crystal undergoes a phase change, in which the two $SbCl_3$ groups become crystallographically equivalent. This results in a number of NQR lines being halved.

At low temperature there will be six ^{37}Cl resonances, in addition to those arising from ^{35}Cl. The ratio of the resonant frequencies for each site is equal to $\gamma(^{37}Cl)/\gamma(^{35}Cl)$, where γ is the magnetogyric ratio, so assignments can be confirmed by observing both frequencies. Similarly, observation of three resonances for each ^{123}Sb (43% abundant, $I = 7/2$) site can confirm the ^{121}Sb assignments. Thus there should be a total of 22 resonances (12 chlorine and 10 antimony) in the NQR spectrum.

The first and simplest interpretation of these gradients was suggested by Townes and Dailey [16], and their theory is still widely used. They showed that for nuclei such as ^{14}N and halogens, the source of the electric field gradient is mainly the valence shell p electrons. For halogens, this leads to the conclusion that there is a linear relationship between resonance frequency and ionic character of the bond to the halogen atom. A useful consequence of this is that halogens attached to the same element have similar frequencies, and that these may be distinctive. For example, ^{35}Cl resonances for Si—Cl bonds are 15—21 MHz, for S—Cl 30—40 MHz and for N—Cl 43—57 MHz; the highest values are for the most covalent bonds. More sophisticated molecular orbital calculations may be used to relate resonance frequencies and net charges on the halogen atoms. This sort of method has been used to show that the chlorine atoms in *cis*-$[PtCl_2(NH_3)_2]$ carry a slightly larger negative charge than those in *trans*-$[PtCl_2(NH_3)_2]$ (−0.65 compared with −0.63), and that the Pt—Cl bonds in the *cis* isomer are correspondingly more ionic [17]. Once such differences have been recognized, the precise frequencies observed become of diagnostic value. In the preceding section, assignment of Al—Cl and Te—Cl bonds in $TeCl_3^+ AlCl_4^-$, or of the bridging and terminal chlorines in $[Ga_2Cl_7]^-$, or distinguishing the chlorines in PCl_4^+ from those in PCl_6^-, all depend on such accumulated knowledge.

In more sophisticated applications, NQR measurements can be used to give geometrical information, which we can illustrate by two examples. In a study of a single crystal of PBr_3O [18], the Zeeman effect (magnetic field splitting) of each line was measured, as a function of crystal orientation. By plotting the zero-splitting loci it was possible to show that the crystal was orthorhombic, with space group *Pnma*. Then, assuming that the principal axis of the bromine field gradients lay along the P—Br bonds, the angles between the bonds were calculated. The three angles were all 107.6—107.7°, but these values are some 2° larger than those determined by X-ray analysis. The results are not highly accurate, but they may provide a starting point for a difficult crystal structure determination.

Determination of internuclear distances is also possible in a very few cases. In a study of $Mn(CO)_5D$ [19] the ^{55}Mn and ^2H quadrupole coupling constants were measured, using a spin-echo double resonance technique, and it was possible to determine the direct dipole—dipole interaction between the two spins. From this the Mn—D distance was calculated to be 1.61 ± 0.01 Å, in excellent agreement with the value of 1.601 ± 0.016 Å found by neutron diffraction. Subsequent work with $Mn(CO)_5H$ has given a value of 1.59 ± 0.02 Å for the Mn—H distance. Such measurements are rare, but they do illustrate the fact that in applying any spectroscopic technique it is possible to obtain information normally provided by other methods. In this particular example, study of nuclear quadrupole resonances has enabled a coupling between two nuclear magnetic dipoles to be measured. In this context, we should remember that nuclear quadrupole coupling constants are also important in microwave spectroscopy (Chapter 4) and Mössbauer spectroscopy (Chapter 7), and may also be determined by NMR experiments under some conditions.

Problems

Note: Magnetic properties of nuclei are listed in Tables 2.1 (p. 31) and 2.2 (p. 32–3)

3.1 Fig. 3.5 shows an ESR spectrum of $[BH_3\cdot]^-$ in solution. Sketch equivalent spectra for $[AlH_3\cdot]^-$ and $[InH_3\cdot]^-$.

3.2 Fig. 3.6 shows an ESR spectrum of a chromium porphyrin complex. Describe the appearance of the spectrum you would expect to observe for a sample enriched to 100% in ^{53}Cr.

3.3 The solution ESR spectra of (a) $[Nb(1,3,5-C_6H_3Me_3)_2]$ and (b) $[Ti(C_6H_6)_2]^-$ are shown below. Account for the patterns that are observed.

(a)

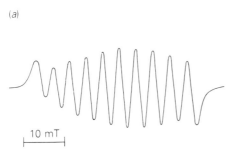

10 mT

(b)

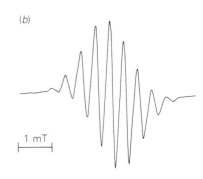

1 mT

Fig. 3.21

3.4 Reaction of Al atoms with benzene at 77 K yields a paramagnetic product. The ESR spectrum of this compound at 4 K shows hyperfine interactions with one Al atom ($I = 5/2$) and with two equivalent protons. At 220 K there are hyperfine couplings to Al and six equivalent protons. Account for these observations.

3.5 Show how ESR derivative spectra having three lines can arise by
(a) hyperfine coupling to two nuclei with spin 1/2,
(b) anisotropy of **g** factors for an asymmetric system,
(c) zero-field splitting for an octahedral high-spin d^7 transition metal ion,
(d) hyperfine coupling to a nucleus with spin 1.
How can these possibilities be distinguished?

3.6 The ESR spectrum below was obtained from a solution of a product of the reaction between VCl_2O and PEt_3. Account for the observed pattern.

Fig. 3.22   10 mT

[The spectrum is redrawn from G. Henrici-Olivé and S. Olivé, *J. Am. Chem. Soc.*, **93**, 4154 (1971).]
Describe qualitatively how the spectrum would change if (a) the solution was frozen (b) PEt_3 was replaced by $AsEt_3$ and (c) vanadium was replaced by niobium.

3.7 By measuring the separations between lines in the spectra illustrated in Figs 3.5 and 3.6, and using the given scales, calculate as many hyperfine coupling constants as possible.

3.8 The ESR spectrum of a frozen solution of $[Rh(py)_4Cl_2]Cl$ (py = pyridine) and tetrabutyl-ammonium chloride after electrolysis is shown here. The relative intensities of the right-hand group of lines are 1:3:5:7:7:5:3:1. What can you deduce from this spectrum?

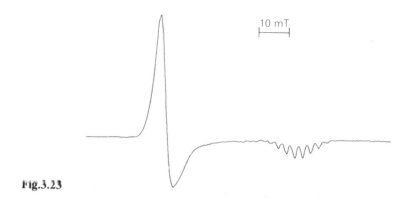

Fig.3.23

[The spectrum is reproduced, with permission, from J.E. Anderson and T.P. Gregory, *Inorg. Chem.*, **28**, 3905 (1989). Copyright (1989) American Chemical Society.]

3.9 The ESR spectrum shown in Fig. 3.24 was recorded for a solution of $[(NH_3)_5Co-O-O-Co(NH_3)_5]^{5+}$. What can you deduce from this spectrum? How might the spectrum be modified if $^{17}O_2$ was used in the preparation of the peroxo-complex?

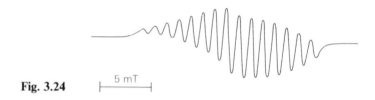

Fig. 3.24

[The spectrum is redrawn from E.A.V. Ebsworth and J.A. Weil, *J. Phys. Chem.*, **63**, 1890 (1959).]

3.10 Describe the patterns due to hyperfine coupling that you would expect to observe in tetrahedral clusters containing (*a*) four Co atoms, (*b*) four Rh atoms and (*c*) two Co and two Rh atoms. Assume that in each case the unpaired electron density is uniformly distributed over the metal atoms.

3.11 On γ-irradiation of powdered $K^+PF_6^-$ the radical $PF_4\cdot$ is formed. What would you expect to see in the ESR spectrum of this species, assuming that it is tetrahedral? How would the spectrum be modified if the structure was trigonal-bipyramidal?

3.12 Describe the ESR spectrum that you would expect to observe for $Si_2H_5\cdot$ in an isotropic medium.

3.13 A solution of potassium in ethylamine at concentrations of 10^{-5} M or above shows four equal resonances, evenly spaced, with a weaker resonance close to the center of the spectrum. At lower concentrations the central line is the dominant feature. What can you deduce from these observations?

3.14 How many lines should be observed in the isotropic ESR spectra of (*a*) a tetragonally-distorted high-spin Fe^{III} complex, with small zero-field splitting, (*b*) a similarly distorted Cu^{II} complex, and (*c*) an octahedral V^{II} complex?

3.15 The second-derivative ESR spectrum shown below was recorded for the dithiadiazolyl radical, 3.VIII. The relative intensities of the multiplets are 1:4:7:7:4:1. Account for the observed pattern.

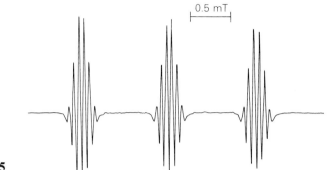

Fig. 3.25

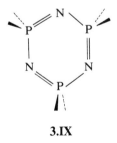

3.VIII

The spectrum is taken, with permission, from G.K. Maclean, J. Passmore, M.N. Sudheedra Rao, M.J. Schriver, P.S. White, D. Bethell, R.S. Pilkington and L.H. Sutcliffe, *J. Chem. Soc., Dalton Trans.*, 1405 (1985).

3.16 $SiCl_4$ reacts with pyridine to give $SiCl_4(py)_2$, and with NMe_3 to give $SiCl_4(NMe_3)$. The pyridine ligands could be *cis* or *trans* in the first case, and the trimethylamine could occupy an axial or an equatorial position in the latter. To what extent could these various possibilities be distinguished by ^{35}Cl NQR spectroscopy?

3.17 The cyclic phosphazenes $P_3N_3Cl_3R_3$ have structures of the type 3.IX.

3.IX

How many isomers are possible, and how far would it be possible to distinguish them by NQR spectroscopy? How would the situation be different for isomers of (*a*) $P_3N_3Cl_4R_2$ and (*b*) $P_3N_3Cl_2R_4$?

3.18 How many resonances might be observed in the NQR spectrum of an equilibrated mixture of $SiCl_4$ and $SiBr_4$, assuming that crystallographic non-equivalence does not lead to splitting of resonances? Would there be any difference if SiI_4 was used instead of $SiBr_4$?

3.19 It is conceivable that arsenic pentachloride could exist in the solid phase as $AsCl_5$, $AsCl_4^+ AsCl_6^-$, $AsCl_4^+ Cl^-$ or As_2Cl_{10}, with two chlorine bridges between octahedral arsenic atoms. How could NQR spectroscopy distinguish between these possibilities?

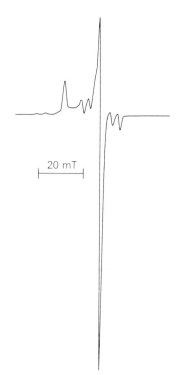

Fig. 3.26

3.20 The palladium atom in the complex PdL^{2+}, where L is 7,16-dimethyl-1,4,10,13-tetrathia-7,16-diazacyclooctadecane, has square-planar coordination to four sulfur atoms. It undergoes possible electrochemical reduction to PdL^{+}, and the ESR spectrum of this product in a MeCN glass at 77 K is shown in the Fig. 3.26. Account for the form of the spectrum.

3.21 What resonances would you expect to see in the NQR spectrum of the complex $In_3Br_3Co_4(CO)_{15}$, with the structure shown?

$$\begin{array}{c}
(CO)_3 \\
Co \\
\end{array}$$

(CO)_3
Co
(CO)₄Co — In · · · · · · In — Co(CO)₄
Br
In — Br
Br
Co(CO)₄

3.X

3.22 One crystalline modification of the yellow allotrope of arsenic has eight lines in its ^{75}As NQR spectrum. What does this indicate about the crystal structure of this form?

3.23 The compound $ICl_2^+ SbCl_6^-$ consists of chains of octahedral $SbCl_6$ and square planar ICl_4 units, linked by chlorine bridges. Assuming that there is one ICl_2SbCl_6 per asymmetric unit, what is the maximum number of resonances you would expect to see in the NQR spectrum? For which of the nuclei present can the asymmetry parameter, η, be determined?

3.24 Reaction of CH_3COBr and $AlBr_3$ in 1:1 ratio yields a crystalline product which at low temperatures shows eight ^{79}Br nuclear quadrupole resonances. On warming the number of lines decreases. If a 1:2 reacting ratio is used, the product has seven ^{79}Br resonances. How can these observations be explained?

Further reading

For accounts of ESR spectroscopy see Refs [1], [3] and [6], and also C.J. Poole Jr., *Electron Spin Resonance*, Interscience, New York (1967).

The Specialist Periodical Reports on ESR (Ref. [4] below) refer to recent books and reviews, and review developments and applications of the technique.

Refs. [9] – [13] provide major accounts of NQR spectroscopy. The first two are very detailed, while the others are shorter reviews of the subject. For examples of applications of NQR in inorganic chemistry, see:

E.A.C. Lucken, *Z.anal.Chem.*, **273**, 337 (1975).

H. Chihara and N. Nakamura, *MTP International Review of Science*, Physical Chemistry Series One, Volume 4, Butterworths, London (1972), p. 125.

Recent results are reported each year in *Spectroscopic Properties of Inorganic and Organometallic Compounds*, Specialist Periodical Reports, The Royal Society of Chemistry, London.

For semi-empirical methods of calculating electric field gradients, see F.A. Cotton and C.B. Harris, *Proc. Natl. Acad. Sci. USA*, **56**, 12 (1966).

References

1 J.E. Wertz and J.R. Bolton, *Electron Spin Resonance, Elementary Theory and Practical Applications*, McGraw-Hill, New York (1972).
2 J.T. Groves and R.C. Haushalter, *J. Chem. Soc., Chem. Commun.*, 1165 (1981).
3 M.C.R. Symons, *Chemical and Biochemical Aspects of Electron-Spin Resonance Spectroscopy*, Van Nostrand Reinhold, New York (1978).
4 Specialist Periodical Reports, *Electron Spin Resonance*, Royal Society of Chemistry, London, annually.
5 Landolt-Börnstein, *Magnetic Properties of Free Radicals*, Vol. 9, Part a (eds H. Fischer and K.-H. Hellwege), Springer-Verlag, Berlin (1977).
6 B.A. Goodman and J.B. Raynor, *Adv. Inorg. Chem. Radiochem.*, **13**, 135 (1970).
7 A. Abragam and B. Bleaney, *Electron Paramagnetic Resonance of Transition Metal Ions*, Clarendon Press, Oxford (1970).
8 L. Kevan and L.D. Kispert, *Electron Spin Double Resonance Spectroscopy*, Wiley, New York (1976); M.M. Dorio and J.H. Freed, *Multiple Electron Resonance Spectroscopy*, Academic Press, New York (1979).
9 T.P. Das and E.L. Hahn, *Nuclear Quadrupole Resonance Spectroscopy*, Academic Press, New York (1958).
10 E.A.C. Lucken, *Nuclear Quadrupole Coupling Constants*, Academic Press, New York (1969).
11 E. Schempp and P.B. Bray, *Nuclear Quadrupole Resonance Spectroscopy*, in *Physical Chemistry, an Advanced Treatise* (ed. D. Henderson), Vol. 4, Academic Press, New York (1970).
12 J.A.S. Smith, *J.Chem.Educ.*, **48**, 39/A77/A147/A243 (1971).
13 A.L. Porte, *Ann. Rep. Progr. Chem., Sect. C*, **80**, 149 (1984).
14 T. Deeg and A. Weiss, *Ber. Bunsenges. Phys. Chem.*, **79**, 497 (1975).
15 T. Okuda, K. Yamada, Y. Furukawa and H. Negita, *Bull. Chem. Soc. Jpn*, **48**, 392 (1975).
16 C.H. Townes and B.P. Dailey, *J. Chem. Phys.*, **17**, 782 (1949).
17 H.H. Patterson, J.C. Tewksbury, M. Martin, M.-B. Krogh-Jespersen, J.A. Lomenzo, H.O. Hooper and A.K. Viswanath, *Inorg. Chem.*, **20**, 2297 (1981).
18 T. Okuda, K. Hosokawa, K. Yamada, Y. Furukawa and H. Negita, *Inorg. Chem.*, **14**, 1207 (1975).
19 P.S. Ireland and T.L. Brown, *J. Magn. Reson.*, **20**, 300 (1975).

4 Rotational Spectra and Rotational Structure

4.1 Introduction

Molecules stand at the boundary between the 'sensible' world inhabited by macroscopic bodies and the quantum world of atoms and sub-atomic particles. This means that we can use a mixture of classical and quantum concepts in dealing with them, though a full description of their behavior is also possible using purely quantum-mechanical concepts. This chapter deals with the phenomena associated with the rotation of molecules in the gas phase, which is quantized but can otherwise be described at least approximately in classical terms. We shall therefore use a semi-classical description of rotating molecules and the changes in energy associated with changes in rotational quantum numbers. The effects of such changes are seen in various different areas of spectroscopy; we shall see how they can be used to give information about the size and shape of molecules and about their atomic compositions. Because instrumental resolution is very high, especially in the microwave region, information about rotations of molecules may be very precise, but often we can only measure one or two parameters for a particular molecule, and this places a severe restriction on the usefulness of rotational spectroscopy as a structural technique. We shall examine the advantages and limitations of the method in this chapter, and illustrate some of its applications. In particular, we shall see how rotational spectroscopy can give us precise values for bond lengths and bond angles in small molecules, and how the study of isotopically-substituted samples can give specific bond lengths even in larger molecules. Rotational spectroscopy has been used for a wide range of molecules, though comparatively few transition metal compounds have been studied, mainly because of lack of volatility.

4.2 Rotation of molecules

4.2.1 Classical rotation

The rotation of a macroscopic body can be described classically in terms of angular momentum about an instantaneous rotation axis. The angular momentum P is equal to the angular velocity ω multiplied by a quantity I, which is the *moment of inertia* about the axis of rotation. The number of axes about which rotation can occur is, of course, infinite, and so the number of possible moments of inertia is also infinite. However, all these possibilities can be dealt with in terms of a second-rank tensor quantity (the inertia tensor), which can be thought of as an ellipsoid whose width in any direction is a measure of the corresponding moment of inertia (Fig. 4.1a). The three principal axes of this ellipsoid are mutually perpendicular; one of them relates to the maximum possible moment of inertia I_C, one to the minimum possible moment I_A, and the third to an intermediate moment of inertia I_B. These three moments are not in

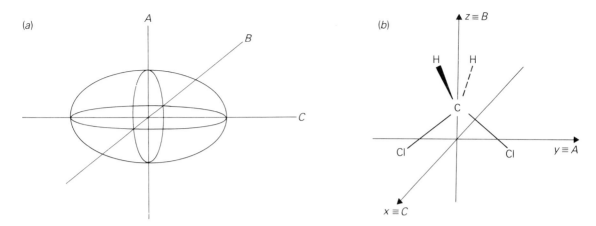

Fig. 4.1 (*a*) Inertia ellipsoid and its three principal axes *A*, *B* and *C*. (*b*) Principal rotation areas *A*, *B* and *C* of CCl_2H_2.

themselves enough to define the tensor completely; a full definition requires in addition the three products of inertia corresponding to the off-diagonal elements of the tensor, but the three *principal moments* we have defined above are enough for our present purposes.

The three principal axes of the ellipsoid are the principal axes of rotation of the molecule, which are labeled *A*, *B* and *C* as shown in Fig. 4.1(*a*). As an example, the three principal axes of CCl_2H_2 are (i) the 2-fold symmetry axis, (ii) an axis lying in the plane of the CCl_2 group, perpendicular to the 2-fold axis and passing through the center of mass of the whole molecule, and (iii) an axis lying in the plane of the CH_2 group, perpendicular to the other two and passing through the center of mass (Fig. 4.1*b*). We may label these as a Cartesian set *z*, *y* and *x*, but for each molecule we must work out which of them is *A*, corresponding to the least moment of inertia I_A, and so on, using the atom masses and positions. In a case like this, I_C is certain to be the moment of inertia about an axis perpendicular to the plane containing the heavy atoms, so $x \equiv C$; the *y*-axis passes closest to the two very heavy chlorine atoms and is therefore *A*, leaving $z \equiv B$. For more complex molecules, or molecules with less symmetry or no heavy atoms, it is often impossible to decide how the rotation axes are arranged without a detailed computation starting with atomic positions in some arbitrary Cartesian axis system.

4.2.2   Quantization of rotational angular momentum

Classically, angular momentum can take any value, but in a body of the size of a typical molecule the angular momentum, *P*, is quantized: its possible values are given by Eqn (4.1), in which *h* is Planck's constant, and the integer *J* is called the rotational quantum number, which must be zero or positive.

$$P = \sqrt{J(J + 1)}h/2\pi \tag{4.1}$$

Although the fundamental quantization is of the angular momentum, spectroscopic measurements in general are concerned with changes in molecular energies, and we may use classical expressions to relate allowed values of the angular momentum to the corresponding molecular energies. Thus, as *P* equals $I\omega$, where *I* is the moment of inertia appropriate to the rotation axis and ω is the angular velocity, and as the kinetic energy of rotation is $(I\omega^2)/2$, we find:

Rotational energy $E_r = P^2/(2I)$
$$= J(J + 1) \cdot h^2/(8\pi^2 I). \tag{4.2}$$

This is written as $BJ(J + 1)$, where B is called the *rotation constant*. It is usual to express B in units of cm^{-1} rather than Joules or ergs, so we divide by hc, giving

$$B = h/8\pi^2 cI. \tag{4.3}$$

The rotational energy is thus limited to the values given by expression (4.2), with J a positive integer or zero. If B is in MHz and I is in atomic mass units $\times$ $Å^2$ the conversion is $B = 505\,391/I$. The gaps between successive rotational energy levels are not identical, but increase steadily with J (Fig. 4.2).

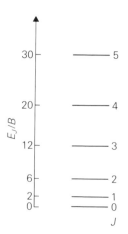

Fig. 4.2 Rotational energy levels for a diatomic molecule.

4.2.3

Rotation constants and moments of inertia

Expression (4.3) shows that the rotation constant B is closely connected to a quantity called the moment of inertia I; but which moment of inertia is involved? For a linear molecule there is only one, as there is no mass off the axis, but for non-linear molecules we need to define three rotation constants, A, B and C corresponding to the three principal moments of inertia of the molecule I_A, I_B and I_C. Some or all of these may be identical, by chance or by reason of the symmetry of the molecule, and we may distinguish three possible situations.

In *spherical tops* the x, y and z axes are related to each other by a set of four 3-fold symmetry operations. Consequently all three principal moments of inertia are equal, and so there is only one rotation constant, B. This arises only in molecules with tetrahedral, octahedral, cubic or higher symmetry, such as SiH_4 (4.I, T_d symmetry) and SF_6 (4.II, O_h symmetry).

In *symmetric tops* two of the three principal moments of inertia are equal to each other. The corresponding rotation constant is called B, while the unique one may be A or C, depending on whether it corresponds to the least (I_A) or the greatest moment of inertia (I_C). A symmetric top belongs to one of the axial point groups (see

4.I   **4.II**

Appendix), in which a symmetry operation of order greater than 2 (an axis of rotational symmetry C_n or a rotation − reflection axis S_n) relates the x and y axes. The component of angular momentum about the z-axis is quantized separately from the total angular momentum, and a second quantum number K is used to express this (Fig. 4.3): K cannot be greater than J. The total energy associated with rotation is then

$$E_r/hc = BJ(J + 1) + (a - B)K^2 \tag{4.4}$$

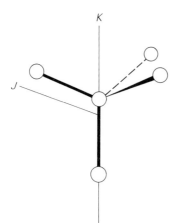

Fig. 4.3 The two rotational quantum numbers, J and K, for a symmetric top.

where a is A or C, whichever is different from B. This means that if A is unique the rotational energies increase with increasing K, whereas if C is unique the energies decrease with increasing K. These two cases are called prolate and oblate symmetric tops respectively; examples are $CH_3C{\equiv}CH$ (4.III, prolate) and BCl_3 (4.IV, oblate). The rotation about the top axis has angular momentum $Kh/2\pi$. This combines vectorially with angular momentum about other axes to give the total angular momentum $\sqrt{J(J + 1)}h/2\pi$.

Occasionally molecules are found to have two moments of inertia which are by

4.III   **4.IV**

chance approximately equal, although this is not required by symmetry. The rotational behavior of such molecules may be treated as if they were symmetric tops.

In *asymmetric tops* all three principal moments of inertia are different, as are the three rotation constants A, B and C. In this case rotational energy cannot be defined by a simple expression like Eqn (4.4), and there are no genuine quantum numbers associated with the rotation except J itself. There are in fact $(2J + 1)$ different energy levels for each value of J; their relative energies depend on the relative magnitudes of A, B and C, and can only be calculated by solution of the wave equation for the rotational motion. Asymmetric rotor problems can be dealt with very easily by use of computer programs designed for the purpose, but are not amenable to the 'back of envelope' calculations that can be used for problems involving symmetric tops, spherical tops and linear molecules.

An asymmetric top molecule *need* have no symmetry at all,* and *must* not have any symmetry axis of order greater than 2 (otherwise it would be a symmetric top).

We should note that for any planar molecule, as for a planar macroscopic object, the moment of inertia about the axis perpendicular to the plane is equal to the sum of the other two, and is thus the greatest (I_C). If the planar molecule is also a symmetric top (such as BF_3 or XeF_4) I_C is then equal to $2I_B$, so that $B = 2C$.

4.2.4

Centrifugal distortion; the semi-rigid rotor

The classical and quantum treatments above assume that the rotating molecules are perfectly rigid and have a constant moment of inertia irrespective of the speed of rotation. In any real body rotation leads to centrifugal forces, which distort the body (Fig. 4.4). As the molecule stretches, the moment of inertia increases, and there is a corresponding decrease in the effective rotation constant, B. For a semi-rigid linear molecule, for instance, we may write the rotational energy as:

$$E_r = BJ(J + 1) - DJ^2(J + 1)^2. \tag{4.5}$$

D is called the centrifugal distortion constant for the linear molecule, and is usually of

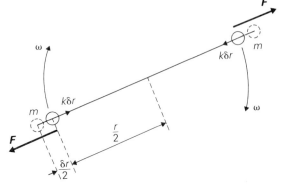

Fig. 4.4 Centrifugal distortion illustrated for a symmetrical diatomic molecule. Rotation at an angular velocity ω generates a centrifugal force F on each atom of mass m, which leads to an extension δr of the bond. This generates a contrary force $k\delta r$, where k is the force constant. The moment of inertia increases from $2m\left(\dfrac{r}{2}\right)^2$ to $2m\left(\dfrac{r+\delta r}{2}\right)^2$.

---

* Strictly speaking, *all* molecules have some symmetry, as the identity operation is a symmetry operation. However, here and elsewhere we use the term 'no symmetry' as it is used in normal speech, rather than in a rigorous mathematical sense.

the order of one millionth of B. Occasionally it is found necessary to include a sextic distortion constant H as well, in a term $HJ^3(J + 1)^3$, in which case D is called the quartic distortion constant. Similar terms are used for spherical tops.

For symmetric tops centrifugal distortion may arise from rotation about the top axis as well as from overall molecular rotation, and three distinct quartic centrifugal distortion constants are needed:

$$E_r = BJ(J + 1) + (A - B)K^2 - D_JJ^2(J + 1)^2 - D_{JK}J(J + 1)K^2 - D_KK^4. \quad (4.6)$$

For asymmetric tops a total of five quartic distortion constants are required. Again, sextic terms can be added, but this is rarely necessary.

4.2.5   Degeneracy of rotational levels

The general rotational level with quantum number J has a degeneracy of $(2J + 1)$. For a linear molecule this degeneracy is related to the orientation in space of the molecule and its rotation axis, and it can be removed by applying a magnetic field. This leads to separation of different component energy levels with the same J but different orientations relative to the external field direction, and is known as the Zeeman effect.

For non-linear molecules rotational energy levels are also degenerate, but the situation is more complex than for linear molecules. In all cases, though, there is at least a $(2J + 1)$ degeneracy due to different space orientations, and this can be removed by application of an external magnetic field. Details of the degeneracies for non-linear molecules can be found in Refs. [1] and [2].

One consequence of the $(2J + 1)$ degeneracy of the rotational energy levels is that the populations of the levels increase as $2J + 1$, and so do the intensities of transitions. This increase is reduced by the Boltzman factor $N_E/N_0 = e^{-E/kT}$. The exponential population decrease becomes significant when E is of the order of kT, which is 209 cm^{-1} at room temperature. As $E = 2BJ(J + 1)$, with a typical B value of 0.1 cm^{-1} (3000 MHz), this occurs when $J(J + 1)$ is about 2000, or J is of the order of 40–50. Rotational spectra and rotational fine structure commonly cover quite wide ranges of J, from zero to 100 or thereabouts.

4.2.6   Nuclear spin statistics

In addition to the $(2J + 1)$ degeneracy referred to above, the various levels may have different statistical weights because of the interaction of the nuclear spins of the atoms making up the molecule with the overall angular momentum. For example, for most centro-symmetric linear molecules in which all nuclei have zero spin, the rotational levels with odd values of J have zero statistical weight, and only even values of J are possible. Note that $^{16}O_2$ is unusual in having a $^3\Sigma_g^-$ ground state, and in this case it is the even-J levels that are missing. If in a linear molecule, XAX, XAAX, X_2 etc., A has zero spin, but X has non-zero spin I, the odd-J and even-J rotational levels have relative weights $(I + 1) : I$ or $I : (I + 1)$ depending on whether the nuclei of X have half-integral spins $(1/2, 3/2, 5/2, etc)$ or integral spins $(1, 2, 3, etc)$, respectively. Thus if X is H $(I = 1/2)$ the odd-J levels have three times the statistical weight of the even-J levels, whereas if X is D $(I = 1)$ the odd-J levels have only half the statistical weight of the even-J levels. If A also has non-zero spin the situation becomes correspondingly more complicated.

Similar but more complex statistical weight effects operate in non-linear molecules. The most important is probably that applying to symmetric tops which have nuclei with non-zero spin off the axis. For example, for a three-fold rotor with three ^{1}H nuclei off the axis, such as SiH$_3$NCS, there is a K-dependent statistical weight variation: levels for which K is a multiple of three have twice the statistical weight of other levels (Fig. 4.5). Details of some of these effects may be found in Refs [1]–[6].

Fig. 4.5 K-structure of the $J = 7 \rightarrow 8$ bunch of transitions in the pure rotation spectrum of the SiH$_3$NCS molecule in its vibrational ground state, showing the eight components split by centrifugal distortion, and an intensity pattern arising from the three spin 1/2 nuclei lying off the axis. Redrawn from Ref. [2], p. 111.

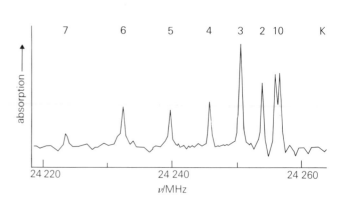

4.2.7 Stark effect

The $(2J + 1)$-fold degeneracy of a rotational energy level may be reduced or removed by the application of an external electric field. The resulting splitting is known as the Stark effect. It is of great importance in practical spectroscopy, as a fluctuating electric field can be applied, and by monitoring the oscillating part of the total signal very weak absorption signals can be detected. The magnitude of the splitting is dependent on the molecular dipole moment and so measurement of the splitting under known conditions allows direct determination of dipole moments for specific rotational and vibrational states of specific isotopic species. The splitting also depends on the rotational quantum numbers of the transition involved, and can be used to assign quantum numbers to lines. A clear distinction can be made, for instance, between lines arising from $K = 0$ states of a symmetric top molecule and those for which K is non-zero. Those for which $K = 0$ have only a weak, second-order Stark effect and the splittings are proportional to the square of the electric field, E^2; those for which K is non-zero show a strong first-order Stark effect, with splittings directly proportional to E. Lines with $K = 0$ can then be identified by comparing spectra recorded with small and large fields. More details can be found in Refs [1]–[6].

4.3 Rotational selection rules

Spectroscopic information is obtainable when a quantized property such as angular momentum changes, giving a measurable change in molecular energy. We must therefore ask about the allowed changes in the rotational quantum number J (and K in the case of the symmetric top). The rules depend on whether changes are taking place in other quantized molecular properties at the same time, *i.e.* whether we are studying a pure rotational transition, a vibrational transition with associated rotational changes or an electronic transition with associated vibrational and rotational changes. All these selection rules are based on symmetry, but it is best at this stage to present the results rather than to attempt to use symmetry to derive or justify them.

Pure rotation spectra

The basic requirement for observation of a pure rotation spectrum is that there should be an oscillating dipole associated with molecular rotation. To a first approximation this is the same as saying that the molecule must have a permanent dipole. One simple way of looking at the interaction of a rotating permanent dipole with radiation is to suppose that the oscillatory electric vector of the radiation resonates with the component of the rotating dipole which fluctuates in the same direction (Fig. 4.6). This is an oversimplified picture, as you will see if you try to work out how the $J = 0$ level can be excited to the $J = 1$ level in this way, but otherwise it is helpful. Only molecules of point groups C_{nv} or C_n can have permanent dipole moments, and these molecules all have rotational spectra. Here n can be any integer from one to infinity, so this includes the low-symmetry cases C_1 and C_s, as well as $C_{\infty v}$.

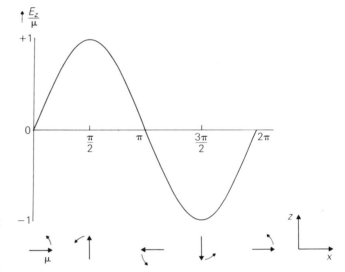

Fig. 4.6 Fluctuating z-component E_z/μ of the permanent dipole moment μ as a molecule rotates through one complete revolution in the xz plane.

Pure rotation spectra of linear molecules involve changes only in rotational quantum numbers, and may be observed using direct absorption or emission of radiation, providing the molecule possesses a dipole. We may picture the interaction of the rotating dipole with the electric vector of the radiation as a simple coupling of two electrical oscillations of the same frequency. In these circumstances the rotational quantum number J can change only by ± 1. For absorption of radiation, ΔJ is $+1$, and the change in rotational energy is

$$B[(J + 1)(J + 2) - J(J + 1)] = 2B(J + 1). \tag{4.7}$$

The pure rotation spectrum of an unsymmetrical linear molecule thus consists of a series of lines separated by $2B$ (Fig. 4.7). These lines appear in the far-IR region (<200 cm^{-1}, <6 THz) or in the millimeter-wave or microwave regions (<1000 GHz or <100 GHz, respectively) depending on the rotation constant B. If there is centrifugal distortion the effective rotation constant B now depends on J, so that the lines are no longer at exactly equal intervals. Instead they occur at positions given by:

$$2B(J + 1) - 4D(J + 1)^3. \tag{4.8}$$

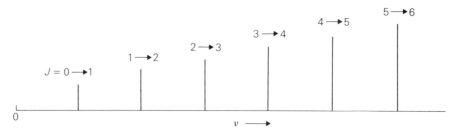

Fig. 4.7 Rotational spectrum of a rigid non-symmetric linear molecule.

It is possible for a molecule to interact with radiation by a non-resonant mechanism, the Raman effect (Section 5.3.2). Here the radiation with energy $h\nu$, which is much greater than that of a typical rotational quantum $h\omega_R$, interacts with the rotating polarizability of the molecule (itself a tensor quantity like the inertia tensor, see Section 4.2.1). The result is the apparent absorption of quanta of energy $h\nu$ and the emission of quanta of energy $h\nu \pm h\omega_R$. The observed rotational Raman spectrum thus consists of a series of emission lines separated from the exciting radiation by the various allowed rotation energies $h\omega_R$. In this case the rotational selection rule is $\Delta J = \pm 2$, so that the ideal line spacing is $4B$, ignoring the effect of centrifugal distortion. The rotational Raman effect is of great significance for the study of homonuclear diatomic molecules, which have no dipole and hence give no absorption spectrum. Unfortunately, the method is applicable only to small molecules at present, as only limited resolution is available. Rotation constants for H_2, N_2, O_2 and the halogens have been measured in this way, but only a few larger molecules, mainly organic, give useful results. More recently various so-called coherent Raman effects, in which two different laser beams irradiate a gas sample simultaneously, have been used. They can give high resolution and signals much stronger than those of the simple Raman effect, but the equipment required is expensive and complex to operate. Further details may be found by looking up references cited for the various techniques (CARS, Stimulated Raman Spectroscopy, Inverse Raman Spectroscopy) in the Glossary, Section 1.6.

The selection rules for pure rotational transitions of symmetric tops are $\Delta J = \pm 1$, $\Delta K = 0$ for direct absorption or emission, and $\Delta J = \pm 1$ or ± 2, $\Delta K = 0$ for the Raman effect. We obtain simple spectra in both cases, with a single series of lines in absorption ($\Delta J = +1$) and two series ($\Delta J = +1$ and $+2$) in the Raman effect. If we can neglect centrifugal distortion, these series have constant spacings of $2B$ or $4B$, and lines for all values of K coincide. If there is centrifugal distortion, separate lines may be observed for the different K values that are possible for each value of J, with frequencies given by:

$$2B(J + 1) - 4D_J(J + 1)^3 - 2D_{JK}(J + 1)K^2 \tag{4.9}$$

for $\Delta J = 1$. The last term corresponds to the separate K structure (see Fig. 4.5); the K splitting is small compared to the separation between bunches. Notice that neither A nor D_K appears in the line position formula, and these constants cannot be obtained from the pure rotation spectrum of a symmetric top.

The pure rotation spectrum of an asymmetric top is very complex, and cannot be reduced to a formula giving line positions. Instead it has to be dealt with by calculation of the appropriate upper and lower state energies (Section 4.2.3). The basic selection rule, $\Delta J = 0, \pm 1$, applies to absorption spectra, and here all three cases may in principle be relevant, as energy levels with greater J are not necessarily above all levels with lower J values. There are other selection rules which depend on the symmetry of the molecule and are described in Refs [1]–[6]. For the rotational Raman effect $\Delta J = \pm 2$ transitions are allowed as well.

It might be thought that a spherical top, with its very high symmetry, could not generate a pure rotation spectrum either by direct resonant absorption or by the Raman effect: to a first approximation this is indeed so. But for molecules with no center of inversion, the effects of centrifugal distortion may lead to a slightly asymmetric distortion of the molecule as it rotates, and hence to a small but non-zero dipole moment. The effect is extremely weak, and obviously increases rapidly with J. The result is that interaction of the rotating dipole with radiation can give a weak pure rotational spectrum, and such spectra have been observed for some spherical tops, such as CH_4, SiH_4 and GeH_4. The selection rule is still $\Delta J = 0, \pm 1$, the case $\Delta J = +1$ giving a series of lines with a spacing of $2B$ in an absorption spectrum (Fig. 4.8), so that B, and hence the bond length, can be obtained from such a spectrum.

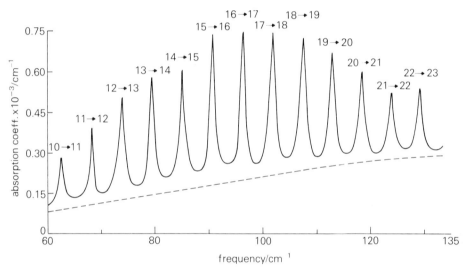

Fig. 4.8 Far-infrared pure rotational spectrum of SiH_4. Redrawn from A. Rosenberg and I. Ozier, *Can. J. Phys.*, **52**, 575 (1974).

4.3.2   Vibration−rotation spectra

If vibrational quantum numbers change in a transition, there will be associated changes in rotational quantum numbers for gases, whose rotation is quantized. We shall go into some aspects of the consequences of these changes in more detail in Section 5.8.3, but must note here the essential features governing the generation of rotational structure in vibration bands. As usual, we must consider molecules of different types in turn.

The stretching vibration of a homonuclear diatomic molecule generates no oscillating dipole, so there is no way in which it can interact directly with radiation. A hetero-nuclear diatomic molecule gives an absorption band, normally observed in the IR region, with rotational fine structure corresponding to the selection rule $\Delta J = \pm 1$. This band consists of a series of lines either side of the vibrational origin with an ideal spacing of $2B$. There is no line at the origin, as $\Delta J = 0$ is forbidden. The case $\Delta J = +1$

Example 4.1

The microwave spectrum of a molecule contains just one group of lines between 60 and 75 MHz, as shown in the first figure below. Which type of molecule (spherical, symmetric or asymmetric top) is most likely to give the pattern of lines shown? The highest frequency line shown has a second order Stark effect, and all others have a first order effect. The highest frequency line of the next group to lower frequency is at 56 503.43 MHz. Deduce as much information as you can from these data.

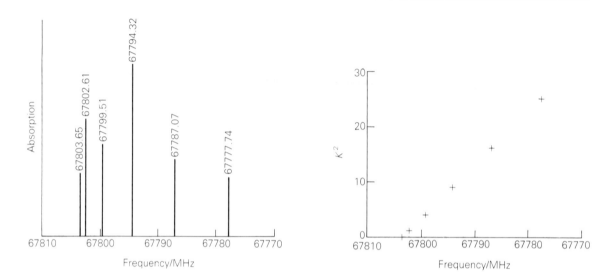

The pattern appears to be that expected for a symmetric top, with lines due to the various K states displaced from the $K = 0$ line (which has a second-order Stark effect) by an amount proportional to K^2. Plotting the frequencies directly against K^2 confirms this, and shows that the highest K value is five. (Right-hand figure above.)

This indicates that this is the $J = 5 \rightarrow 6$ bunch. If we neglect centrifugal distortion for the moment, the apparent B value is 67 803.65/(2 × 6) = 5650.30 MHz. This is confirmed by the separation between the bunches of lines, 67 803.65 − 56 503.43 = 11 300.22 MHz, which is close to $2B$.

We now use Eqn (4.9) to define the ground state rotation constant, B_0, and centrifugal distortion constants, and D_J and D_{JK}. Taking the highest frequency line in each bunch as having $K = 0$, we have

$$\nu_{5,0} = 12B_0 - 864D_J = 67\ 803.65 \text{ MHz and}$$
$$\nu_{4,0} = 10B_0 - 500D_J = 56\ 503.43 \text{ MHz.}$$

Whence, $B_0 = 72D_J + 5650.304$ and $B_0 = 50D_J + 5650.343$.

Thus, $D_J = 1.77$ kHz and $B_0 = 5650.432$ MHz.

We then use the separation between the $K = 0$ and $K = 5$ lines to give

$$2D_{JK} \times 6 \times 5^2 = 67\ 803.65 - 67\ 777.74 \text{ MHz.}$$

Therefore $D_{JK} = 86.37$ kHz.

The higher intensity of the $K = 3$ line suggests that there are three off-axis spin-1/2 nuclei, giving the characteristic ..weak..weak..strong.. intensity pattern, as shown in Fig. 4.5.

gives lines to higher frequency, called the R-branch, and the case $\Delta J = -1$ gives a P-branch to lower frequency.

Two complications arise. First, centrifugal distortion leads to variation in the effective rotation constant with J, giving terms containing J^3 in the expression for line positions, as in Eqn (4.8). The second complication is generally more important. We have to consider the effect of the vibration on the effective rotation constant, B. This can be thought of as arising from the vibrational averaging of the internuclear distance r in the following way. A rigid diatomic molecule has a fixed value of r, and the moment of inertia is determined exactly by r and the atomic masses involved, so there is no variation of I_B or B. As the molecule in fact has a zero-point vibration even in the ground vibrational state, the instantaneous internuclear distance varies over a range, and we have to take a quantum-mechanical average (expectation value) of $1/r^2$ to give the correct effective value of B in the ground state. This is called B_0. In a higher vibrational state the vibrational amplitude is greater, and the averaging process gives a different effective value of B, called B_1 for the first vibrational excited state, B_2 for the second, and so on. A general expression can be written for the vibrational variation of B.

$$B_v = B_e - \alpha(v + 1/2). \tag{4.10}$$

Here B_e is called the equilibrium rotation constant; it corresponds to the rotation constant of a hypothetical molecule with a fixed internuclear distance of r_e, the equilibrium internuclear distance. The quantity α can be called a vibration–rotation constant. Thus in general the two different vibrational states involved in a transition have different values of B_v. For a fundamental transition the difference is equal to α in Expression (4.10) above. In the upper state $v = 1$ and $J = J'' + 1$, and so the combined vibration/rotation energy is $G_v + B_1(J'' + 1)(J'' + 2)$, where G_v is the purely vibrational energy, and J'' is the J quantum number in the lower vibrational state. In the lower state $v = 0$ and $J = J''$, and so the energy is $B_0 J''(J'' + 1)$. Thus line positions in the R-branch are given by:

$$\begin{aligned} \text{Transition energy} &= G_v + 2B_1(J'' + 1) + (B_1 - B_0)J''(J'' + 1) \\ &= G_v + 2B_1 + (3B_1 - B_0)J'' + (B_1 - B_0)J''^2. \end{aligned} \tag{4.11}$$

The variation of B with vibrational state thus introduces a term in J^2 to the expression, which normally outweighs the J^3 term arising from centrifugal distortion. The corresponding expression for P-branch line positions is

$$\text{Transition energy} = G_v - (B_1 + B_0)J'' + (B_1 - B_0)J''^2. \tag{4.12}$$

Suitable manipulation of measured transition energies (such as a combination of P- and R-branch line positions) allows us to derive B_0, B_1 and their difference directly from the spectrum. If centrifugal distortion is significant the above expressions must be modified by inclusion of the appropriate terms in the derivations.

For linear polyatomic molecules there are two possible types of infrared bands, depending on whether the dipole change in the vibration is parallel or perpendicular to the axis of the molecule. For a parallel band, the same rules apply as for a heteronuclear diatomic molecule. For a perpendicular band, $\Delta J = 0$ transitions are allowed as well. In either case the P- and R-branches have the same structure as above, and upper- and lower-state rotation constants can be derived in the same way. The $\Delta J = 0$ transitions for a perpendicular band give a Q-branch, which has a frequency equal to G_v in the

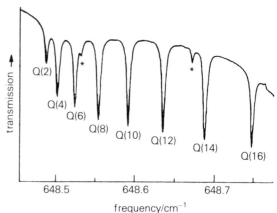

Fig. 4.9 Part of the Q-branch of the bending vibration of $^{13}CO_2$, recorded using a diode laser source spectrometer. The lines marked with asterisks arise from other isotopic species. Redrawn from M.J. Reisfeld and M. Flicker, *J. Mol. Spectrosc.*, **69**, 330 (1978).

ideal case. However, if the effects of the change of B with v are appreciable the branch may be resolved (Fig. 4.9) into a series of lines with frequencies given by:

$$\text{Transition energy} = G_v + (B_1 - B_0)J''^2. \tag{4.13}$$

For a symmetric top, as for a polyatomic linear molecule, there are two types of band, depending on whether the dipole change in the vibration is along the top axis or perpendicular to it. Parallel bands of symmetric tops have the rotational selection rules $\Delta J = 0$ or ± 1, $\Delta K = 0$, and so have structures similar to those of perpendicular bands of linear molecules, but with the added complication that a distinct sub-band is present for each value of K. If D_{JK} is negligible these effectively coincide, and we observe clusters of lines, one for each value of J in the P-branch and in the R-branch, and a single very narrow Q-branch (Fig. 4.10a). If D_{JK} is larger, or if other more complicated effects are important, separate lines may be observed for each possible set of values of J and K. Analysis is very similar to that described above for a linear molecule, giving B for both vibrational states, and perhaps values for D_J and D_{JK}.

Perpendicular bands of a symmetric top present an entirely different picture, at least superficially. The rotational energy of a symmetric top molecule in a vibrational state of e symmetry contains an extra term due to the interaction of the rotational angular momentum with the angular momentum associated with the vibration:

$$E_{vr} = G_v + BJ(J + 1) + (A - B)K^2 - 2A\zeta kl \tag{4.14}$$

where ζ is a Coriolis coupling constant*, l is the angular momentum quantum number associated with the vibration, and k is equal to K, with the proviso that the sign of kl depends on whether the two angular momenta are in the same sense or opposite senses. The effect of this extra term is to give two levels for each value of K, one with positive kl, the other with a negative kl value. The selection rules for perpendicular bands are $\Delta J = 0, \pm 1, \Delta K = \pm 1$, and this results in separate sub-bands for each value of $K\Delta K$, with their origins separated by $2(A - A\zeta - B)$. Each sub-band has the same J structure

---

* The Coriolis coupling term arises from the effects of Coriolis coupling between the two components of a degenerate vibration in a rotating molecule; see Ref. [1] for a detailed description of the coupling and its effects.

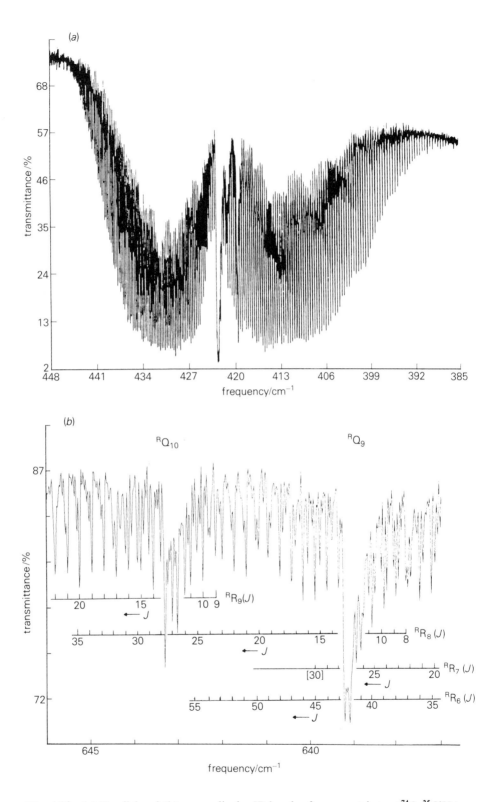

Fig. 4.10 (*a*) Parallel and (*b*) perpendicular IR bands of a symmetric top ($^{74}Ge^{35}ClH_3$). Taken from H. Bürger, K. Burczyk, R. Eujen, and A. Rahner and S. Cradock, *J.Mol.Spectrosc.*, **97**, 266 (1983) and S. Cradock, H. Bürger, R. Eujen and P. Schulz, *Mol. Phys.*, **46**, 641 (1982).

as for a perpendicular band of a linear molecule (Fig. 4.10*b*), and if resolved can give upper- and lower-state B values. The sub-band origins give the value of $A - A\zeta$, as B is known, but neither A nor $A\zeta$ can be determined separately in this way.

The rotational fine structure of a vibration band of an asymmetric top is complex, and depends on the direction of the dipole change in the vibration relative to the three principal inertial axes. Three different band types are possible in principle, labelled a-type, b-type and c-type for dipole changes along the A, B and C axes, respectively. Analysis is possible, giving the three rotation constants in favorable cases, as the line separations are often dependent on all three constants rather than being at simple intervals of $2B$, *etc.* An example is B_2H_6, for which analysis [7] of various IR bands has given A, B and C for the ground state; the use of these to define the structure is discussed in Section 4.6.1.

In IR spectra of spherical tops the allowed vibrational transitions are all of the same symmetry species (Section 5.8.3), though others can appear by reason of perturbations. The structure of a band due to the triply-degenerate vibration of a symmetric top (t_2 symmetry for a tetrahedral molecule) is like that of a perpendicular band of a linear molecule, with $\Delta J = 0, \pm 1$, giving rise to P-, Q- and R-branches. As for symmetric tops, however, there are additional terms in the upper-state energy analogous to $2A\zeta kl$ in Expression (4.14), and the separation of successive lines is not a simple multiple of B, but includes a contribution from a Coriolis coupling term $B\zeta$ as well.

Thus the rotational fine structure of a vibration band can give information about the rotation constant B for a linear molecule and for a symmetric top, and about all three rotation constants for an asymmetric top. The A rotation constant for a prolate symmetric top (or the C rotation constant for an oblate symmetric top) and the B rotation constant for a spherical top cannot be obtained from analysis of the structure of a single fundamental band. Similar information is in principle also obtainable from the analysis of the rotational structure of Raman bands. Even if no resolved detail is present in a vibrational band because of insufficient instrumental resolution, the band contour can give information about the symmetry of the molecule and the vibration; we shall return to this in Chapter 5.

4.3.3   Electronic spectra

Analysis of the rotational fine structure of electronic bands is also possible, and much information about unstable diatomic molecules has been derived in this way. The rotation constant for a diatomic molecule may be calculated for each of many vibrational states of each electronic state involved in the transition (see Section 4.4.4). Much less information has been forthcoming for larger molecules. This is partly because we are looking at small differences between high energy transitions, and it is very difficult to attain the resolution needed to observe the rotational fine structure of an electronic band. More fundamentally, the fine structure is often not defined because the excited state has a very short lifetime, with a consequent uncertainty in the energy levels because of the Uncertainty Principle. Hence we find broad unresolved bands in the electronic spectra of most polyatomic molecules. We shall see in Section 6.7.1 how the use of lasers can help to reduce the complexity of electronic bands, by specifically exciting single upper states, but they can do nothing to reduce the effects of dissociation and related effects such as predissociation.

4.4 Instrumentation

4.4.1 Pure rotational absorption spectroscopy

As mentioned in Section 4.3.1, most pure rotation spectra are observed by direct absorption of electromagnetic radiation in the far-IR, millimeter wave or microwave regions of the spectrum, corresponding to wavelengths from 0.05–60 mm (200 to 0.16 cm^{-1}, 6000 to 5 GHz). Over this wide range it is necessary to use several different techniques for generation and detection of the radiation.

In the far-IR region, say 200–20 cm^{-1} (6000–600 GHz), there are only broad-band sources of radiation, besides a few lasers which operate at fixed frequencies and are not generally useful. No tunable monochromatic sources are available. Most spectra are obtained using interferometers (Section 5.4), though grating spectrophotometers have been used. It is possible to obtain spectra with quite high resolution by the standards of normal mid-IR spectroscopy in this region, but even so it is not easy to achieve a resolution better than about 0.01 cm^{-1}. Pure rotation spectra of light molecules (such as the hydrogen halides, NH_3 and H_2O) fall in this region, and have been extensively studied.

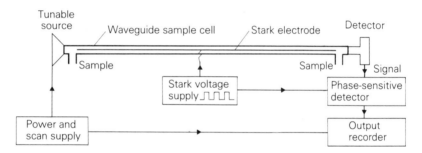

Fig. 4.11 Schematic diagram of a microwave spectrometer.

In the microwave region (5–100 GHz, say) it is comparatively easy to generate highly-monochromatic radiation that can be tuned over a wide range using a single source, and an even wider range if several sources are used. A typical microwave spectrometer (Fig. 4.11) thus consists of a set of tunable sources (klystrons or backward-wave oscillators), an absorption cell and a detector. As absorption in this region is usually very weak, a typical cell length is 1–2 m. The radiation is best passed through the apparatus using wave-guide techniques rather than conventional optical methods (mirrors or lenses) and the need to match the size of the waveguide to the wavelength of the radiation means that several sets of matching guides are used. As each source/waveguide combination is useful over a frequency range of less than one octave it can become quite tedious covering the whole range from 5 to 100 GHz, spanning more than four octaves. A microwave measurement has the enormous advantage over a typical IR measurement of extremely high precision, since a typical line position of 30 GHz (3×10^{10} Hz) is measurable to ±10 kHz, or one part in 3 million! There are two points here; the lines are extremely narrow, and their positions can also be measured with extreme precision because the source is very stable. Almost every molecule except the lightest has some part at least of its pure rotation spectrum in the microwave region, and almost all the information we have about rotation spectra comes from such measurements.

The region between the far-IR and microwave regions, corresponding to frequencies between about 100 and 1000 GHz, is much more difficult to study. There are no tunable sources that emit directly in this region, but it is possible to generate harmonics from the output of a klystron or backward-wave oscillator operating below 100 GHz to give some power in this, the millimeter wave region. It is not easy to work with such polychromatic radiation, but a few groups have managed to overcome the problems and so obtain useful spectra in this most interesting region. One advantage of working with millimeter waves rather than IR radiation is that the frequencies can still be measured precisely. An advantage over the microwave region is that higher-J transitions are observed, so that the effects of centrifugal distortion can more readily be measured and accounted for. As mentioned in Chapter 1, Fourier transform methods can be used in the microwave and millimeter wave regions, but only a few groups have applied them as yet.

4.4.2   High resolution Raman spectroscopy

As we shall see in Section 5.4, a basic Raman spectrometer consists of a powerful source of monchromatic light, usually a laser, an efficient monochromator to separate the weak Raman-scattered light from the much stronger Rayleigh-scattered light of the original frequency, and a detector. In pure rotational spectroscopy, for which gaseous samples are used, the signals are inherently weak and very powerful lasers are used as sources. The monochromator must be extraordinarily efficient to prevent any unshifted radiation reaching the detector, while the weakness of the scattering makes it hard to achieve the highest possible resolution by employing very narrow slits. The advantages of photographic recording are hard to beat, and until recently most rotational Raman spectra had been obtained using photographic methods, which allow the accumulation of extremely weak signals over many hours. Modern Raman spectrometers use photo-multipliers to detect signals (often cooled with liquid N_2 to reduce noise levels) coupled with efficient double or even triple monochromators to allow operation close to the exciting line. The technique is most important for homonuclear diatomic molecules, which give no rotation spectrum by direct absorption.

It is also possible to use Raman techniques to study the rotational structure of vibrational bands, and it is one of the main advantages of Raman spectroscopy that the same apparatus can be used for both types of study. However, the vibrational bands are usually even weaker than the pure rotation lines, and few really high-resolution spectra have been obtained. It is apparently impossible to operate with an effective resolution of better than about 0.1 cm^{-1}, although coherent Raman techniques give a considerably higher resolution, at the expense of more complexity in the apparatus. Again, photographic recording was until recently the best method of accumulating weak signals.

4.4.3   High resolution mid-IR spectroscopy

If rotational structure is to be resolved in a vibration band it is usually necessary to use very high resolution, though of course rotational fine structure can be seen for light molecules even with quite crude instruments. There are a few very high-resolution grating instruments operating especially in the higher-frequency portion of the IR region, but it is now possible to purchase an interferometer with an effectively constant resolution of better than 0.005 cm^{-1} over the whole mid-IR range ($5000-500$ cm^{-1}). It appears that for simple symmetrical molecules there are significant advantages in working with IR spectra rather than with the relatively few lines obtainable in a typical microwave spectrum. The large number of lines in one vibration–rotation band means that the ground-state constants such as B_0 and the centrifugal distortion constants D_J and D_{JK} can be specified more precisely than from two or three microwave lines, which is often all that can be measured, even though the precision of each individual line position is much greater in the microwave region (± 10 kHz) than in the IR (± 0.001 cm^{-1} = ± 30 MHz). This is not just a statistical effect, but arises because the precision of the higher-order terms such as centrifugal distortion constants, which affects that of B_0, is much greater if transitions with a wide range of J and K values (for a symmetric top) can be measured. An example may be found in Ref. [8], which describes the analysis of a high-resolution IR spectrum of SiBrH$_3$ to give a precise value for D_J, and hence a more reliable value of B_0 than could be calculated from a microwave spectrum.

It is worth pointing out at this stage that an instrumental resolution of 0.005 cm^{-1} is comparable with the Doppler width of IR transitions, due to the random thermal motions of molecules in gases. There is therefore little advantage to be expected from any further increase in instrumental resolution, unless steps are taken to reduce Doppler width, for example by using molecular beams to reduce the random motion of sample molecules.

4.4.4   High resolution electronic spectroscopy

Again, the attainment of high resolution in an electronic spectrum depends to a large extent on new techniques, though conventional monochromators with resolutions of better than 0.01 cm^{-1} in the visible region have been built. More modern advances have depended on the development of tunable lasers and double resonance techniques, some of which we examine below (Section 4.4.5). Much work has been done on the electronic spectra of stable and unstable diatomic molecules, and an enormous amount of information is available on the rotation constants of the electronic ground states of such molecules, as well as of their excited states. For example, the rotational fine structure of some vibrational bands in the $A\,^2\Pi \rightarrow X\,^2\Pi$ emission spectrum of SbO have been analyzed to give an accurate bond length, $r_e(\text{SbO}) = 1.8227$ Å, in the $X\,^2\Pi_{\frac{1}{2}}$ lower state [9].

4.4.5   Tunable laser and double resonance experiments

This heading refers to a wide range of experiments that are essentially more sophisticated than a simple absorption measurement, though it is certainly possible to use a tunable laser to scan a conventional absorption spectrum for any sample that absorbs in its range of operation. More interesting experiments include IR/microwave double resonance, in which a fixed frequency IR laser is used to irradiate a sample within a vibration–rotation band. This perturbs some particular ground-state and upper-state levels so that if other transitions involving these levels are observed in the microwave region, for instance, changes in the intensities, linewidths and so on may be noticeable (Fig. 4.12). In particular, if the IR laser irradiation is chopped or otherwise modulated at some specific frequency the perturbation will also fluctuate at this frequency. This allows very sensitive phase-sensitive detectors to be used to identify transitions that involve one or other of the perturbed levels and reject all others.

The same technique can be used when the irradiation is in the electronic spectrum. Thus phase-sensitive detection combined with a modulated tunable visible or ultraviolet laser can give signals deriving from only a single upper-state or lower-state level, which is specific with respect to electronic, vibrational and rotational quantum numbers. This allows much simplification of the observed spectrum.

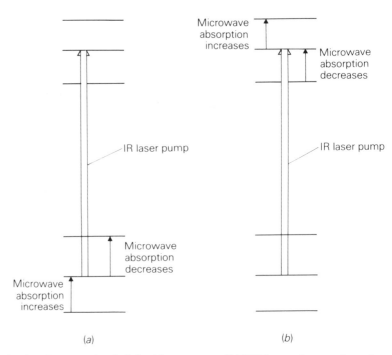

Fig. 4.12 Some simple microwave/optical double resonance (MODR) experiments. In each case an IR laser pumps sample molecules from a particular rotational level of the vibrational ground state to a level of an excited state. In (*a*) microwave lines of the ground state molecule are detected by their change in intensity, and in (b) lines of the excited state are detected. Note that some lines increase in intensity, and that others decrease.

4.5 Using the information in a spectrum

4.5.1 Fingerprinting

It is, of course, possible to use rotation spectra or rotational detail in other spectra simply as a method of recognizing or identifying samples, and extensive lists of microwave line positions have been published to facilitate these processes. The method is particularly applicable in the microwave region, where the narrow instrumental and spectral linewidths mean that lines even less than 1 MHz apart can easily be distinguished, while the chance of two different samples giving lines within the accuracy of measurement of each other is small. Even this small chance of accidental coincidence can be much reduced if two or three lines are checked, and samples can be identified as known compounds with a high degree of certainty in this way. In the same way even quite

Table 4.1 Molecules and ions detected in space by their pure rotation spectra

---

(*a*) Diatomic species
CO, CN, CS, SO, SiO, NO, NS, CH, CH^+

(*b*) Triatomic species
H_2O, HCN, HNC, OCS, H_2S, N_2H^+, SO_2, HNO, HC_2, HCO, HCO^+, HCS^+

(*c*) Tetratomic species
NH_3, CH_2O, HNCO, CH_2S, HNCS, $C \equiv C - C \equiv N$

(*d*) Some selected larger species
$HC \equiv C - CN$, CH_3OH, HCOOH, $CH_3C \equiv CH$, CH_3CN, CH_3CHO, NH_2CHO,
$CH_2 = NH$, CH_3SH, CH_3NH_2, $HCOOCH_3$, CH_3OCH_3, CH_3CH_2OH, $H(C \equiv C)_nCN$
($n = 2\text{-}5$)

---

small amounts of some impurities in a sample can be detected and identified. On the other hand, the method is limited in applicability as it is restricted to volatile species and to molecules with a dipole moment. Other methods are much less useful, though for small molecules such as the hydrogen halides it is easy to spot the regular patterns of strong lines in their IR and far-IR spectra.

One very important area is the identification of molecular species in astronomical sources. A surprising number of polyatomic molecules and ions (Table 4.1) have been

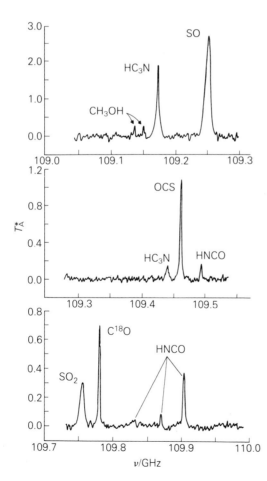

Fig. 4.13 Portions of emission spectrum of the source of Orion A, showing lines assigned to severn different molecular species between 109 and 110 GHz. The lines include transitions for ground and excited vibrational states of $HC \equiv C - C \equiv N$. The resolution is 1 MHz and the ordinate is in the form of an antenna temperature, which is corrected for local background and other effects. Redrawn from P.F. Goldsmith, R.L. Snell, S. Deguchi, R. Krotov, and R.A. Linke, *Astrophys. J.*, **260**, 147 (1982).

identified in the so-called molecular clouds, which appear to contain regions with moderate effective temperatures (tens or even hundreds of Kelvins) which may represent an early stage in the condensation of interstellar matter into stars. An example of an emission spectrum from such a source is shown in Fig. 4.13.

4.5.2 **Determination of rotation constants**

Much the most important use of rotation spectra and spectra containing rotational detail is in the measurement of rotation constants, which can in principle be used to define moments of inertia, and hence internuclear distances. We shall see later that there are in fact severe limitations to the usefulness of this information, but for the moment we must note what can and cannot be measured.

For a linear molecule, there is only one rotation constant, which can be measured from the pure rotation spectrum, from a vibration–rotation band or from an electronic spectrum. As the line positions depend upon other parameters as well as B_0 it is necessary to measure more than one line position to obtain an accurate measure of the rotation constant, and this can be a limiting factor.

For a symmetric top, only B can be determined easily from a pure rotation spectrum or a vibration–rotation band in absorption, and the same problem can arise. For example, B_0 for GeFH$_3$ is just over 10 GHz, so that the first line appears just above 20 GHz, and is the only datum measurable using a spectrometer operating up to 40 GHz. But this line position is in fact equal to $2B_0 - 4D_J$, and we cannot obtain B_0 without estimating the distortion constant. This is one reason why the use of a high-resolution mid-IR interferometer may give more accurate results, though a combination of data from IR and microwave spectra will usually be best of all. If a microwave line position is measured to ± 100 kHz at 20 GHz, the relative precision of the measurement in experimental terms is ± 1 part in 200 000, which is comparable with the expected relative magnitudes of D_J and B_0. In the above example, for instance, B_0 is 0.333 9513 cm^{-1}, while D_J is 0.285×10^{-6} cm^{-1}. These values were actually determined from the millimeter wave spectrum [10], from which the effects of D_J could be explicitly measured, as pure rotation lines up to $J'' = 12$ were observed, near 360 GHz.

It is possible in principle to determine A for a symmetric top if the Raman spectrum is studied with sufficient resolution, but this has been done only for a few simple molecules. In other cases A has been found from high-resolution IR spectra by making use of a breakdown of the simple rotational selection rules arising from perturbations, or couplings between close-lying levels. However, these occur only by chance in particular systems.

For an asymmetric top, all three rotation constants can in principle be determined from pure rotation spectra or from vibration–rotation spectra, but it is not always appreciated that the accuracy of some of the constants may be low even though they are derived from very precise microwave spectra. This arises in part because of the problems of centrifugal distortion, as mentioned above for symmetric top molecules, but also because the measured line positions may well depend only weakly on one of the rotation constants, which is then correspondingly uncertain.

The rotation constant for a spherical top can only be determined from the 'forbidden' pure rotation spectrum or from a detailed analysis of the fine structure of a vibration–rotation band. An analysis based on the use of combination differences (differences between observed transition energies involving the same upper state level but different lower-state levels) is most effective in this case.

4.5.3 Determination of centrifugal distortion constants

Centrifugal distortion constants are not of prime importance in the determination of molecular structure but it may be necessary to determine them in order to interpret an observed spectrum. To determine a quartic distortion constant it is generally necessary to have very precise data, preferably over a good range of the quantum numbers (J, and K for a symmetric top). Otherwise there is no particular difficulty in obtaining reliable values for D_J and D_{JK} for a symmetric top, or D for a linear molecule. D_K has very little influence on the form of the spectrum for a symmetric top, and is correspondingly difficult to determine.

4.5.4 Isotopic substitution

Altering the isotopic mass of an atom in a molecule may alter some or all of the moments of inertia. We shall see in Section 4.6 that such changes can be used to define precisely atom positions within a molecule. Here we note that we can use the observation of a change in rotation constant to confirm the presence of a particular type of atom in the molecule. The technique has been most extensively used in studying unstable species such as diatomic molecules and ions for which elemental analysis is impracticable. For example, if we wish to confirm that the species responsible for a given spectrum is indeed an oxide, it is sufficient to show that the lines in the spectrum shift (and may split) when $^{18}O_2$, either pure or mixed with normal $^{16}O_2$, is used instead of normal oxygen in the preparation. It is also usually

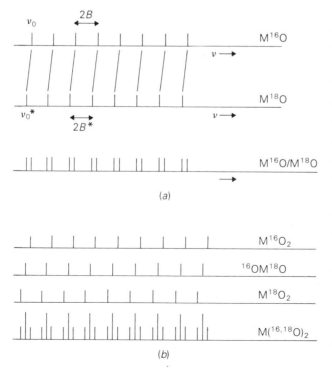

Fig. 4.14 Mass effects on rotational fine structure in vibrational or electronic spectra of partial or total substitution of ^{18}O for ^{16}O in the preparation of (a) a species MO; as B^* is less than B, the separations between the pairs of lines will increase to higher frequency: (b) a species MO_2 containing two equivalent oxygen atoms. In (b), only the effects of mass are considered; the fact that ^{16}O and ^{18}O have nuclear spin $I = 0$ means that for the symmetrical species alternate lines will not be observed because odd-J levels have zero statistical weight (see Section 4.2.6). All the lines shown for the mixed isotopic species occur.

possible to use the exact change in rotation constant to determine whether the species contains a single non-oxygen atom, while the presence of more than one oxygen atom may be revealed if $^{18}O^{16}O$ is present in the reactant oxygen, as well as the symmetrical species (see Fig. 4.14). The effect of a change in isotopic mass is, of course, small if the substituted atom is close to the center of mass, and zero if it is at the center of mass, so that ^{13}C-substitution will not reveal the presence of carbon in CO_2, for example. Similarly, the rotation constants of BCl_3 are not affected by the boron isotopic mass, as the boron atom lies at the center of mass in this planar molecule.

4.6 Using rotation constants to define molecular structures

4.6.1 General; r_0 and r_e structures

The primary use of a measured rotation constant is for the calculation of the corresponding moment of inertia. This depends on the molecular structure, because it is the sum of the products of atomic masses, m_i, and the squares of the displacements, r_i, of the atoms from the appropriate axis.

$$I = \sum_i m_i r_i^2 \qquad (4.15)$$

For a diatomic molecule there is only one structural parameter, the bond length, and the single moment of inertia is sufficient to determine this if the atomic masses are known with the necessary accuracy. (In the past the relative moments of inertia of isotopic molecules were used to determine the relative atomic masses, but this method has been superseded.) At one time the factor limiting the precision of bond-length determination was the uncertainty in the value of Planck's constant, h, used in the conversion of the rotation constant B to I using Eqn (4.3). This is no longer a limitation, as the constant is known with adequate precision [$h = 6.626\ 18(11) \times 10^{-34}$ J s].

A complication arises when we consider what the exact meaning of the moment of inertia I_0 derived from the ground-state rotation constant B_0 might be. It clearly does not correspond to a single fixed internuclear distance, as the zero-point vibration ensures that the rotation constant is an effective value averaged over the whole range of distances over which the atoms move. The internuclear distance r_0 calculated from I_0 is thus only an effective (average) value of the internuclear distance, and in particular it differs from the equilibrium distance, r_e, even if the vibration is harmonic. We can only calculate r_e for a diatomic molecule if we determine B_e by finding the values of B in two different vibrational states (see Section 4.3.2), convert this to I_e, and then use Eqn (4.15) to give r_e. If this is done by analyzing a vibration–rotation band or by observing microwave spectra of two different vibrational states, we find that the resulting r_e values are independent of the isotopes involved. Values of r_0, on the other hand, vary with isotopic mass, as this alters the effect of the zero-point averaging (see Table 4.2).

Table 4.2 Bond lengths in N_2/Å

Isotopic species	r_0	r_e
$^{14}N_2$	1.100 105	1.097 700
$^{14}N^{15}N$	1.100 043	1.097 702
$^{15}N_2$	1.099 985	1.097 700

CHAPTER 4: ROTATIONAL SPECTRA AND STRUCTURE

For more complicated molecules these problems are worse. We may extend Eqn (4.10) to cover any number of vibrations by writing

$$B_v = B_e - \sum_i (v_i + 1/2)\alpha_i^B \qquad (4.16)$$

so that the measured rotation constant in the ground state, B_0, can only be converted to B_e if *all* the vibration–rotation constants α_i^B are known. For a non-linear molecule there is a set of constants α_i^A, α_i^B and α_i^C relating to the three different rotation constants A, B and C. Even for an asymmetric linear triatomic molecule, with only two structural parameters (the two bond-lengths), the single rotation constant gives only one moment of inertia, and this is insufficient to define both parameters. As we have seen for a diatomic molecule, the r_0 values for different isotopic species are inconsistent with each other, because of the effects of vibrational averaging in the ground state, and it is only if we can obtain B_e values for two different isotopes of the molecule that we can combine the information to give an equilibrium structure. This requires finding rotation constants in ground and excited states for all three vibrations of the triatomic molecule for both isotopic species, a total of eight experimental observations to define two bond lengths. It is rare for this amount of information to be collected, and we must remember that any attempt to combine B_0 information from different isotopic species is at best an approximation. In awkward cases (as for instance in N_2O, where one of the N atoms lies very close to the center of mass) it may lead to results inconsistent with any possible structure. In this case, the moment of inertia I_0 is actually less for the isotopic species with ^{15}N in the central position than for the normal species with two ^{14}N atoms, purely because of the difference in zero-point averaging effects. In any case, isotopic substitution of an atom near the center of mass affects the moments of inertia (and hence the rotation constants) very little, so its position will not be given accurately by this method.

For a symmetric linear triatomic like CO_2, of course, there is only one structural parameter, and this can be determined from the rotation constant for a single isotopic species, as for a diatomic. Even so, it is necessary to collect four pieces of experimental information to correct B_0 to B_e, in order to define r_e.

Non-linear molecules have more rotation constants than linear ones, and more information is in principle available for each isotopic species, though some of it may not be easy to obtain. The discrepancies between B_0 and B_e types of information are important, and we may have to make an inordinate number of measurements in order to calculate a few reliable structural parameters.

A case in point is $COCl_2$ [11], for which three rotation constants (A, B and C) have been determined experimentally for the ground vibrational state and for the first excited states of all six vibrations, a total of 21 experimental determinations. From these, the equilibrium rotation constants A_e, B_e and C_e can be found. But after all this effort, these data contain only two independent pieces of information, as $I_C = I_A + I_B$ for a planar body; this is indeed found to be so for the equilibrium moments of inertia, though not for those derived from the rotation constants of any of the vibrational states (ground or excited). We thus have insufficient data to define the structure, as this requires three parameters (C=O and CCl distances, and an angle).

Another case is that of B_2H_6. Four parameters are required to define the structure, and so we must combine data from at least two isotopic species, as we can only obtain three rotation constants in each case [12]. As each isotopic molecule has $3N - 6 = 18$

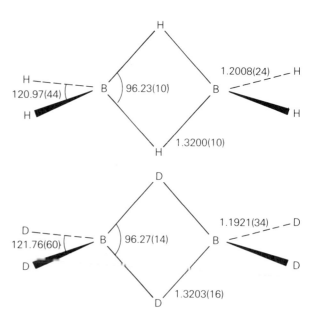

Fig. 4.15 Mixed-boron-isotope r_0 structures of B_2H_6 and B_2D_6, derived from ground-state rotation constants of $^{10}B_2H_6/^{11}B_2H_6$ and $^{10}B_2D_6/^{11}B_2D_6$, respectively. Distances are in Å and angles in degrees, and estimated standard deviations are given in terms of the least significant digit, in parentheses. Notice that the terminal atom positions are rather different, but that the central BH_2B units are almost identical. The data are taken from Ref. [12].

different vibrations (Section 5.2.2), it is clearly an immense labor to accumulate enough data to correct these rotation constants to give A_e, B_e and C_e for each isotopic species. In these circumstances, it is usual to compromise by using B_0 values for different isotopic species to define a 'mixed-isotope r_0' structure. In this case we can deal separately with B_2H_6 and B_2D_6, and for each of them we can derive a 'mixed-boron-isotope' r_0 structure using data for the ^{10}B and ^{11}B forms (Fig. 4.15). Unfortunately, such structures have even less well-defined character than the r_0 structure for a single isotopic species, but they are often the best that can be attained. In any such case it is probably best to accumulate more information than is strictly required and use a least-squares fitting technique to adjust structural parameters to reproduce the observed rotation constants as precisely as possible. This has the advantage of revealing any incompatibilities in the data and may help one to avoid proposing ridiculous structures.

Another technique that is helpful here is the use of predicate observations. Again a least-squares fitting method is used, but each structural parameter (bond length or angle) is assigned an initial value and an uncertainty. The least-squares fit is then applied using these initial values and uncertainties as additional observations, so that the final set of structural parameters depends on the initial guesses as well as the spectroscopic observations. An example of the use of this technique is a study of the structure of methyl stannane using microwave data [13] for CH_3SnDH_2 and CH_3SnD_2H, with several Sn isotopes detectable at natural abundance. The lack of isotopic substitution data for the methyl group makes it impossible to define the structure fully without making assumptions, but including reasonable values for the CH bond distance (1.083 ± 0.005 Å) and the SnCH bond angle ($110.5 \pm 0.5°$) makes least-squares refinement possible, to give the structure shown in Fig. 4.16. The big advantage of the method is that it guards against reaching a solution that contains unrealistic bond lengths or angles. A big snag is that it is perfectly possible to use the method in circumstances where there is insufficient spectroscopic information to define the structure adequately, so that the final result is effectively determined almost entirely by the initial assumptions.

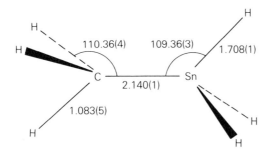

Fig. 4.16 Isotopic average structure of methyl stannane derived from microwave spectra of CH_3SnDH_2 and CH_3SnD_2H, using assumed methyl group geometry (bond length and angle) as predicate observations. Distances are in Å and angles in degrees, and estimated standard deviations are quoted in parentheses, in terms of the least significant digit. Data are taken from Ref. [13].

4.6.2

Partial structures; r_s structures

It may not be possible to collect enough information to define a structure completely, yet there is certainly some structural information hidden in a collection of rotation constants for a set of isotopic species. In these circumstances we may be able to derive some structural parameters, often very precisely, without finding a full structure. An example of this is the determination of a single bond length from the rotation constants for three isotopic species. Isotopic substitution of an atom allows us to specify its position relative to the center of mass, using the difference in the moments of inertia, and if this is done for two atoms the bond length (r_s) is determined. For a linear molecule or a symmetric top in which the substituted atom lies on the z-axis, the z-coordinate of the atom can be calculated as

$$z^2 = \frac{\Delta I}{\Delta m} \cdot \frac{M^*}{M} \tag{4.17}$$

where ΔI is the change in moment of inertia on substitution, Δm is the change in mass on substitution, M is the total mass of the original molecule and M^* is the total mass of the substituted molecule. Enough data were available for CH_3SnH_3 [13] (see Section 4.6.1) to give an r_s SnH bond length of 1.713(5) Å, and a corresponding angle CSnH of 110.01(3)°. These results are in reasonable agreement with those determined by the least-squares analysis with predicate observations (Fig. 4.17). The discrepancy in the bond angle is probably due to neglect of the $B_0 \rightarrow B_e$ corrections; this neglect often leads to apparent differences in bond lengths and angles involving H and D. Similarly, microwave data for $^{12}C/^{13}C$ and $^{78}Se/^{80}Se$ pairs of molecules of HNCSe allows the determination of r_s CSe as 1.719 Å, though the other atoms (N, H) cannot be located without making further assumptions [14].

The precision of these calculations is high, as only the accurately determined moments of inertia and the precisely known atomic masses are involved. This apparent precision is, of course, partly illusory if moments of inertia are derived from B_0 constants rather than B_e constants; only the latter will give true substitution coordinates, r_s. In fact some of the effects of vibrational averaging included in the rotation constants B_0 cancel out when they are combined in this way, and 'mixed-isotope r_0' or 'r_0/r_s' parameters are certainly much closer to the r_e values than single-isotope r_0 parameters. Attempts have been made to allow for various other effects of changing isotopic mass, giving parameters labelled r_c and r_m, which combine data from different isotopic species to give even better approximations to r_e. True r_s parameters are geometrically consistent; 'r_0/r_s', r_c and r_m parameters are nearly but not quite

Example 4.2

The B_0 rotation constants for the $^{70}Ge^{79}Br$, $^{70}Ge^{81}Br$ and $^{74}Ge^{79}Br$ isotopic species of $GeBrH_3$ are 2438.52, 2410.15 and 2375.86 MHz respectively. Deduce the Ge—Br r_s bond length. (Isotopic masses are ^1H, 1.007825, ^{70}Ge 69.92425, ^{74}Ge 73.92118, ^{79}Br 78.91839, ^{81}Br 80.91642).

....

Using the first two rotation constants and Eqn (4.3), we obtain moments of inertia I and $I*$ for the parent and substituted species:

I = 2438.52/505391 atomic mass units $\times$ $Å^2$

= 207.2532 amu $Å^2$;

$I*$ = 209.6928 amu $Å^2$;

so ΔI = 2.4396 amu $Å^2$.

The molecular weights, M and $M*$, are

M = 151.86612 amu

and $M*$ = 153.86415 amu,

so Δm = 1.99803 amu.

Using Eqn (4.17) we obtain

$z_{Br}{}^2$ = 1.23707 $Å_2$

Whence $|z_{Br}|$ = 1.11224 Å.

Using the first and last rotation contants in the same way, we obtain:

I = 207.2532 amu $Å^2$,

$I*$ = 212.7193 amu $Å^2$,

ΔI = 5.4660 amu $Å^2$,

M = 151.86612 amu,

$M*$ = 155.86305 amu,

Δm = 3.99693 amu

$z_{Ge}{}^2$ = 1.40354 $Å^2$,

and $|z_{Ge}|$ = 1.18471 Å

The two heavy atoms must be on opposite sides of the center of mass, so

r_s(Ge—Br) = 2.29695 Å

The precision of this result may be estimated by considering likely uncertainties of ± 0.03 MHz for each rotation constant, i.e. 1 part in 80 000. Corresponding uncertainties in I and $I*$ are then ± 0.0026 amu $Å^2$ and the probable error in ΔI is $\sqrt{2}$ times this, ± 0.0037 amu $Å^2$. This is 1 part in 660 and 1 in 1400 for the two ΔI values. Each z_s has half the relative error of the corresponding ΔI, as we take square roots, and the final error in r_s is obtained by combining the z_s errors. The Ge—Br distance should be quoted as 2.2970 ± 0.0009 Å.

geometrically consistent. The equations required to derive r_s structural parameters from isotopic changes in moments of inertia were given by Kraitchman [15] and the method was developed further by Costain [16].

4.6.3   Use of spectroscopically calculated corrections; r_z structures

The measured value of a vibration−rotation constant, α, for any vibration can be expressed as the sum of harmonic contributions, which can be calculated on the basis of the assumption of harmonic oscillator dynamics, and an anharmonic contribution, which cannot. Consequently, one way to avoid measuring all the α values for all vibrations of a whole set of isotopic molecules is to use the results of a normal coordinate analysis (see Section 5.14) to calculate the harmonic contributions to the α values. Taking the harmonic contributions only, we define a new B value, labeled B_z, as

$$B_z = B_0 + \tfrac{1}{2}\Sigma\alpha_{\text{harm}} \tag{4.18}$$

in place of B_e as defined in Section 4.3.2. Using B_z rotation constants we can then generate a set of moments of inertia that we may label I_z and a structure or partial structure labelled r_z. This approximates to the equilibrium (r_e) structure better than the r_0 structure does, but still differs from it in respect of the anharmonic contributions to α^B which we have ignored in the calculation of B_z. It has the great advantage that only one (ground state) rotation constant needs to be measured for each isotopic species, as the harmonic contributions to α^B are calculated from the normal coordinate analysis. We can also use B_z data to define substitution structures with more confidence than if B_0 data were used, though there is still no guarantee that the neglect of anharmonic contributions will be justified in this case. An r_z structure has been calculated for B_2H_6 and B_2D_6 from their rotational constants (Section 4.6.1) and an approximate r_e structure

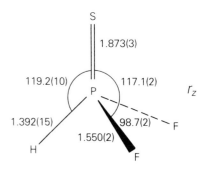

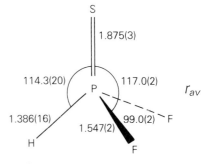

Fig. 4.17 Structures for PF_2HS derived from rotation constants with calculated harmonic vibrational corrections (r_z) and from simultaneous analysis of the same rotational data with electron diffraction data (r_{av}). Parameters, which are taken from Ref. [17], are given in Å or degrees, with estimated standard deviations in parentheses.

derived by estimating the remaining anharmonic contributions [12]. Another great advantage of the r_z structure is that it can be shown to be exactly equivalent to the r_α^0 structure determined in an electron diffraction experiment (Section 8.3), again using vibrational corrections calculated from the normal coordinate analysis. This means that data from the two techniques can be compared and even combined in defining a final joint structure, sometimes known as an r_{av} (for average) structure. The r_z and r_{av} structures found for PF_2HS [17] are shown in Fig. 4.17. It is clear that the structures are closely similar, except for the HPS angle, which has very little effect on the rotation constants.

Problems

Isotopic masses:

^1H	1.007 825 0	^{19}F	18.999 405
^6Li	6.015 125	^{35}Cl	34.968 853
^7Li	7.016 004	^{37}Cl	36.965 903
^{12}C	12.000 000	^{79}Br	78.918 39
^{13}C	13.003 354	^{81}Br	80.916 42
^{14}N	14.003 074	^{138}Ba	137.905 00
^{15}N	15.000 109	^{200}Hg	199.968 33
^{16}O	15.994 914	^{202}Hg	201.970 64
^{18}O	17.999 162		

4.1 The microwave spectrum of the vapor over molten ^6LiF shows a regular series of lines whose first member is at 89 740.46 MHz. What species is likely to produce such a simple spectrum? Calculate its bond length.

4.2 Where would the first line be expected in the microwave spectrum of ^7LiF, if the first line in the spectrum of ^6LiF occurs at 89 740.46 MHz?

4.3 The relative populations of rotational energy levels depend on the degeneracy $(2J + 1)$ and the Boltzmann factor $(\exp[- E_J/kT])$. Calculate the relative populations of the levels of ^6LiF for $J = 0, 10, 20 \ldots 70$ at a temperature of 1500 K, when kT takes the value 3.135×10^7 MHz.

4.4 The vapor of $^{138}Ba^{16}O$ shows microwave lines assigned to the $J = 0 \rightarrow 1$ transitions of molecules in the vibrational state $v = 0$ and 1 and frequencies of 18 702.04 MHz and 18 617.98 MHz. Calculate B_e and r_e for this molecule.

4.5 The only microwave line of $^{74}GeFH_3$ below 40 GHz is the $(J, K)(0, 0) \rightarrow (1, 0)$ transition at 20 023.21 MHz. Other $K = 0$ transitions are found in the millimeter wave spectrum, that with $J'' = 9$ appearing at 200 198.22 MHz. Calculate B_0 and D_J. The line with $K = 8, J'' = 9$ is at 200 029.16 MHz. Calculate D_{JK}.

4.6 Assuming bond lengths $rCH = 1.10$ Å, $rC—C = 1.40$ Å and $rCN = 1.15$ Å, calculate the moments of inertia of the cyano-acetylenes $H—(C\equiv C)_n—CN$ with $n = 1, 2, 3$ and hence calculate their rotation constants B.

4.7 Calculate rotation constants for CO_2, CS_2 and OCS, assuming bond lengths of 1.2 Å for C=O and 1.6 Å for C=S, and masses of 12, 16 and 32 amu for carbon, oxygen and sulfur. What effect does replacing ^{12}C by ^{13}C have on these rotation constants?

4.8 The relative populations of rotational levels at room temperature are governed by the degeneracy $(2J + 1)$ and the Boltzmann factor $\exp[- E_J/209]$ where E_J is in cm^{-1}. Calculate the relative populations for $J = 10, 20, 30 \ldots 100$ for a molecule with $B = 0.2$ cm^{-1}. By plotting a graph of population against J estimate the value of J for which the population is a maximum.

4.9 The first two lines in the pure rotation spectrum of $^{12}C^{16}O$, observed by millimeter wave spectroscopy, are found at 115 271 201 kHz and 230 537 974 kHz. Derive values for the rotation constant B_0 and the centrifugal distortion constant D. Calculate the effective internuclear distance in the ground state molecule, r_0.

4.10 The pure rotation transitions of O_2 from $J = 17$ to $J = 19$ levels are observed at the following frequency shifts in the Raman spectra of the various isotopic species: $^{16}O_2$ 106.1427 cm^{-1}; $^{16}O^{18}O$ 100.2399 cm^{-1}; $^{18}O_2$ 94.3867 cm^{-1}. Given that the centrifugal distortion constants for the three species are 4.85×10^{-6}, 4.7×10^{-6} and 3.84×10^{-6} cm^{-1}, respectively, calculate values of B_0 and hence r_0 for each isotopic species. Why do these values differ from each other? The values of α, the coefficient in Eqn (4.10) which allows correction of B_0 to B_e, are 0.0159, 0.0146 and 0.0133 cm^{-1}, respectively; show that the three isotopic species have values of r_e, the equilibrium internuclear distance, which are not significantly different from one another.

4.11 What information about the bond lengths and angles of the following molecules could be determined from spectroscopic measurements? In which cases would you expect to observe a pure rotation spectrum?

XeF_4O SiF_4 SCl_2O_2 SF_2O_2 SCl_2 $GeH_3C{\equiv}CH$ PF_5 CH_3SSiH_3

What other techniques would be useful in cases where rotational spectroscopy gave insufficient information to define the structure?

4.12 What do the following molecules have in common, which is relevant to the form of their rotation spectra? Sketch the pure rotation spectrum you would expect for any one of them, showing both the overall pattern for several different J values and the detail associated with one initial J value.

SiH_3I $SF_5C{\equiv}CH$ $[Ni(\eta^5{-}C_5H_5)NO]$ PCl_3

4.13 Calculate the three rotation constants for the molecule $^{14}N^{16}O_2{}^{35}Cl$. Assume $r(N{=}O)$ 1.202 Å, $r(N{-}Cl)$ 1.840 Å and angle ONO 130.6°.

4.14 Ground-state rotation constants $(B_0/$MHz$)$ for the hydrogen-bound complex, $H_3N{\cdots}HCl$, which is a symmetric top, are as follows:

Isotopic species
$^{14}N{\cdots}^{35}Cl$	4243.2593 ± 0.0016
$^{14}N{\cdots}^{37}Cl$	4168.8107 ± 0.0007
$^{15}N{\cdots}^{35}Cl$	4098.3113 ± 0.0012

Using Eqn (4.17), calculate the r_s N$\cdots$Cl distance in the ground-state molecule.

4.15 Ground-state rotation constants $(B_0/$MHz$)$ for various isotopic species of CH_3HgCN are as follows:

$^{12}CH_3{}^{200}Hg^{12}C^{14}N$	1964.836
$^{13}CH_3{}^{200}Hg^{12}C^{14}N$	1928.247
$^{12}CH_3{}^{202}Hg^{12}C^{14}N$	1964.476
$^{12}CH_3{}^{200}Hg^{13}C^{14}N$	1938.143

Calculate the two different Hg—C bond lengths.

4.16 If r_e for BrF is 1.756 Å and r_0 is 1.759 Å, calculate the positions of the lowest frequency lines in the rotational spectra of (a) ^{79}BrF in its vibrational ground state, (b) ^{81}BrF in its ground state and (c) ^{79}BrF in its first vibrationally excited state.

Further reading

Refs [1] and [2] are general accounts of spectroscopic methods that can give information about the rotations of molecules, and Refs. [3]–[6] deal specifically with rotation spectroscopy.

References

1 A. Herzberg, *Molecular Spectra and Molecular Structure*, Van Nostrand, New York: Vol. I, *Diatomic Molecules*, 2nd edn (1950); Vol. II, *IR and Raman Spectra of Polyatomic Molecules* (often known as *IR and Raman Spectra*) (1945); Vol. III, *Electronic Spectra of Polyatomic Molecules* (1966).
2 J.M. Hollas, *High Resolution Spectroscopy*, Butterworths, London (1982).
3 C.H. Townes and A.L. Schawlow, *Microwave Spectroscopy*, McGraw-Hill, New York (1955).
4 T.M. Sugden and C.N. Kenney, *Microwave Spectroscopy of Gases*, Van Nostrand, London (1965).
5 W. Gordy and R.L. Cook, *Microwave Molecular Spectra*, Interscience, New York (1970).
6 H.W. Kroto, *Molecular Rotation Spectra*, Wiley, London (1975).
7 J. Harper and J.L. Duncan, *J.Mol.Spectrosc.*, **100**, 343 (1983).
8 H. Bürger, H. Beckers and J. Kauppinen, *J.Mol.Spectrosc.*, **108**, 215 (1984).
9 W.J. Balfour and R.S. Ram, *J.Mol.Spectrosc.*, **105**, 246 (1984).
10 S. Cradock and J.G. Smith, *J.Mol.Spectrosc.*, **98**, 502 (1983).
11 Y. Yamamoto, T. Nakanaga, H. Takeo, C. Matsumura, M. Nakata and K. Kuchitsu, *J.Mol.Spectrosc.*, **106**, 376 (1984).
12 J.L. Duncan and J. Harper, *Mol.Phys.*, **51**, 371 (1984).
13 J.R. Durig, C.M. Whang, G.M. Attia and Y.S. Li, *J.Mol.Spectrosc.*, **108**, 240 (1984).
14 J. Vogt and M. Winnewisser, *Ber.Bunsenges.phys.Chem.*, **88**, 448 (1984).
15 J. Kraitchman, *Am.J.Phys.*, **21**, 17 (1953).
16 C.C. Costain, *J.Chem.Phys.*, **29**, 864 (1958).
17 P.D. Blair, S. Cradock and D.W.H. Rankin, *J.Chem.Soc., Dalton Trans.*, 755 (1985).
18 J.A. Duckett, A.G. Robiette and I.M. Mills, *J.Mol.Spectrosc.*, **62**, 34 (1976).
19 M. Kreglewski and P. Jensen, *J.Mol.Spectrosc.*, **103**, 312 (1984).
20 L. Fusina, I.M. Mills and G. Guelachvili, *J.Mol.Spectrosc.*, **79**, 101 (1980) and references therein.
21 L.H. Scharpen and V.W. Lawrie, *J.Chem.Phys.*, **49**, 221 (1968).
22 P.A. Baron and D.O. Harris, *J.Mol.Spectrosc.*, **49**, 70 (1974).
23 A.P. Cox and R. Varma, *J.Chem.Phys.*, **44**, 2619 (1966).
24 J. Pfab, *Chem.Phys.Letters*, **99**, 465 (1983).

5 Vibrational Spectroscopy

5.1 Introduction

Vibrational spectroscopy has been extensively used by inorganic chemists since about 1940, particularly with the development of infra-red (IR) and Raman spectrometers that are reliable and easy to use. It is uniquely wide in its applicability: every molecular or covalently bonded sample has some form of vibrational spectrum, and it is easy to study gases, liquids, solids and solutions. Spectra can even be obtained under extreme conditions of low or high temperature or pressure, and by various subtle means it is also possible to study short-lived species.

Above all, IR spectrometers are almost universally available, and so IR spectroscopy is one of the first techniques the inorganic chemist will call upon when studying a new sample. Raman spectroscopy is also important, even though it is less widely used. In this Chapter we show that the range of information which may be obtained from the use of the more common techniques is very wide; we also explain the ways in which less conventional sources of information about molecular vibrations may provide valuable data.

5.2 The physical basis; molecular vibrations

5.2.1 Vibrational motions and energies

It is possible to use vibrational spectroscopy as a tool for identifying and measuring concentrations of samples without needing to know anything about how or why the spectra arise. However, much more information can be obtained if we have a reasonable level of understanding of this physical basis. The basic picture we need is a purely mechanical one. We regard a molecule as composed of massive particles (atoms) joined by connections which are much lighter (ideally mass-less) (Fig. 5.1). We can identify the connections with the bonding electrons. The motions of such a body can be

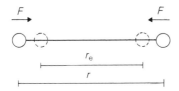

Fig. 5.1 Vibrating diatomic molecule showing the equilibrium positions of atoms as dashed circles, instantaneous positions as solid circles, and restoring forces F.

calculated classically using Newton's Laws of Motion, if we assume for the moment that the restoring forces in the connections are of the form

$$F(x) = -kx \tag{5.1}$$

where x is a displacement from equilibrium. The internal motions are those that involve no overall translation or rotation of the molecule. They can be resolved into various oscillations about the equilibrium structure. These are the characteristic or *normal vibrations* of the molecule, each with a characteristic vibration frequency

$$\omega_i = \frac{1}{2\pi}\sqrt{\frac{k_i}{\mu_i}} \tag{5.2}$$

where k_i is the force constant [see Eqn (5.1)] and μ_i a reduced mass for the particular motion. Classically, such motions can involve any amount of energy, but because of the atomic scale involved the energy is quantized. The quanta of vibrational energy ε_i are related to the classical vibration frequency by

$$\varepsilon_i = h\omega_i \tag{5.3}$$

where h is Planck's Constant.

Because of the Uncertainty Principle, it is not possible for a molecule to have zero vibrational energy; if it had, we could specify both the positions and the momenta of the atoms precisely. All molecular vibrations therefore have some zero-point energy, and this can be shown to be equal to one half of a quantum, so that the total vibrational energy of a molecule with only one normal vibration is

$$E_v = h\omega(v + 1/2) \tag{5.4}$$

where v is the vibrational quantum number, a positive integer or zero.

5.2.2 Number of vibrational motions

A set of N atoms moving in 3-dimensional space has $3N$ degrees of freedom. This is true both for independent atoms and for an N-atom molecule. For all but linear molecules, external molecular motions − 3 translations and 3 rotations − account for 6 of these. This leaves $3N − 6$ vibrational motions (or modes of vibration). A linear molecule has only two rotations, and so it has $3N − 5$ vibrational modes. Thus a

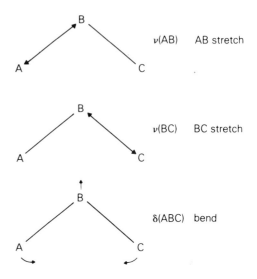

Fig. 5.2   Three vibrational modes of a bent triatomic molecule ABC.

diatomic molecule ($N = 2$), which must of course be linear, has only one mode of vibration, the bond-stretching motion. A bent triatomic molecule ABC has $3 \times 3 - 6 = 3$ modes (Fig. 5.2). It is also useful to note that, in general, a molecule has one bond stretching mode per bond, although in systems with rings or cages the stretching motions are not always described so simply.

5.2.3   Non-ideal restoring forces; anharmonicity

So far we have assumed a simple (Hooke's Law) restoring force which is proportional to displacement [Eqn (5.1)]; this implies that the potential energy V has the form

$$V(x) = 1/2\,k\,x^2. \tag{5.5}$$

This represents a quadratic potential function. It can be analyzed in two different ways: we can either use classical mechanics, or we can use the Schrödinger equation, which is the wave-mechanical equivalent. In either case, we find that the motion is simple harmonic, with fixed frequency ω and equally-spaced energy levels as described by Eqn (5.4) and shown in Fig. 5.3(a). However, Function (5.5) cannot possibly be a

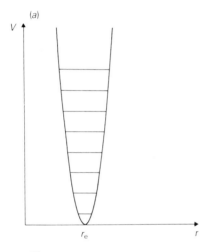

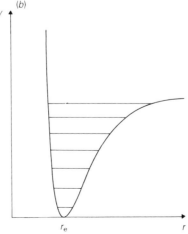

Fig. 5.3 Potential wells for diatomic molecules: (a) parabolic potential and (b) Morse potential.

true representation of the potential energy associated with a real displacement $x = r - r_e$ of a bond between two atoms. It is obvious that as r decreases below the equilibrium bond distance and the atoms 'penetrate' each other, the potential energy must increase dramatically so that some singular behavior (in the mathematical sense) must occur before r reaches zero. On the other hand, as r increases, the potential energy does not increase without limit; it must approach asymptotically a finite value, the dissociation energy. This is the energy required to separate the atoms or groups of atoms at either end of the bond by a great distance, so that there is no bonding between them.

A reasonable approximation to the expected behavior of a 'real' bond is given by the Morse function,

$$V(x) = D_e[1 - \exp(-ax)]^2 \tag{5.6}$$

where D_e is the dissociation energy and a is a constant. The Morse function yields vibrational energy levels given by Eqn (5.7), and these have the advantage that they are closely related to those for the simple harmonic case [Eqn (5.4)]

$$E_v^{\text{Morse}} = h[\omega_e(v + 1/2) - x_e(v + 1/2)^2]. \tag{5.7}$$

The energy levels for a Morse function are illustrated in Fig. 5.3(b).

It can be shown that any function capable of representing the variation of potential energy with displacement will lead to energy levels representable by a power series in $(v + 1/2)$

$$E_v^{\text{general}} = h[\omega_e(v + 1/2) - x_e(v + 1/2)^2 - y_e(v + 1/2)^3 - \ldots]. \tag{5.8}$$

The simple harmonic and Morse oscillators are thus particular cases of this general anharmonic oscillator for which the power series has been truncated after the first and second terms, respectively.

For the anharmonic oscillator, then, the vibrational energy levels associated with a particular vibration are not equally spaced. The relative magnitudes of the coefficients of the power series vary with the sort of vibration. For bond stretches involving heavy atoms, where $h\omega_e$ is of the order of a few hundred cm^{-1}, hx_e may be of the order of 1 cm^{-1} or less. On the other hand, for bond stretches involving hydrogen atoms the anharmonic term is often larger in proportion, up to 100 cm^{-1} compared with $h\omega_e = 2000$-3500 cm^{-1}. For angle bending motions anharmonicities are often small, and may be negative, in which case the level separations increase as v increases. We shall meet other consequences of anharmonicity later.

Note that two forms of Eqn (5.7) will be found in other texts: one uses x_e (as here) and the other uses $\omega_e x_e$ as the coefficient of $(v + 1/2)^2$.

5.3   Observing molecular vibrations

5.3.1   Absorption in the infrared

We know that chemical bonds involve electrostatic attraction between positively charged atomic nuclei and negatively charged electrons. Displacements of atoms during a molecular vibration therefore lead to distortions in the electric charge distribution of the molecule, which can be resolved into dipole, quadrupole, octopole, *etc.* terms in various directions. Molecular vibrations therefore lead to oscillations of electric charge, with frequencies governed by the normal vibration frequencies of the system. An oscillating molecular dipole, for instance, can then interact directly with the oscillating electric vector of electromagnetic radiation of the same frequency, leading to resonant absorption of radiation. The quantum energy of the radiation ($h\nu$) is equal to the quantum energy of the oscillator ($h\omega$) if $\nu \equiv \omega$, so that resonant absorption introduces exactly enough energy to raise the oscillator from a level v to a level $(v+1)$ [Fig. 5.4(a)]. Electric dipole interactions are thus responsible for absorption of radiation in the IR region of the spectrum. In this range the frequencies are comparable with those of molecular vibrations, which are generally between 10^{11} and 10^{13} Hz, corresponding to 30-3000 cm^{-1}.

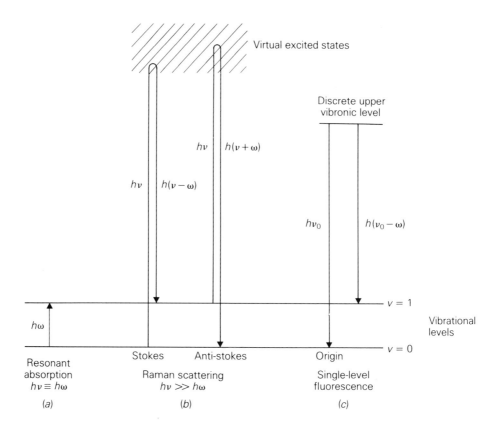

Fig. 5.4 Energy levels involved in (a) resonant absorption (IR spectroscopy), (b) Raman scattering and (c) single-level fluorescence.

It is important to realize that not all molecular vibrations lead to oscillating dipoles. We shall see in Section 5.7 how arguments based on symmetry may be helpful in deciding which vibrations will interact with IR radiation, and which will not. Interactions with oscillating quadrupoles, *etc.* are also possible, but are so much weaker than dipole interactions that they can usually be ignored. For magnetic species, such as free radicals, and some transition metal complexes, oscillating magnetic dipoles may in principle interact with the magnetic vector of the radiation, but these interactions too are comparatively weak.

| 5.3.2 | **Raman scattering** |

IR spectra arise when radiation of energy corresponding to vibrational transitions in a molecule is absorbed. But there is another way in which molecular vibrations can be excited by radiation, a way which does not involve direct or resonant absorption. The radiation involved is normally of frequency much greater than that of any molecular vibration. When radiation of frequency ν interacts with a molecular vibration of frequency ω, one quantum of the incident radiation may disappear, with creation of a quantum of frequency $(\nu \pm \omega)$. We may think of the initial photon as being absorbed, so that the molecule is excited to a 'virtual' state, whose lifetime is so short that there is effectively immediate re-emission of a photon whose energy may be different from $h\nu$ by a quantum of vibrational energy. The process, which is a very weak one, is illustrated in Fig. 5.4(b). Conservation of energy requires a corresponding change in the vibrational energy of the molecule. The resulting spectrum is called a Raman spectrum: it contains lines whose frequencies differ from that of the incident radiation by the frequencies of molecular vibrations. The lines to low frequency of the incident radiation are called Stokes lines, and those at higher frequency are called anti-Stokes lines; the latter arise from vibrationally-excited states. Scattered radiation of lower frequency is in general more important, because the proportion of molecules in excited vibrational states is governed by the Boltzmann distribution [Eqn (5.9)], and will be lower than that in the lowest state.

$$N_v/N_0 = \exp\left(-\frac{E_v - E_0}{kT}\right) \tag{5.9}$$

A conventional vibrational Raman spectrum therefore consists of a series of Stokes emissions separated in frequency from the exciting radiation by the various vibration frequencies of the sample.

The Raman scattering process is clearly not a resonant interaction, and it results in comparatively weak spectra. The interaction comes about through the polarization of the electron cloud of the molecule by the oscillating electric vector of the incident quantum (Fig. 5.5). The important molecular property connected with Raman spectra is therefore the polarizability, and a vibration will give rise to a Raman line if it leads to a change in the polarizability of the molecule. We shall see in Section 5.7 that symmetry considerations allow us to decide which vibrational modes will lead to Raman scattering.

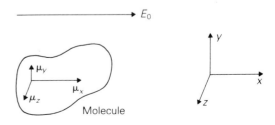

Fig. 5.5 x, y and z components of the dipole moment induced in a molecule by an external electric field E_0 parallel to the x-axis. They are related to the components of the polarizability tensor α_{xx}, α_{xy} and α_{xz}, respectively.

CHAPTER 5: VIBRATIONAL SPECTROSCOPY

Information from electronic spectra

Electronic spectra often contain information about the vibration frequencies of both the lower and the upper electronic states of the species involved, as described in Chapter 6. This is particularly true if an emission spectrum can be observed: absorption spectra arise mainly from the vibrational ground-state because of the Boltzmann distribution of population, unless the sample is already excited in some other way. Much detailed information about the vibration frequencies of different electronic states of diatomic molecules and some simple polyatomic molecules has been obtained in this way. For larger molecules it is not possible to disentangle the information about the electronic ground state from that relating to the excited state. In recent years, however, the introduction of tunable lasers as excitation sources has allowed laser-induced fluorescence spectra to be obtained. In these experiments, one specific vibronic (vibrational/electronic) level of the upper state is excited; fluorescence can then occur

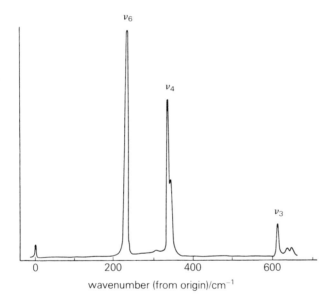

Fig. 5.6 Induced emission (luminescence) spectrum of Cs_2MnF_6 at 5 K. The origin of the band is at the left, and the peaks are assigned to three of the internal vibrations of the MnF_6^{2-} ion, with small splittings in some cases due to the crystal lattice. Redrawn from 'The Extraction of Vibrational Information from Electronic Spectra of Transition Metal Complexes', C.D. Flint, *Advances in Infrared and Raman Spectroscopy*, **2**, 53 (1976).

to many different vibrational levels of the electronic ground state, as shown in Fig. 5.4(*c*). Ground-state vibration frequencies are obtained as shifts from the exciting frequency or from the vibrational origin of the electronic band. There is then no confusion between ground-state and excited-state information. The selection rules depend on the symmetry properties of the electronic transition. Much of the work reported so far with this technique involves organic samples, but the application to inorganic species should be a useful addition to other sources of information. The method is applicable to gases and crystals, but vibrational relaxation in liquids tends to be too efficient to preserve the simplicity inherent in fluorescence from a single excited vibronic level. The technique is discussed in some detail in Ref. [1], p. 395, and an example of a spectrum showing fluorescence from a crystalline sample is shown in Fig. 5.6.

Inelastic scattering of neutrons and electrons

We have just seen that a non-resonant photon will normally leave an encounter with a molecule unchanged in energy, but it may very rarely lose energy to the molecule and emerge as a Raman-shifted photon of lower energy. In the same way, we may think of the ideal encounter of a particle with a molecule as an elastic collision in which no energy is transferred. However, rare inelastic collisions may also occur, resulting in the transfer of specific vibrational quanta of energy from the particles to the molecules. We may, therefore, take a beam of neutrons or electrons, all having the same, known velocity, and pass it through or reflect it from a sample. Analysis of the velocity distribution will reveal a 'vibrational spectrum' in the form of a spectrum of energy-losses by the particles. Any particle could in principle be used; most results have been obtained with neutrons or electrons, each of which has various advantages and disadvantages for particular samples.

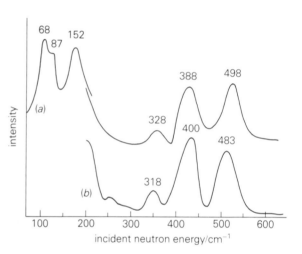

Fig. 5.7 Inelastic neutron scattering spectra of (a) Ni(π-C$_3$H$_5$)$_2$ and (b) Pd(π-C$_3$H$_5$)$_2$. The bands of the Ni compound found at 152, 328, 388 and 498 cm^{-1} have been observed in either IR or Raman studies, but the two peaks at lowest frequency have not. These are assigned to torsional motions of the ligands, which are very weak in the Raman effect, though formally allowed. Redrawn from J. Howard and T.C. Waddington, *J.Chem.Soc., Faraday II*, **74**, 879 (1978).

Neutrons interact comparatively weakly with matter, but Inelastic Neutron Scattering (INS) is capable of revealing both rotational and vibrational transitions in solid samples. There are effectively no vibrational selection rules, so the technique is particularly useful for detecting vibrations that do not appear in either IR or Raman spectra. The spectra of π-allyl complexes of nickel and palladium shown in Fig. 5.7 include some bands seen only in the IR spectra, some seen only in the Raman spectra, and some seen in neither. More information about the technique and its applications can be found in Ref. [2].

Electrons interact much more strongly with matter. Consequently, gases can be studied, but in most cases only electronic excitations are observed. The technique has been extensively used to study vibrations associated with surfaces and thin film samples, usually in an electron microscope. Vibrational excitation has only been reported for a few simple molecules in the gas phase, as shown in Fig. 5.8, and vibrational structure may also be seen on some bands arising from electronic transitions. Applications of the technique, commonly known as Electron Energy Loss Spectroscopy (EELS), to studies of solid surfaces are beyond the scope of this book. The subject is discussed more fully in Refs [3] and [4].

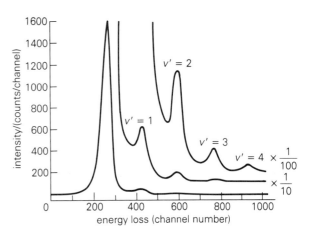

Fig. 5.8 Electron energy loss spectrum of gaseous N_2, showing peaks due to the excitation of a few vibrational quanta. The energy scale is 1.75 meV per channel, and the large peak between channels 200 and 300 represents elastic scattering (*i.e.* no energy loss). Adapted, with permission, from Ref. [4].

5.4 Spectrometers

For many years the standard IR spectrometer has been the dispersive scanning spectrophotometer. Non-dispersive interferometers, though available, were not widely used prior to the advent of modern computer technology. In the past few years interferometers have displaced scanning spectrophotometers in many applications. A third type of spectrometer, fundamentally simpler than the other two, uses a tunable monochomatic source; instruments using IR-emitting diodes have been produced, but are not yet able to replace interferometers.

Dispersive instruments use prisms or gratings to separate radiation of different frequencies, by refraction or diffraction. Photographic recording of IR radiation is extremely difficult, so scanning spectrophotometers are used instead. However, only one resolution element is detected at once, rather than all being recorded simultaneously as in a spectrograph; this makes the scanning spectrophotometer an inherently inefficient device. Nevertheless, despite its mechanical complexity, its limited energy throughput and its inefficiency, it has dominated the market until comparatively recently. These cheap, reliable and effective instruments have now been challenged by the alternative type, the interferometer.

In the interferometer no dispersing element is used, and no separation of different wavelengths is necessary. All frequencies are measured simultaneously, by recording an interferogram, from which the spectrum is then obtained by Fourier transformation (see Section 1.5). Modern interferometers covering the mid-IR region (4000-200 cm^{-1}) include fast mini-computers required for the Fourier transformation, and are almost as easy to operate as scanning spectrophotometers. The availability of the computer means that spectrum manipulation, storage and retrieval for replotting or comparison with standards are all easy, and this may be felt to justify the higher cost of interferometers. However, the cost differential is rapidly diminishing, and as some of these benefits can be obtained simply by the addition of a micro-computer to a spectrophotometer, the choice between spectrophotometer and interferometer is becoming more and more a matter for the individual spectroscopist. Each type of instrument has its particular advantages and disadvantages, and it seems likely that both types will continue to be available for the foreseeable future.

A basic Raman spectrometer consists of an intense source of exciting radiation, usually a laser, operating in the visible region on a single frequency, with a visible-light spectrophotometer to measure the Raman-shifted emission from the sample. The sensitivity of the instrument may be increased by increasing the power of the laser or by decreasing the level of stray light in the monochromator; double- or even triple-grating monochromators are generally used to ensure that the Raman signals, often as weak as 10^{-6} or less of the intensity of the exciting radiation, are not swamped by unshifted radiation by-passing the monochromator. All radiation in the spectrum lies in the visible region, so glass optics can be used, and this offers an important advantage over the IR region, where glass is unusable, and mechanically weaker salt optics or front-silvered mirrors must be employed.

Recently it has been shown that Raman spectra may be obtained from a wide variety of samples using a combination of an interferometer capable of operating in the near-IR region and a near-IR laser. The advantages include those associated with the interferometer (see above) and the associated computer, those associated with the high-power directed beam of the laser, which may also be operated in pulses rather than continuously, and also the virtual elimination of fluorescent background signals because of the long wavelength of the laser radiation.

Further information about scanning IR spectrophotometers, interferometers and Raman spectrometers may be found in Refs [5], p. 17; [6], p. 9; and [7], p. 132, respectively.

5.5 Sample handling

5.5.1 Infrared

One of the factors responsible for the wide use of IR spectroscopy is the ease with which various samples may be studied. Glass and silica are opaque in the IR region below 4000 cm^{-1}, so that other materials must be used as windows for sample cells. For most purposes the salts NaCl and KBr are adequate. These can be used down to 600 and 350 cm^{-1}, respectively; for lower frequencies, CsI can be used down to 200 cm^{-1} but it is softer and much more expensive. Spectra of these typical window materials are shown in Fig. 5.9. All these materials are water-soluble, but this is only rarely a limitation, because intense IR absorption by H_2O itself makes it very difficult to use IR spectroscopy to study aqueous systems.

Samples are most readily studied as liquids or solutions. Thin films of liquid between two plane-faced salt windows are simplest to prepare, and a variety of designs of spaced-window cells, with fixed or variable sample thickness, are commercially available. The effects of solvent absorption in a solution may be partially compensated by comparing the transmittance of matched cells, one containing solution and the other pure solvent, ideally in a double-beam spectrometer. However, it is often necessary to use several solvents so that regions obscured by bands of one are free of interference in the spectrum of another.

Large solid samples are also easy to study, though the strength of the IR absorption may require the preparation of thin sheets of crystalline or amorphous materials. Powders are slightly less easy to handle, and coarse, irregular solid samples give great problems because of scattering of IR radiation by solid particles of sizes comparable

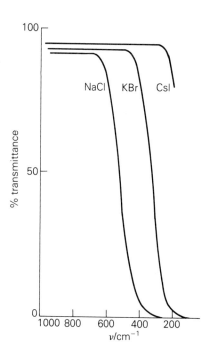

Fig. 5.9 Low frequency IR spectra of window materials.

with the wavelengths of the radiation (2-40 $\times$ 10^{-6} m). Both powders and coarser materials are usually finely ground, and then intimately mixed with a liquid whose refractive index is close to that of the solid. The resulting mull, which should be optically clear, is usually examined as a thin film, pressed between two salt windows. A heavy paraffin fraction (Nujol) is often used as mulling agent, and its few absorption bands can usually be readily distinguished from those of the sample (Fig. 5.10). Ideally the regions where these interfering signals occur should be covered using a second

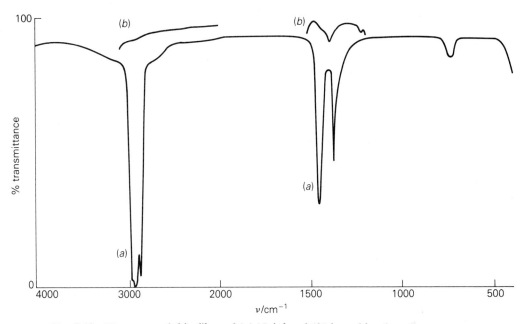

Fig. 5.10 IR spectra of thin films of (*a*) Nujol and (*b*) hexachlorobutadiene.

SECTION 5.5: SAMPLE HANDLING

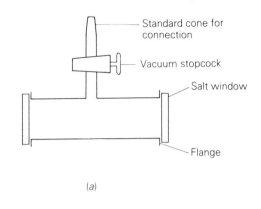

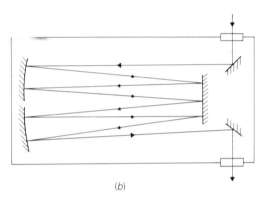

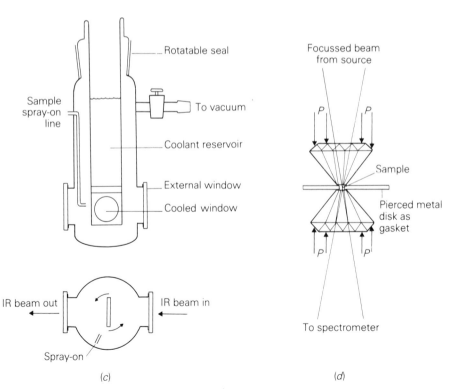

Fig. 5.11 Some arrangements for IR sampling: (*a*) gas cell, (*b*) long path gas cell, (*c*) apparatus for low temperature studies of solids, and (*d*) diamond pressure cell.

sample prepared with a mulling agent that has no absorptions in these regions, such as hexachlorobutadiene.

An alternative form of sample for solids is the pressed disk. This may be of pure sample, but more often it is diluted with a salt, usually KBr. The resulting disk needs no windows, and is also conveniently stored for future reference. Scattering may be a problem unless the sample is very finely ground. Another problem is that reactions such as halide exchangé may occur during disk preparation, and it is sometimes necessary to evacuate the die during pressing to prevent oxidation of samples under the extreme pressure.

Gases require the most elaborate sampling arrangements. For most purposes, glass tubing cells with flat salt windows sealed to flanged ends [Fig. 5.11(a)] are used. Care must of course be taken with gases that may react with the glass, salt windows, or sealants. For samples with very low vapor pressures, or for investigation of minor components of a gas stream, a greater path-length is helpful, and this may best be achieved using a folded-path cell with internal mirrors to direct the IR beam several times through the sample. Various cells, such as that shown in Fig. 5.11(b), give paths of up to 20 m.

Sometimes samples need to be heated, cooled, compressed, treated with gases, irradiated and so on. The apparatus required for these experiments usually includes salt windows, and is designed to fit the sample compartment of a commercial IR spectrometer. One important experimental arrangement of this type, shown in Fig. 5.11(c), allows samples to be condensed at very low temperature (10-20 K) on a salt window. This makes it possible to study samples isolated at high dilution in an inert matrix of a solid such as N_2 or Ar. We shall see in Section 5.16.3 that such studies can be important even for stable substances, and can allow spectroscopic identification of species such as free radicals and other reactive fragments that are normally short-lived.

It is also possible to record vibrational spectra of solid samples under high and varying pressure. Under these conditions many solids undergo phase changes, and the nature of the accompanying changes in structure can be studied. Salt windows are generally too weak mechanically to be included in high-pressure apparatus, and the cell shown in Fig. 5.11(d) has diamond windows. Needless to say, small samples are usually used, and additional optics are needed to concentrate the beam of the IR spectrometer. In such experiments the signal-to-noise ratio tends to be very low, and interferometers, which allow repeated scanning, are more useful than spectrophotometers.

5.5.2   Raman

As glass tubes are ideal for obtaining Raman spectra, sampling of liquids is much easier than for IR spectroscopy. Moreover, one important advantage of Raman over IR spectroscopy is that water has very weak Raman scattering, and vibrational spectra of aqueous solutions are almost always studied most readily using Raman techniques. The small size of the laser beam also makes it easy to investigate small solid samples, which generally require no container unless protection from air is needed. For gases, however, the weakness of the Raman effect combined with low sample density makes spectra hard to obtain without special care. It is almost essential to use a small cell with internal mirrors to give multiple traversal of the sample by the laser beam, and the highest attainable laser power should be employed. Typical arrangements for recording Raman spectra of liquids, solids and gases are shown in Fig. 5.12.

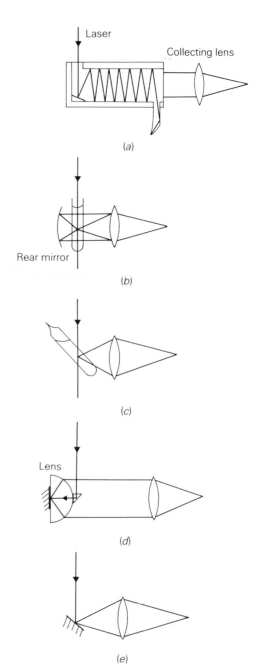

Fig. 5.12 Some arrangements for Raman sampling: (*a*) gas cell with internal reflecting surfaces, (*b*) and (*c*) liquid samples, (*d*) liquid or solid sample using hemispherical lens, and (*e*) solid sample.

Problems arise in recording Raman spectra of fluorescent samples, as the fluorescence is often much more intense than the Raman signal. If this behavior is inherent in the sample, very little can be done about it except to change the exciting wavelength to one that reduces the effect. But if it is due to a small proportion of an impurity it can usually be dealt with by careful purification or by exposing the sample to the laser beam for some time (a process called photo-bleaching). A related problem is experienced with colored samples. If an appreciable proportion of the laser radiation is absorbed by the sample, local heating will usually result in decomposition at the point of irradiation if the sample is a solid, though efficient convective mixing will often prevent

CHAPTER 5: VIBRATIONAL SPECTROSCOPY

overheating in liquid samples. One way of overcoming this is by spinning a large-area sample of the solid in the laser beam, so that any individual spot is irradiated for a short time only, and is able to cool before being brought back into the beam.

5.6   Effects of phase on spectra

We shall see in Section 5.8.3 that changes in rotational quantum numbers accompany the vibrational transitions for gases. At high resolution these result in fine structure; at lower resolution, various characteristic band envelopes may be observed. We only find these in the spectra of gases; rotation is not quantized in condensed phases, and the characteristic rotational envelope of a gas-phase band is absent. This has some advantages, as the rotational envelope results in a spreading of the intensity of a band over a

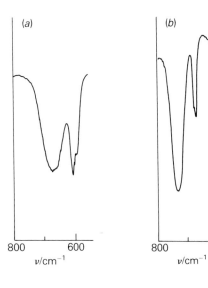

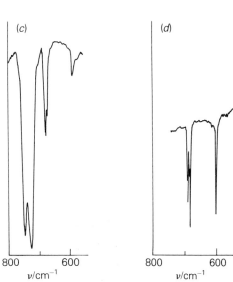

Fig. 5.13   Parts of IR spectra of a sample (*a*) as a gas, (*b*) as an amorphous solid, (*c*) as a crystalline solid and (*d*) isolated in a matrix of solid argon. In the gas phase the bands are broad, but have distinctive shapes. The amorphous solid gives fairly broad bands (as does the liquid), but the crystalline solid has sharper bands, and splittings, which depend on the symmetry in the crystal, are observed. The matrix spectrum has very sharp lines, with small splittings, which are effects of the matrix. The frequencies for the gas and matrix are more or less the same, but the bands are shifted in other phases.

wide range, and may lead to overlapping of bands due to different vibrations. A weak band close to a strong one may become difficult to detect in the gas phase, unless its own rotational envelope contains a prominent feature such as a Q-branch close to the band origin. In condensed phases bands are often narrower, so that nearby vibrations are less likely to be obscured, while the peak heights are greater than in the gas phase. This effect is at its most extreme in matrix-isolated samples, where very sharp pure vibrational peaks are usually observed. Modes very close together may then be distinguished, and useful spectra are obtainable with very small samples. Fig. 5.13 illustrates how the spectrum of a sample may depend on its phase.

In liquids and solutions there are several factors leading to broadening of the vibrational band, even in the absence of rotation envelopes. In neat liquids and solutions there is little short-range order, and at any instant different molecules are under the influence of randomly-differing local environments. The random differences in environment result in a range of different vibration frequencies, and so each vibration band is broadened. This broadening can also arise from the time-variation of the local environment of a typical molecule, and it is possible to use measured bandwidths to probe the timescale of fluctuations of local environment. Sometimes specific intermolecular complexes are formed, either between two sample molecules or between sample and solvent molecules. For instance, hydrogen-bond formation in liquid water lowers the OH stretching frequencies by several hundred cm^{-1}, and it would be quite unrealistic to use the liquid water frequencies as characteristic of the isolated triatomic molecule.

In glassy solids the effects of random differences in local environment are even greater than for liquids; where fairly strong interactions occur, some bands may be very broad indeed. Even in the same sample, however, other modes whose frequencies are insensitive to the local environment may well give quite narrow bands, and glasses (best formed by condensation from the gas phase onto a cold surface) should not be ignored as possible samples for the study of vibrational spectra.

In crystalline solids the effects of randomness of environment and of quantized rotation are absent, and very sharp vibration bands may be observed, especially in the Raman effect. In the IR, problems arise because of the effects of surface reflection; these reach a maximum in the vicinity of an absorption band, and distort its apparent profile (the Christiansen effect). The preparation of the samples, by fine grinding, must be undertaken very carefully, if high quality spectra are to be obtained.

The spectra of solid samples often contain more bands than expected on the basis of the molecular symmetry of single, isolated molecules. These arise because the symmetry of the free molecule is reduced to that of the environment, which must be very low in the case of a random glassy solid, and may be low even in a crystalline sample. This effect of site symmetry in a crystal is then compounded with the effects due to the presence of more than one molecule in the unit cell, which is the basic vibrating unit in a crystal. If there are m molecules in the unit cell, each of the internal motions of the molecule will give rise to m components. The activities of these components depend on their symmetries just as the activities of internal vibrations of the molecules are related to their symmetries (see Section 5.7).

The effects of low site symmetry are the same as those to be expected from distortion of the symmetrical molecule to a less-symmetrical form. First, modes of the symmetrical form that are inactive in a particular spectrum (IR or Raman) may become active in the

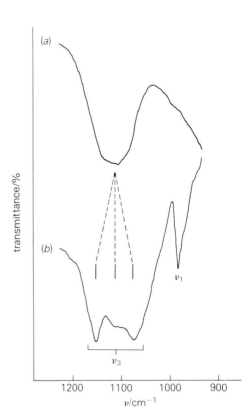

Fig. 5.14 IR spectra of SO_4^{2-} (a) in solution and (b) in crystalline $MgSO_4 \cdot 7H_2O$. The mode which is inactive in solution (ν_1) becomes active in the solid, while the triple degeneracy of ν_3 is lifted.

lower symmetry of the site. For example, the totally-symmetric stretching mode of the sulfate ion is IR inactive in an environment of high symmetry, as in solution; it becomes active in crystalline phases in sites of lower symmetry. Secondly, degeneracy may be removed by reduced site symmetry, so that two (or three) distinct vibration bands occur rather than one. For the sulfate ion, the triply-degenerate modes can show splitting into two or three components. These features are illustrated in Fig. 5.14. Site effects are well-defined in crystals, but are by no means confined to regular solids, and may be observed even in melts, as in molten mixtures of alkali sulfates, where the random distribution of M^+ ions around the sulfate anions may constitute a low-symmetry environment [8].

A further complication in the case of a crystalline solid is the possible appearance of lattice modes, which involve motions of whole molecules rather than of their constituent atoms. A full factor group analysis of the motions of the entire cell is required to account for these, and we do not go into this process here. Full details are given in Ref. [9]. The modes are usually of rather low frequency, and their effect can often be ignored except in the far IR region and at low frequencies in Raman spectra. It is possible for such lattice modes to combine with higher-frequency modes giving weak additional features near strong fundamental bands, or near the expected positions of weak or missing fundamentals. These can complicate the assignment of spectra of crystals, and care is needed in deciding on the numbers of active fundamentals in crystal spectra for this reason.

5.7   Vibrational spectra and symmetry

5.7.1   Fundamental vibrational selection rule

We have seen in Section 5.2 that vibrational energy is quantized, with the result that each vibrational mode has a characteristic frequency. The possible energies associated with a vibration are $h\omega(v + 1/2)$, where v, a quantum number, is a positive integer or zero. For the simple forms of interaction of radiation or particles with molecular vibrations, the quantum numbers change only by ± 1, and so the vibrational spectrum consists of a set of signals at frequencies given by

$$h\omega\left[(v + 1 + 1/2) - (v + 1/2)\right] = h\omega. \tag{5.10}$$

These signals correspond precisely to the characteristic vibrations of the molecule. If $v = 0$ they are called fundamental bands. As levels with $v > 0$ are normally not significantly populated, these bands account for most of the intensity in an absorption spectrum. The IR absorption spectrum and the Stokes Raman scattering spectrum both correspond to $v \to v + 1$ transitions, so the fundamental vibrational selection rule is normally stated as $\Delta v = +1$. We should remember that $\Delta v = -1$ is also permitted, but that this corresponds to the emission of radiation in the IR or to the anti-Stokes lines in the Raman spectrum, which will be weak, as the population of molecules with $v > 0$ will be small.

The main exceptions to the $\Delta v = \pm 1$ selection rule occur in electronic spectra (see Section 6.6), and in the resonance Raman effect (Section 5.17), where progressions due to transitions with $\Delta v = 0, 1, 2, 3 \ldots$ may be observed.

5.7.2   Symmetry selection rules

In vibrational spectroscopy we use electromagnetic radiation to investigate the transitions between molecular vibrational levels. But there are some molecular vibrations which cannot interact with this radiation. For instance, the stretching motion of a homonuclear diatomic molecule such as N_2 generates no oscillating dipole, and so direct interaction with the oscillating electric vector by the normally-dominant dipole mechanism is impossible. This vibration is therefore inactive in the IR spectrum. Similarly, some modes of some molecules may involve no change in polarizability, and so they are Raman-inactive. It is useful to employ the symmetry properties of molecules and molecular vibrations in dealing with this matter, and we may state the selection rules for infra-red activity and for Raman activity of molecular vibrations in symmetry terms as follows:

1   a molecular vibration will give rise to a fundamental band in the IR spectrum if, and only if, the symmetry of the vibration is the same as that of one or more of the vectors x, y, and z [see Fig. 5.15(a)].
2   a molecular vibration will give rise to a fundamental band in the Raman spectrum if, and only if, the symmetry of the vibration is the same as that of at least one component of the polarizability tensor, α, of the molecule [see Fig. 5.15(b)].

We must therefore consider first how to determine the symmetry of a vibration, and secondly what the consequences of these symmetry selection rules may be. If you are not familiar with the use of symmetry please read the Appendix before continuing.

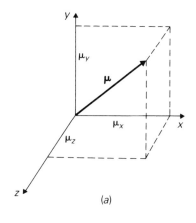

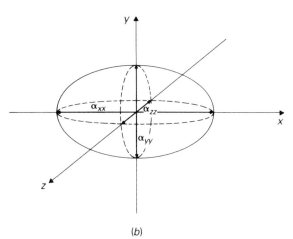

Fig. 5.15 (*a*) Dipole μ (a vector quantity), showing its components along three Cartesian axes, *x*, *y* and *z*. (*b*) Polarizability ellipsoid (a tensor quantity), showing its three major axes, α_{xx}, α_{yy} and α_{zz}.

5.7.3

Symmetry of vibrations

The symmetry of any property of a molecule may be determined by seeing how it behaves when operated on by the various symmetry elements that make up the overall symmetry point group of the molecule (see Appendix). For simplicity, this behavior is best described in terms of the characters $\chi_v(R)$ of the vibration (v) with respect to each of the symmetry elements (R). The row of characters for a particular vibration, Γ_v, is called a representation, and it should be identical with one of the irreducible representations of the point group. These express all the possible behaviors of attributes of molecules of the particular point group involved, and are collected in the character table. The symmetry of the vibration can then be read from the character table, as a conventional symbol (see Appendix) for the symmetry species.

For example, consider the molecule H_2O. This has symmetry elements E, $C_2(z)$, $\sigma_v(xz)$ and $\sigma_v(yz)$, making up the point group C_{2v}, whose character table is:

C_{2v}	E	$C_2(z)$	$\sigma_v(xz)$	$\sigma_v(yz)$		
a_1	$+1$	$+1$	$+1$	$+1$	z	x^2, y^2, z^2
a_2	$+1$	$+1$	-1	-1	R_z	xy
b_1	$+1$	-1	$+1$	-1	x, R_y	xz
b_2	$+1$	-1	-1	$+1$	y, R_x	yz

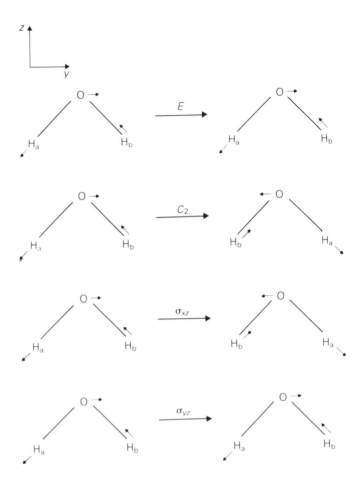

Fig. 5.16 Four symmetry operations in the point group C_{2v}, operating on the asymmetric stretching vibration of H_2O.

The four irreducible representations are given the labels a or b, depending on the sign of the character for C_2 ($+1$ or -1, respectively); a_1 is as always the totally symmetric symmetry species. To determine the representation appropriate to a particular vibration (the antisymmetric stretch) we apply the symmetry operations to a diagram (Fig. 5.16) in which the motions of the atoms are represented by arrows. The character is derived as the trace (sum of diagonal elements) of the transformation matrix that expresses the result of the operation on the motion represented by the arrows (see Appendix). Here each symmetry operation will either leave the motion unchanged or reverse its direction, so the transformation matrices are simply $+1$ or -1, respectively.

The identity operator (E) leaves the motion unchanged, so $\chi_v(E)$ is $+1$. Equally, operation of the reflection $\sigma_v(yz)$ leaves all arrows unmoved, so $\chi_v(\sigma_{yz})$ is $+1$. Operation of the C_2 rotation axis shifts H_a to where H_b was, and vice versa, so the corresponding pair of arrows ends up reversed compared to the starting pair. So now $\chi_v(C_2)$ is -1. Finally, we see that $\chi_v(\sigma_{xz})$ is -1, so

$$\begin{array}{ccccc} & E & C_2 & \sigma_{xz} & \sigma_{yz} \\ \Gamma_v = & +1 & -1 & -1 & +1, \end{array}$$

CHAPTER 5: VIBRATIONAL SPECTROSCOPY

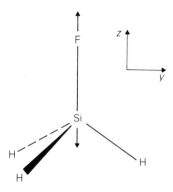

Fig. 5.17 Si—F bond stretching vibration of SiFH$_3$.

and we deduce that the vibration is of symmetry species b_2. The character table shows that the y axis also has b_2 symmetry, so this vibration will give a dipole change along y, and will be IR active (*i.e.* give rise to a fundamental band in the IR spectrum). The character table also shows that the component α_{yz} of the polarizability tensor has b_2 symmetry, so the vibration will also be active in the Raman spectrum.

For most point groups we can usually deduce the direction of any dipole change associated with a vibration without going into details of the representation. Reference to the column of the character table showing the Cartesian axis directions then gives the symmetry species directly. Thus for SiFH$_3$, the stretching vibration of the SiF bond must give rise to a dipole change along the C_3 axis, conventionally the z axis (see Fig. 5.17). The character table for C_{3v}, the point group concerned, shows that this vibration must therefore be totally symmetric, of a_1 symmetry species, and that it will also be Raman active. We already know it to be IR active because there is a dipole change during the vibration.

With practice, the assignment of vibrations to symmetry species quickly becomes automatic, particularly for common point groups having only one or two symmetry elements. Thus in a molecule with a single mirror plane (point group C_s), it is quite easy to decide whether or not a particular vibration is symmetric with respect to the plane. It can thus be assigned to one (a') or other (a'') of the two possible symmetry species. The systematic treatment given in the example can be used in case of doubt.

We sometimes find that the representation generated is not the same as any of the irreducible representations of the character table, or that we are unable to define the character according to the procedure used above. This usually indicates that the vibration we have chosen is not consistent with the symmetry of the molecule, and must be taken together with others to form symmetry-adapted vibrations.

For example, if we try to assign characters to the vibration of H$_2$O in which bond O—H$_a$ stretches, we can see (Fig. 5.18) that χ_v is $+1$ for elements E and σ_{yz}; but for C_2 and σ_{xz} the arrow representing the motion is neither left unmoved nor simply reversed, but ends up where a similar arrow relating to a stretching motion of the O—H$_b$ bond would have been. This indicates that we must consider the stretching motions of both bonds together. We may proceed in two ways.

(1) We must take as our basis the two vectors representing the two bond stretching motions, so that the transformation matrices showing the effects of the symmetry operations will be 2×2 matrices; only the trace of the matrix is needed to generate the

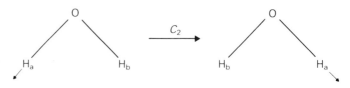

Fig. 5.18 Effect of the C_2 operation on the stretching motion of one bond in H_2O.

character χ_v. For the operations E and σ_{yz} both vectors are unmoved, so the transformation matrix is $\begin{pmatrix} 1 & 0 \\ 0 & 1 \end{pmatrix}$, with a trace of $+2$. For the operations C_2 and σ_{xz} each vector is shifted to where the other was originally, so the transformation matrix is $\begin{pmatrix} 0 & 1 \\ 1 & 0 \end{pmatrix}$, with a trace of 0. The representation Γ_v is then $+2$ 0   0 $+2$, which is not one of the irreducible representations of the character table; it is a reducible representation. It is, however, equal to the sum of the two irreducible representations $\Gamma(a_1)$ and $\Gamma(b_2)$, and we deduce that the two vibrations associated with OH bond stretching give rise to a_1 and b_2 group vibrations. The character table then shows that both are IR and Raman active.

(2) We may know from experience that whenever we have two equivalent bonds related by symmetry (a 2-fold axis C_2, a mirror plane σ or an inversion center i), their motions combine to give symmetric and antisymmetric stretching vibrations. We can then decide what is the symmetry species of each combination.

Here, the symmetric combination (where both bonds stretch in phase) clearly must be a vibration mode of a_1 symmetry species. The antisymmetric mode, where one bond lengthens as the other contracts, has been shown above (at the beginning of the present Section) to be of b_2 symmetry species.

Finally, consider an example of a molecule with a center of inversion, i. Cyanogen ($N{\equiv}C{-}C{\equiv}N$) has two equivalent CN bonds, and we must treat their stretching motions together (see Fig. 5.19). The symmetric stretching mode does not give rise to a dipole change, and hence is inactive in the IR. The symmetry species here is called Σ_g^+, as the molecule is linear. The antisymmetric stretch gives rise to a dipole change along z, the molecular axis, and we deduce that this vibration is IR active, with symmetry species Σ_u^+. Here the sub-scripts g and u refer to the behavior with respect to the inversion center, g (from German *gerade*, even) meaning symmetric, and u (*ungerade*, uneven) meaning antisymmetric. The Raman activities are the reverse of the IR activities, the symmetric mode being Raman active, the antisymmetric mode being inactive. This is an instance of a general rule, the Rule of Mutual Exclusion. For a molecule with an inversion center, i, only *ungerade* vibrations may be IR active, and only *gerade* vibrations may be Raman active. Therefore no vibration may be active in both spectra.

In some point groups there are symmetry species which are associated with neither IR nor Raman activity. A vibration of this type is the a_2 torsion (twisting) motion of

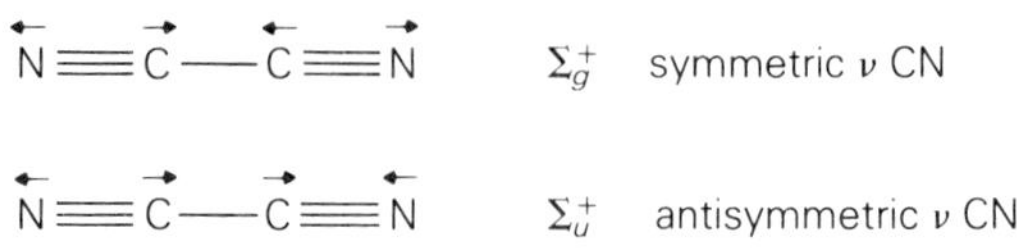

$$N{\equiv}C{-}C{\equiv}N \qquad \Sigma_g^+ \quad \text{symmetric } \nu \text{ CN}$$

Fig. 5.19 Two $C{\equiv}N$ stretching motions of cyanogen, C_2N_2.

$$N{\equiv}C{-}C{\equiv}N \qquad \Sigma_u^+ \quad \text{antisymmetric } \nu \text{ CN}$$

methylsilane (CH_3SiH_3), which has C_{3v} symmetry. In this vibration the two MH_3 groups twist about their common axis in opposite directions [Fig. 5.20(a)]. Such a motion can be detected neither by IR nor by Raman spectroscopy. In principle it can be detected by inelastic neutron scattering or by electron energy-loss spectroscopy [Section 5.3.4], for which there are effectively no symmetry selection rules.

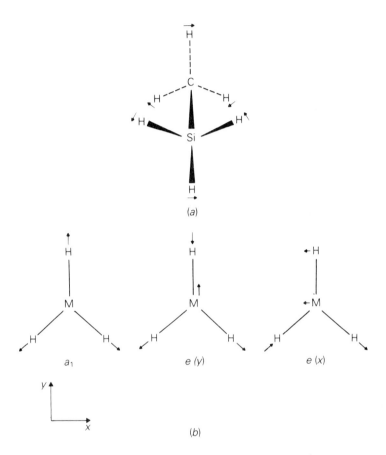

Fig. 5.20 (a) Torsional vibration of CH_3SiH_3. (b) Symmetric (a_1) and degenerate antisymmetric (e) M—H stretching motions of an MH_3 group, viewed along the C_3 (z) axis.

We have seen that two equivalent bonds always have symmetric and antisymmetric stretching vibrations. In the same way, there are three stretching modes associated with a group of three equivalent bonds (in an MH_3 group, say), while four bonds give rise to four modes, and so on. For point groups including one rotation axis of order 3 or more, the z-axis is unique, being defined by convention to coincide with this rotation axis, but the x and y axes must be taken together, so that motions along x and y are equivalent, or degenerate. The three equivalent M—H bonds of and MH_3 group in a molecule with a 3-fold axis through M then give rise to a symmetric combination of all three stretching motions (symmetry species a_1) and a degenerate pair of antisymmetric combinations of stretching motions, with symmetry species e [see Fig. 5.20(b)]. In a four-fold group, as in the square-pyramidal XeF_4O molecule, the four equivalent XeF bond stretches combine to give a totally symmetric (a_1) component, a second non-degenerate component of b_2 symmetry and a doubly degenerate pair of components of symmetry species e (Fig. 5.21). In the C_{4v} point group only a_1 and e vibrations are IR active, but a_1, b_2 and e vibrations are all observed in the Raman spectrum.

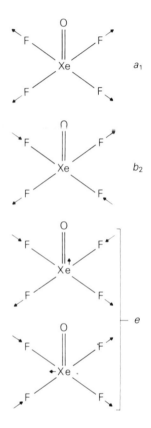

Fig. 5.21 Xe—F stretching vibrations of XeF_4O.

5.7.4   Symmetry of an entire set of normal vibrations

We can now tell how many vibrations a molecule has and decide their symmetry species, and so determine their IR and Raman activities. For small molecules it is often possible to write down descriptions of various vibrations until the total of $3N - 6$ is reached, decide the symmetry selection rules for each in turn, and so decide how many modes should appear in the IR and Raman spectra. But for more complicated molecules we need to have a systematic way of working out how many of the molecular vibrations will be active.

The set of three translational degrees of freedom for each of a set of N atoms is modified when the atoms make up a molecule. Thus the $3N - 6$ internal motions ($3N - 5$ for a linear molecule) become distinct from the translations and rotations of the molecule as a whole, the six (or five) external motions. However, the symmetry properties of the set of $3N$ degrees of freedom remain unchanged. If we can generate the representation of this complete set of degrees of freedom for the isolated atoms,

CHAPTER 5: VIBRATIONAL SPECTROSCOPY

and then subtract the representation of the external motions, we would obtain the representation Γ_{vib} of the whole set of molecular vibrations. This is a reducible representation, which must then be reduced, or decomposed, to the sum of a set of irreducible representations. This set will finally give us the numbers of vibrations of each symmetry species. Using PF_5 (5.I) as an example, we shall proceed first to build up the reducible representation Γ_y corresponding to the whole set of $3N$ degrees of freedom, then to reduce it, and only then remove the external motions to leave Γ_{vib}.

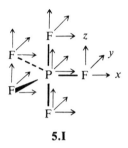

5.I

(1) We use as our basis the set of vectors representing all possible components of motion of individual atoms, making a total of $3N = 15$ vectors. Our transformation matrices are therefore 15×15 in size, and the advantage of only needing to consider the trace to derive the character χ is apparent! These $3N$ vectors are drawn on the atoms in 5.I.

(2) If an atom moves when the symmetry operation is applied, the three vectors representing its motions contribute zero to the trace of the transformation matrix, and hence to the character.

(3) For any atom that does not move when the operation is applied, there is a particular contribution to the character, and this is the same for every such atom for a particular type of symmetry operation. These contributions, which do not depend on the point group or even the particular orientation of the symmetry element, are listed in Table 5.1. They are called χ_{xyz}, the character of the general x, y, z basis vector for the particular operation. For example, any C_2 axis will leave one component vector unchanged and reverse the other two, so the corresponding transformation matrix is of the form,

$$\begin{pmatrix} -1 & 0 & 0 \\ 0 & -1 & 0 \\ 0 & 0 & 1 \end{pmatrix}$$

with a trace of -1; any mirror plane will reverse one component and leave the other

Table 5.1 Characters Representing Symmetry Operations on x, y and z Displacement Vectors

Symmetry operation (R)		$\chi_{xyz}(R)$
E	(identity)	3
i	(inversion)	-3
σ	(reflection)	$+1$
$C(\theta)$	(rotation through angle θ)	$1 + 2\cos\theta$
$S(\theta)$	(rotation–reflection, angle θ)	$-1 + 2\cos\theta$

two unchanged, so the trace is $+1$. The values of χ_{xyz} for the various symmetry operations of the D_{3h} point group are:

R	E	$2C_3$	$3C_2$	σ_h	$2S_3$	$3\sigma_v$
χ_{xyz}	$+3$	0	-1	$+1$	-2	$+1$.

(4) We then calculate the character $\chi_y(R)$ for each symmetry operation as $n_u\chi_{xyz}(R)$, where n_u is the number of atoms in the molecule left unmoved by the operation. The overall reducible representation Γ_y is then given by the row of characters $\chi_y(R)$, one for each type of symmetry operation in the point group. For PF_5 we have:

R	E	C_3	C_2	σ_h	S_3	σ_v
n_u	6	3	2	4	1	4
$\chi_y(R)$	18	0	-2	4	-2	4 .

(5) The collection of characters Γ_y is a reducible representation, and is then reduced using the following equation. This is done once for each symmetry species, to give the reduced form of the representation Γ_y:

$$n_i = 1/h \Sigma g_r \chi_y(R) . \chi_i(R).\tag{5.11}$$

Here n_i is the number of times the ith symmetry species appears in Γ_y, h is the order of the point group (equal to the total number of symmetry operations, and given at the head of the character table), g_r is the number of equivalent symmetry operations of type R, and $\chi_i(R)$ is the character listed for operation R for this symmetry species in the character table. The summation runs over all the classes of symmetry operations of the group. We may illustrate the reduction for PF_5, using the character table for D_{3h} given in Table 5.2, by showing the calculation required for the $a_1{}'$ symmetry species:

$$n(a_1') = \tfrac{1}{12} [1.18.1 + 2.0.1 + 3.-2.1 + 1.4.1 + 2.-2.1 + 3.4.1] = 2.$$

Table 5.2 Character Table for Point Group D_{3h}

$h = 12$	E	$2C_3$	$3C_2$	σ_h	$2S_3$	$3\sigma_v$			
a_1'	$+1$	$+1$	$+1$	$+1$	$+1$	$+1$			$x^2 + y^2,\ z^2$
a_2'	$+1$	$+1$	-1	$+1$	$+1$	-1	R_z		
e'	$+2$	-1	0	$+2$	-1	0	(x, y)		$(x^2 - y^2, xy)$
a_1''	$+1$	$+1$	$+1$	-1	-1	-1			
a_2''	$+1$	$+1$	-1	-1	-1	$+1$	z		
e''	$+2$	-1	0	-2	$+1$	0	(R_x, R_y)		(xz, yz)

Out of all possible motions of the six atoms, there are therefore two of $a_1{}'$ symmetry. In the same way we may find the numbers of motions of other symmetry species, and so obtain the reduced representation:

$$\Gamma_y = 2a_1' + 1a_2' + 4e' + 3a_2'' + 2e''.$$

(6) From the character table, the irreducible representations Γ_t and Γ_r corresponding to the translations and rotations are found; they are indicated by the symbols x, y, z, R_x, R_y and R_z. By removing these from the representation of all motions we are left

CHAPTER 5: VIBRATIONAL SPECTROSCOPY

with Γ_{vib}, the list of the numbers of vibrations of each symmetry species for the molecule. For PF_5 we find Γ_t is $1e' + 1a_2''$, Γ_r is $1a_2' + 1e''$; subtracting these from Γ_y we obtain $\Gamma_{\text{vib}} = 2a_1' + 3e' + 2a_2'' + 1e''$.

To check that we have the correct total number of vibrational degrees of freedom, we count one for each a (or b) species vibration, and two for each e (doubly degenerate) vibration. This gives a total of 12 for PF_5, which is equal to $3N - 6$ for our six-atom molecule. The character table shows us that only e' and a_2'' vibrations, corresponding to dipole changes in the x/y and z directions, respectively, can be IR active, so we expect five fundamental bands in the IR spectrum of PF_5. Only a_1', e' and e'' vibrations will be Raman active, so we expect six fundamental bands in the Raman spectrum, only three of which (due to the three e' vibrations) should coincide in frequency with IR bands.

Notice that, in this degenerate point group, the 12 vibrational degrees of freedom result in only 8 distinct vibration frequencies, 4 of which correspond to doubly-degenerate modes, and that neither the IR nor the Raman spectrum contains bands due to all eight modes. Vibrations of a_2' and a_1'' symmetry are inactive in both IR and Raman spectra. PF_5 does not have any modes of these symmetry species, but other molecules with D_{3h} symmetry, such as $Fe(CO)_5$, do have one or more of these totally inactive modes.

We can repeat the above analysis for any other suggested structure for PF_5, such as the square pyramid of C_{4v} symmetry. The final result in this case is that such a molecule should have six IR bands, all with coincident Raman bands, plus three Raman bands with no IR counterparts. We can therefore, at least in principle, distinguish the D_{3h} and C_{4v} structures by the numbers of bands in the spectra.

5.8   Assignment of bands to vibrations

5.8.1   Introduction

For some purposes we do not need to know which modes of vibration are responsible for which IR or Raman bands, but sometimes we must go further, and assign the bands in the spectrum to particular modes. Clearly this is much easier if we can associate bands in the spectrum with the various possible symmetry species. In some cases overall selection rules leave little doubt. For PF_5 only modes of e' symmetry are active in both IR and Raman spectra; any Raman bands coinciding in frequency with IR bands can confidently be assigned to these modes. The only IR bands without Raman counterparts should be due to a_2'' modes. That leaves us with modes of a_1' and e'' symmetry, which are forbidden in the IR but active in the Raman. To distinguish between these symmetry species we need more information, and it is extremely helpful to measure polarization of Raman scattering from fluids and band contours of IR (or Raman) bands of gases.

5.8.2   Raman polarization

When radiation has been scattered, its polarization properties may be the same as those of the incident light, or they may have changed. There are various possibilities, depending on the relation between the directions of incident and emitted radiation and the polarization properties of the incident light, but we will not go into details [which

Example 5.1

Determine the number and symmetry species of the normal vibrational modes of $Ni(CO)_4$, and their activities in infrared and Raman.

....

Nickel carbonyl, $Ni(CO)_4$, belongs to the point group T_d. It has nine atoms, and so there are 27 degrees of freedom. To work out the reducible representation of these, the tetrahedron is best seen in relation to a cube: the nickel atom goes at the center, and each of the CO groups lies along a different diagonal. The symmetry elements for T_d are then easily identified: the three-fold axes lie along the diagonals, and the two-fold and S_4 axes go through the centers of the faces, and the σ_d planes coincide with the face-diagonals (see figure).

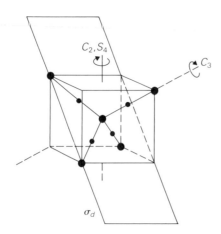

The reducible representation is then obtained as follows:

Identity (E) – all 27 degrees of freedom unchanged – entry 27.

C_3 rotation – x, y, z vectors permuted for axial atoms; all other atoms shifted – total 0.

C_2 rotation – all atoms shift except Ni, for which one vector is unchanged, two are reversed – entry – 1.

S_4 only Ni atom unshifted – two vectors permuted, other one reversed – entry – 1.

σ_d 5 atoms lie in each such plane; with two vectors permuted, the other unchanged – entry 5.

The reducible representation then looks like this:

E	$8C_3$	$3C_2$	$6S_4$	$6\sigma_d$
27	0	– 1	– 1	5

We can reduce this using Eqn (5.11) and the character table given in the appendix. The order of the group is 24. We then have the following number of degrees of freedom in the various symmetry species:

a_1: $1/24 [27.1.1 + 8.0.1 + 3. - 1.1 + 6. - 1.1 + 6.5.1] = 2$
a_2: $1/24 [27.1.1 + 8.0.1 + 3. - 1.1 + 6. - 1. - 1 + 6.5. - 1] = 0$
e: $1/24 [27.1.2 + 8.0. - 1 + 3. - 1.2 + 6. - 1.0 + 6.5.0] = 2$
t_1: $1/24 [27.1.3 + 8.0.0 + 3. - 1. - 1 + 6. - 1.1 + 6.5. - 1] = 2$
t_2: $1/24 [27.1.3 + 8.0.0 + 3. - 1. - 1 + 6. - 1. - 1 + 6.5.1] = 5$

So the degrees of freedom transform as $2a_1 + 2e + 2t_1 + 5t_2$. From the character table we see that the three translations transform as t_2 and the three rotations as t_1. The vibrations therefore transform as $2a_1 + 2e + t_1 + 4t_2$.

To work out more about the sort of motion that these represent, we can treat a simple tetrahedral species MX_4 in exactly the same way; we find that the vibrational modes transform as $a_1 + e + 2t_2$. A similar analysis using vectors along the M—X bonds to represent the bond stretching modes in tetrahedral MX_4 shows that the stretching modes transform as $a_1 + t_2$. Since the stretching modes of the C—O bonds in $Ni(CO)_4$ will behave in exactly the same way as the M—X bonds in MX_4, we know now that the vibrational modes for $Ni(CO)_4$ can be further factorized as follows: the NiC stretches transform as $a_1 + t_2$; the CO stretches transform as $a_1 + t_2$; the NiC_4 bends transform as $e + t_2$. This accounts for 13 of the 21 vibrational modes. The remaining eight, $e + 2t_1 + t_2$, must represent bending of the CO groups. The e and t_2 components are exactly analogous to the corresponding deformations of the NiC_4 core. The t_1 deformations can be related to rotations of the NiC_4 core: they can be represented by rotation of this core within a rigid sphere containing the four oxygen atoms.

---

can be found elsewhere (Ref. [7], p. 62)]. However, we can measure a depolarization ratio ρ for each band in the Raman spectrum of a fluid sample. This is the ratio of the intensity recorded with a polarizing element (polaroid screen or half wave plate) set, first parallel, and then perpendicular to the plane of polarization of the incident radiation. The maximum value ρ_{max} is governed by the scattering geometry, and corresponds to a depolarized band, while any lower value corresponds to a polarized band. The detailed theory is complicated, but the use of the depolarization ratio lies in a simple rule: totally symmetric vibrations give rise to polarized Raman bands; vibrations that are not totally symmetric give rise to depolarized bands. Fig. 5.22 shows how this can be used to identify modes having similar frequencies.

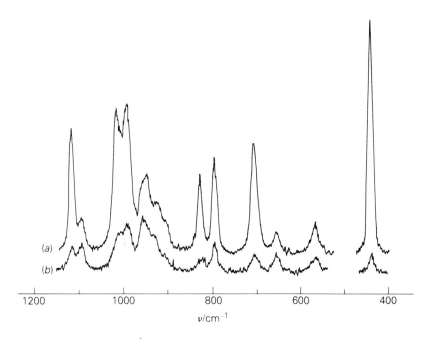

Fig. 5.22 Raman spectra of liquid η^1-cyclopentadienylsilane, $C_5H_5SiH_3$, recorded with the analyzing half wave plate set (a) parallel and (b) perpendicular to the plane of polarization of the incident radiation. The bands at 1090, 950 and 660 cm^{-1} are clearly depolarized, and can be assigned to a'' modes. The other bands are all polarized, but their depolarization ratios are not all equal.

For a given spectrometer and sample geometry, ρ_{max} is more or less constant, and may be taken as known. Thus if the value of ρ for a band is found to be less than ρ_{max} the band may be assigned to a totally-symmetric mode. The depolarization ratio is only required to be zero for totally symmetric modes of molecules of cubic or icosahedral point groups. For totally symmetric modes of molecules of lower symmetry, ρ is greater than zero, and in some cases may be impossible to distinguish from ρ_{max}. The ratio also tends to be higher for totally symmetric modes that involve angle bending than for pure stretching modes.

We can now complete the assignment for PF_5, by distinguishing between the a_1' and e'' modes. The two a_1' modes will be polarized. The third Raman band without an IR counterpart must be of e'' symmetry: it will be depolarized, as will be the bands due to the e' modes. The final set of observations to be expected for the D_{3h} structure is shown in Table 5.3. Fig. 5.23 illustrates the observed spectra. The C_{4v} structure, with three modes of a_1 symmetry, should give three polarized Raman bands coinciding with three of the IR bands, and thus would be clearly distinguished from the D_{3h} structure.

For powders or microcrystalline samples, multiple external reflections among grains destroy polarization, so that we cannot obtain useful data. Work with oriented single crystals can give detailed information; specialized texts (see Ref. [7], p. 63) should be consulted to enable the best use to be made of such experiments.

Table 5.3 Vibrational Modes of PF_5

Symmetry species	IR		Raman	Polarization	
a_1'	-		2	p	$\rho < \rho_{max}$
e'	3	coincident	3	dp	$\rho = \rho_{max}$
a_2''	2		-	-	
e''	-		1	dp	$\rho = \rho_{max}$

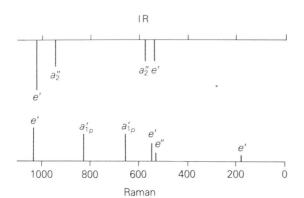

Fig. 5.23 Bands observed in vibrational spectra of PF_5. Polarized Raman bands are labeled p.

Polarized IR measurements are in principle useful for crystals and other oriented samples [10]. Dipole-induced absorption is due to a dipole change in a particular direction relative to the molecular orientation in the crystal, so that rotation of the sample leads to changes in relative band intensities if polarized incident radiation is used. Fig. 5.24 shows how such measurements can be used in the assignment of bands to particular vibrational modes.

CHAPTER 5: VIBRATIONAL SPECTROSCOPY

Band contours in gases

In gases, the rotational energies of molecules are quantized; vibrational transitions may be accompanied by changes in rotational quantum numbers, and hence in rotational energy. A gas-phase vibrational band is therefore usually complicated. It does not just consist of a single $\Delta J = 0$ transition from $v = 0$ to $v = +1$ for each rotational level; it

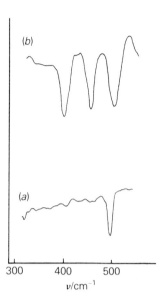

Fig.5.24 IR spectra of a single crystal of $Rh(CO)_2$(acetylacetonate), recorded with the plane of polarization (*a*) parallel to and (*b*) perpendicular to one of the crystal axes. The striking difference in the spectra allows us to deduce that the mode just below 500 cm^{-1} causes a dipole change perpendicular to the axis, whereas the three modes giving the bands visible in (*a*) cause dipole changes along the axis. Redrawn from D.M. Adams and W.R. Trumble, *J.Chem.Soc., Dalton Trans.*, 690 (1974).

Table 5.4 Branches due to Rotational Structure in Vibration Bands

ΔJ	Label	Line position/B
-2	O	$-4J+2$
-1	P	$-2J$
0	Q	0
$+1$	R	$+2J+2$
$+2$	S	$+4J+6$

includes lots of lines due to transitions with $\Delta J \neq 0$, where J is the general rotational quantum number. The resulting set of lines forms a band envelope. The various vibration bands for a given molecule have distinct band envelopes that depend on the symmetries of the vibrations concerned, because the rotational selection rules are different for different vibrational symmetries. In some cases the individual lines making up the envelope may be resolved (see Chapter 4).

The five most important values encountered for ΔJ are -2, -1, 0, $+1$ and $+2$, and these give rise to characteristic branches in vibration bands which are labelled O, P, Q, R and S (see Table 5.4). For the simplest possible case, where the rotation constants are the same in the lower and upper vibrational states, the corresponding patterns of rotation fine-structure lines are also given in the Table. They are simply derived from the term $BJ(J + 1)$ for the rotational energy. Note that in this case all $\Delta J = 0$ lines coincide, and so Q branches are often line-like features close to the vibration transition energy ν_{vib} or ν_0. For the other branches characteristic spacings of rotation lines are predicted, $\pm 2B$ for $\Delta J = \pm 1$ (R and P branches) and $\pm 4B$ for $\Delta J = \pm 2$ (S and O branches). In general, S and O branches occur only in Raman spectra, whereas P, Q and R branches are found in both IR and Raman spectra (see Fig. 5.25).

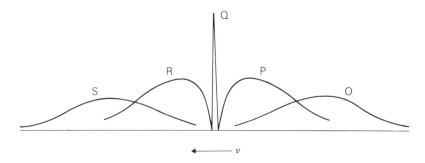

Fig. 5.25 Rotational branches in a general vibrational band.

The details of rotational line positions in vibration bands are exceedingly complex, and beyond the scope of this book. The underlying rules, and some of the consequences, are set out in Refs [1] and [11]. There are, however, some simple rules that enable us to deduce the symmetry of a vibration from the band envelope of the corresponding IR or Raman band, even at rather limited resolution, where the $2B$ spacing of components of P and R branches, for instance, is not resolved.

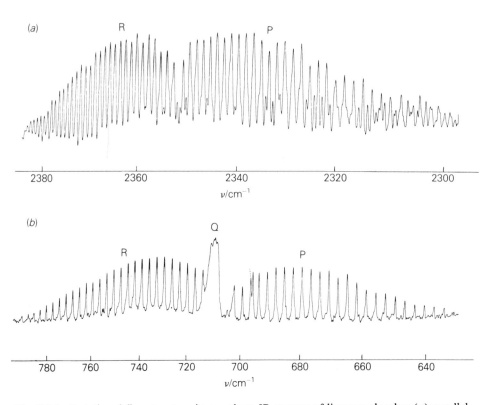

Fig. 5.26 Rotational fine structure in gas phase IR spectra of linear molecules: (a) parallel band of CO_2 (the asymmetric stretch) and (b) perpendicular band of HCN (the bend). Adapted, with permission, from *Tables of Wavenumbers for the Calibration of IR Spectrometers*, IUPAC, Butterworths, London (1961).

CHAPTER 5: VIBRATIONAL SPECTROSCOPY

A simple case is that of the linear molecule. Here we can distinguish two distinct types of IR envelope, which correspond to vibrations causing dipole changes along the linear axis (parallel band) or perpendicular to the axis (perpendicular band). For a parallel band, arising from an asymmetric stretching mode of Σ symmetry, the rotational selection rule is $\Delta J = \pm 1$ only; there is therefore no Q-branch, for which $\Delta J = 0$. On the other hand a perpendicular band, due to a doubly-degenerate bending mode of Π symmetry, has $\Delta J = 0$ or ± 1, so a Q-branch appears (see Fig. 5.26). IR bands of linear molecules are therefore readily associated with Σ and Π symmetry species. In the Raman spectrum, $\Delta J = 0$ and ± 2 transitions are allowed for both Σ and Π vibrations, but for Π vibrations *only* $\Delta J = \pm 1$ is allowed as well. It is thus more difficult to distinguish Σ and Π vibrations using the Raman bands, as we would have to detect the presence of P and R branches with spacings of $2B$ as well as the O and S branches with spacings of $4B$. The polarization measurement is likely to be more useful in this instance, and is the only method applicable if the individual lines in the O, P, Q, R and S branches cannot be resolved.

Non-linear symmetric tops which have a rotation (or rotation–reflection) axis of order three or more present a more complicated problem, as we must now consider two rotational quantum numbers. J is still the overall quantum number, but the component of angular momentum along the top axis is also quantized, with quantum number K. There are still only two types of IR band, depending on whether the associated dipole change is parallel or perpendicular to the top axis. For parallel bands, due to vibrations of some a symmetry species (a_2'' in PF_5; a_1 in $SiFH_3$ . . .), the rotational selection rules are $\Delta J = 0, \pm 1$; $\Delta K = 0$ except that when $J = 0$, $\Delta J = 0$ is forbidden. The overall structure is superficially similar to that of a perpendicular band of a linear molecule, but with P, Q and R sub-branches for each value of K for which there is a significant population. As the Q sub-branches superimpose, the result is usually a band with distinct P, Q and R branches. For perpendicular bands, due to vibrations of e symmetry, such as e' in PF_5, the rotational selection rules are $\Delta J = 0, \pm 1$; $\Delta K = \pm 1$, and the P, Q and R sub-branches due to different initial K values are now separated because of the change in K. The usual result is a broad band with overlapping P and R sub-branches, and a regular series of Q sub-branches, one for each value of $K \Delta K$. These Q branches are often distinct, but sometimes they merge into the rest of the band. Thus again a distinction is in principle possible between bands due to vibrations of different symmetry species (Fig. 5.27). The selection rules for Raman spectra are different, but parallel and perpendicular vibrations of symmetric tops may be distinguished by Raman band contours in the same way as described above for IR bands, although this may not be enough to allow us to assign symmetry species to all bands.

For asymmetric tops, which have three different moments of inertia, there is no symmetry element of order greater than two, and up to three different symmetry species of vibration may give rise to IR bands. These correspond to dipole changes along the x, y and z Cartesian axes. If these axes coincide with the three principal inertial axes A, B and C (see Chapter 4), either because of symmetry constraints, or just accidentally, we find that very simple band contours arise. These may be labelled a-type, b-type or c-type if the dipole change is along the A, B or C axis. The band contours illustrated in Fig. 5.28 may be taken as typical, though the details vary with the magnitudes of the three moments of inertia. Thus c-type bands always have sharp distinct Q branches on an indistinct background with no apparent gap between P and R branches; b-type bands have a central dip with no distinct Q-branch (or sometimes a

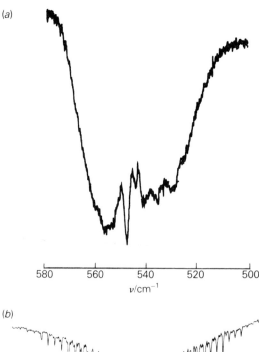

(a)

580 560 540 520 500
ν/cm^{-1}

(b)

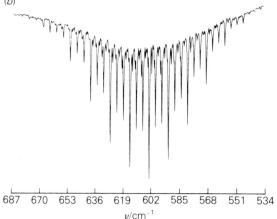

687 670 653 636 619 602 585 568 551 534
ν/cm^{-1}

Fig. 5.27 Parallel and perpendicular IR bands of symmetric top molecules: (a) parallel band of $SiClH_3$ under fairly low resolution, and (b) perpendicular band of $GeClH_3$ with moderate resolution. A high resolution spectrum of a parallel band similar to (a) is shown in Fig. 4.10.

Q-branch clearly divided into two, again leaving a distinct central gap); and a-type bands have more or less clear P, Q and R branches.

For the point groups C_{2v}, D_2 and D_{2h}, the symmetry means that x, y and z axes coincide with the A, B and C axes. We then have to work out for each molecule which of the Cartesian axes corresponds with which inertial axis, as this depends on the details of the mass-distribution. In molecules with point groups D_2 or D_{2h} the center of gravity is fixed at the intersection of the three 2-fold axes. We can work out the moments of inertia about the three relevant axes, and use the rule that $I_A < I_B < I_C$. For C_{2v} molecules only one of the inertial axes is defined by symmetry. To define the other two the center of gravity must be located, and it may often be easier to use a combination of simple mechanical principles and information from the spectrum to deduce the exact correspondence between the axis sets.

As an example, we may take gaseous $SiCl_2H_2$, whose IR spectrum (Fig. 5.29) contains

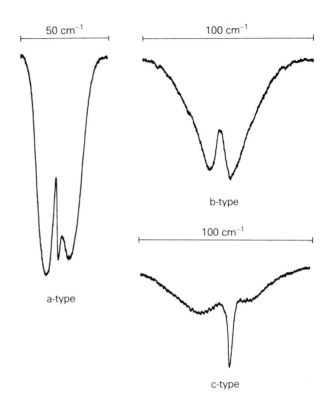

Fig. 5.28 Typical a-type,
b-type and c-type IR bands.

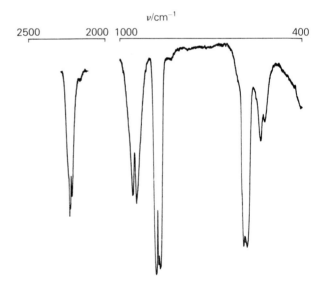

Fig. 5.29 Part of the IR
spectrum of gaseous SiCl₂H₂.
The highest frequency band
consists of overlapping c-type
and b-type components; of
the other four bands, the
outer two are b-type, and the
inner two are a-type.

bands with three distinct types of envelope, as expected for a molecule with C_{2v}
symmetry. There are 2 a-type bands, 4 b-type bands and 2 c-type bands, some of which
overlap. For a five-atom molecule there are nine modes, and their symmetry species
can be determined as follows:

SECTION 5.8: ASSIGNMENT OF BANDS TO VIBRATIONS   207

$$
\begin{array}{lllll}
R & E & C_2(z) & \sigma_{xz} & \sigma_{yz} \\
\Gamma_\gamma \quad = & 15 & -1 & 3 & 3
\end{array}
$$

$$n(a_1) = \tfrac{1}{4}[15.1 \quad -1.1 \quad +3.1 \quad +3.1] = 5$$
$$n(a_2) = \tfrac{1}{4}[15.1 \quad -1.1 \quad -3.1 \quad -3.1] = 2$$
$$n(b_1) = \tfrac{1}{4}[15.1 \quad -1.-1 \quad -3.1 \quad +3.1] = 4$$
$$n(b_2) = \tfrac{1}{4}[15.1 \quad -1.-1 \quad +3.1 \quad -3.1] = 4$$
$$\Gamma_t + \Gamma_r = a_1 + a_2 + 2b_1 + 2b_2$$
$$\Gamma_{vib} = 4a_1 + a_2 + 2b_1 + 2b_2.$$

Only a_1, b_1 and b_2 vibrations give IR bands; four of these should be of one type (a_1) and two each of the other types. The C-axis must be perpendicular to the plane of the three heavy atoms, and so we deduce that the c-type bands are due to b_1 vibrations. As the spectrum includes four b-type bands, we assign these to a_1 vibrations and the a-type bands to b_2 vibrations (Fig. 5.30). Note that the a_2 vibration is observed in the Raman spectrum of the liquid, which also confirms the assignment of the four b-type bands in the IR spectrum of the gas to a a_1 vibrations, as they coincide with partially polarized Raman bands.

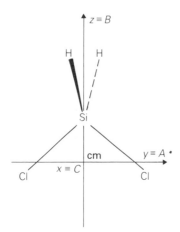

Fig. 5.30 Relationship between the Cartesian axes (x, y, z) and the inertial axes A, B, C of $SiCl_2H_2$. The center of mass is marked cm.

For many molecules the Cartesian axes do not coincide uniquely with the principal inertial axes, and hybrid band contours arise. For any planar molecule, the C-axis is automatically perpendicular to the plane, so that any IR-active out-of-plane motions give rise to c-type bands. Many non-planar molecules such as ethyl germane (5.II) have their heavy atoms lying in a plane of symmetry, to which the C-axis is perpendicular,

5.II

and their a'' modes therefore also have c-type contours. But if there is no molecular symmetry other than the plane, the x and y axes are not uniquely determined by the symmetry and *in-plane* vibrations and may give a/b-hybrid bands depending on the relative directions of the A and B axes and the dipole change. These bands will, however, still be distinguishable from c-type bands, and the essential symmetry distinction can still be made. Thus the spectrum of ethyl germane (Fig. 5.31) includes bands that seem to have a and b band shapes, and both types must be assigned to a' modes. In other cases, other hybrid band shapes may arise. If the molecule has no symmetry element except the identity operation, all vibrations are of the same symmetry species. The dipole changes may take any orientation with respect to the inertial axes, and bandsmay have a-, b- or c-type contours or any hybrid.

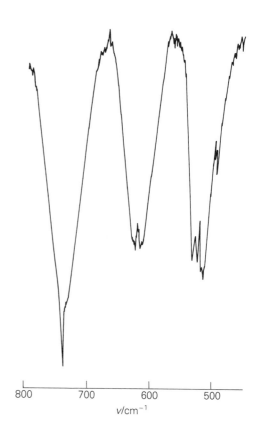

Fig. 5.31 Part of the gas phase IR spectrum of ethyl germane. The band at 740 cm^{-1} is clearly c-type, and can be assigned to an a'' mode. The other two bands must arise from vibrations of a' symmetry, but they have markedly different shapes, one closely resembling a b-type band, and the other resembling an a-type band.

Analogous distinctions between band envelopes may be made for Raman spectra of gaseous asymmetric tops, but they are not commonly used. For molecules of C_{2v}, D_2 or D_{2h} symmetry the four different symmetry species giving rise to Raman activity correspond to four distinct band envelopes, called A, B_a, B_b and B_c-types (Ref. [1], p. 244), while for other point groups various hybrids of these may arise.

Finally, for spherical tops, with three identical moments of inertia, only one symmetry species gives rise to IR activity, as x, y and z axes are triply degenerate. In Raman spectra totally symmetric modes are uniquely identified in that they give rise only to Q-branches ($\Delta J = 0$), whereas vibrations of any other symmetry give $\Delta J = 0$, ± 1 and ± 2 branches.

Intensities of allowed fundamentals

A fundamental band in an IR spectrum may only be observed if there is an interaction between the electric vector of the incident radiation and the oscillating dipole due to the corresponding vibration. The intensity of the band therefore depends on the strength of that interaction, which is proportional to the square of the magnitude of the oscillating dipole $d\mu/dQ$. Here Q represents the changing 'coordinate' involved in the vibration, analogous to $(r - r_e)$ for a diatomic oscillator. Symmetry selection rules correspond to symmetry requirements that for a forbidden band, $d\mu/dQ$ must be zero, at least at the equilibrium configuration, where Q is zero. For an allowed band, the value of $d\mu/dQ$ is not restricted, and it may accidentally be zero. The magnitudes of dipole moment derivatives vary widely, and it is not safe to assume from the failure to observe a band that the corresponding mode is symmetry-forbidden. Thus the lower-frequency stretching mode of HCN, mainly associated with $C \equiv N$ stretching, gives rise to IR absorption so weak that it cannot effectively be observed by conventional means. We shall see below that some particular bond-stretching or angle-bending motions have characteristically high intensities. There is a wealth of empirical lore in this area, but the underlying physical reasons for the observed intensities are not well understood. Attempts to calculate intensities have been made; Ref. [12] is a review of this area.

In a Raman spectrum, the intensity of a band depends on the squares of the derivatives of components of the polarizability tensor, $d\alpha/dQ$. In the absence of symmetry selection rules that require it to be zero, the value of the derivative is not restricted. The Raman scattering due to different vibrations of a particular molecule is generally less variable than the intensity of the different IR bands, but the general level of intensity of Raman spectra varies from molecule to molecule much more than that of IR spectra. For example, some simple organic solvents such as C_6H_6 and CCl_4 give very strong Raman spectra, whereas water gives only very weak scattering, but for each solvent the various allowed bands have similar intensities. The absence of a band from a reasonably well-defined Raman spectrum is thus more likely to indicate a genuine symmetry selection rule than the absence of an IR band. But mistakes have been made when trying to decide between alternative structures on the basis of the presence or absence of particular bands in IR and Raman spectra. It is better to rely on the whole pattern of observed bands, including polarization and band contour data.

Mode numbering

We find it helpful to number modes so that we can refer to them simply in the form v_n. Unfortunately, it has not been possible to find a fool-proof numbering system applicable to all molecules, but two simple rules are well-established.

1   The order adopted for modes of different symmetry species takes the more highly symmetric modes before those of lower symmetry.

2   Modes of a particular symmetry species are numbered consecutively from highest to lowest frequency; thus v_1 is always the highest-frequency mode of the totally symmetric species.

The major source of discrepancy between different authors is rule 1, because the order of decreasing symmetry cannot be defined unambiguously. The problem can be illustrated for the point group C_{2v}. The b_1 and b_2 vibrations are equally symmetric, and an arbitrary choice has to be made, first as to which set is to be labeled b_1, and secondly

Table 5.5 Mode Numbering for $SiCl_2H_2$

Vibration type	a_1	a_2	b_1	b_2
SiH stretch	v_1		v_6	
SiH deformation	v_2	v_5	v_7	v_8
SiCl stretch	v_3			v_9
SiCl deformation	v_4			

Note: it is possible to order the two symmetry species b_1 and b_2 in two ways, and this affects the mode numbering (see text).

as to whether to maintain the same order in mode numbering. Table 5.5 shows the numbering of the nine vibrations of $SiCl_2H_2$ as an illustration of the use of the two rules.

There is a more complicated example in the point group D_{3h}. In this point group, symmetry species which are symmetric with respect to the horizontal plane, σ_h, are labeled $'$; those antisymmetric to the plane are labeled $''$. Some authors order the symmetry species a_1', a_2', e', a_1'', a_2'', e'', regarding the $'/''$ distinction as most important. Others treat the symmetry with respect to the 3-fold axis as more important, ordering the species as a_1', a_1'', a_2', a_2'', e', e''. This clearly results in a different numbering of all modes except those of symmetry species a_1' and e''. Neither approach is obviously more correct than the other.

We find a deviation from the two rules for triatomic molecules. The bending mode is always labeled $\dot{v}_2$, whatever the molecule. For molecules AX_2, whether linear like CO_2 or bent like H_2S, the symmetric stretch is of higher symmetry than the antisymmetric stretch and so is labeled v_1, even though it is lower in frequency. For molecules XAY, like OCS or ClNO, the two stretching modes are of the same symmetry, and so are numbered in order of decreasing frequency.

5.8.6 Non-fundamental transitions

In real life, the business of deducing molecular symmetry from vibrational spectra may be complicated by transitions which do not correspond to the $v = 0 \rightarrow v = +1$ vibrational selection rule. Various types occur, and they may be categorized according to the changes in vibrational quantum numbers v_i of the set of molecular vibrations (see

Fig. 5.32 Vibrational energy levels and types of vibrational transitions: (*a*) Fundamental transitions v_i, v_j, (*b*) Hot bands $(n + 1)v_i - nv_i$, $(m + 1)v_j - mv_j$, (*c*) Hot bands $(nv_i + v_j) - nv_i$, $(v_i + mv_j) - mv_j$, (*d*) First overtones $2v_i$, $2v_j$, (*e*) Second overtones $3v_i$, $3v_j$, (*f*) Combination $v_i + v_j$, (*g*) Difference $v_i - v_j$.

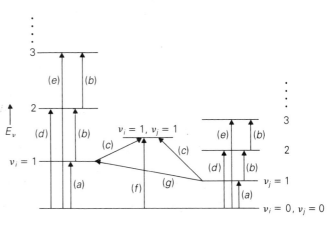

Table 5.6 and Fig. 5.32). The list given is not exhaustive, but includes two important types of non-fundamental transitions. Those with $\Sigma(\Delta v) > 1$ are formally forbidden under the fundamental vibrational selection rule. In general they give very weak bands in IR or Raman spectra unless special factors apply. Their appearance is connected with anharmonicity, which leads to a breakdown of the selection rules derived assuming simple harmonic motion. This is true of overtones, combinations and difference bands.

Table 5.6 Types of Vibrational Transitions

(a) Formally allowed ($\Delta v = 1$ for only one mode)

v_i	Δv_i	v_j	Δv_j	Name	Symbol
0	+1	0	0	Fundamental	v_i
n	+1	0	0	Hot band	$(n+1)v_i - nv_i$
0	+1	m	0	Hot band	$v_i + mv_j - mv_j$

(b) Formally forbidden

v_i	Δv_i	v_j	Δv_j	Name	Symbol
0	+2	0	0	First overtone	$2v_i$
0	+3, etc.	0	0	Second overtone etc.	$3v_i$, etc.
0	+1	0	+1	Combination	$v_i + v_j$
0	+1	+1	−1	Difference band	$v_i - v_j$

Hot bands, on the other hand, are formally allowed, because Δv is +1 for only one mode, and zero for all others. Hot bands differ from fundamentals only in that the initial state of the transition is a vibrationally excited state, and not the vibrational ground state of the molecule. They are in effect the fundamentals of molecules that are already vibrationally excited. Such molecules are in general less abundant than ground state molecules, because of the Boltzmann distribution of thermal energy. The ratio of the number of molecules in an excited state to those in the ground state, assuming dynamic thermal equilibrium at room temperature, is $e^{-E_v/209}$, where E_v is the excess energy in cm^{-1}. When E_v is 1000 cm^{-1}, this ratio is 0.005, a proportion which is small enough to be ignored. However, when E_v is 200 cm^{-1}, the ratio is 0.34; therefore a sample of a compound with a vibration of frequency lower than about 300 cm^{-1} has a significant proportion of its molecules in vibrationally excited states. In the presence of such low frequency vibrations, no band in the IR or Raman spectrum truly represents a single transition. Each band is in fact made up of components due to $\Delta v = +1$ transitions arising from the corresponding modes of the various vibrationally excited molecules. One of the effects of anharmonicity is that these transitions do not all have exactly the same frequency, and so a band may have a very complex form. However, it is sometimes possible to distinguish a progression of Q branches, such as is shown in Fig. 5.33.

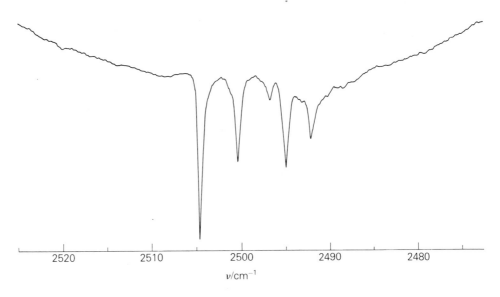

Fig. 5.33 An N—D stretching vibration of gaseous PF_2ND_2 gives this IR band, which shows several hot bands, arising from at least two low frequency modes. The fundamental Q branch is at 2505 cm^{-1}.

The details of hot bands may be hard to analyze, but it is the formally forbidden transitions, due to overtones, combinations and differences, that cause most confusion. In Raman spectra, these are usually much weaker than fundamentals. All fundamentals for a given molecule are usually of comparable intensity, so it is unlikely that a forbidden band will be mistaken for a weak fundamental. This is more of a possibility in IR spectra. For example, in the IR spectrum of HCN there is a band in the region where the lower-frequency stretching mode would be expected. Comparison with the Raman spectrum of the gas shows that the strong Raman band in this region is some 20 cm^{-1} lower in frequency than the observed IR band. The two clearly do not correspond; the Raman band is assigned to the fundamental, and the IR band to the second overtone of v_2.

Once the fundamental $\Delta v = 1$ selection rule is breached, the IR and Raman activities of overtones and combinations are governed by the same symmetry selection rules that apply to fundamentals. Before we can apply these, however, we must define the symmetry of a vibrational transition rather more precisely.

A transition involves two distinct vibrational states of the molecule, and we can define the symmetry species of each of these. The symmetry of the transition is then defined as the direct product of the symmetries of the initial and final states involved, i and f

$$\Gamma_t = \Gamma_i \times \Gamma_f. \tag{5.12}$$

The direct product representation Γ_t is generated by multiplying together the corresponding characters of the representations Γ_i and Γ_f. For fundamental, overtone or combination transitions, the initial state i is the vibrational ground state, which is always totally symmetric, with all characters equal to $+1$. Thus the symmetry of the transition is the same as that of the final state. For overtones and combinations we need to generate the representation Γ_f as the direct product of the representations corresponding to each vibration excited in the state f.

For example, if an a_2 and a b_2 vibration of a C_{2v} molecule are both singly excited, we calculate the representation of the final state like this:

R	E	C_2	σ_{xz}	σ_{yz}
Γ_f	$+1 \times +1$	$+1 \times -1$	-1×-1	$-1 \times +1$
$=$	$+1$	-1	$+1$	-1

This corresponds to b_1 symmetry, and so the combination of an a_2 vibration and a b_2 vibration generates a b_1 transition.

Similarly, for double excitation of a b_1 vibration:

$\Gamma_f =$	$+1 \times +1$	-1×-1	$+1 \times +1$	-1×-1
$=$	$+1$	$+1$	$+1$	$+1$

This corresponds to a_1 symmetry, and so the first overtone of a b_1 vibration is an a_1 transition. Note that for all point groups without degeneracy all first overtones are totally symmetric.

In point groups with one or more axes of order three or more, the direct products obtained in this way may have to be reduced in order to discover the symmetry species of the components of the overtone or combination. For instance, for a combination of two vibrations of e symmetry species in C_{2v}:

R	E	C_3	σ_v
$\Gamma_f =$	$+4$	$+1$	0

This reduces to $a_1 + a_2 + e$.

For a combination all three components exist, though only a_1 and e components give rise to IR transitions from the ground state. For an overtone, only a_1 and e components (the so-called symmetric product components) actually exist. Details for other point groups are given in Refs [11] and [13].

Overtones and combinations are usually weak in IR and Raman spectra. The symmetry selection rules operate as for fundamentals, in respect of activity and Raman polarization and IR band contours. They can be valuable in making a complete vibrational assignment. Vibrations that are not observed directly may still be involved in overtone or combination bands, from which the unknown frequencies may be deduced.

Overtones and combinations sometimes give rise to unusually strong bands in IR or Raman spectra by stealing intensity from a nearby fundamental of the same symmetry. This phenomenon, known as Fermi resonance, occurs because the two vibrationally excited states mix. Before mixing, one is the upper state of a fundamental and the other the upper state of an overtone or combination band. After mixing, both states become partly fundamental and partly overtone or combination (Fig. 5.34). As a result of the mixing, the intensity of the fundamental is shared between the two bands involved. The energies of the upper states, and hence of the transitions, are also affected. The mean frequency does not change, but the upper states move further apart, and so do the frequencies of the bands. Fermi resonance thus introduces additional strong bands, which may be mistaken for fundamentals; it also affects the apparent frequencies of fundamentals involved.

For example, in the Raman spectrum of CO_2 we expect only one fundamental, due to the symmetric stretch, of symmetry species Σ_g^+. Instead, we observe two intense

CHAPTER 5: VIBRATIONAL SPECTROSCOPY

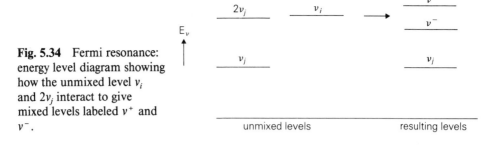

Fig. 5.34 Fermi resonance: energy level diagram showing how the unmixed level ν_i and $2\nu_j$ interact to give mixed levels labeled ν^+ and ν^-.

unmixed levels   resulting levels

Raman bands, one at 1285 and the other at 1388 cm^{-1}. The $\varPi_u$ bending fundamental is at 667 cm^{-1} in the IR, and its overtone should appear near 1334 cm^{-1}, almost mid-way between the two observed Raman bands. There is a $\varSigma_g^+$ component of the overtone, and this component must have mixed with the $\varSigma_g^+$ fundamental to produce two bands of similar intensity, each being more-or-less half fundamental and half overtone. A similar case of Fermi resonance occurs in SiH$_3$NCS, where the first overtone of the C=S stretch at 1033 cm^{-1} interacts with the N=C stretch near 2070 cm^{-1} to give two IR bands of comparable intensity (Fig. 5.35). Remember that Fermi resonance is only possible when a fundamental and a second-order band have the same symmetry and are close together in energy.

For purely harmonic vibrations, overtone or combination frequencies would be exact sums of the constituent fundamental frequencies. In real oscillators the effect of anharmonicity is to change (usually to reduce) the frequency of an overtone or

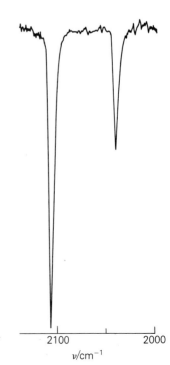

Fig. 5.35 IR spectrum of SiH$_3$NCS, showing the effect of Fermi resonance. The C=S stretching fundamental is at 1033 cm^{-1}, so its overtone is expected near 2066 cm^{-1}, which is close to the observed mean frequency of the two bands shown, 2074 cm^{-1}.

ν/cm^{-1}

combination [see Section 5.2.3]. The general formula for vibrational energy levels (to second order, for two vibrations) is

$$
\begin{aligned}
E_{v1,v2} = {} & h\omega_1(v_1 + 1/2) - hx_{11}(v_1 + 1/2)^2 \\
& + h\omega_2(v_2 + 1/2) - hx_{22}(v_2 + 1/2)^2 \\
& - hx_{12}(v_1 + 1/2)(v_2 + 1/2).
\end{aligned}
\tag{5.13}
$$

The fundamental transition energies ($v = 0 \rightarrow v = 1$) are thus $h\,[\omega_1 - 2x_{11} - x_{12}]$ and $h\,[\omega_2 - 2x_{22} - x_{12}]$, and the overtone energies ($v = 0 \rightarrow v = 2$) are $h\,[2\omega_i - 6x_{ii} - x_{12}]$, which is not twice the fundamental energy. The combination $v_1 + v_2$ occurs at $h\,[\omega_1 + \omega_2 - 2x_{11} - 2x_{22} - 2x_{12}]$, which is not equal to the sum of the fundamental energies. The measurement of these five transition energies, however, allows us to calculate all five of the constants (ω_i, x_{ii}, x_{12}) in Eqn (5.13), and enables harmonic vibration frequencies ω_e to be deduced from anharmonic transition energies.

Difference bands ($v_i - v_j$) behave exactly like the corresponding combination bands ($v_i + v_j$) as far as symmetry selection rules, band contours and polarization are concerned. However, there is one very important distinction. The energy levels involved in difference bands are the same as those involved in fundamental transitions. Consequently the difference frequency should be precisely equal to the difference between the frequencies of the corresponding fundamentals. For the same reason, difference bands cannot be intensified by Fermi resonance.

Hot bands, as mentioned earlier, show the effects of anharmonicity in the formation of sequences of bands; as they are already formally allowed transitions there is no enhancement in intensity due to Fermi resonance, though mixing of upper-state levels may affect the regularity of hot-band sequences.

5.9 Structural information from vibrational spectra

We have seen how the symmetry of a molecule determines the number of bands expected in IR and Raman spectra, so that we can use the numbers of bands observed to decide on the actual symmetry of a molecule, such as PF_5. We have also seen how Raman polarization data and gas-phase band contours may be used to make complete assignments of the observed bands to vibrations of particular symmetry species. The task of assignment may be complicated by the absence of some expected bands with fortuitously low intensity, or by the presence of formally-forbidden overtone or combination bands, which may have appreciable intensity if the conditions for Fermi resonance are satisfied. Now we must consider in more detail some of the ways we may use vibrational spectra to give structural information about inorganic substances.

There are a number of different levels at which the information content of a vibrational spectrum can be analyzed to allow identification of a sample to be made, or to allow conclusions to be drawn about its structure. At one level, the spectrum as a whole may be treated as a fingerprint, which can be used simply to recognize the product of some reaction as a known compound. This requires the existence of a file of standard spectra. At the other extreme, the numbers of bands observed may be used to deduce the symmetry of the molecule responsible for the spectrum, and the force constants corresponding to all the internal vibrations can be deduced from the frequencies themselves using a normal co-ordinate analysis. At intermediate levels, deductions may be drawn about the presence or absence of specific groups or assemblies

Example 5.2

Reaction between SiH_3I and $Hg(SCF_3)_2$ gave $S(CF_3)(SiH_3)$, but this decomposed in the presence of the mercury salt to give a mixture of volatile products whose IR spectrum is given below. Deduce what you can about the decomposition products from the frequencies and shapes of the bands.

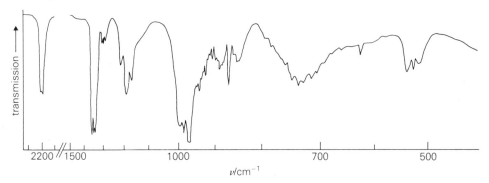

Redrawn from *Infra-red Spectroscopy and Molecular Structure*, ed. M. Davies, Elsevier, Amsterdam (1963).

The spectrum contains bands with well-marked contours, which suggests that the products are all small molecules. At just above 900 cm^{-1} there is a partly-obscured band showing well-defined rotational detail with the 'strong, weak, weak' pattern of intensity alternation that is characteristic of a molecule with a three-fold axis in which the only atoms off the axis are hydrogens. The only candidate for such a molecule in this system is $SiFH_3$. The infrared spectrum of $SiFH_3$ should contain bands due to Si—H stretching, near 2200 cm^{-1}; SiH deformation (parallel and perpendicular; near 1000 cm^{-1}); Si—F stretching (parallel; near 900 cm^{-1}), and SiH_3 rocking (perpendicular; near 750 cm^{-1}). (See Tables 5.7 and 5.9.) All these bands are observed in the spectrum, and so we can conclude that one major product of the decomposition is $SiFH_3$. A comparison with the published spectrum confirms this. If we write the equation for decomposition as giving $SiFH_3$, we find a clue as to the possible other product: $S(CF_3)(SiH_3) \rightarrow SiFH_3 + CF_2S$.

CF_2S would belong to the point group C_{2v}; its IR spectrum should show bands with the characteristic contours of types a, b and c; we would expect to find the S—C stretch around 1300 cm^{-1}, C—F stretches around 1200 cm^{-1}, and deformation modes at lower frequencies; the expected modes and contours are: S—C stretch, $a_1:v_1$ (a-type); C—F stretch, $a_1:v_2$ (a-type) and $b_2:v_5$ (b-type); CF_2 bend, $a_1:v_3$ (a-type); CF_2 rock, $b_2:v_6$ (b-type); CF_2 wag, $b_1:v_4$ (c-type).

The spectrum observed can be accounted for in terms of this compound. The very strong a-type band at 1365 cm^{-1} is assigned to v_1, the S—C stretch. The weak band near it, of the same shape and at 1312 cm^{-1}, is a second order band intensified by Fermi resonance. Similarly, the complex envelope at around 1200 cm^{-1} can be interpreted as two overlapping b-type bands; one is the fundamental v_5 and the other a second order band intensified by Fermi resonance with v_5. The c-type band at 610 cm^{-1} is assigned to the out-of-plane deformation mode, v_4, and the a-type band at 526 cm^{-1} to the CF_2 deformation mode, v_3; further work revealed a b-type band at 422 cm^{-1}, assigned to the CF_2 rocking mode. Plausible assignments of the two second order bands involve the 'missing' fundamental v_2 in combination with v_3 ($a_1 \times a_1 = a_1$; a-type at 1305 cm$^{-1} \approx 526 + 779$) and with v_6 ($a_1 \times b_2 = b_2$; b-type at 1188 cm$^{-1} \approx 422 + 766$) suggesting that v_2 lies near 770 cm^{-1}, where it is hidden by the $SiFH_3$ band at 745 cm^{-1}

of atoms by looking for characteristic group frequencies and intensities. The effects of isotope substitution on the observed spectrum may give valuable information about the numbers of atoms of a specific element present, and about any symmetry relationship between their positions. We shall consider these methods of analysis of vibrational spectra in more detail in the following sections.

5.10   Fingerprints

5.10.1   Comparison with standard spectra

This is a traditional approach, and probably accounts for most vibrational spectra actually recorded. Comparing two spectra, which involves taking into account band shapes and intensities as well as frequencies, is one of the tasks the human eye/brain seems to do rather well. Strenuous efforts have been made to program such recognition abilities into microcomputers and several packages for manipulating and comparing spectra are available. But it is still hard to beat the experienced eye, which can automatically focus its attention on the most significant features of a spectrum. Unless the range of possibilities is very limited, or only simple questions are asked ('Is this compound X?' or 'Is this sample of compound X pure?'), large numbers of standard spectra are needed for comparison. Such standard spectra have been available as drawings for many years, and computer-compatible forms are now becoming available. Most general collections are of spectra of organic samples, so that local collections covering special groups of compounds are of more immediate use to the inorganic chemist. Some collections are commercially available [14].

5.10.2   Detection and identification of impurities

If our compound has already been characterized, any bands which are present in its vibrational spectrum but are not found in that of a pure sample may be assigned to impurities. It is, of course, important to ensure that reference and sample spectra are recorded under the same conditions of phase, temperature, concentration and instrument resolution so that the comparison can be as detailed and reliable as possible.

Detection of impurities, then, is often quite easy. If we are only monitoring a process of purification, it may not be necessary to identify impurities, but unless we do, we cannot be sure how sensitive the check on purity is. An immediate complication is that an impurity is only a minor component of a mixture, and so some regions of its spectrum may be difficult to study. In particular, if product and impurity are chemically similar, their spectra are likely to be very similar too, and most of the impurity bands may be masked by those due to the desired product. Usually, we can find at least one band to use as a measure of the amount of impurity still present: if $S(SnBu_3)_2$ is made from $SnBu_3Cl$, the residual chloride can be detected by the only distinct band, that due to the Sn—Cl stretching mode. In extreme cases, no such band can be identified, and purity must be determined by other methods. It is important to remember that the absence of impurity bands in a spectrum does not mean that the sample studied was pure.

Band shapes can also be useful for identifying unknown impurities, particularly in the gas phase. The width of a band gives some idea of the size of the molecule involved.

In extreme cases, resolution of rotational fine structure can identify a molecule such as HCl unequivocally.

Quantitative analysis of mixtures

Vibrational spectroscopy has been used successfully for quantitative analysis, but considerable care is needed to ensure accurate results. This is particularly true with conventional IR spectrophotometers, which have limited photometric accuracy. With modern ratio-recording double-beam spectrophotometers and FT interferometers this problem is much less severe, and ordinate repeatability of 0.1% transmission is now routinely available. In the IR, the usual presentation of the spectrum in terms of transmittance T means that peak height is not linearly related to the intensity of absorption. Some spectrometers can be operated so as to produce a spectrum of absorbance A, where peak height is a direct measure of intensity. In terms of the light passing through the sample, I, compared with the initial light intensity I_0,

$$T = \frac{I}{I_0} \times 100 \tag{5.14}$$

whereas

$$A = \log_{10} \frac{I_0}{I}. \tag{5.15}$$

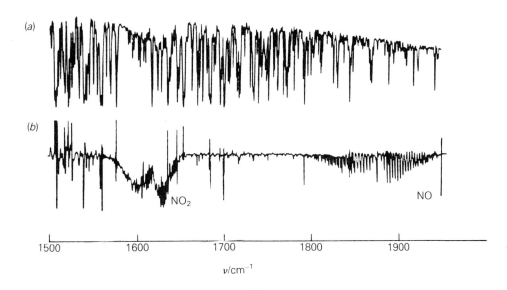

Fig. 5.36 (a) IR spectrum of a sample of air, contaminated with 5 ppm of NO_2 and 10 ppm of NO. The spectrum, obtained with a cell with a path length of 81 m, is dominated by lines due to water. Spectrum (b) is the ratio of (a) with a spectrum of uncontaminated air. Bands attributable to NO and NO_2 can clearly be seen. Adapted, with permission, from P.L. Hanst, A.S. Lefohn and B.W. Gay, *Appl. Spectrosc.*, **27**, 188 (1973).

SECTION 5.10: FINGERPRINTS 219

In the Raman, the intensity of a peak due to a given species is directly related to its concentration.

Another problem is that each IR or Raman band of each compound has its own characteristic intensity; concentrations of different compounds cannot be directly compared by a simple comparison of relative intensities of bands in a mixture. However, the intensity of a particular band does vary linearly with concentration in a mixture unless some strong chemical interaction occurs between components, in which case mixtures of the same components in known concentrations must be used as standards. Once this has been done it is possible to make single-frequency readings to give quantitative concentrations of constituents of a mixture. With suitable sampling arrangements a standard spectrometer can be used to make a wide variety of different measurements. Thus trace gases can be detected in air using a long-path gas cell and/or high pressure (Fig. 5.36), hydrogen atoms can be detected in polycrystalline silicon [15], inorganic fillers can be identified and measured in plastics (Fig. 5.37), and $SiMe_3$ end groups can be determined quantitatively in silicones (Ref. [16], p. 282), all with a standard IR instrument. Species present in aqueous solutions can be studied, most conveniently by Raman spectroscopy, so that, for example, the equilibria involved in the aqueous carbonate/bicarbonate/CO_2 system can be understood [17]. Many other examples of studies of electrolytes in aqueous solution by Raman spectroscopy have been reviewed [18].

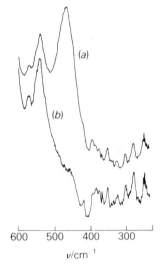

Fig. 5.37 Fourier transform IR spectra of (*a*) polythene containing 0.04% by weight of SiO_2 as a filler, and (*b*) pure polythene. The strong band near 470 cm^{-1} is clearly due to the filler, and its intensity can be used to determine the amount of filler in the composite material. Redrawn from D.R. Battiste, J.P. Butler, J.B. Cross and M.P. McDaniel, *Anal.Chem.*, **53**, 2232 (1981).

5.11   Group frequencies

So far we have considered each molecule and its spectrum in isolation. Indeed, no relationship between the vibration frequencies of different molecules is logically necessary. In practice, however, it is found that there are similarities between the frequencies of molecules containing similar groups of atoms, and this sort of information can be applied to the study of new compounds. Such group-frequency correlations have been extensively developed for organic compounds, and similar correlations can be drawn up for inorganic species [19]. These correlations arise because a normal mode, although formally involving the whole molecule, may actually involve significant motions of only a few atoms constituting a discrete group, and hence have a frequency primarily determined by the group and only slightly affected by the nature of the rest of the molecule. In their simplest form, group frequency correlation tables allow the identification of bonded atom pairs, because bond stretching frequencies are the most characteristic. The correlations can be extended to answer a wide variety of chemical questions. We may be able to determine the number of atoms in a group, and so to distinguish MX_2 from MX_3 groups. We can distinguish monodentate from bidentate sulfate, or terminal from bridging CO ligands, and we can use variations in the CO stretching frequencies in metal carbonyls to make deductions about the electronic natures of the other ligands. The processes involved are described and illustrated below.

5.11.1   Characteristic bond stretching frequencies

The vibration frequency associated with a particular bond depends primarily on the masses of the bonded atoms and the force constant of the bond, and to a lesser extent on the influence of the rest of the molecule. We shall see below how these effects can be treated more formally, but we can deduce some general principles at this stage:
1   Characteristic stretching frequencies are lower for heavier atoms;
2   Characteristic stretching frequencies are lower for weaker bonds, where the restoring forces are smaller;
3   Stretching frequencies vary over a fairly narrow range in a set of related compounds.
The lightest atom, hydrogen, gives rise to some of the most characteristic group frequencies. Virtually all bond stretches involving hydrogen bonded to one other atom occur in the range ~ 4000 to $1700\ cm^{-1}$ (H—F to H—Pb) and the position in this range correlates well with the mass of the other atom and the strength of the bond. Thus in going down any Main Group of the Periodic Table (p-block elements), both the increase of atom mass and the decrease in bond strength lead to a lowering of the stretching frequency. If we go from left to right along a row of p-block elements, the effect of increasing mass is outweighed by the increasing bond strength, and frequencies increase (see Table 5.7).

Among transition metal hydrides the mass of the metal atom has rather little effect; frequencies are in the range 2200-$1700\ cm^{-1}$, the precise value depending on the metal and the other ligands involved. More subtle correlations valid over limited classes of compounds may be found useful here.

The stretching frequencies of bonds not involving hydrogen are generally lower.

Table 5.7   Typical M—H Bond Stretching Frequencies for p-block Elements

B 2300	C 3000	N 3400*	O 3700*	F 3962*
Al 1800	Si 2200	P 2350	S 2600	Cl 2836
Ga 1850	Ge 2100	As 2200	Se 2350	Br 2559
	Sn 1850	Sb 1900	Te 2070	I 2230

*Note: normally reduced considerably by hydrogen bonding in condensed phases.

Except for multiple bonds with very high force constants or for single bonds between light first-row elements (e.g. C—F, B—O), all fall below 1000 cm^{-1}. Group frequency correlations have been established for the most common and widely-studied elements, and an experienced worker soon learns to recognize diagnostic bands relevant to his own field of interest. It must be remembered, however, that the ranges within which particular bond stretches appear may overlap, and circumstances sometimes arise in which a band may be considerably displaced from its expected position. The most likely reason for this is vibrational coupling. If there are groups with similar vibration frequencies close to each other in the molecule, and two or more of the vibrations are of the same symmetry, the group modes will mix; mixing displaces the normal modes to higher and lower frequencies. In these circumstances the observed bands represent mixtures rather than particular group vibrations, and the analysis of the spectrum based on group frequencies no longer applies. The coupling is generally greatest if the displacement vectors lie along the same axis, and the linking atom is relatively light. This coupling between group modes of the same symmetry should not be confused with Fermi resonance between a fundamental and an overtone or combination of the same symmetry [see Section 5.8.6 above]; in the case of coupling between fundamental group modes there is no violation of the $\Delta v = 1$ vibrational selection rule.

An important group frequency is that due to the CO stretching motion of carbonyl ligands in transition metal complexes. The average stretching frequency for a terminal carbonyl ligand in a complex correlates with the electron-richness of the metal. Back-bonding from the d-orbitals of the metal into the π^* antibonding orbitals of the ligand weakens the CO bond and lowers the stretching frequency from its value in free CO. Many examples can be found in the literature. Here we shall note the unambiguous data for the series of complexes $[Co(CO)(NO)(PCl_xPh_{3-x})_2]$ (Table 5.8). It is clear that electronegative chlorine substituents on the phosphine ligands decrease the electron density on the central Co atom, thereby decreasing the $d \rightarrow \pi^*$ back-bonding and raising the CO and NO bond stretching frequencies. Note that solvent effects on CO stretching frequencies in metal complexes are often appreciable, and comparisons should always be made between samples of similar concentration in the same solvent.

Table 5.8 NO and CO Stretching Frequencies for $Co(CO)(NO)(PCL_xPh_{3-x})_2$

x	$v(CO)/cm^{-1}$	$v(NO)/cm^{-1}$
3	2044.5	1793.2
2	2023.6	1771.7
1	1987.3	1747.2
0	1956.7	[1714]*

*Note: different solvent, see text.

5.11.2   Characteristic deformation (angle bending) frequencies

Angle bending modes may give rise to characteristic frequencies just as stretching modes do, but bending frequencies are much more variable. For example, the angle bending motions of germyl halides, GeH$_3$X, consist of two deformation modes and one rocking mode, numbered ν_2, ν_5 and ν_6, respectively. The observed frequencies are shown in Table 5.9, and observed IR spectra of some germyl halides are shown in Fig. 5.38. It will be seen that ν_5 is at almost precisely the same frequency in each case, but that both ν_2 and ν_6 vary in frequency with the mass of the halogen. The characteristic parallel band contour for ν_2 (a_1) and perpendicular contours for ν_5 and ν_6 (e) leave no

Table 5.9 Frequencies (cm^{-1}) of Deformation and Rocking Modes of Germyl Halides

Mode	Fluoride	Chloride	Bromide	Iodide
ν_2	874	848	833	812
ν_5	874	874	871	872
ν_6	643	602	578	558

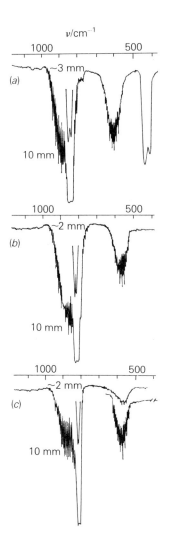

Fig. 5.38 IR spectra of gaseous germyl halides GeH$_3$X with (a) X = Cl, (b) X = Br and (c) X = I showing the two deformation bands near 850 cm^{-1} and the rocking band near 600 cm^{-1}. The extra band at lowest frequency in (a) is due to the Ge—Cl stretch. The pressures of gas used are marked on each spectrum. Taken, with permission, from D.E. Freeman, K.H. Rhee and M.K. Wilson, *J.Chem.Phys.*, **39**, 2908 (1963).

doubt about the correct assignments, and so we may say that the germyl rocking frequency characteristically lies between 550 and 650 cm^{-1}.

The correlation extends beyond this limited set of compounds. All GeH$_3$-compounds give a characteristic strong IR band near 850 cm^{-1}, attributable to a deformation mode of the germyl group; the exact frequency varies with the mass of the attached group.

5.11.3 Characteristic patterns of group frequencies

A clearly-defined group frequency like v(CO), which gives bands in a region where few other fundamentals are likely to occur, can be very important in determining how many examples of the group occur in each molecule and the symmetry relationships between them.

We have already noted that there is one stretching mode for each bond in a molecule. Thus in the absence of complications due to symmetry, we may in principle count the number of CO ligands in a complex simply by counting the CO stretching bands in the vibrational spectrum. If there are symmetry operations relating equivalent groups, two other factors have to be considered.

First, symmetry selection rules govern the activity of the various stretching motions in IR and Raman spectra. It may therefore be necessary to record both spectra to check for modes inactive in one but active in the other. On the other hand, such rules give us additional information about the structure of the complex.

Secondly, if there is a rotation axis relating three or more CO ligands, the number of bands will be less than the number of ligands because some modes will be degenerate. Thus a complex [M(CO)$_3$L] has only two v(CO) bands (a_1 and e) in its vibrational spectrum, so long as the ligand L preserves the 3-fold symmetry of the M(CO)$_3$ group.

In some examples, both factors are important. In the carbonyl stretching region of the IR spectrum of tetrahedral Ni(CO)$_4$ there is only one band, assigned to the triply-degenerate t_2 mode. The fourth stretch, of a_1 symmetry, is only Raman-active.

The symmetry species of a set of equivalent bond stretches may be worked out very simply. The method is the same as was used to determine the OH stretching modes of H$_2$O in Section 5.7.3. We do not always have to work through the whole process though, because the patterns for two, three, four or more symmetry-related modes are reproducible from one molecule to another. However, the labels for the symmetry species depend on the point group, and may be deduced from the appropriate character table.

In a *cis*-octahedral complex [M(CO)$_4$L$_2$] (5.III), the two CO groups *trans* to each other must logically be treated together, as must the other two, which are mutually *cis*.

5.III

We must not try to treat all four CO groups together, as the two pairs are not equivalent. When the four CO ligands are treated as two sets of two, it is immediately clear that as each pair is related by the C_2 axis, there must be an a_1 and a b component from each pair. Moreover, as the two pairs of bonds define two planes at right angles to each other, one of the two b components is b_1 and the other b_2, in the overall C_{2v} symmetry of the complex. We therefore expect two a_1 modes, one b_1 and one b_2 in the CO stretching region. All four will be both IR and Raman active; the two a_1 modes are in principle distinguishable from the others by giving rise to polarized Raman bands in solution or in the gas phase. In contrast, the four CO groups of trans-$[M(CO)_4L_2]$ (5.IV) are all symmetry-related, giving a_1', b_2' and e' stretching modes, only one of which one (e') is IR-active and all three are Raman active.

5.IV

We can extend this kind of approach to make a complete analysis of the vibrations of quite complex molecules by making use of local symmetry to factorize the modes. Let us consider cis-$[M(CO)_4(PH_3)_2]$. The overall point group is C_{2v} if we suppose that the PH_3 groups rotate freely. The four (CO) stretches have just been described; the two (MP) stretches can be analyzed in the same way, and the deformations can be treated similarly. The internal modes of the PH_3 groups present more of a problem. They can be analyzed using local symmetry. The local symmetry of M—PH_3 is C_{3v}, and the PH_3 modes are like those of the CH_3 group in CH_3X. There are a_1 and e PH stretches, a_1 and e PH_3 deformations, and e PH_3 rocks. In CH_3X, the a_1 modes give rise to dipole changes parallel to the C—X axis, and the e modes give rise to dipole changes perpendicular to that axis. In the complex we can use a similar basis to classify the PH_3 modes as symmetric (giving dipole changes parallel to the M—P bond) and antisymmetric (giving dipole changes perpendicular to the bond). We then combine the local modes of the two PH_3 groups within the framework of the overall molecular point group, which is C_{2v}. Because there are two PH_3 groups, we have two 'symmetric' PH stretches, made up by combining the dipole changes along the M—P bonds. If we represent the dipole changes by arrows, as in Fig. 5.39, it is clear that the two local modes can be combined in phase, giving a molecular mode of a_1 symmetry, or out of phase, giving a mode of b_2 symmetry.

The antisymmetric PH stretches are doubly degenerate in the local symmetry; in the overall point group of the molecule, this degeneracy is broken. The associated dipole changes must be perpendicular to the M—P bonds, but the dipole arrows may lie either in the PMP plane or perpendicular to it (see Fig. 5.39), and the two sets are not equivalent. If we take the in-plane modes first, the two dipole arrows can be taken in phase, giving us another a_1 mode, or out of phase, giving us another b_2 mode. The

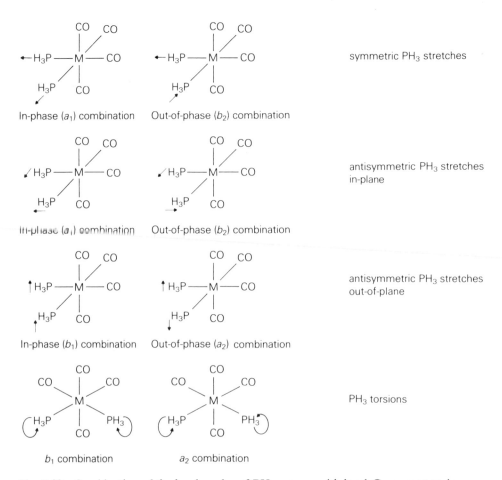

Fig. 5.39 Combination of the local modes of PH_3 groups, with local C_{3v} symmetry, in a molecule with overall C_{2v} symmetry. The local dipole changes associated with local stretching and deformation modes are represented by straight arrows, and local torsional motions are represented by curved arrows.

two modes have the same symmetries as the two derived from the internally symmetric PH_3 stretches, but they involve quite different types of motion. Taking the out-of-plane dipole arrows together, we obtain b_1 and a_2 modes. So we can classify the six PH stretches we expect as a_1 and b_2 derived from internally symmetric modes and usually labelled v_s, and a_1, a_2, b_1 and b_2 derived from internally antisymmetric modes and usually labelled v_a. Both $a_1(v_s)$ and $a_1(v_a)$ should in principle give rise to polarized Raman lines, but the line from the v_s mode is likely to be more strongly polarized.

We can analyze the PH_3 deformation and rocking modes in exactly the same way. There are six PH_3 deformation modes: a_1 and b_2 based on symmetric and a_1, a_2, b_1 and b_2 based on antisymmetric local motion. The rocking modes are locally antisymmetric, and so they combine to give four modes of a_1, a_2, b_1 and b_2 symmetry. The remaining local modes are torsional. They derive from molecular rotation of CH_3X, and can be worked out by representing the local motion by arrows curved round the M—P bonds. In C_{2v} these combine as a_2 and b_1. In this way we have been able to analyze and classify all the internal modes of the PH_3 groups.

CHAPTER 5: VIBRATIONAL SPECTROSCOPY

5.11.4   **Group frequencies and types of ligand binding**

Many ligands have several different modes of binding to other atoms. A classic example is carbonyl, which is known in a number of different environments, including terminal, bridging and triple-bridging (5.V). Bridging CO ligands normally have lower stretching frequencies than terminal ligands in complexes with the same metal and similar overall electron density. The known range of frequencies in terminal sites is very wide (2130-1700 cm^{-1}) but frequencies of bridging ligands lie between 1900 and 1780 cm^{-1}.

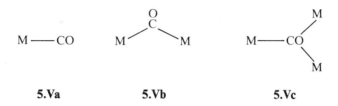

5.Va   **5.Vb**   **5.Vc**

A CO band above 1900 cm^{-1} is therefore almost certainly due to a terminal CO; a band below this frequency may be due to a bridging ligand, but it could also be ascribed to a terminal ligand with some unusually severe reduction of the CO bond strength through $d \rightarrow \pi^*$ back-bonding. Only if the complex also shows carbonyl bands well above 1900 cm^{-1} can one be confident that such electronic effects are absent, and safely assign low-frequency CO stretches to bridging ligands. The IR spectrum of a solution of $Ru_3(CO)_{12}$, for example, has CO stretching bands at 2060, 2030 and 2010 cm^{-1} only; $Fe_3(CO)_{12}$ shows bands at 2043, 2020, 1997 and 1840 cm^{-1} under the same conditions [19]. We will not assign these spectra in detail, but it is now generally accepted that the ruthenium complex has three $Ru(CO)_4$ units held together only by metal–metal bonding, whereas the iron complex has bridging CO ligands, as well as many terminal ones.

Halogen atoms may also act as bridging ligands, sometimes forming simple dimers such as (Al_2Cl_6), and sometimes giving extended polymers such as $[BeCl_2]_x$. By studying a range of compounds of known structure, it is sometimes possible to define characteristic M—X stretching frequencies for terminal and bridging ligands, and to use the observed spectrum of a new compound to establish the presence or absence of bridging ligands.

Polyatomic ligands may have several potential donor sites, and we can sometimes identify various possible modes of attachment by their characteristic group frequencies. The monothioacetate ligand, for instance, can attach to a metal:

1   through oxygen only, to one metal atom (5.VIa);
2   through sulfur only, to two or to one metal atom (5.VIb and c);
3   through both oxygen and sulfur to a single metal atom (5.VId); or
4   through oxygen to one metal atom and through sulfur to another (5.VIe).

Of these cases, only (2) leaves the ligand with an uncomplexed >C=O group, which gives rise to a characteristic CO stretching band near 1600 cm^{-1} in the IR spectrum. The C=S stretch is less characteristic, being weaker and in a more crowded region of the spectrum, but a band near 950 cm^{-1} may be taken as indicating case (1). Cases (3) and (4), involving chelating and bridging ligands, lead to slightly reduced frequencies

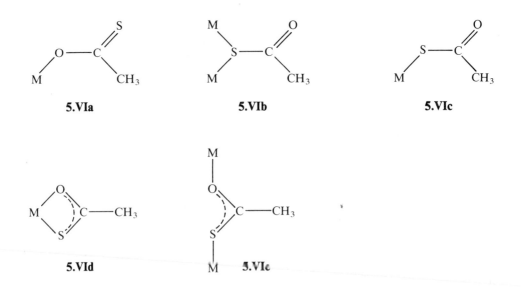

5.VIa   5.VIb   5.VIc

5.VId   5.VIe

for both bond stretches, down to about 1500 cm^{-1} for C=O and about 900 cm^{-1} for C=S, and it is not usually possible to distinguish them in this way.

The perchlorate ion can occur as isolated, undistorted ions, or as distorted but essentially unattached ions, loosely attached through one or two oxygen atoms to one or two metal ions, or it may take part in hydrogen bonding to water molecules in hydrated salts. The IR spectrum of the isolated undistorted ion shows only one band, v_3, in the Cl=O stretching region at 100 cm^{-1}; the Raman spectrum shows v_1 as well, near 900 cm^{-1}. Distortion of the ion makes v_1 IR active, and also lifts the triple degeneracy of v_3. If the ion is attached through one oxygen, so that its symmetry is reduced from T_d to C_{3v}, v_3 splits into two components (a_1 and e), while if the ion is attached to one or more metal atoms through two of its oxygens the degeneracy is entirely removed and three components of v_3 (a_1, b_1 and b_2) appear, as well as v_1. Distortions due to the formation of hydrogen bonds are usually smaller than those due to co-ordination to metal ions. For perchlorate, it appears that at least in aqueous solution a symmetrically-hydrated ion is formed, as only a single v_3 bands is seen; for other oxyanions, however, distortions attributed to hydrogen-bond formation have been claimed on the basis of splittings of v_3 bands.

5.12   Use of isotopes in interpreting vibrational spectra

As we have seen, vibration frequencies depend on the masses of moving atoms. If we substitute an atom in a molecule by an isotope of different mass, we alter the frequencies of some modes – but only those in which the substituted atom moves significantly. This can be a considerable help in assigning bands to particular vibrations, especially where several bands are close together in frequency. If the substitution is only partial (involving only one, say, of a group of symmetry-related atoms) the resulting reduction of molecular symmetry may alter the selection rules. This can allow us to draw conclusions about the number of atoms involved and the symmetry of the group. A particularly important case involves deuterium substitution in hydrogen-containing species.

H/D substitution

The large relative mass increase here leads to large decreases in vibration frequencies, by a factor of up to 0.717 ($1/\sqrt{2}$). Thus all M—H stretching frequencies decrease by several hundreds of cm^{-1}, and this can be used to confirm the group-frequency assignment of such modes, and also to remove M—H bands from regions where they may hide other bands. This is particularly important for the region from 2400 to 1600 cm^{-1}, which covers stretching modes of many multiple bonds including those of CO ligands in metal carbonyls. For example, there are four IR bands near 2000 cm^{-1} in the IR spectrum of [Co(CO)$_4$H] [20]. Only the lowest of these shifts appreciably on deuterium-substitution (see Table 5.10), and so this is assigned to the Co—H stretching mode. The other three high-frequency bands must be due to CO stretches.

Deformation modes involving hydrogen also drop dramatically in frequency on isotopic substitution. This can again serve both to confirm their assignment and to allow other modes in the same region to be distinguished. However, mixing of modes of the same symmetry and similar frequency often occurs, so that deuterium substitution causes shifts of more bands than might have been expected. In [Co(CO)$_4$H], for instance, there is only one expected deformation that mainly involves motion of the H atom; it is doubly degenerate. In the observed spectrum we find that several low frequency bands shift on deuterium substitution. This shows that several normal vibrations contain contributions from this local or internal mode. These bands are listed in Table 5.10. We shall see in Section 5.14 how such frequency shifts may be used to provide precise information about the way in which the local modes contribute to the normal modes of vibration.

Table 5.10   Some IR bands of Co(CO)$_4$H

Co(CO)$_4$H	Co(CO)$_4$D	Assignment	
2121	2120		a_1
2062	2058	ν(CO)	a_1
2043	2043		e
1934	1396	ν(CoH, D)	a_1
703	600		
505	482	deformations involving	
403	393	H-atom motion	
331	296		

Heavy-atom isotope substitution

For atoms other than hydrogen, the relative changes in mass on isotopic substitution are small, and frequency shifts are less dramatic. Indeed, except for first-row elements, band shifts are rarely more than one or two cm^{-1}, and particular precautions must be taken if they are to be observed at all. We must eliminate any sources of broadening contributed by the instrument or sample, to ensure that narrow peaks rather than broad bands are present in the spectra. Gaseous samples may give clear Q-branches, whose positions can be measured precisely, and solutions in non-interacting solvents usually give fairly narrow bands of well-defined shape which can also be useful. Pure liquids and solids often give broad or asymmetric bands and so it is very difficult to measure frequencies precisely. For volatile samples it is sometimes helpful to use matrix-isolation techniques [Section 5.16.3], as the molecules trapped in an inert matrix are generally unable to rotate freely, and give simple sharp vibration bands of well-defined frequency.

Both naturally-occurring and artificial isotopic mixtures can give useful information. Thus the normal 3:1 mixture of chlorine isotopes leads to a characteristic isotope pattern for a single M—Cl bond stretch, with a stronger higher-frequency band due to ^{35}Cl and a weaker, lower-frequency band due to ^{37}Cl. The relative shift $(\Delta v/v)$ will be less than $0.5(\Delta m/m)$, which is readily resolvable if the bands are narrow [Fig. 5.40(a)].

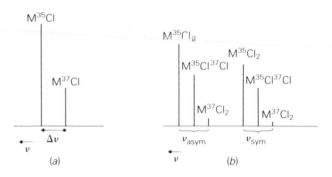

Fig. 5.40 Schematic illustration of band patterns expected from natural isotopes of chlorine for (a) an M—Cl stretching mode and (b) asymmetric and symmetric stretching modes of an MCl_2 group.

A group MCl_2 gives a more complex pattern, as there are three possible isotopic species: $M^{35}Cl_2$ (most abundant), $M^{37}Cl_2$ (least abundant) and also $M^{35}Cl^{37}Cl$. The first two species contain equivalent M—Cl bonds, and so give bands due to symmetric and asymmetric stretches. The bonds in $M^{35}Cl^{37}Cl$ are not equivalent, and its bands usually lie between those of the other two species. The relative abundances of the three species are $9(M^{35}Cl_2) : 6(M^{35}Cl^{37}Cl) : 1(M^{37}Cl_2)$, and we often observe two patterns, each of three bands of intensity-ratio 9:6:1 in decreasing frequency-order, as shown in Fig. 5.40(b). It is often not possible to distinguish the weakest component, and only the relative intensities $(9:6 = 3:2)$ serve to distinguish this case from that of a group containing only one chlorine, for which the intensity ratio is 3:1.

Some elements, such as fluorine and phosphorus, have only one stable isotope, and isotopic substitution cannot be used conveniently to identify vibrations involving them. In other cases, the naturally-occurring element consists mainly of one isotope, and enriched samples must be used. Among the first-row elements, ^{13}C, ^{15}N and ^{18}O are relatively rare, but are obtainable in a variety of chemical forms at various levels of isotopic enrichment. An example of the use of isotopically-enriched ^{15}N is illustrated in Fig. 5.41, in which a band just above 1000 cm^{-1} in the IR spectrum of K[OsO$_3$N] is seen to shift to lower frequency, confirming that it is due to an Os—N stretch.

In recent years, isotopically-enriched samples of various metals and other heavy elements have become available. They have proved to be most useful in identifying bands due to metal—ligand vibrations in complexes, in the same way that samples containing isotopically-substituted ligands allow identification of ligand modes [21]. Fig. 5.42 shows IR spectra of [NiBr$_2$(PEt$_3$)$_2$], made with ^{58}Ni and ^{62}Ni: analysis of isotope shifts enables Ni—Br and Ni—P stretching modes to be identified.

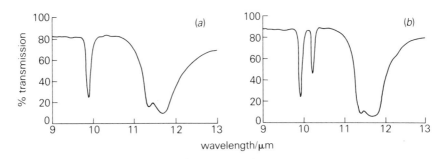

Fig. 5.41 IR spectra of $K[OsO_3N]$ with (a) normal ^{14}N and (b) 30% ^{15}N/70% ^{14}N, showing that the band just above 1000 cm^{-1} (10 microns) shifts to lower frequency and is assignable to the Os—N stretch. The spectra are recorded on a scale linear in wavelength, as is usual in the older literature. Redrawn from J. Lewis and G. Wilkinson, *J.Inorg.Nucl. Chem.*, **6**, 12 (1958).

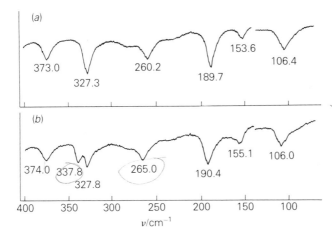

Fig. 5.42 IR spectra of $NiBr_2(PEt_3)_2$ (a) with ^{62}Ni and (b) with ^{58}Ni. Note particularly the bands at 337.8 and 265.0 cm^{-1} for the ^{58}Ni species. These shift to lower frequency for the ^{62}Ni species, for which the higher-frequency band coincides with an unshifted band near 328 cm^{-1}; they are attributed respectively to the asymmetric Ni—Br and Ni—P stretches. The other bands, which show much smaller shifts or remain unaffected by the isotopic substitution, are assigned to modes of the PEt$_3$ groups. Redrawn from K. Shobotake and K. Nakamoto, *J.Am. Chem.Soc.*, **92**, 3332 (1970).

Even the natural level of ^{13}C abundance (1%) may allow weak bands due to 'heavy' molecules to be observed. For example, transition metal carbonyls show weak IR absorptions to low frequency of the very intense ^{12}CO stretching bands. These weak bands are attributed to mono-substituted ^{13}CO species. It may be possible, by considering the possible sites for ^{13}C substitution and the symmetries of the resulting mono-substituted species, to draw definite conclusions about the structure of a complex. In other cases it is possible to substitute progressively pure ^{13}CO for normal ^{12}CO in a complex. The numbers of bands in the spectra of initial, intermediate and final samples will often give an indication of the numbers and relative positions of CO ligands. The application of this method to $Fe(CO)_4$ is described in more detail in Chapter 10.

Isotope studies have been very helpful in studying the reactions of molecular oxygen with metal atoms or halogen atoms. These reactions are carried out in flowing low-pressure gas streams. The products can be trapped in an inert matrix by freezing the mixture of reactants, products and an inert carrier-gas such as N_2 or Ar on a window held at 10-20 K. The reaction of metal atoms with O_2 can lead to the formation of a number of small molecules, and in general more than one of these will be formed in any particular experiment. With halogen atoms X the initial products are usually XOO or XOOX, though the stable form of ClO_2 is known to be OClO. The use of enriched isotopic mixtures of $^{16}O_2$ and $^{18}O_2$ allows such species to be identified as follows.

XOO has two stretching modes, one loosely described as $v(OO)$, the other as $v(XO)$. The use of a 1:1 mixture of $^{16}O_2$ and $^{18}O_2$, with no $^{16}O^{18}O$, can only give two isotopic forms of XOO. As O—X stretching frequencies are much lower than O—O frequencies, the two modes can easily be distinguished. If the precursor O_2 is passed through an electrical discharge so that dissociation and recombination leads to the formation of $^{16}O^{18}O$ as well as the symmetrical species, the products of reaction with X will also include $X^{16}O^{18}O$ and $X^{18}O^{16}O$, which are different molecules; their $v(OO)$ frequencies will probably be quite similar, but their $v(XO)$ frequencies will differ considerably. We therefore expect the two stretching regions of the spectrum to show the patterns illustrated in Fig. 5.43.

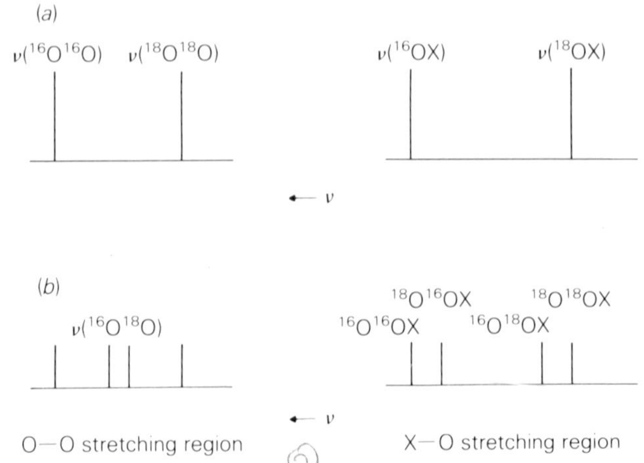

Fig. 5.43 Expected band patterns in the regions associated with O—O and X—O stretching vibrations of XOO, (a) using an unscrambled $^{16}O_2/^{18}O_2$ sample and (b) using a scrambled $^{16}O_2/^{16}O^{18}O/^{18}O_2$ sample.

For XOOX, on the other hand, the species $X^{16}O^{18}OX$ is the only one that contains a mixture of isotopes, so that a sample made from scrambled dioxygen will give three $v(OO)$ peaks rather than four, and six $v(XO)$ peaks. The alternative OXO compound has two oxygen atoms related by symmetry, and so there are only three distinct species. It would give no $v(OO)$ bands, but two $v(XO)$ bands for each species.

The numbers of bands expected in the $v(OO)$ and $v(XO)$ regions for each species, for both unscrambled and scrambled O_2 precursors, are listed in Table 5.11. Fig. 5.44 shows IR spectra for a product formed from fluorine atoms and dioxygen. The pattern

Table 5.11 Expected IR Bands for Possible Products of Reactions of Halogen Atoms with Dioxygen

Product	$^{16}O_2/^{18}O_2$		$^{16}O_2/^{16}O^{18}O/^{18}O_2$	
	$\nu(OO)$	$\nu(OX)$	$\nu(OO)$	$\nu(OX)$
XOO	2	2	4	4
XOOX	2	4	3	6
OXO	0	4	0	6

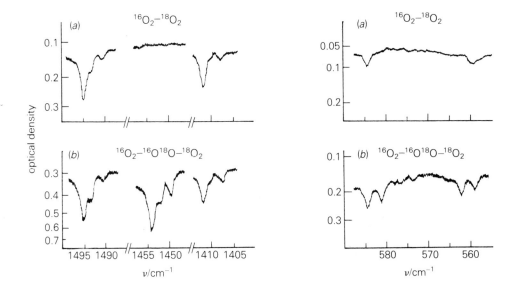

Fig. 5.44 IR spectra of matrix-isolated FO_2 produced by reaction of fluorine atoms with (a) unequilibrated $^{16}O_2$ and $^{18}O_2$ and (b) an equilibrated mixture containing $^{16}O^{18}O$ as well as the symmetric species. Note the two intermediate peaks in the low-frequency (O—F stretching) region that confirm that the F atom is bound asymmetrically, so that the structure is F—O—O. The two bands expected for the O—O stretching bands of the mixed isotope species are too close to appear separately near 1454 cm^{-1}; the additional peaks in the O—O stretching region, which are observed for all isotopic species, are probably due to multiple-site effects. Taken, with permission, from P.N. Noble and G.C. Pimentel, *J.Chem.Phys.*, **44**, 3641 (1966).

of bands shows clearly that FOO has been formed in this case. Under different conditions, FOOF may be formed and identified, and similarly reactions of chlorine and bromine have also been studied.

5.13 Complete empirical assignment of vibrational spectra

We have at last assembled all the information we need to allow the complete assignment of the vibrations of a moderately complicated molecule. We have seen how the vibration frequencies may be observed and how each may be allocated to a particular symmetry species. We have shown how group frequencies derived from experience with other molecules may be used to suggest assignments for particularly characteristic

Example 5.3

The figure (*a*) below shows the Raman spectrum of gaseous O_2 enriched 55% with ^{18}O. Figure (*b*) shows the resonance Raman spectrum band assigned to the O—O stretching mode in the spectrum of oxyhemocyanin, oxygenated with a similarly enriched sample of O_2. It is known that the coordination site of O_2 in hemocyanin contains two copper atoms. What can you deduce about the geometry of coordinated O_2 in oxyhemocyanin?

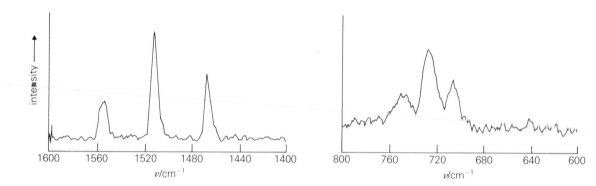

Redrawn from T.J. Thamann, J.S. Loehr and T.M. Loehr, *J. Am. Chem. Soc.*, **99**, 4187 (1977).

....

The three bands in each spectrum are assigned to the O—O stretching modes of $^{16}O_2$, $^{16}O^{18}O$, and $^{18}O_2$. The relative areas under the three components in the two spectra are very similar. Moreover, the central peak in the spectrum of oxyhemocyanin is not significantly broader than the two outer peaks. If O_2 in oxyhemocyanin were coordinated in a way that left the two oxygen atoms in different environments – if, for instance, the O_2 molecule were coordinated end-on or at an angle to the Cu—Cu axis – then molecules containing one atom of ^{16}O and one of ^{18}O could coordinate in two different ways; the band due to $^{16}O^{18}O$ should show two components, or (if the two stretching modes were too similar for full resolution) should at least appear broader than the bands due to coordinated $^{16}O_2$ and $^{18}O_2$. In the absence of any such splitting or broadening, we may conclude that the O_2 molecule is symmetrically coordinated, with the two O atoms in equivalent sites. Of course a small degree of asymmetry could not be detected in this way; the minimum detectable asymmetry could only be determined by calculations.

The frequency of the O—O stretching mode has been reduced on coordination from near 1500 cm^{-1} to about 700 cm^{-1}. In free dioxygen the O—O bond order is two, and this has clearly been reduced, probably to one, in the copper complex. This suggests that the coordinated dioxygen has a hydrogen peroxide-like structure, with a Cu—O—O—Cu linkage. The O—O stretching frequencies in simple dioxygen species are:

Species	Vibrational frequency/cm^{-1}
O_2^+	1858
O_2	1555
O_2^-	1145 (KO$_2$)
O_2^{2-}	738 794 (Na$_2$O$_2$ anhydrous)

bands, and how the use of isotopes may assist us in the deduction of molecular structure and of the types of atoms involved in each mode. We have also seen how the expected vibrations of a model molecule may be organized into symmetry species, and how in simple cases this leads to very simple correlations between the vibration frequencies observed and the expected modes of the model. We must now try to bring together all this information in the assignment of the spectrum of a more complex molecule, attempting at the same time to come to some definite conclusions about its structure.

We will consider $GeCl_2H_2$ and $GeBr_2H_2$, both of which have been studied using IR and Raman spectroscopy. Each contains five atoms, giving nine normal vibrations. Generation of the reducible representation and reduction assuming C_{2v} symmetry (5.VIIa) leads us to expect $4a_1 + 1a_2 + 2b_1 + 2b_2$ vibrations. All of these are Raman active and all except the a_2 mode are also IR active. The list of observed bands (Table 5.12) shows that all the IR bands observed coincide with Raman bands (within 20 cm^{-1}, which is close enough considering that IR and Raman spectra are of different phases), and the number of Raman bands is indeed nine. At this stage we cannot exclude the *cis* square planar alternative structure (5.VIIb), which also has C_{2v} symmetry. This would have $4a_1 + 1a_2 + 1b_1 + 3b_2$ modes, all Raman active and all but one IR active. It is possible to exclude the *trans*-form (5.VIIc), which is centrosymmetric (point group D_{2h}), and should have no IR/Raman coincidences. This structure would give only three Raman-active modes ($2a_g + 1b_{3g}$), rather than the nine bands observed.

Returning to the expected tetrahedral form, we can assign the polarized Raman bands to three of the four expected a_1 modes. The corresponding IR bands have b-type contours. Without calculating moments of inertia about three Cartesian axes we

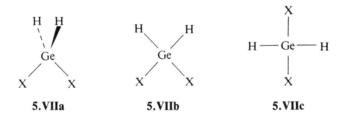

| | 5.VIIa | 5.VIIb | 5.VIIc |

Table 5.12 Observed IR and Raman Bands of Dihalogenogermanes

GeCl$_2$H$_2$ IR(g)	Raman(l)	GeBr$_2$H$_2$ IR(g)	Raman(l)		Assignment
2150 c	2155 *dp*	2138 c	2144 ?	b_1	GeH stretch
2135 b	2132 *p*	2121 b	2114 *p*	a_1	
854 b	840 *p*	848 b	826 *p*	a_1	GeH$_2$ scissors
779 a	772 *dp*	757 a	739 *dp*	b_2	GeH$_2$ wag
—	648 *dp*	—	640 *dp*	a_2	twist
524 c	533 *dp*	489 c	496 ?	b_1	GeH$_2$ rock
435 a	420 ?	322 a	311 ?	b_2	GeX stretch
410 b	404 *p*	290 b	285 *p*	a_1	
not studied	163 *dp*?	not studied	104 *p*	a_1	GeX$_2$ scissors

cannot be sure which band-type corresponds to which of the three IR-active symmetry species, but we now know that the B-axis is associated with a_1 vibrations, and is therefore the z axis. The axis of greatest moment of inertia must be x, as it is perpendicular to the plane containing the three heavy atoms. This is confirmed by the observation that one of the GeH stretching modes (near 2100 cm^{-1} in each case) gives rise to a c-type band. We can therefore make the symmetry assignment of observed bands on the basis:

a_1	IR active, b-type	Raman polarized
a_2	IR inactive	Raman depolarized
b_1	IR active, c-type	Raman depolarized
b_2	IR active, a-type	Raman depolarized

There are, as expected, two c-type and two a-type IR bands, all coinciding with depolarized Raman bands, and one Raman band without a counterpart in the IR spectrum. Three of the four a_1 modes have already been identified, and the remaining a_1 mode must be that at lowest frequency; the Raman band is not discernably polarized for $GeCl_2H_2$, although it is for $GeBr_2H_2$, and no IR band contours are available. All the bands have now been assigned to symmetry species, and all we need to do to complete the assignment is to associate each band with a particular form of vibration, best expressed in terms of local motions such as Ge—H bond stretching.

The nine motions include four bond stretching modes, one for each bond, but these must be combined into symmetry-adapted group modes. The two equivalent Ge—H bond stretches in the xz plane give symmetric and antisymmetric combinations of a_1 and b_1 symmetry respectively, while two equivalent Ge—X bond stretches in the yz plane give combinations of a_1 and b_2 symmetry. The other five motions are all angle-

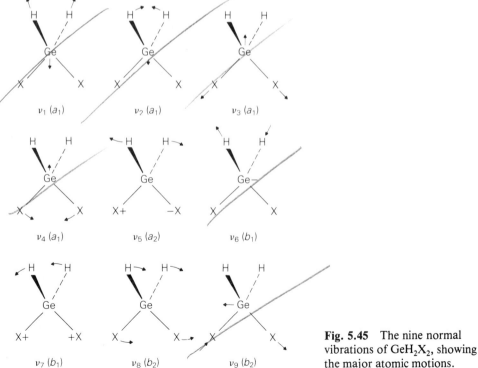

Fig. 5.45 The nine normal vibrations of GeH_2X_2, showing the major atomic motions.

CHAPTER 5: VIBRATIONAL SPECTROSCOPY

bending modes. Removal of the $2a_1 + 1b_1 + 1b_2$ stretches which have already been identified shows that they have symmetries $2a_1 + 1a_2 + 1b_1 + 1b_2$. Both the GeH_2 and the GeX_2 groups have symmetric bending modes of a_1 symmetry. In principle these could mix to some extent, but in practice they can be considered separately as their energies must be very different. The other three modes, each of different symmetry, must involve motions of all four peripheral atoms. As the hydrogen atoms are the lightest they must move most, and it is usual to describe all three as deformation modes of the GeH_2 group (Fig. 5.45).

Group frequency considerations suggest that both GeH stretching modes should be near 2100 cm^{-1}, so the lower-frequency c-type band near 500 cm^{-1} must be a GeH_2 deformation, as is the a_2 mode near 645 cm^{-1}. The two bands near 760 cm^{-1} and 850 cm^{-1} are affected only slightly in frequency by the substitution of Br for Cl, suggesting that they too are essentially GeH_2 deformation modes. This leaves the two bands, near 400 cm^{-1} for the dichloride and near 300 cm^{-1} for the dibromide, as the GeX stretches, and the lowest frequency band in each case as the GeX_2 deformation. The final assignment is then as shown in Table 5.12.

5.14   Normal co-ordinate analysis

We have seen that it is sometimes possible to sort out the observed vibrational bands according to their symmetry, and that group frequency and similar empirical arguments may make it reasonable to associate bands with particular vibrational modes. It is always necessary, however, to remember that the bands are not in fact associated with local, group modes, but with normal modes of the whole molecule. Often particular normal modes are essentially identical to particular group modes, as we have seen for $GeCl_2H_2$ and $GeBr_2H_2$, but in other cases substantial mixing of local modes occurs. This is always to be expected when modes of the same symmetry and similar frequency occur close together in a molecule. In such circumstances we may not describe a band as being due to a single local mode, and the empirical methods we have used so far break down.

For example, we expect GeF_2H_2 to have two modes of b_2 symmetry, giving rise to two a-type bands in the IR spectrum of the gas. Two such bands are observed, at 813 and 720 cm^{-1}, compared with 779 and 435 cm^{-1} in the spectrum of $GeCl_2H_2$. As fluorine is much lighter than chlorine, the antisymmetric GeF_2 stretch is at a considerably higher frequency than the analogous mode of $GeCl_2H_2$, and is therefore closer in frequency to the GeH_2 rocking mode. The consequence is that neither of the two bands observed for the difluoride should be assigned to either of the expected group modes, the GeF_2 antisymmetric stretch and the GeH_2 rock; both are due to whole-molecule modes with contributions from both group modes. The relative contributions can be determined by normal co-ordinate analysis.

Again, we may consider the two stretching motions of each of the linear triatomic species OCS, CO_2 and NCO$^-$. For OCS it is reasonable to argue that the higher frequency motion ($v_1 = 2062$ cm^{-1}) corresponds to the CO stretch and the lower frequency ($v_3 = 859$ cm^{-1}) to the CS stretch, as the two terminal atoms have very different masses. At the other extreme, for CO_2, the terminal atoms are the same, and so we must have complete mixing to give two symmetry-adapted normal modes, whose harmonic frequencies are $v_1 = 1351$ cm^{-1} and $v_3 = 2396$ cm^{-1} (Ref. [11], p. 222).

Example 5.4

The infrared spectra of gaseous SiF_2H_2 and SiF_2D_2 are given below. Make as complete a vibrational assignment as you can.

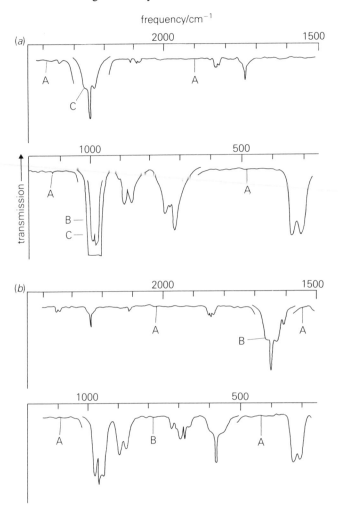

Spectrum (a) is SiF_2H_2 at pressures A 80, B 20 and C 4 mm Hg; spectrum (b) is SiD_2F_2 at A 50 and B 7 mm Hg. Redrawn from S. Cradock, E.A.V. Ebsworth and A.G. Robiette, *Trans. Faraday Soc.*, **60**, 1502 (1964).

....

SiF_2H_2 and SiF_2D_2 have C_{2v} symmetry. The vibrational modes, with their symmetry classes, are given in Table 5.5. In this case, we need to discuss corresponding modes for SiF_2H_2 and SiF_2D_2. The symmetry classes of modes do not change on isotopic substitution, but the order of the frequencies may. For instance, Si—F stretching modes may well be lower than SiH deformation frequencies but higher than SiD deformations. If we followed the mode-numbering rules strictly, we should have to use different numbering schemes for the two molecules. To avoid this confusion we use a common scheme, but mark the mode labels ν_1, ν_2, *etc.* with a prime when they refer to SiF_2D_2, to indicate non-standard numbering.

We expect to see four bands due to the totally symmetric modes and these will have b-type contours. There should also be two c-type bands due to the b_1 modes and two a-type bands due to b_2 modes. The a_2 mode is IR-inactive. At first sight the spectrum of SiF_2H_2 is

CHAPTER 5: VIBRATIONAL SPECTROSCOPY

not consistent with these expectations. We can only see one c-type band, one a-type band, and two b-type bands of moderate or greater intensity, and there is a broad, strong band near 700 cm^{-1} with a most peculiar contour, one that does not match any of the expected band shapes for a C_{2v} molecule. However, the spectrum of SiF_2D_2 is much easier to assign. The two Si–D stretching modes appear near 1650 cm^{-1}; the stronger, c-type, band is assigned to ν_6', and the weaker, b-type band due to ν_1' can be distinguished on the low frequency side. The a-type band at 960 cm^{-1} and the b-type band at 878 cm^{-1} are due to the Si–F stretching modes, ν_8' and ν_3' respectively; they appear at about the same frequencies in SiF_2H_2. Similarly, the b-type band at 317 cm^{-1} is assigned to the SiF_2 bending mode; it too barely shifts in frequency from SiF_2H_2 to SiF_2D_2. There remain the SiD_2 deformation modes; we expect one a-type, one b-type and one c-type band, and that is what is observed: ν_2' (b-type) at 707 cm^{-1}, ν_9' (a-type) at 680 cm^{-1} and ν_7' (c-type) at 579 cm^{-1}. None of these three bands has counterparts in the spectrum of SiF_2H_2.

Looking back at the spectrum of SiF_2H_2, we can begin to understand the apparent anomalies. The two Si–H stretching modes, partly resolved in SiF_2D_2, overlap so completely that only the c-type contour due to ν_5 can be observed. Between 800 and 1000 cm^{-1} we only observe one strong a-type and one strong b-type band. Multiplying the frequencies of ν_2' (the a_1 SiD_2 deformation) and ν_8' (the b_2 SiD_2 deformation) by 1.4, which should give us rough values for the corresponding frequencies for SiF_2H_2, leads to frequencies of about 990 cm^{-1} for the a_1 mode and about 950 cm^{-1} for the b_2 mode. This means that they would come close to the Si–F stretching modes of the same symmetry; coupling and intensity borrowing might well lead to a failure to observe the separate modes. If you look carefully at the spectrum near 900 cm^{-1}, you can see a weak band that probably represents ν_3. This leaves us with the problem of ν_7, the c-type SiH_2 deformation mode, clearly visible at 579 cm^{-1} in the spectrum of SiF_2D_2. Multiplying this frequency by 1.4 we obtain 780 cm^{-1} for this mode in SiF_2H_2. This is close to the broad and peculiar envelope centered at around 730 cm^{-1}. The c-type band in the spectrum of SiF_2H_2 is obviously perturbed by some factor we have not allowed for. The most likely candidate is the a_2 mode, which though forbidden can perturb allowed modes if it almost coincides with them. There is a method of calculating the frequencies for an isotopically substituted molecule fairly precisely from those of another isotopic species. By this method, we obtain the set of fundamental frequencies for SiF_2H_2 and SiF_2D_2 that are given in the table below. All the tentative suggestions made above are substantiated. Furthermore, the IR spectrum of solid SiF_2H_2 shows not two but four strong bands in the region between 800 and 1000 cm^{-1}; in the condensed phase the frequency-shifts and the restriction on molecular motion eliminate the coupling and intensity borrowing that make the spectrum of gaseous SiF_2H_2 so puzzling at first sight.

Mode	Symmetry species	Band contour	SiF_2H_2 frequency/ cm^{-1}	SiF_2D_2 frequency/ cm^{-1}
ν_1	a_1	b	2246	1616
ν_2	a_1	b	982	707
ν_3	a_1	b	870	878
ν_4	a_1	b	322	317
ν_5	a_2		ca.730?	ca.532?
ν_6	b_1	c	2251	1651
ν_7	b_1	c	ca.730	579
ν_8	b_2	a	981	960
ν_9	b_2	a	903	680

In the NCO^- ion the terminal atoms have similar masses and the force constants $f(CO)$ and $f(CN)$ are expected to be much the same. We might therefore expect to find two local modes of rather similar frequency. Instead we find two normal modes of very different frequencies, showing that again we have substantial mixing of the local modes, even though this is not required by symmetry. Normal co-ordinate analysis aims to provide a quantitative analysis of such mixing of local or group modes, and so to improve our understanding of the vibrational motions of a molecule. It also gives us a lot of other information, derived mathematically from the analysis.

The analysis is based on Newton's Laws of Motion applied to the motion of individual atoms. Account is taken of their kinetic energies, which involve the masses of the atoms and their instantaneous velocities, and potential energies, which are derived from the instantaneous displacements of the atoms from their equilibrium positions in an assumed force field. As the classical vibration frequency for a diatomic oscillator is equal to the quantized energy gap in units of h (Planck's constant), we can derive the normal vibration frequencies from a purely classical analysis. For most purposes, a purely harmonic force field [see Section 5.2.1] is adequate. This is lucky, as the mathematics required to manipulate the equations becomes much more complicated if non-quadratic terms are introduced. Here we cannot go into the details of a normal co-ordinate analysis, as they are complicated even using harmonic force fields; the method is well described elsewhere [22, 23].

It is important to realize that a normal co-ordinate analysis is almost always an 'under-determined' problem mathematically. There is not enough information available to allow a unique solution to be found, and the particular solution adopted will depend on the various assumptions made in the course of the analysis, as well as on the observed frequencies. Considerable care is therefore needed in interpreting the results of a normal co-ordinate analysis.

There is also the problem of anharmonicity. As we cannot use an anharmonic force field in the analysis, it may seem preferable to work with idealized harmonic frequencies ω_e rather than the observed vibration frequencies, which include anharmonic contributions. This requires the collection of much more experimental information so that the harmonic and anharmonic terms can be separated, and yet still does not take account of the anharmonic terms in the force field. On the whole, it seems best to use the observed vibration frequencies in the analysis, recognizing that the results are of only limited validity.

The results of a normal co-ordinate analysis are of several kinds. First, there are the calculated frequencies themselves; by adjusting the force field it should be possible to make these identical to the observed frequencies. Secondly, there is the resulting force field, the various terms of which can be taken as force constants for the motions used initially to define the problem. Such terms are often found to be more-or-less transferable from one problem to another, and so have some claim to significance beyond the level of the particular problem in which they were derived.

The third kind of result is derived from the force field and the mathematical solution of the equations of motion for each particular normal mode. We can define the potential energy distribution in a normal mode, which shows how the various local modes mix in the vibration. For example, in GeF_2H_2 we might find that the 813 cm^{-1} mode is 80% GeH_2 rock, 20% GeF_2 stretch, with the reverse proportions in the 720 cm^{-1} mode.

The mean square amplitudes of vibration for any atom or, more importantly, the mean square amplitudes of relative motion for any atom pair may also be calculated.

Such information is very useful in the analysis of electron diffraction patterns of gases (Chapter 8), and in other experiments where measurements must be conducted upon large collections of molecules in different instantaneous states. Centrifugal distortion constants and Coriolis coupling constants can also be calculated, and the values compared with those obtained from analysis of rotational spectra or rotation fine-structure of vibration bands (see Chapter 4). Details of the calculation of such physically observable quantities can be found in Ref. [22], p. 212.

Once again, we must remember that the results are not determined only by the experimental vibration frequencies, but by various assumptions made in the course of setting up and solving the problem. It is possible, however, to use the observed values of these physically observable quantities to check the validity of some of these assumptions, and hence to derive a more reliable force field.

Additional information that can be used to narrow down the range of possible solutions in a normal co-ordinate analysis can be obtained using isotopically substituted samples. As we have seen in Section 5.12, isotopic substitution can lead to changes in vibration frequencies. The distribution of the isotope shifts among the various bands gives independent evidence about the force field, which is assumed to be invariant, while the mass-dependent kinetic energy terms alter. Each isotopic substitution gives in principle one extra piece of information about the force field, and for several small molecules enough isotopic species have been studied to allow the force field to be specified fully. Again, the influence of anharmonicity is not removed by these studies, and may even be worse if the anharmonicity is significantly different for the isotopic species, as it is with M—H and M—D stretches.

5.15 Changes of structure with phase

Some compounds have different structures in different phases. N_2O_5, for instance, is molecular as a gas, but ionized as NO_2^+ and NO_3^- in the crystal. Such structural changes are of course accompanied by changes in vibrational spectra. The IR spectrum of gaseous N_2O_5 is like that of the glass formed by rapid condensation, but completely different from that of the annealed and crystalline solid, whose spectra shows bands characteristic of NO_2^+ and NO_3^-. Other structural changes, such as association or hydrogen bonding, are reflected in vibrational spectra, and can be followed by vibrational spectroscopy, as can processes of solvation.

Changes in conformation may also occur as samples are condensed or concentrated. An extreme example is B_2F_4 [24], whose IR and Raman spectra have been interpreted in terms of planar molecules of D_{2h} symmetry in the solid phase and a structure with a dihedral angle of 90° between the BF_2 end groups in the gaseous and liquid states.

One situation commonly encountered involves a molecule with two conformers, which are easily interconverted, with one marginally more stable than the other. Changing the temperature alters the relative proportions of the conformers; the vibrational spectra therefore show bands whose relative intensities vary with temperature, as shown in Fig. 5.46.

5.16 Vibrational spectroscopy of unstable molecules

For stable samples we rarely need to obtain spectra in a hurry, and commercial spectrometers generally operate on timescales ranging from a few minutes to several hours.

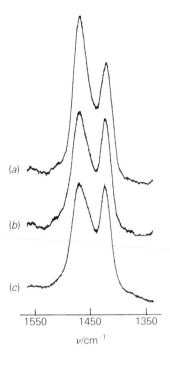

Fig. 5.46 Raman spectra of PMeO(OME)$_2$ showing a variation of the relative intensities of bands assigned to two different conformers as the temperature of the liquid sample changes. Trace (*a*) was recorded at 150 K, (*b*) at 200 K and (*c*) at 295 K. Taken, with permission, from B.J. van der Veken, M.A. Herman and A.J. Barnes, *J.Mol.Struct.*, **99**, 197 (1983).

The most modern interferometers can obtain spectra in shorter times, but even so they cannot deal with samples that decay in seconds, let alone milli-, micro- or nano-seconds. Such samples require the use of special techniques which may be described loosely as flow, flash and freeze methods.

5.16.1   **Flow**

If an unstable species is generated in a static situation, such as in a bulb containing gases or a beaker containing a solution, its decay simply leads to a steadily-decreasing concentration at all points in the apparatus. If, on the other hand, it is continuously generated in a flowing system — gas or liquid travelling at a uniform rate down a tube — its subsequent decay establishes steady-state concentrations at points down-stream from the region of generation. At such points we can observe the spectrum of the unstable product together with that of any stable products, carrier materials and so on. The flow method thus transforms the time-variation of concentration into a distance-variation and enables us to make an essentially static measurement (Fig. 5.47). The distance scale of the decay of this steady-state concentration depends on the ratio of the flow rate to the rate of decay.

The problem with this method is that we must generate the species of interest continuously — often by mixing two gases or solutions — in a way that can be matched to the timescale of the measurement. Thus even here we may want to reduce the total time required to record the spectrum. In the past, much of the work with flowing gases used photographic recording of electronic emission or absorption spectra. These give only limited vibrational information, and then only for small molecules. More modern laser techniques allow much more subtle experiments to be applied to both small and

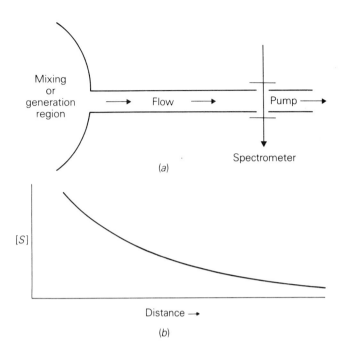

Fig. 5.47 (*a*) Schematic diagram of a flow experiment. (*b*) Variation in concentration [*S*] of an unstable species along the flow tube.

[*S*]

Distance →

(*b*)

Fig. 5.48 Q-branch of the bending vibration of FO_2, prepared in a flow-tube reaction. The band is centered near 578 cm^{-1}, and the whole spectrum shown covers only 1.5 cm^{-1}. Under the resolution of Fig. 5.26(*b*) the whole Q-branch would appear as a single line! Adapted, with permission, from C. Yamada and E. Hirota, *J.Chem.Phys.*, **80**, 4694 (1984).

large unstable molecules in flow systems. An example is shown in Fig. 5.48, which shows the Q-branch of v_2 of FO_2, prepared in a flow tube by reaction of fluorine atoms and dioxygen molecules. The spectrum was recorded using a tunable IR diode laser.

5.16.2 Flash

Given the availability of very fast spectroscopic techniques, using either photographic recording or pulsed lasers, we can consider working with a non-flowing sample and simply recording a spectrum at some interval after sudden generation of the species, but before its decay is complete. This is the technique of flash photolysis, and again it can be applied to gaseous or to liquid samples. Originally a high-intensity flash-lamp was used to generate the unstable species, while the spectrum was recorded using a second, weaker flash. Modern flash photolysis experiments often use lasers because of their superior intensity, their monochromaticity and their directed beams. Unfortunately a monochromatic beam is of only limited use for spectroscopy unless it is tunable. Moreover, IR absorption spectra are difficult to record in this way, as photographic recording is effectively useless in the IR. Interferometers can be used, though the decay of the sample following the generating flash means that quite elaborate multiple interferograms must be recorded under computer control. These must then be combined to generate an interferogram in which each point corresponds to the same time-delay after the flash. Each flash and interferogram only contribute one point to the composite interferogram, and thousands may be required. It it not a trivial experiment!

Over narrow spectral ranges, many flash-generation/laser-detection experiments may be devised and carried out, and the use of electric-field tuning (Stark effect), magnetic-field tuning (Zeeman effect) and double-resonance effects can lead to the identification and precise measurement of many vibration−rotation lines of short-lived species. However, it is not at present a routine matter to extend such measurements over an appreciable portion even of one vibration−rotation band. The production of the $CH_3 \cdot$ radical by flash photolysis of CH_3I, for instance, has been monitored using a tunable IR laser set at a fixed frequency of 608.301 cm^{-1}, coinciding with the strong Q-branch of the v_2 band. The decay of the transient absorption was monitored over a few hundred microseconds, following a 30 ns photolysis flash [25].

5.16.3   Freeze

A third response to the problem of recording spectra of short-lived species is to increase their lifetimes, by cooling the sample in an inert environment. We then have effectively unlimited opportunity to study the sample using conventional instruments. A wide variety of methods has been developed for the preparation of such samples, which are described as matrix-isolated. In many cases, where the inert environment is provided by a frozen gas (Ar or N_2 are often used), the temperature needed may be only a few degrees above absolute zero. Such temperatures are best attained by the use of micro-refrigerators using helium as the working fluid. These micro-refrigerators are commercially available, and avoid the explosion hazard of liquid hydrogen and the expense of liquid helium. In any case, high-vacuum equipment is needed to ensure adequate thermal insulation. The technical details of matrix-isolation experiments and the results obtained by these methods are described in Refs [26]−[29].

Preparation of matrix-isolated samples generally follows one of two basic routes. One involves generating a short-lived species in the gas phase, by photolysis, thermolysis, chemical reaction, discharge, *etc.* The products are then co-condensed with a large excess (typically a thousand-fold) of the inert matrix material. In the other method, the unstable material is generated in the cold matrix, by photolysis, or by chemical reaction of a stable precursor with a photolytically-generated atom.

Spectra of samples obtained using an extremely simple means of generation, evaporation from a solid, are shown in Fig. 5.49. Here the vapors above heated Na_3PO_4

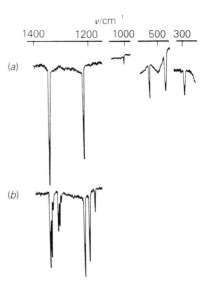

Fig. 5.49 IR spectra of NaPO₃ condensed with a large excess of Ar from (*a*) normal ^{16}O-containing Na_3PO_4 and (*b*) Na_3PO_4 prepared from oxygen with 28% ^{18}O. The spectra are shown by detailed calculation to be consistent with a planar trigonal PO_3^- ion with Na co-ordinated to two of the oxygen atoms. Adapted, with permission, from S.N. Jenny and J.S. Ogden, *J.Chem.Soc., Dalton Trans.*, 1465 (1979).

have been co-condensed with argon. The novel monomeric metaphosphate, $NaPO_3$, was identified in the matrix by its IR spectra. Both normal ^{16}O and enriched $^{16}O/^{18}O$ samples were used. The structure of $NaPO_3$ is planar, like that of KNO_3, and the sodium atom is coordinated to two of the oxygen atoms (5.VIII).

Fig. 5.50 shows IR spectra of products formed when $SiMe_3N_3$ is photolyzed in a matrix. Loss of N_2 is presumed to lead to the formation of a nitrene, $SiMe_3N{:}$, but this

5.VIII

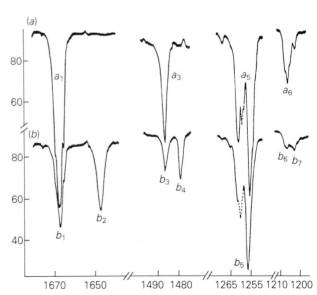

Fig. 5.50 IR spectra of the products formed when $SiMe_3N_3$ is photolyzed in a matrix, (*a*) using normal $SiMe_3N_3$ and (*b*) using a sample enriched in ^{15}N. Note particularly the shift of part of the intensity under the envelope labelled a_1 to lower frequency (b_2). This suggests that these bands should be assigned to C = N stretches, which are expected to absorb in this region, and to show a large $^{14}N/^{15}N$ shift. Adapted, with permission, from R.N. Perutz, *J.Chem.Soc., Chem.Commun.*, 762 (1978).

is unstable, and rearranges to give the observed product. The IR spectra are consistent with it being $SiHMe_2(N=CH_2)$. In particular, a band near 1670 cm^{-1}, which shifts to lower frequency when a ^{15}N-enriched sample is used, is assigned to the $C=N$ stretching mode.

Recently the use of photolysis in matrices with plane polarized light and the subsequent recording of the polarized infrared spectra of the products has given important additional information about the structures of unstable intermediates. Elegant and closely-analyzed work has shown that $Mn_2(CO)_9$, formed by photolysis of $Mn_2(CO)_{10}$, probably contains a semi-bridging carbonyl group; in the complicated photolysis of $Fe_2(CO)_9$, it has been possible to deduce a great deal about the forms of the intermediate $Fe_2(CO)_8$ and about the photochemical processes by which it is formed and reacts. The reader is referred to the original papers for details (Refs [30, 31]).

A further example of the use of matrix isolation is discussed in Section 10.16, and many other examples will be found in Refs [26] – [29].

5.17 Resonance Raman spectroscopy

A Raman spectrum is normally excited using light of a frequency that is not absorbed by the sample. The probability of absorption is very low, and excitation is to a virtual excited state whose lifetime is very short and whose energy is correspondingly poorly defined. If we use light whose absorption probability is high, the effects of local heating by the laser beam are usually so great that the sample decomposes. If we can avoid this local heating, possibly by spinning the sample rapidly under the laser spot so that different parts of the sample are heated in turn, we find that stronger Raman signals are observed as the exciting light is absorbed by the sample. This effect is the resonance Raman effect [32, 33]. The energy levels involved are shown in Fig. 5.51, and the process is also discussed in Section 6.8.7.

The intensity of the effect is high, but in general only a few vibrational modes are involved, and often only one is detectable. It is usual to find a progression indicating significant transition probabilities to many vibrational levels of the modes concerned. The particular vibrations excited are often found to be closely associated with the electronic transition within whose band the exciting radiation falls, and this can sometimes be useful in assigning electronic transitions. The use of a tunable laser may be helpful, as excitation profiles, which show the intensity of resonance-enhancement as a function of exciting frequency, may allow overlapping electronic bands to be distinguished from each other. In terms of vibrational spectroscopy there are three major advantages over simple Raman spectroscopy.

1 The intensity is enhanced, enabling weakly scattering or very dilute samples to be used.

2 The spectrum is greatly simplified, as only a few vibrations are enhanced.

3 The observation of a progression enables the vibrational potential function to be specified over a wide range of the quantum number v.

The resonance Raman effect is observable with a normal Raman spectrometer, and does not depend on any special properties of laser excitation. The intensity enhancement relies on the electronic transition being allowed by symmetry, so the vibrations whose intensities are enhanced are almost always totally symmetric (Section 5.7).

The technique has enabled spectra to be obtained from the metallic cores of some important biological molecules such as hemoglobin and some enzymes, where the low

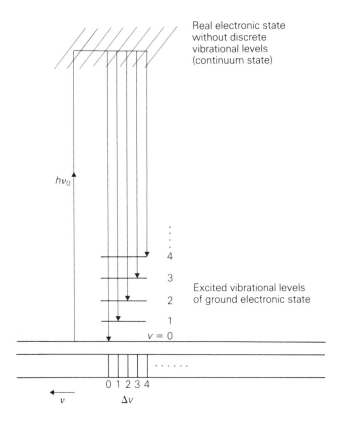

Real electronic state
without discrete
vibrational levels
(continuum state)

hv_0

4
3
2
1
$v = 0$

Excited vibrational levels
of ground electronic state

0 1 2 3 4

v Δv

Fig. 5.51 Energy level
diagram illustrating the
formation of a progression
of Stokes lines in a
resonance Raman
spectrum.

concentration of the metal and the low solubility of the substance itself make it difficult
to study vibrational spectra in any other way. The IR or normal Raman spectrum of
such a large species is also, of course, very complex, and the resonance Raman spectrum
is more readily interpreted because of its simplicity. We shall illustrate the effect with
two simpler examples. The tetrahedral chromate ion, CrO_4^{2-}, has only one totally symmetric vibration, and it gives a very simple resonance Raman spectrum, with a clear
progression up to the 9th overtone (Fig. 5.52). The broadening of the higher bands

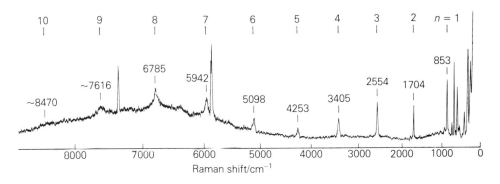

10   9   8   7   6   5   4   3   2 $n = 1$

6785   5942

~7616

~8470   5098   3405   2554   1704   853
                  4253

8000   7000   6000   5000   4000   3000   2000   1000   0
Raman shift/cm^{-1}

Fig. 5.52 Resonance Raman spectrum of solid K_2CrO_4. A clear-cut progression of peaks due
to the totally symmetric stretching mode of the anion CrO_4^{2-} is seen, up to the ninth or
tenth member. The frequency scale is non-linear. Excitation was by UV laser line within the
absorption band whose low-frequency wing in the blue/violet region is responsible for the
yellow color of the salt. The sample was rotated rapidly to reduce the effects of local heating
by absorbed laser light. Taken, with permission, from W. Kiefer and H.J. Bernstein,
Mol.Phys., **23**, 835 (1972).

reflects progressively shorter lifetimes for the final highly-excited vibrational states. Excitation in this case was with a laser line in the ultraviolet, well within the absorption band responsible for the yellow color. The second example is $K_4[Ru_2OCl_{10}]$, **5.IX**,

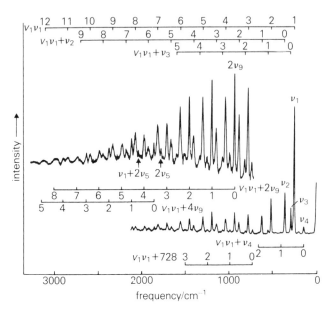

5.IX

which has a linear bridging oxygen and overall D_{4h} symmetry in the anion. With this symmetry it has four totally symmetric vibrations, and all are seen in the resonance Raman spectrum (Fig. 5.53). However, only the symmetric RuORu stretch (labelled ν_1) produces progressions. A few peaks due to the excitation of overtones of vibrations of lower symmetry (*e.g.* $2\nu_9$) can also be seen, but the main feature of the spectrum is a set of progressions in ν_1, with or without excitation of one of the other a_{1g} modes. This may be taken to imply that the electronic band, within whose absorption profile the excitation occurred, involves a transition between states having different Ru—O bond lengths, and probably involves excitation of electrons having some Ru—O bonding character. We shall see in Chapter 6 how such arguments can sometimes help in the assignment of electronic transitions.

Fig. 5.53 Resonance Raman spectrum of $K_4[Ru_2OCl_{10}]$ at about 80 K. All four of the totally-symmetric modes expected on the basis of D_{4h} symmetry appear below 500 cm^{-1}, and progressions in the symmetric Ru—O—Ru stretching frequency ν_1 dominate the rest of the spectrum. Taken, with permission, from J.R. Campbell and R.J.H. Clarke, *Mol.Phys.*, **36**, 1133 (1978).

Problems

5.1 The reduced mass μ for a vibrating molecule AB is given by the expression

$$\frac{1}{\mu} = \frac{1}{m_A} + \frac{1}{m_B} \, .$$

Calculate the force constants for HBr and NO (vibration frequencies 2559 and 1876 cm^{-1}, respectively).

5.2 Calculate the difference in stretching frequency for the natural isotopes of Cl bonded to a single atom of (a) ^{6}Li; (b) ^{60}Co; (c) ^{238}U, assuming vibration frequencies of 1000, 400 and 200 cm^{-1} respectively.

5.3 Use Eqns (5.4) and (5.9) to calculate the relative populations at room temperature of molecular states with $v = 0$, 1 and 2 for a vibration frequency of 600 cm^{-1}.

5.4 The CH stretching fundamental of HCN gives an IR band at 3312 cm^{-1}, and its first overtone appears at 6522 cm^{-1}. Calculate the harmonic frequency ω_e and the anharmonicity x_e for this vibration.

5.5 Determine the symmetry species of each of the vibrations of SCl$_2$.

5.6 To which point groups do the molecules SO$_3$, ClO$_3$ and XeO$_3$ belong? Determine the numbers of bands you would expect to observe in the M—O stretching region of the IR and Raman spectra of each of these compounds.

5.7 Determine the symmetry components arising from the CO stretching vibrations of a symmetrical octahedral molecule M(CO)$_6$, and state which are active in the Raman and IR spectra. Show how these results are modified by the existence of about 1% ^{13}C in natural samples.

5.8 The IR spectrum of CS$_2$ shows a band at 397 cm^{-1} with a PQR contour, and a band at 1523 cm^{-1} which lacks a Q-branch. The IR spectrum of OCS contains PR bands at 2021 and 642 cm^{-1} and a PQR band at 466 cm^{-1}. How do you interpret these observations?

5.9 Show how a study of the IR and Raman spectra could enable a distinction to be made between *syn-* and *anti*-planar structures of N$_2$F$_2$.

5.10 The infrared and Raman spectra of matrix-isolated CrF$_2$, TiF$_2$ and CrF$_3$ are summarized below. What can you deduce from them about the geometries of the molecules?

Compound	IR/cm^{-1}	Raman/cm^{-1}
CrF$_2$	654, 255	565 pol
CrF$_3$	749, 177, 125	668 pol
TiF$_2$	740, 643, 171	no data

5.11 Determine the point groups and the numbers of infrared and Raman bands that would be expected for the *cis* and *trans* isomers of SCl$_2$F$_4$.

5.12 Show how the IR and Raman spectra of molecular B(OH)$_3$ could be used to distinguish between planar (C_{3h}) and non-planar (C_{3v}) forms.

5.13 The infrared bands around 1450–1500 cm^{-1} in the spectrum of BF$_3$ that are assigned to the asymmetric B—F stretching mode show marked splitting because of the presence of significant amounts of both ^{10}B and ^{11}B in natural boron, yet the Raman band at 888 cm^{-1}, assigned to the symmetric B—F stretch, shows no such isotopic effects. What can you deduce from these observations?

5.14 Given the assignment of fundamentals for SiF_2H_2 in Example 5.4, calculate the frequencies of all first overtones $2v_i$ and binary combinations $v_i + v_j$, and their symmetries.

5.15 Given the frequencies of v_2, v_5 and v_6 of $GeClH_3$ in Table 5.9, and $v_3 = 423$ cm^{-1}, find all the overtones and combinations (with any number of constituents) in the frequency range 2000–2500 cm^{-1}; the Ge—H stretching modes v_1 and v_4 lie in this region and may be perturbed by any combinations nearby. Specify the symmetry of each overtone or combination.

5.16 Sb_4 has been shown to be a significant constituent of vaporized antimony, and may be assumed to have the same structure as P_4. How many vibrational modes would you expect from $^{121}Sb_4$ trapped in a matrix? How many modes would you expect from $^{121}Sb_3^{123}Sb$? Determine the symmetry species and spectroscopic activities of the vibrations for each molecule.

5.17 The Raman spectrum of a colored sample shows many more lines than expected, with prominent peaks at 154, 240, 306, 456, 604, 708 and 750 cm^{-1}. Suggest an explanation for these observations and derive molecular vibration frequencies for the sample.

5.18 Raman spectra of solutions of $HgCl_2$ and $HgBr_2$ in methanol each show a single band due to the solute, at 324 and 205 cm^{-1}, respectively. A solution containing both solutes shows additional bands at 344 and 234 cm^{-1}. Suggest an assignment for all four bands.

5.19 In the electronic spectra of $[Re_2X_8]$ (X = Cl, Br, I), there are bands assigned to $\delta \rightarrow \delta^*$ and $\pi \rightarrow \delta^*$ transitions; the $\delta \rightarrow \delta^*$ transitions involve electrons from the δ-bonding orbital which is associated with the metal-metal bond; the δ^* orbital is antibonding between the metal atoms, while the π-level contains metal-halogen bonding character. If the resonance Raman spectra are excited by irradiation in the envelope of the band assigned to the $\delta \rightarrow \delta^*$ transition, the following bands are intensified and show progressions:

X	Band frequency/cm^{-1}
Cl	275
Br	276
I	257

If the spectrum is excited by irradiation in the envelope of the band assigned to the $\pi \rightarrow \delta^*$ transition, the resonance Raman spectra show intensification of the same bands, and in addition the following additional peaks, with progressions whose lengths are given in brackets:

X	Band frequency/cm^{-1}
Cl	361 (2)
Br	211 (6)
I	151 (4)

Comment on these observations in relation to the suggested assignments of the electronic bands.

5.20 In the infrared spectrum of ethene, $v(C=C)$ forbidden; bands due to CH deformation modes occur below 1500 cm^{-1}. Some complexes of transition metals containing co-ordinated ethene show an infrared band near 1600 cm^{-1}. Explain the occurrence of this band. How could you find other evidence to support this explanation?

5.21 Reaction of molecular H_2 with $W(CO)_3(PR_3)_2$ (R = cyclohexyl or isopropyl) gives a new compound thought to be $W(CO)_3(H_2)(PR_3)_2$. The IR spectrum of this product shows a band at 2690 cm^{-1}; a sample generated using HD in place of H_2 gives a band at 2360 cm^{-1} in the IR. The corresponding band for a sample containing D_2 is obscured by $v(CO)$ bands. Bands assigned to $v(W—H)$ modes appear below 1600 cm^{-1}. Suggest an assignment of the high-frequency bands, and account for the observations of only one such band in the sample generated using HD.

5.22 Show how the numbers of IR bands in the different parts of the CO stretching region could be used to distinguish between cluster-structures (5.X) and (5.XI).

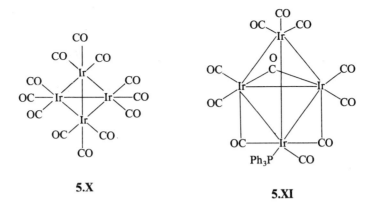

5.X

5.XI

5.23 Show how a study of the IR and Raman spectra of PF_3H_2 in the liquid state would enable you to decide between the three possible structures based on trigonal bipyramidal coordination of phosphorus.

5.24 Photolysis of a mixture of $AsCl_3$ and ozone in rare gas matrices at very low temperature generated a new species in addition to $AsCl_3$. Vibrational bands of the new species were identified as set out below:

IR/cm^{-1}	Raman/cm^{-1}
996	997
435	434
—	373
248	247

The band at 996 cm^{-1} shifted to 949 cm^{-1} when $^{18}O_3$ was used. How far can you assign the spectra on the assumption that the product is $AsCl_3O$? What additional measurements would confirm this identification?

5.25 An aqueous solution of mercury(I) nitrate gives a very strong Raman band at 169 cm^{-1}, as well as bands due to nitrate anions. Solid mercury(I) chloride gives a strong Raman band at 167 cm^{-1}, with another weaker band at 270 cm^{-1}, while mercury(II) chloride gives only a single band near 320 cm^{-1} (as vapor or in solution in water or methanol). Suggest an assignment of the Raman bands of mercury(I) chloride, and comment on the structural implications for both chlorides.

5.26 Photolysis of nickel atoms and methane in an argon matrix using radiation from a mercury arc lamp gave a product with the following infrared bands.

Product from $^{12}CH_4$/cm^{-1}	$^{13}CH_4$/cm^{-1}	$^{12}CD_4$/cm^{-1}
2950, 2860	2940, 2855	2197
1945	1945	1407
1120	1120	892
642	641	not observed
555	542	not observed

What can you deduce from these data about the nature of the product?

5.27 When a small amount of BI_3 is added to gaseous BF_3 the IR spectrum of the mixture shows bands due to BF_3 at 1446, 691 and 480 cm^{-1} (frequencies for ^{11}B) together with additional bands at 1410 cm^{-1} (b-type band contour), 1185 cm^{-1} (a-type), 610 cm^{-1} (a-type), 529 cm^{-1} (c-type), 318 cm^{-1} (b-type) and 270 cm^{-1} (a-type). Addition of more BI_3 gives a mixture whose IR spectrum contains new bands assigned to BI_3 and others at 1295 cm^{-1}

(b-type), 758 cm^{-1} (a-type), 403 cm^{-1} (c-type), 366 cm^{-1} (b-type); 250 cm^{-1} (a-type) and 125 cm^{-1} (b-type). Assign all the bands and comment on the species responsible.

5.28 In the low-frequency region of the resonance Raman spectrum of [Fe(OEP)(CN)(py)] (OEP = octaethylporphyrin, py = pyridine), there are bands at 449 and 191 cm^{-1}. The former shifts to 446 cm^{-1} when ^{13}CN is used, whereas the latter is unaffected; when the complex is prepared from pyridine-d_5, the former band is unaffected but the latter shifts to 188 cm^{-1}. What can you deduce from these observations?

5.29 The infrared and Raman spectra of NF$_3$O and NF$_3$S give the following fundamental frequencies (polarized Raman bands are indicated by 'pol'). Assign the spectra to the appropriate vibrational modes, and consider the implications of the assigned frequencies to simple representations of the bonding.
NF$_3$O: 1689(pol), 880, 740(pol), 548(pol), 513, 398 cm^{-1}
NF$_3$S: 1512(pol), 812, 768(pol), 520, 430, 340 cm^{-1}.

5.30 The infrared spectrum of matrix-isolated CrF$_4$ contains only two bands, at 784 and 303 cm^{-1}. What can you deduce from this about the molecular geometry of CrF$_4$? How might you be able to obtain additional evidence to support your deductions?

5.31 The infrared spectrum of gaseous and the Raman spectrum of liquid WF$_4$O are set out below. Use them to decide the probable molecular geometry.

IR/cm^{-1}	Raman/cm^{-1}
1055 m	1057 s pol
733 m	732 vs pol
698 vs	
	631 vw br
	328 br sh
298 vw	301 mw
	298 w
248 w	248 w pol
236 m	234 w

(s = strong, m = medium, w = weak, v = very, br = broad, sh = shoulder, pol = polarized)

5.32 Studies of adducts of BrF$_5$ with strong fluoride acceptors such as SbF$_5$ give solid products believed to contain the cation BrF$_4^+$, and have identified the vibrational bands associated with this cation. The infrared spectra of alkali metal salts of the anion BrF$_4^-$ have given the vibrational frequencies listed below. Use these frequencies to determine the probable geometries of the ions and to assign the spectra.

BrF$_4^+$		BrF$_4^-$	
IR/cm^{-1}	Raman/cm^{-1}	IR/cm^{-1}	Raman/cm^{-1}
	736 sh		
730 s		542 vs	
	723 vs		
690 vs	704 m		523 vs
606 mw	606 s		449 s
419 m	426 mw	317 s	
not observed	385 mw		246 m
369 vw	369 mw	183 w	
not observed	219 w		

5.33 Raman bands due to BiI$_4^-$ have been identified at the following frequencies: 141 s pol; 129 s dp; 107 m pol; 80 w; 53 vbr. What can you deduce from this information about the probable shape of the anion?

Further reading

General and theoretical

Ref. [11] is *the* authority on every aspect except modern instrumentation and developments. Refs [22] and [23] contain modern and older treatments − both are useful.

P. Gans, *Vibrating Molecules*, Chapman and Hall, London (1971). This is a useful treatment of several aspects, including a comprehensive bibliography.

Infrared techniques

Refs [5] and [6] are examples of the many books dealing with instrumental techniques and their application in this field.

Raman spectroscopy

Ref. [7] is a masterly treatment, with a well-selected bibliography.

The following may also be useful:

T.R. Gibson and P.J. Hendon, *Laser Raman Spectroscopy*, Wiley-Interscience, London (1970).
J.G. Grasselli, M.K. Snavely and B.J. Bulkin, *Chemical Applications of Raman Spectroscopy*, Wiley-Interscience, New York (1981).
M.C. Tobin, *Laser Raman Spectroscopy*, Wiley-Interscience, New York (1971).

Vibrational spectroscopy of inorganic substances

S.D. Ross, *Inorganic IR and Raman Spectra*, McGraw-Hill, London (1972).
K. Nakamoto, *Infrared Spectra of Inorganic and Co-ordination Compounds*, 2nd edn, Wiley, New York (1970).

See also Ref. [19].

References

1 J.M. Hollas, *High Resolution Spectroscopy*, Butterworths, London (1982).
2 R.K. Thomas, Chap. 6 in *Molecular Spectroscopy*, Specialist Periodical Reports, The Chemical Society, London (1979), Vol. 6.
3 H. Ibach, *Electron Energy Loss Spectroscopy*, Springer-Verlag, Berlin (1977).
4 S. Trajmar, J.K. Rice and A. Kuppermann, *Advances in Chemical Physics*, Wiley, New York (1970), Vol. 18.
5 A. Lee Smith, *Applied Infrared Spectroscopy*, Wiley-Interscience, New York (1979).
6 P.R. Griffiths, *Chemical Infrared Fourier Transform Spectroscopy*, Wiley-Interscience, New York (1975).
7 D.A. Long, *Raman Spectroscopy*, McGraw-Hill, New York (1977).
8 W.C. Child, D.H. Smith and G.M. Begun, *C.R. Conf. Int. Spectrosc. Raman* 7th, 268 (1980).
9 B. Schrader, Chap. 6 in *Molecular Spectroscopy*, Specialist Periodic Reports, The Chemical Society, London (1978), Vol. 5.
10 C. Turrell, *Infrared and Raman Spectra of Crystals*, Academic Press, New York and London (1972).
11 G. Herzberg, *Infrared and Raman Spectra*, Van Nostrand, New York (1945).
12 D. Steele, Chap. 3 in *Molecular Spectroscopy*, Specialist Periodic Reports, The Chemical Society, London (1978), Vol. 5.
13 H.H. Jaffé and M. Orchin, *Symmetry in Chemistry*, Wiley, New York (1965).

14 *DMS Working Atlas of IR Spectroscopy*, Butterworths, London (1972); *Sadtler Collection of IR Spectra*, Philadelphia; Aldrich Library of IR Spectra (1970).

15 D.S. Ginley and D.M. Haaland, *Appl. Phys. Letters*, **38**, 271 (1981).

16 A. Lee Smith (ed.), *Analysis of Silicones*, Wiley-Interscience, New York (1974).

17 B.G. Oliver and A.R. Davis, *Can. J. Chem.*, **51**, 698 (1973).

18 D.E. Irish and M.H. Brooker, Chap. 6 in *Advances in IR and Raman Spectroscopy*, Heyden, London (1976), Vol. 2.

19 D.M. Adams, *Metal−Ligand and Related Vibrations*, Edward Arnold, London (1967).

20 W.F. Edgell and R. Summitt, *J. Am. Chem. Soc.*, **83**, 1772 (1961).

21 N. Mohan, A. Müller and K. Nakamoto, Chap. 5 in *Advances in IR and Raman Spectroscopy*, Heyden, London (1975), Vol. 1.

22 S. Califano, *Vibrational States*, Wiley, New York and London (1976).

23 E.B. Wilson, J.C. Decius and P.C. Cross, *Molecular Vibrations*, McGraw-Hill, London (1955).

24 J.N. Gayles and J. Self, *J. Chem. Phys.*, **40**, 3530 (1964).

25 G.A. Laguna and S.L. Baughcum, *Chem. Phys. Letters*, **88**, 568 (1982).

26 H.F. Hallam (ed.), *Vibrational Spectroscopy of Trapped Species*, Wiley, London (1973).

27 A.J. Downs and S.C. Peake, Chap. 9 in *Molecular Spectroscopy*, Specialist Periodical Reports, The Chemical Society, London (1973), Vol. 1.

28 B.M. Chadwick, Chap. 4 in *Molecular Spectroscopy*, Specialist Periodic Reports, The Chemical Society, London (1975), Vol. 3.

29 S. Cradock and A.J. Hinchcliffe, *Matrix Isolation*, Cambridge University Press, Cambridge (1975).

30 I.R. Dunkin, P. Harter and C.J. Shields, *J. Am. Chem. Soc.*, **106**, 7248 (1984).

31 S.C. Fletcher, M. Poliakoff and J.J. Turner, *Inorg. Chem.*, **25**, 3597 (1986).

32 J. Behringer, Chap. 3 in *Molecular Spectroscopy*, Specialist Periodical Reports, The Chemical Society, London (1975), Vol. 3.

33 R.J.H. Clark, Chap.4 in *Advances in IR and Raman Spectroscopy*, Heyden, London (1975), Vol. 1.

6 Electronic and Photoelectron Spectroscopy

6.1 Introduction

Our present understanding of quantum mechanics is largely based on the study of electronic spectra. The spectra of atoms, for instance, with their series of lines, provided the pioneers of quantum mechanics with incontrovertible evidence of the quantum nature of matter at the atomic level. The electronic structures of atoms can only be described in quantum terms, and are now well understood.

In the same way, the electronic spectra of molecules can give information about their electronic structures, and have helped to establish our present understanding of chemical bonding in molecules. We shall see, however, that electronic spectra rarely give us direct information about molecular structures. This is partly because molecular electronic spectra are very different from the line-spectra of atoms, consisting in most cases of broad bands, and partly because the spectra are usually very complicated, so that assignment is often impossible. This complexity arises simply because a typical molecule has a lot of electrons in different energy levels, together with a large number of unoccupied levels, and excitation can occur from each of the occupied levels to many or all of the unoccupied levels. There are also practical difficulties in obtaining electronic spectra outside a fairly narrow range of frequencies covering the visible and near-ultraviolet (UV) regions from 14 000 to 50 000 cm^{-1} (corresponding to wavelengths of 700 to 200 nm). Very few molecules have electronic transitions below this range, in the IR region, but *all* molecules have most if not all of their electronic transitions in the far-UV region, above 50 000 cm^{-1}. Note that the electron volt, which is often used as an energy unit, is equivalent to 8067 cm^{-1}.

For these and other reasons more useful information about molecular structures can be obtained by photoelectron spectroscopy, where the excited electron is ejected completely instead of remaining in a higher-energy level of the molecule. The energy required to remove the electron depends only on the energy of the level it originally occupied relative to that of an unbound electron, so photoelectron spectra are in principle much simpler than electronic excitation spectra. Much of this chapter will therefore be concerned with the two major types of photoelectron spectroscopy (PES), though we shall note how electronic excitation spectroscopy can be used to give structural information about some molecules. The spectra of transition-metal complexes in the visible and near-UV regions have been studied in considerable detail, and may be most informative. They can also be used to detect, identify and measure the concentrations of such compounds.

6.2 Excitation and ejection of electrons

The interaction of a photon and a molecule, like other physical processes, is subject to the laws of conservation of energy, of linear momentum and of angular momentum. These restrict some of the conceivable processes whereby excitation could occur, while

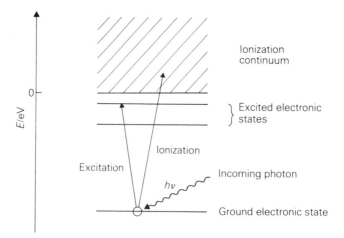

Fig. 6.1 Photoexcitation and photoionization processes.

allowing others to occur with very few restrictions. These processes are illustrated in Fig. 6.1.

Thus excitation of an electron within the energy levels of a molecule requires the supply of energy from an incoming photon. In some cases angular momentum (as orbital angular momentum of the excited electron) must also be supplied, but not linear momentum. The incoming photon, on the other hand, has both energy and linear momentum, so that resonant absorption of the photon (involving transfer of the whole of the photon energy to the molecule) must also involve the transfer of the linear momentum of the photon. This leads to an increase in the translational momentum of the molecule, which is not quantized. Non-resonant absorption of a photon, in which both surplus energy and surplus linear momentum are carried away by a photon of lower energy, is a much less favorable process. It leads to the Raman effect, which, as we have already seen in Chapter 5, is many orders of magnitude less efficient than direct resonant absorption.

In a photoionization process, on the other hand, any energy from the incoming photon over and above that needed to ionize the molecule is carried away as kinetic energy of the electron ejected from the molecule; conservation of linear momentum is generally accomplished by recoil of the ion formed. This means that resonant photo-ionization offers no particular advantages and photoelectron spectroscopy is best done using a monochromatic light source with more than sufficient energy to ionize the electrons of interest. This has the advantage that the surplus energy, appearing as kinetic energy of the electrons, can be measured by electrostatic analysis of the electron velocities.

6.3 Electron energy levels in atoms and molecules

In order to appreciate the various processes involved in electronic spectroscopy we need to have a reasonable picture of the levels, both occupied and unoccupied, available to electrons in molecules. It is best to start with the levels in atoms, and some of the most successful quantitative descriptions of the behavior of electrons in molecules use an approach based on the Linear Combination of Atomic Orbitals (LCAO).

Each of the solutions found for the Schrödinger equation (quantum mechanical wave equation) for a one-electron atom can be regarded as relating to one of a set of orbitals, which is characterized by a unique combination of quantum numbers and is given a corresponding symbol, such as $1s$, $2p$, etc. Each orbital has a distinct wavefunction, which describes the distribution of an electron occupying the orbital, and an energy which such an electron must possess. Each electron in an atom must have exactly the energy appropriate to one or other of the orbitals.

There are n^2 distinct solutions of the wave equation for each value of n, and the corresponding orbitals are said to constitute a shell. Within each shell the orbitals are divided into sub-shells according to the value of the angular momentum quantum number l, which can take values from zero to $(n - 1)$; there are $(2l + 1)$ orbitals in the sub-shell corresponding to a given value of l. This leads to the familiar sets of s, p, d, f, etc. orbitals in each shell.

For a one-electron atom, all the orbitals of a given shell have the same energy relative to the energy of a free electron; this is proportional to n^{-2}, and to the nuclear charge, Z, as increasing the nuclear charge makes the electron more stable. For an atom with many electrons, we can still use the one-electron atom wavefunctions as approximate descriptions of the behavior of the electrons, though we should modify the functions to account for the mutual repulsion of the electrons. In particular, this repulsion means that orbitals in different sub-shells no longer have the same energy. Within any one shell the energies are always $E_s < E_p < E_d \ldots$, while the principal quantum number n still orders orbitals of the same value of l, so that $E_{1s} < E_{2s} < E_{3s} \ldots$, $E_{2p} < E_{3p} < E_{4p} \ldots$ etc. It is possible however for orbitals of different principal quantum numbers to overlap in energy, so that $E_{3d} > E_{4s}$, for example.

One of the most rigorous rules of quantum mechanics is the Pauli Exclusion Principle, an important consequence of which is that each orbital can hold no more than two electrons. We can speak of an orbital as being full if it contains two electrons, half-filled if it has a single electron, or empty.

For a typical many-electron atom, such as Fe, we may divide the orbitals into three major categories. At lowest energies we find the so-called core levels, corresponding to filled shells and sub-shells $(1s^2)$, $(2s^2 2p^6)$ and $(3s^2 3p^6)$. (The superscripts denote the total numbers of electrons in the sub-shell.) There is then a set of valence levels, whose binding energies are generally quite low (5-25 eV). Some of these are full, some half-filled and some empty. For Fe the valence levels would include the $3d$, $4s$ and $4p$ sub-shells, which could hold a total of 18 electrons if filled. The nuclear charge of Fe is 26, and the core levels contain 18 electrons, so there are only 8 electrons in the valence levels. Lastly there is a set of unoccupied (empty) levels (sometimes called Rydberg levels or virtual levels) consisting of the sub-shells $4d$, $4f$ and all orbitals whose principal quantum number is 5 or more. This last set is in principle infinite in number. At electron energies even higher than the virtual levels we find a continuum of unbound states, corresponding to an ion and a free electron. Excitation of electrons into this continuum corresponds to ionization.

For a ground state atom, the core levels are completely filled, the valence levels are partially filled, and the virtual levels and the ionization continuum are unoccupied. Electronic excitations are then possible, from core or valence levels to valence, virtual or continuum levels, as shown in Fig. 6.2. Core-level electrons may be excited into unoccupied or half-filled valence levels, or into virtual levels, or they may be ionized, by excitation into the continuum: only the last of these possibilities concerns us here.

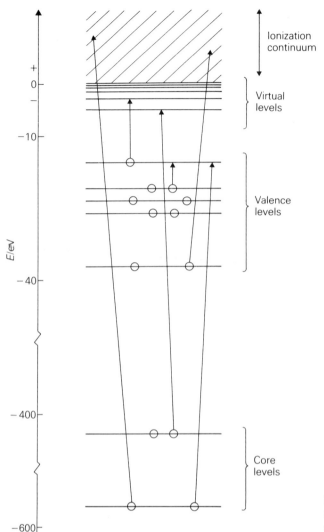

Fig. 6.2 Core, valence and virtual levels, showing various possible excitation and photoionization processes.

The high binding energies of core levels in most atoms means that X-ray photons are required to ionize core electrons, giving rise to X-ray photoelectron spectra (XPS or XPES). Valence-shell electrons may also undergo transitions of three types, depending on their destination; in this case all are observable in practice. Valence–valence excitations and valence–virtual excitations may be considered together, and give rise to valence-electron excitation spectra. The excitation of valence electrons into the ionization continuum gives rise to valence-electron photoelectron spectra.

So far we have considered the transitions possible for individual atoms. Exactly the same possibilities exist for molecules. Not only that, but as far as the core levels are concerned the fact that atoms have combined to form molecules is almost irrelevant. The characteristic core levels of the constituent atoms remain in the molecules, with only very slight shifts in energy (of the order of a few eV) due to the electrostatic charge on each atom. Only the valence levels are greatly affected by molecule formation, as this is where the bonding actually occurs. Thus intravalence excitation and valence-electron ionization transitions contain information characteristic of molecules, whereas core-electron ionizations relate to the atoms making up the molecules. All atoms other than hydrogen can be detected and identified using X-ray excited photo-electron spectroscopy, which is therefore sometimes called Electron Spectroscopy for Chemical Analysis (ESCA). Moreover, the slight shifts in core levels, called chemical shifts, can be used to differentiate atoms of the same element in different environments, either within one molecule or in a mixed sample. The technique is therefore very powerful in the examination of systems that are chemically inhomogeneous, or even varying with time.

The remainder of the present Chapter will deal with core-level photoelectron spectroscopy, valence-level photoelectron spectroscopy and electronic excitation spectroscopy in turn.

6.4   Core-level photoelectron spectroscopy

As we have seen, the removal of a core-electron requires an energy characteristic primarily of the atom concerned, so this technique offers us the ability to identify the constituent atoms of any sample.

6.4.1   Experimental method

To generate an XPS spectrum we need a source of X-rays, ideally monochromatic. The usual X-ray source consists of a metal anode which is bombarded by a beam of high-velocity electrons, and gives off the characteristic X-ray spectrum of the metal concerned. This comprises a few intense lines and a much weaker continuous background. A filter that transmits only the most energetic K line and absorbs less energetic X-rays then leaves an almost monochromatic X-ray beam, though the linewidth is of the order of 1-2 eV, and the 'line' may in fact consist of two or more components. A crystal monochromator may be used to narrow the line and eliminate unwanted components, but the increase in complexity and the reduction in intensity that are inevitably involved mean that this is only done if very high resolution is required (see below). The lines most often used are the K_α lines of Mg (1254 eV) and Al (1487 eV).

The XPS spectrum consists of a record of the number of electrons ejected as a function of the energy carried by the electron, so the electrons must be detected after an analysis of their energies. There are several different types of electron detector in use, most of which rely on the production of bursts of electrons when a single fast-moving electron strikes a surface. In this way each electron arriving at the detector generates an electrical pulse which can be amplified and counted. Energy analysis of electrons is equivalent to velocity analysis, and is generally done by the application of a controlled electrostatic field which deflects electrons with a particular velocity into the detector. Further details of X-ray photoelectron spectrometers may be found in Refs [1]–[3].

The sample arrangements depend on the physical nature of the sample, of course. By far the largest proportion of XPS studies are carried out using solid samples. These present few problems with the requirement for a high vacuum (though volatile solids may need to be cooled), and they give the highest possible density of sample, and therefore signal strength, but it must be remembered that the atomic environment includes neighboring molecules or ions. Gases and liquids are harder to cope with, but useful spectra have been obtained for both.

6.4.2   Core-level photoionization process

The interaction of an X-ray photon with an atom, leading to ejection of a core electron, can be treated as a simple inelastic collision. The energy transferred from the photon to the electron is more than enough to remove it from the atom. The interaction is strong, as it involves the electric field of the radiation and the electric charge of the electron. Moreover, it is not inhibited by the need to conserve both energy and momentum, because both are shared between the ejected electron and the resulting ion. In the simplest possible photoionization event the whole of the photon energy is transferred to the electron, so we may write

$$h\nu = \mathrm{BE} + \mathrm{KE} \tag{6.1}$$

where $h\nu$ is the energy of the incident photon, BE is the binding energy of the ejected electron and KE is its kinetic energy. Thus if we know $h\nu$ and measure KE we can derive the electron binding energy very simply. Core electrons are not normally involved in the bonding of the molecule, and so when they are ejected the resulting ions are usually formed in their vibrational ground states. In any case, typical vibration frequencies of less than 0.5 eV (~ 4000 cm^{-1}) cannot normally be resolved, except under special circumstances when a monochromator is used to reduce the source linewidth from ~ 1.2 to ~ 0.2 eV. In a few cases, such as that illustrated in Fig. 6.3, traces of vibrational structure have been observed, but this effect is rarely useful. So for the rest of this section we can ignore processes in which the ion is excited, either immediately or during the ejection of the electron.

What we record in an X-ray photoelectron experiment is a spectrum of kinetic energies of ejected electrons, each corresponding to the binding energy of one core level in the sample. For a sample containing only one type of atom these signals would

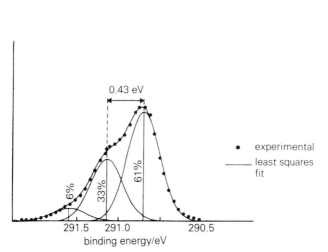

Fig. 6.3 Carbon $1s$ core level line of gaseous methane, showing structure arising from excitation of the symmetric CH stretching mode of the ion CH$_4^+$. The peak has been resolved into three components by a computer fitting procedure, which is often valuable in analyzing composite peaks. The two satellites which are shown correspond to the formation of ion states with one and two quanta of vibration. •, experimental points; —, least squares fit. Redrawn from U. Gelius, *J. Electronic Spectr.*, **5**, 985 (1974).

correspond to the expected s, p, d ... sub-shells of the core shells. For a sample with different atoms the result is simply a superposition of those expected for each of the constituent atoms. Thus metallic cobalt gives signals due to removal of electrons with binding energies 926 eV $(2s)$, 794 and 779 eV $(2p)$, 101 eV $(3s)$ and 60 eV $(3p)$, while any cobalt oxide would also give signals due to the oxygen $1s$ core level with binding energy 540 eV, as shown in Fig. 6.4. Any one of the Co signals is adequate to identify Co in a sample, as most energies can be uniquely associated with one sub-shell of one element. Similarly, the signal at 540 eV uniquely identifies oxygen. Chemical shifts are generally of a few eV or less, and overlap of lines due to different elements is unlikely, as each element contributes at most a few lines in the range 50-1200 eV, while the instrumental linewidth is of the order of 2 eV. The $1s$ level binding energies of elements increase roughly as Z, the nuclear charge, and can of course only be used for elements lighter than that used in the X-ray source. Nevertheless, every element except H and He has at least one line in the range 50-1200 eV which can be used for its identification. Table 6.1 gives some convenient lines for a number of elements.

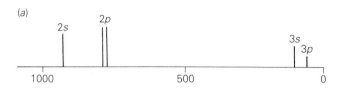

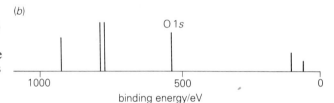

Fig. 6.4 Schematic X-ray photoelectron spectra of (a) metallic cobalt and (b) an oxide of cobalt, showing the characteristic core level lines of cobalt in both cases and the oxygen $1s$ line in (b).

binding energy/eV

Table 6.1 Some core level binding energies (eV) accessible with Al K_α (1487 eV) excitation

	Li	Be														B	C	N	O	F	Ne
$1s$	55	111														188	284	399	532	686	867

	Na	Mg														Al	Si	P	S	Cl	Ar
$2p$	31	52														73	99	135	164	200	245
																74	100	136	165	202	247
$2s$	63	89														118	149	189	229	270	320
$1s$	1072	1305																			

	K	Ca	Sc	Ti	V	Cr	Mn	Fe	Co	Ni	Cu	Zn	Ga	Ge	As	Se	Br	Kr
$3p$	18	26	32	34	38	43	49	56	60	68	74	87	103	122	141	162	182	214
													107	129	147	168	189	223
$3s$	34	44	54	59	66	74	84	95	101	112	120	137	158	181	204	232	257	289
$2p$	294	347	402	455	513	575	641	710	779	855	931	1021	1116	1217	1323	1436	—	—
	297	350	407	461	520	584	652	723	794	872	951	1044	1143	1249	1359	1476	—	—
$2s$	377	438	500	564	628	695	769	846	926	1008	1096	1194	1298	1413	—	—	—	—

Spin complications

We noted above that the 2*p* levels of cobalt gave rise to two lines in the XPS spectrum, separated by some 15 eV, and other examples of this are shown in Table 6.1. This is a consequence of electron spin. We know that an atom with a single *p* electron has two states of slightly different energy depending on whether the spin angular momentum (quantum number *s*) is aligned parallel or anti-parallel to the orbital angular momentum (quantum number *l*). Exactly the same sort of behavior is found for an atom with a single *p*-level vacancy. In this case *l* is 1 and *s* is 1/2, and so the atom has two distinct states, as *j*, the vector sum of *l* and *s*, may be 3/2 or 1/2. Similarly for *d* levels there are two components, with *j* values 5/2 and 3/2, and for *f* levels the *j* values for the two components are 7/2 and 5/2. The lines from *s* levels are of course single.

In molecules with unpaired valence electrons, spin−spin coupling between valence and core electrons leads to further small splittings. Thus for the NO molecule a splitting of 1.5 eV occurs in the N 1*s* line near 415 eV, and for the O_2 molecule a splitting of 1.1 eV appears in the oxygen 1*s* line near 453 eV (Fig. 6.5). Such splittings are too small to detect unless a monochromator is used.

Fig. 6.5 Spin−spin splitting of the oxygen 1*s* core level line in the X-ray photoelectron spectrum of gaseous O_2. Redrawn from U. Gelius, E. Basilier, S. Svensson, T. Bergmark and K. Siegbahn, Uppsala University Institute of Physics, UUIP-817 (1973).

6.4.4   Depth effects in solid samples

When the target atom in an XPS experiment is in a gas-phase molecule, the ejected photoelectron can escape freely, though even here it may lose some energy as it departs, exciting or even ionizing valence electrons in the process. However, electrons ejected from solids are likely to be captured by the surrounding matter, unless the distance to the surface is very small. Even for electrons originating in atoms near the surface various energy-loss processes reduce the main signal and generate a spectrum of bands to lower energies. The analysis of such spectra can give information about the sample, but for our present purposes we are concerned only with the unperturbed signal, which must be derived from atoms near the surface. How near depends on the electron energy. If the binding energy is low the ejected electrons carry most of the photon energy and can escape from perhaps a depth of 100 Å, while low-energy electrons ejected from levels of binding energy of 1000 eV or more may only reach the surface from depths as small as 1 Å: this effectively includes only the surface layer of atoms. We can therefore choose the depth to which we probe by using appropriate electron binding energies.

Such studies show that even freshly-prepared samples are often strikingly contaminated. Thus metal surfaces which have been exposed to air, even at room temperature, are usually oxide-coated (Fig. 6.6), and the oxides present are not always those we might anticipate from a knowledge of the stability of bulk oxide samples. Conversely, exposure of samples to the high vacuum within the XPS spectrometer may lead to surface alteration, especially loss of water from hydrated samples. A very common surface contamination is due to a film of pump-oil acquired from the diffusion pumps, which gives a prominent carbon $1s$ signal.

Such unrepresentative surfaces can be stripped away by bombardment of the sample with electrons or ions, and this is often done as a routine precaution to ensure that the surface examined is typical of the bulk material. It can, however, be useful to have some surface contamination, such as the pump oil, to provide a calibration point. This enables the effects of sample-charging, which alters the velocity of the departing electron [see below, Section 6.4.5], to be allowed for. As an alternative calibrant, some unreactive material may be sputtered onto the surface; gold is a favored material, giving $4f$ signals near 85 eV, as well as several others down to $4s$ near 760 eV.

As XPS is useful for the study of surfaces it can give a good deal of information about species adsorbed on catalysts. Such work is important, but it is outside the scope of this book.

6.4.5   Chemical shifts

Once an atom has been identified, it is in principle possible to define its effective charge by observing the chemical shift between its core lines and those of a similar atom in a standard environment. In practice the measurement is difficult, and many misleading shifts were reported before the possible pitfalls were appreciated.

The problems may be divided into three groups.
1   Instrumental factors, including the breadth of the exciting line and imperfections (aberrations) in the electron-velocity analyzer, which can be overcome in principle by improvements in the design, for example by inclusion of a monochromator.

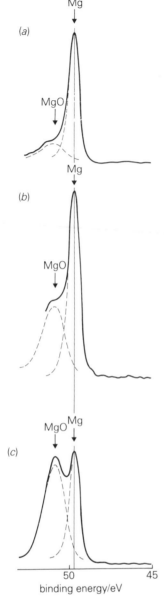

(a)

Mg

MgO

(b)

Mg

MgO

(c)

MgO Mg

50   45
binding energy/eV

Fig. 6.6 Magnesium $2p$ core level lines for (*a*) a 'clean' metal surface, and (*b*) and (*c*) two oxidized surfaces, showing the growth of a separate signal due to Mg^{2+} ions in MgO. Redrawn from K. Seigbahn, Uppsala University Institute of Physics, UUIP-880 (1974), p. 50.

2   Sample effects, including the problems of charging of solid samples, and the difficulty of defining a reference level from which the binding energy can be measured for an insulating solid. These are best dealt with by use of an internal standard, such as the C($1s$) line of hydrocarbon contaminants, or a line of some chemically-unreactive metal such as Au sputtered on to the sample. Any reputable XPS study should report the standard used, and the energy adopted for the standard.

3   The final problem is that various energy-loss processes may accompany or follow the photoionization event, reducing the energy of the ejected electron. There is no cure for this, but fortunately it is usually found that such processes do not occur in some fraction of photoionization events, so that the observed spectrum contains at least

CHAPTER 6: ELECTRONIC AND PHOTOELECTRON SPECTROSCOPY

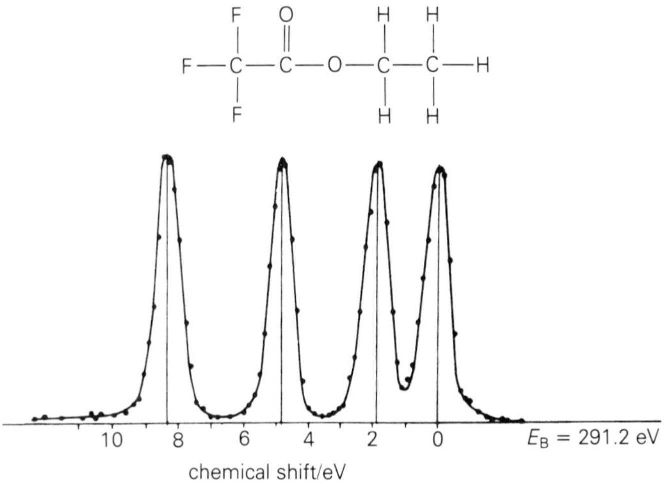

Fig. 6.7 Four distinct carbon 1s core level lines from a sample of ethyl trifluoroacetate. A monochromator was used to reduce the width of the Al K_α source radiation. Redrawn from U. Gelius, E. Basilier, S. Svensson, T. Bergmark and K. Siegbahn, *J. Electronic Spectr.*, **2**, 405 (1974).

some intensity in an unshifted line, accompanied by various 'satellites' corresponding to the energy loss processes.

Once we have adequate instrumentation and have adopted an internal standard, what do we find? Quite simply, the core energy levels for atoms of a particular element depend on its environment. We can, therefore, distinguish between different compounds, or between chemically distinct atoms in a single compound. However, the peaks due to atoms in different sites frequently coincide by chance, as the shift range (up to about 10 eV) is not very large compared with the ultimate linewidth (~ 0.2 eV). Positively charged atoms attract the core electrons more strongly than neutral or negative ones, so metal atoms in oxides, salts and so on have higher binding energies than similar atoms in pure metal samples (Fig. 6.6). Even for carbon compounds distinct signals are observed, depending on the numbers of electronegative atoms attached to the C atom concerned (Fig. 6.7). Such chemical shifts can clearly give useful information about the oxidation state of an element in a sample, though the small shift range makes it hard to define non-overlapping regions characteristic of each oxidation state.

For example, the molybdenum $3d$ binding energies observed [4] for $Mo(CO)_4(PPh_3)_2$, $MoCl_2(CO)_3(PPh_3)_2$ and $MoCl_4(PPh_3)_2$, in which the formal oxidation states of the metal are 0, 2 and 4, are 230.9, 232.7 and 235.3 eV, respectively. Thus a high oxidation state is associated with a high binding energy, as expected. However, the binding energy for the Mo^V complex $MoCl_3O(PPh_3)_2$ is only 235.5 eV, and oxidation states IV and V cannot be distinguished reliably. This could be taken to show significant back-donation from O to Mo, which would reduce the effective charge on Mo. Such arguments are appealing, but must not be carried too far without careful theoretical analysis.

6.4.6   Comparison of chemical shifts with theory

There are three fundamentally different approaches to the calculation of core-level binding energies. In one, we calculate the orbital energies of the electrons in the original molecule, and equate the orbital energy with the binding energy − an assumption known, loosely, as Koopmans' Theorem. The absolute binding energies are of course very large, and even if we do a very expensive calculation we should not expect our calculated orbital energies to match them exactly. But we may hope that the small differences in binding energies that correspond to the chemical shifts between different samples will be correctly reproduced. This is only a hope, and this way of calculating chemical shifts is probably the least satisfactory.

A second approach is to use some theoretical model to calculate effective atomic charges. The chemical shifts are then related to the differences in charge in different samples. Here we are dealing much more directly with the small differences between samples, and considerable success has been claimed for groups of closely-related molecules. An example, using a very simple model of the bonding in nickel(II) complexes in which the charge is calculated simply from the electronegativity difference between the central atom (Ni) and the attached ligand atom, is illustrated in Fig. 6.8. Here we see how the observed chemical shifts in binding energy correlate very well with the calculated charge on Ni. This approach, which relates chemical shifts to a single effective charge parameter, requires that *all* core levels on a given atom shift together, so that the $2s-2p$ gap, for instance, remains constant. This seems to be effectively true in practice, and the success of a charge-based calculation is a reflection of how well the actual differences in charge are reproduced [2].

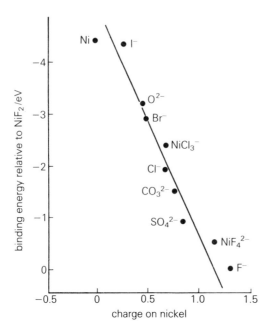

Fig. 6.8 Relative binding energies of nickel $2p$ electrons plotted against charge on nickel. Redrawn from L.J. Matienzo, L.I. Yiu, S.O. Grim and W.W. Swartz, *Inorg.Chem.*, **12**, 2762 (1973).

The third approach attempts to avoid the approximation of Koopmans' Theorem by deriving binding energies as the differences in *total* energies between the molecule and the highly excited positive ions formed by removing core electrons. The problems of calculating the total energies of such ions are by no means trivial, and important advances in techniques of calculation have been made in the process of developing satisfactory methods. It is particularly important to allow for the electronic relaxation of the ion − adjustments of all other electrons − though the timescale is too short to allow vibrational relaxation. This approach is only worth attempting if full *ab initio* methods are used, as semi-empirical methods such as CNDO, INDO and MNDO are incapable of dealing with ions in states which have incompletely filled core orbitals. Ideally we should include configuration interaction (CI), but this is rarely possible except for small molecules.

The second approach has on the whole been most successful, as it is possible to use 'cheap' methods such as CNDO to calculate effective atomic charges. Inclusion of corrections for relaxation effects improves the accuracy of its predictions. This leads to the relaxation potential model, which can be used to rationalize observed chemical shifts, to predict the chemical shifts of new systems and, it is hoped, to increase our understanding of chemical bonding.

6.5   Symmetry and molecular orbitals

Core levels can be regarded as essentially atomic in character, and a core electron can be clearly identified in terms of the particular atom and the type of atomic orbital with which it is associated. In contrast, valence electrons of molecules are best discussed in terms of molecular orbitals. A particular molecular orbital may seem to be associated primarily with one bond, or with a single atom, but in principle it must be associated with the whole molecule. The use of symmetry to classify molecular orbitals is almost essential for a detailed understanding of electronic spectroscopy. We can only describe this briefly here; further background material can be found in the Appendix, and full details can be found in Refs [5]−[9].

Assigning a particular molecular orbital to its symmetry species is like assigning a vibration to its symmetry species (Section 5.7): we need to consider the effects of the various symmetry operations on the sign and orientation of the wavefunction. One problem is that we do not necessarily know what a particular wavefunction looks like, and it is here that the LCAO approach to the construction of molecular orbitals is very useful. We begin by classifying the valence-shell atomic orbitals (AOs) of the constituent atoms, in the symmetry of the molecule, as the set of molecular orbitals (MOs) has the same distribution of symmetry species as the set of contributing atomic orbitals. Then we can attempt to construct molecular orbitals, bearing in mind that an orbital of any particular symmetry species arises from combinations only of atomic orbitals of that same symmetry species. The Appendix on symmetry includes details of the assignment of atomic orbitals to symmetry species.

A simple example will illustrate these points. Consider the H_2O molecule, with symmetry C_{2v} (Fig. 6.9). The oxygen atom has four valence-shell AOs ($2s$, $2p_x$, $2p_y$, $2p_z$) and each hydrogen atom has one AO ($1s$). There are thus a total of six constituent AOs, and we shall find six MOs formed from them. The symmetry species of the oxygen AOs may be deduced by considering the effect of each of the four symmetry operations E, C_2, σ_{xz}, σ_{yz} on the respective wavefunctions. The results are: $2s$, a_1; $2p_x$, b_1; $2p_y$, b_2; and $2p_z$, a_1. The two hydrogen-atom $1s$ AOs must be taken together, and generate a symmetric and an anti-symmetric combination of symmetry species a_1 and b_2, respectively.

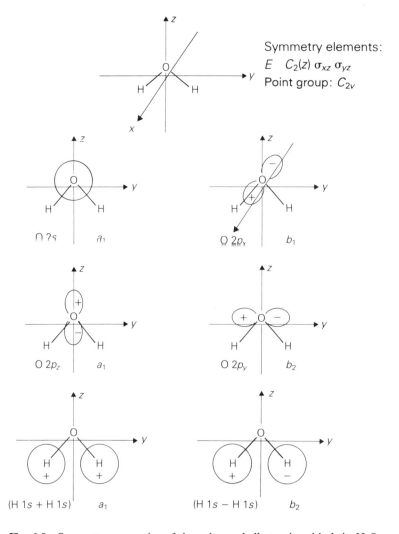

Fig. 6.9 Symmetry properties of the valence shell atomic orbitals in H_2O.

The six atomic orbitals combine to give six molecular orbitals, and these must belong to the same set of symmetry species, and we can write the reduced representation:

$$\Gamma_{mol} = 3a_1 + 1b_1 + 2b_2.$$

Clearly the MO of b_1 symmetry can have no contribution from any AO except O $2p_x$, and can be properly considered as a localized oxygen lone-pair orbital. As there is no possible overlap with H-atom orbitals, the MO has the same energy as the oxygen AO except for a small charge effect. The two MOs of b_2 symmetry are formed by overlap of the O $2p_y$ and H ($1s - 1s$) orbitals, each of b_2 symmetry. The overlap of two orbitals on different centers generates a bonding MO of lower energy and an anti-bonding MO of higher energy.

In principle all three MOs of a_1 symmetry have contributions from all three constituent a_1 orbitals, and we cannot easily deduce the final pattern of energies and bonding characters. In this case, the O $2s$ AO has a much lower energy than either the O $2p_z$ AO

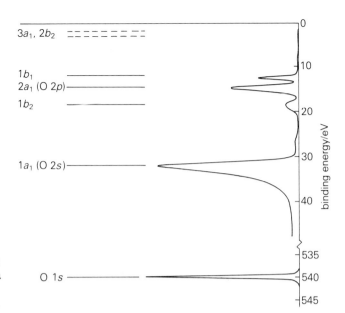

Fig. 6.10 Photoelectron spectrum and energy level diagram for H_2O. Redrawn from K. Siegbahn, Uppsala University Institute of Physics, UUIP-880 (1974).

or the symmetric combination of H $1s$ AOs, and is more closely held to the oxygen core than the O $2p_z$ AO. It is therefore reasonable to suggest that one of the MOs of a_1 symmetry is largely O $2s$ in character, while the O $2p_z$ and H $(1s + 1s)$ orbitals combine to generate bonding and anti-bonding MOs. The final pattern of energy levels is shown in Fig. 6.10. The result is that we have six MOs, two bonding, of symmetry species a_1 and b_2, two non-bonding, of symmetry species a_1 and b_1, and two anti-bonding, of symmetry species a_1 and b_2. The eight electrons in the valence shell occupy the bonding and non-bonding levels, leaving the anti-bonding levels unoccupied.

Of course, this is a very simple case, and it is often necessary to consider many more AOs and MOs. OF_2, for instance, has six more AOs (four valence-shell orbitals on each fluorine, rather than one for each hydrogen in H_2O) and hence six more MOs. In fact these can legitimately be regarded as fluorine 'lone pair' orbitals quite distinct from the bonding, non-bonding and anti-bonding MOs which are analogous to those found for H_2O. Similarly, $O(CF_3)_2$ presents us with yet another increase in complexity, but the additional orbitals can be regarded, at least to a first approximation, as 'CF_3 group' orbitals associated with CF bonding or with F atom lone pairs, again superimposed on a basic set of 'skeletal' MOs analogous to those found for H_2O.

The energies of the MOs cannot, of course, be derived purely from symmetry arguments. They may be calculated by various empirical, semi-empirical or non-empirical methods [10]−[12].

6.6 Valence-electron photoelectron spectroscopy

6.6.1 Instrumental

Valence electrons, with binding energies below about 40 eV, may be photoionized with X-rays in the same way as core electrons. However, there are distinct advantages in using lower-energy photons for the excitation. The most commonly-used light source is a discharge lamp, generally with a collimating capillary which allows a pencil of light to escape into the sample region without permitting sample to flow back into the discharge.

Various gases may be used, but almost all spectra are recorded using He discharge lamps. At a comparatively low energy density excited neutral He atoms give rise to quite strong emission of He(I) lines, of which by far the strongest is He(I$_\alpha$) at 58.4 nm (21.22 eV), corresponding to the electronic transition He $1s^12p^1 \rightarrow$ He $1s^2$. At lower gas pressures and higher energy density ionization of He in the discharge becomes more important, and He(II) emission lines, due to transitions of He$^+$ ions, become progressively stronger. Again, the line at lowest energy is most important. It is called He(II$_\alpha$), and corresponds to the transition He$^+2p^1 \rightarrow$ He$^+1s^1$. Its energy (40.8 eV) is considerably higher than that of the He(I$_\alpha$) line.

The great advantage of using lower-energy photons rather than X-rays is the greater resolution attainable. This can be attributed to the narrower intrinsic width of the source emission line, which is less than 0.01 eV, and to the lower energy of the ejected electrons. The resolving power of an electrostatic analyzer is related to the electron energy, so that $\Delta E/E$ is effectively constant. If we can construct an analyzer with $\Delta E/E = 0.001$, this contributes >1 eV for the electrons ejected with >1000 eV kinetic energy using X-ray radiation, but only 0.01 eV for 10 eV electrons produced using 21 eV [He(I)] excitation. In practice a resolution of 0.02 eV (20 meV) is attainable with a relatively simple design using UV sources, and this is adequate to show vibrational features (characteristically ~ 0.1 eV).

UV-excited photoelectron (UPS) spectrometers use electrostatic velocity analyzers, often of a cylindrical design, which are cheap and simple to construct. Details of experimental arrangements are described in Refs [13] – [15].

6.6.2   **Vibrational structure of PE bands**

Only valence shell MOs that are occupied by at least one electron can be studied by photoelectron spectroscopy (PES), but for these we can deduce the binding energy and also in some cases the bonding or anti-bonding character of the orbital. The binding energy of the electron in an orbital (BE) is found from the kinetic energy of the ejected electron (KE) and the photon energy as before:

$$h\nu = \text{BE} + \text{KE} + \Delta E_{\text{vib}}. \tag{6.2}$$

Here ΔE_{vib} is a new term equal to the difference in vibrational energy between the ion and the original molecule, measured in each case from the ground vibrational level. As most of the molecules will be in or near the ground state before ionization, ΔE_{vib} is rarely negative, but may be positive or zero, depending on whether the ion is vibrationally-excited or not. There is no simple selection rule for Δv in this case, and in some cases extensive vibrational progressions are found. These can help us in making assignments and can tell us about the vibrations of molecular ions.

To understand what causes vibrational fine structure, it is simplest to start with a diatomic molecule. Photoionization is like electronic excitation; the main difference is that the upper state is an ion. We can therefore draw potential functions for the ground-state molecule, and for the molecular ion produced by loss of an electron, on the same diagram. For the ground-state molecule, the spread of internuclear distances is represented by the ground-state wavefunction, with a maximum probability near the center. By the Franck−Condon Principle, ionization occurs so fast that the internuclear distance does not have time to change; the ion is produced with the internuclear distance that was appropriate for the molecule, in what is called a 'vertical' transition.

To understand the consequences of this constraint, we must look at the potential function of the ion. We will assume for the time being that this has a well-defined minimum. If we remove an electron from an anti-bonding orbital, the ion will be more tightly bound than the molecule was. The minimum in the potential function will therefore be at a shorter internuclear distance, and the vibration frequency will be higher than for the molecule. We can see from Fig. 6.11 that excitation is to vibrational levels of the ion whose wavefunctions give high probabilities for the inter-atomic distance found in the neutral molecule, and so overlap the ground-state wavefunction. There will be a series of levels satisfying this criterion, and so we observe a progression, in the frequency of the ion. Since we have removed an anti-bonding electron the vibrational frequency of the ion is greater than that of the molecule.

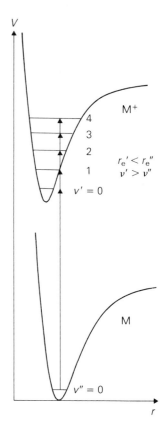

Fig. 6.11 Ionization of a diatomic molecule by removal of an electron from an anti-bonding orbital. The ions may be formed in several different vibrational states, and so the photoelectron spectrum shows a band with a series of vibrational lines.

If we remove an electron from a bonding orbital, the bond in the ion is weaker than that in the molecule. The minimum in the potential function is at a greater internuclear distance, and the vibration frequency is lower. Once again, ionization from the

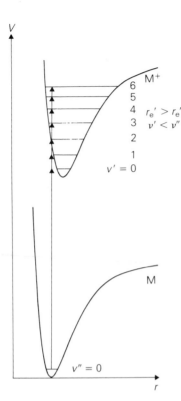

Fig. 6.12 When a diatomic molecule is ionized by loss of an electron from a bonding orbital, the ion may be formed in one of several vibrational states. The corresponding band in the photoelectron spectrum shows a vibrational progression, and the vibrational frequency is less than that of the molecule.

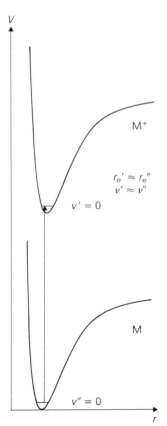

Fig. 6.13 Ionization of a diatomic molecule by loss of a non-bonding electron. Most ions are formed in their vibrational ground state, and the corresponding band in the photoelectron spectrum consists of one sharp line, possibly with one or two weak vibrational satellites.

ground-state of the molecule leads to the formation of ions in a series of vibrational states, and the UPS band shows a progression with $\nu_{ion} < \nu_{molecule}$ (see Fig. 6.12).

If we remove an electron from a non-bonding level, the potential functions of ion and molecule should have minima at very similar internuclear distances. If the forms of the two potential functions are similar too, we may well find that the ground-state level for the ion lies almost exactly over that for the molecule in the energy diagram (Fig. 6.13). Ionization from the ground-state of the molecule then produces ions, most of which are in their vibrational ground-state, and we see a single sharp line with no progression (corresponding to the $0 \rightarrow 0$ transition). In practice few UPS bands are quite so starkly simple; we usually see the vestiges of a progression. But it is generally easy to recognize a peak due to ionization from an essentially non-bonding level. As the bonding or antibonding character increases, the progression lengthens, and the intensity maximum moves from the $(0 \rightarrow 0)$ peak. The progression associated with ionization of a strongly bonding level may show a maximum (the vertical ionization

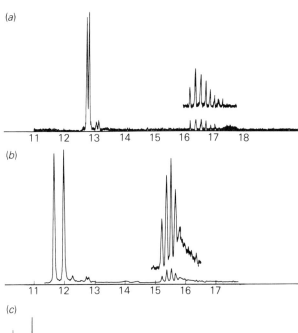

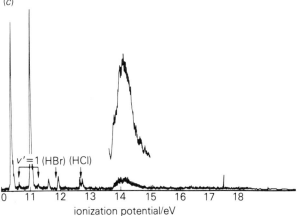

Fig. 6.14 He(I) photo-electron spectra of (a) HCl, (b) HBr and (c) HI, showing the increasing spin−orbit coupling in the first sharp band, which is due to halogen non-bonding electrons in π-type orbitals. Weak vibrational satellites can be seen in each spectrum, and the spectrum of HI shows clearly that small amounts of HCl and HBr are also present. Redrawn from Ref. [13].

ionization potential/eV

energy) well down the band, and we may not detect the $(0 \rightarrow 0)$ level (the adiabatic ionization energy).

We can therefore find out a lot about the character of the orbital from which ionization occurs by studying the form of the UPS band. The UPS spectrum of HCl [Fig. 6.14(a)], for instance, shows two bands below 21.22 eV. The one at lower energy is a sharp doublet with essentially no progression, just one pair of weak satellites. The other shows a long progression with a frequency of 1610 cm^{-1}, a reduced v(H—Cl). The doublet is assigned to ionization from the π-non-bonding orbitals of Cl, split by spin-orbit coupling [see Section 6.6.4 below]; the progression is due to ionization from the H—Cl bonding level.

The spectra of O_2 and N_2 in this region are also informative. The spectrum of O_2 [Fig. 6.15(a)] shows four bands. The one at lowest energy shows a progression in an increased O—O stretching frequency, showing that ionization has occurred from an anti-bonding orbital. The remaining bands are complicated because O_2^+ may be produced in quartet or doublet states, but all show clear progressions with reduced O—O stretching frequencies. The energy-level scheme is given in Fig. 6.16(a). The second band, with its very long progression of vibrational peaks, is assigned to both the quartet and doublet states formed by removal of an electron from the π_u level, while the third and fourth bands are assigned respectively to the quartet and doublet

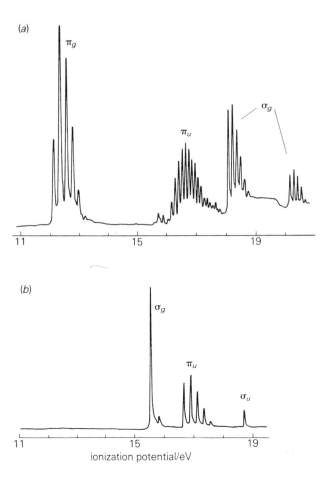

Fig. 6.15 He(I) photo-electron spectra of (a) O_2 and (b) N_2.

ionization potential/eV

states formed by removal of an electron from the $\sigma_g(2p)$ level; both levels are expected to be strongly bonding, as suggested by the lower vibration frequencies of the ions. The spectrum of N_2 [Fig. 6.15(b)] looks very different. Of course, there is no band corresponding to the lowest energy band of O_2, because the π_g level of N_2 is empty; but in marked contrast to the spectrum of O_2, the first and third bands show sharp and dominant $(0 \to 0)$ peaks and clearly correspond to ionization from non-bonding levels. Moreover, the second band shows a progression with a reduced $N \equiv N$ stretching frequency, and is assigned to ionization from the π_u bonding level. The first and third bands represent ionization from $\sigma_g(2p)$ and $\sigma_u(2s)$ levels. In O_2 these are respectively bonding and antibonding, and $\sigma_g(2p)$ is below $\pi_u(2p)$. In N_2, $\sigma_g(2p)$ is raised in energy and becomes less bonding through mixing with $\sigma_g(2s)$ of the same symmetry, which is filled and lies below it: $\sigma_u(2s)$ is lowered in energy and becomes less antibonding by mixing with the empty $\sigma_u(2p)$ which lies above it [see Fig. 6.16(b)]. Mixing is greater for nitrogen than for oxygen because the gap between $2s$ and $2p$ is smaller. The σ-character of the highest-filled level in N_2 is confirmed by the 'end-on' coordination found in most complexes of N_2 with transition metals.

With polyatomic molecules there are many different vibrational modes, and so analysis of what might happen on ionization becomes conceptually much harder. In practice, many bands show no resolved structure. The band maximum then gives us the vertical ionization energy, and the band origin gives the adiabatic ionization energy. In the UPS spectra of some quite complex molecules we may see resolved progressions. When we do, we conclude that the vibrational coordinate associated with the progression is very much affected by the orbital from which the electron was

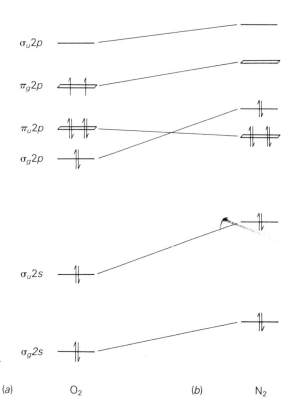

Fig. 6.16 Energy level diagrams for the valence shell molecular orbitals of (a) O_2 and (b) N_2.

removed. For example, in the spectra of many acetylenes, the bands due to ionization from the π-bonding levels show progressions with reduced $C \equiv C$ stretching frequencies. In the spectrum of ammonia (Fig. 6.17), the band due to ionization from the lone pair orbital shows a long progression in a reduced NH_3 symmetric deformation frequency. This reminds us that the angle in NH_3 is largely determined by the lone pair, and the ground state of NH_3^+ is essentially planar.

If the potential function of the ion shows no minimum, but a continuous decrease in energy with increasing internuclear distance, there are no discrete vibrational levels; ΔE_{vib} can take a continuous range of values, giving a broad band. A similar situation will arise if the ion-state formed, although itself vibrationally bound, is 'predissociated' by overlapping a dissociating state at similar energy and distance. In other cases the ion state is only weakly bound, so that only a few discrete levels exist, and the band shows a few lines followed by a broad continuous portion. There are thus different reasons for observing bands without structure, but when we are able to observe vibrational structure, it can be most helpful.

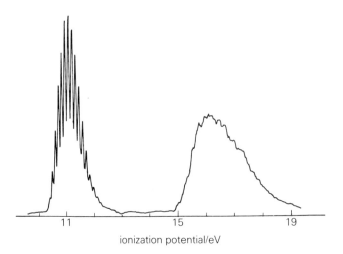

ionization potential/eV

Fig. 6.17 He(I) photo-electron spectrum of NH_3.

6.6.3   Spin–orbit coupling

If an electron is removed from a p-orbital of a rare gas atom the interaction of the electron spin angular momentum (quantum number s) with the orbital angular momentum (quantum number l) generates two possible states for the ion, depending on whether the angular momenta add or subtract. The overall quantum number j can be $\frac{3}{2}$ or $\frac{1}{2}$ and the difference in energy between these states is related to the spin–orbit coupling constant for the atom. Similar effects operate for d orbitals, but not for s orbitals, which have no orbital angular momentum.

In exactly the same way, an atom such as a halogen atom in a covalent molecule can show spin–orbit coupling. The magnitude of the coupling constant increases with nuclear charge: values for the hydrogen halide ions in their ground states are HF 0.03 eV, HCl 0.08 eV, HBr 0.33 eV, HI 0.66 eV (Fig. 6.14). We find a simple splitting into two states only for molecules in which there is a single halogen atom, lying on an axis of at least 3-fold symmetry. In molecules with more than one halogen atom more complicated patterns appear. In linear molecules, only Π states give a spin–orbit splitting, and this can aid in the assignment; similarly in non-linear molecules only E states (where the electron has been removed from a molecular orbital of e symmetry) show the effect. The magnitude of the splitting in a given band gives an indication of the contribution of the halogen p-orbitals to the corresponding molecular orbital. Thus for BrF the first band shows a splitting of about 0.32 eV and is clearly due to a level largely localized on Br; in IBr (Fig. 6.18) the first two bands show splittings of 0.58 and 0.36 eV and are assigned to lone-pairs mainly localized on I and Br, respectively. As a level becomes more delocalized the magnitude of spin–orbit coupling falls, as in the series RI where the constant has values of 0.66 (R = H), 0.63 (R = CH_3) and 0.56 eV (R = CMe_3). The decrease presumably arises because the level of the lone pair on the halogen atom mixes with other levels of e symmetry in the alkyl halides. For R = SiH_3 the value is 0.55 eV, showing a greater delocalization effect for the silyl compound than for the methyl. We cannot be sure that this is a sign of $(p \to d)\pi$ bonding in the Si compound; the delocalization could involve the SiH bonding or anti-bonding levels of e symmetry.

Similar effects can be seen for other heavy atoms, such as Se, Hg and Xe, though usually only in linear molecules.

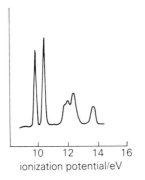

10   12   14   16
ionization potential/eV

Fig. 6.18 He(I) photoelectron spectrum of IBr. The spin–orbit splittings of the two bands at lowest IP suggest that they are due to electrons localized mainly on iodine and on bromine, respectively (see text). Redrawn from S. Evans and A.F. Orchard, *Inorg. Chim. Acta*, **5**, 81 (1970).

6.6.4   Jahn–Teller distortions

Removal of an electron from a degenerate level leaves an ion in a degenerate electronic state. The Jahn–Teller effect concerns the spontaneous removal of this degeneracy by a distortion of the ion. Thus the first band of CH_4 arises from removal of an electron from one of the three t_2 orbitals, forming CH_4^+ in the triply-degenerate $^2T_2$ state; this distorts spontaneously. With the degeneracy removed, there are three maxima in the spectrum, and so the band appears to be broad. Distortion of doubly-degenerate states gives two maxima, but often these are not resolved, and we may only observe a broadening of the band. Jahn–Teller distortions may occur in molecules which also show spin–orbit coupling, giving more complex patterns. An analogous effect in linear molecules is called the Renner–Teller effect.

The photoelectron spectrum of a molecule is closely related to the set of occupied valence orbitals. For simple molecules, some PE bands may show discrete structure due to vibrational excitation of the ion; this may be interpreted in terms both of the bonding/non-bonding/anti-bonding character of the electron removed and of the vibration frequencies of the ion. All this information is related to molecular structure, but it is rarely possible to come to definite conclusions about the constitution, shape, size or conformation of a sample purely from its PE spectrum. Of course, at the simplest possible level the PE spectrum may be used as a fingerprint, allowing us to identify a pure sample or a constituent of a mixture by comparison with known spectra. This process is particularly effective where small molecules are involved, for the band-shapes and vibrational structure are often characteristic. For more complex molecules in whose spectra vibrational detail is not resolved, assignments are difficult to make. Bands often overlap, and the only methods available to help making assignments are relating the intensities of bands, comparing spectra of related molecules, possibly observing spin–orbit coupling, or making comparisons with theory.

Band intensities are often used in making assignments, but they have to be interpreted with care. Though there should be some correlation between the degeneracy of a fully occupied level and the intensity of the band derived from ionization from that level, many factors (including band overlap) obscure or destroy the validity of this generalization. With the spectra of transition metal complexes, a comparison of spectra obtained with He(I) and with He(II) radiation is often used to help in making assignments. Ionization from orbitals with d character is normally supposed to be stronger in spectra excited with He(II) than with He(I) radiation. Recently it has become possible to record band intensity profiles with a range of excitation energies between 20 eV and 100 eV using synchrotron radiation; these profiles show that the relationship between ionization energy and band intensity is not a simple one. The spectra of $M(CO)_6$ (M = Cr, Mo, W) show a very pronounced intensification of the bands due to ionization from the predominantly metal-d orbitals of t_{2g} symmetry at photon energies that appear to correspond to absorptions in the high energy electronic spectra [16]. Studies of this kind show much promise in helping to understand the PE spectra of moderately complex inorganic and organometallic compounds, but at present the empirical assignment of these spectra is fraught with uncertainty. The most fruitful studies have been based on calculated energy levels. As with core levels, there are various approaches to the computation of valence level binding energies. Here, however, the appeal of empirical and semi-empirical methods is stronger, as they are able to handle valence levels explicitly, without the need to include core levels. The size of a semi-empirical calculation is thus much less than that of an *ab initio* calculation for the same molecule, even if only first row elements are concerned. However, the results of the comparison of valence level PES with these simpler calculations have not been encouraging. Not only do the semi-empirical methods (CNDO, INDO, MNDO, *etc.*) fail to reproduce the absolute values of binding energies, but they also often predict the wrong orders of levels. They can hardly, therefore, be taken as satisfactory aids to assignment.

There are at least two problems involved, apart from the particular failings of the assumptions built into each method. These arise from the different effects of electron correlation in the molecule and the ion (which of course have different numbers of electrons), and the effects of spontaneous relaxation of the electrons remaining in the ion as the ejected electron leaves. Both effects are more important for valence levels than for core levels, because the separations between levels are much smaller. These problems can only be resolved by more elaborate and sophisticated calculations. Allowing for electron correlation requires use of configuration interaction (CI) methods, which enlarge the scale of the computation enormously, while the relaxation effect is best dealt with by calculating the difference between overall energies for molecule and for ion (ΔSCF, ΔINDO, etc.). Calculations for transition metal species present particularly difficult problems. It is hard to calculate the energies of ions with several unpaired electrons, and the effects of electron correlation and of relaxation are particularly significant in transition metal compounds.

6.7   Valence excitation spectroscopy

6.7.1   Experimental methods

Most of what is commonly called electronic spectroscopy is concerned with transfer of electrons between valence shell molecular orbitals. For atoms and diatomic molecules, emission spectroscopy using discharge excitation is very useful, but for larger molecules most work has been done using absorption techniques. More recently, the use of lasers to excite specific upper state levels has allowed emission spectra to be studied. This technique is known as laser-induced fluorescence spectroscopy.

Emission electronic spectroscopy is the simplest possible form of spectroscopy. The sample is a low pressure gas excited thermally, electrically or by a microwave or radio-frequency discharge, and acts as the light source. We then only need a dispersing element and a detector to complete the experiment. Unfortunately, only diatomic molecules are easy to study in this way, as larger molecules almost invariably fragment on excitation, but a vast amount of information has been obtained for atoms and diatomic molecules. Most work has been done in the visible and near-UV regions (wavelengths 700-200 nm; frequencies 14 000-50 000 cm^{-1}) where photographic recording is easy, and glass or quartz optics are ideal. Both prism and grating spectrographs are simple to construct and operate, and many decades of experience has resulted in the development of reliable instruments. Recording spectrophotometers are more elaborate, but are also available commercially. Modern Fourier transform interferometers, designed to operate mainly in the IR, can also be used up to 16 000 cm^{-1} or so, covering part of the visible region, and instruments designed to cover the entire visible region should be available soon. These instruments combine constant resolution and high precision of frequency measurement with great photometric accuracy.

For absorption spectroscopy we need a light source as well as the sample, dispersing element and detector. Suitable continuum emission is obtained from a filament lamp in the visible region and from a high-pressure H_2 (or better, D_2) lamp in the near-UV region. Detection is usually electrical, with a photomultiplier. Prism and grating spectrophotometers are commercially available, and absorption spectra of solids, liquids, solutions and gases are all readily obtained.

For the laser-induced fluorescence (LIF) experiment we need to have a tunable laser source, most commonly a dye laser, pumped by a high-power UV laser. This may well be pulsed rather than continuous in its output, which allows us to use a gated detector,

Example 6.1

Explore the assignment of the UVPE spectra of $Cr(CO)_6$, $W(CO)_6$, $[Cr(CO)_5(NMe_3)]$ and $[W(CO)_5(NMe_3)]$, shown in the figure below.

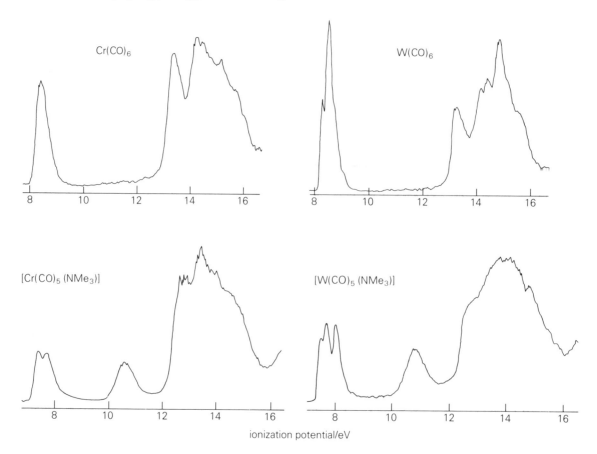

ionization potential/eV

The figure is taken, with permission, from B.R. Higginson, D.R. Lloyd, J.A. Connor and I.H. Hillier, *J. Chem. Soc., Faraday Trans. II*, **70**, 1418 (1974).

....

The molecular orbitals in the hexacarbonyls are likely to fall into four groups: the metal t_{2g} orbitals; the π-bonding and the σ-donor orbitals derived mainly from CO; the orbitals that make up the internal ligand σ-bonding system; and the deep-lying molecular orbitals consisting mainly of oxygen $2s$ atomic orbitals from the CO ligands. The He(I) spectra show bands in three broad regions. The band corresponding to ionization of the most loosely bound electrons must correspond to ionization from the t_{2g} orbitals, which are mainly metal d orbitals. In the spectrum of $W(CO)_6$ there is some sign of multiplet splitting in this band, derived from spin–orbit coupling. Between 13 and 16.5 eV there is a complicated system of overlapping bands that can only be assigned with the help of molecular orbital calculations. Assignments in the third region are similarly difficult. So without calculations it is only possible to identify the HOMO in these spectra.

In the spectra of the substituted derivatives there are similar difficulties with overlapping bands at ionization energies above 12 eV. However, there are two interesting features. First, the band at lowest ionization energy in each spectrum shows additional splitting. This would be expected as the local symmetry of the metal and its ligands changes from O_h in the hexacarbonyls to C_{4v} in [M(CO)$_5$(NMe$_3$)]. In C_{4v}, the t_{2g} orbitals of the regularly octahedral molecules split into e and b_2 components. The extra splitting is clear in both spectra, though again the details can only be understood with the help of calculations. In each spectrum there is an additional band, not present in the spectra of the hexacarbonyls, at about 11 eV. It is reasonable to assign this to ionization from the metal–N σ orbital. The remaining assignments need the help of molecular orbital calculations.

---

discriminating between prompt fluorescent emission and delayed phosphorescence. In the simplest experiment the total emission is monitored as a function of excitation frequency, giving an excitation spectrum. This is closely related to an absorption spectrum, but not identical to it, because the 'spectrum' recorded relates only to those molecules that fluoresce rather than losing energy in other ways. Fig. 6.19 shows the absorption and emission processes.

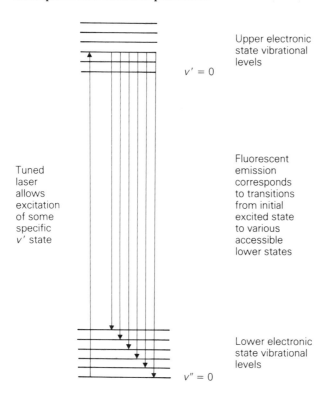

Upper electronic state vibrational levels

$v' = 0$

Tuned laser allows excitation of some specific v' state

Fluorescent emission corresponds to transitions from initial excited state to various accessible lower states

Lower electronic state vibrational levels

$v'' = 0$

Fig. 6.19 Levels and transitions involved in a simple laser-induced fluorescence experiment.

A more elaborate LIF experiment uses a simple monochromator to disperse the emission, so that each separate transition is detected separately. We now have a 'two-dimensional' spectrum, as we can record the spectrum of the emission (in one dimension) for a range of excitation frequencies. As explained in Section 5.3.3, this can result in a very simple spectrum, due to emission from a single vibrational level of the electronically-excited state to various vibrational levels of the electronic ground state, as shown in Fig. 6.19. This gives us information about some of the vibrations of the ground state (see, for example, Fig. 5.6). Scanning the excitation frequency reveals different upper-state vibrational levels, giving information about the vibrations of the upper state, and about the electronic nature of this state. To obtain the greatest possible simplification of the spectrum, the sample must be cooled (as a gas by expansion through a supersonic nozzle in a high-pressure carrier gas such as He or Ar, or as a solid or matrix-isolated sample) so that the initial excitation populates only a single level at a time.

As well as emitting and absorbing light, samples of solids also reflect light, and it is possible to obtain reflectance spectra in various ways. A specular reflectance spectrum is obtained from a polished flat surface, whereas a diffuse reflectance spectrum is obtained from a rough surface or a loose powder sample. Both of these techniques give information about the solid surface rather than the bulk. The resulting spectrum is similar to a conventional transmission spectrum, as radiation that is absorbed is not reflected. An example is shown in Fig. 6.20.

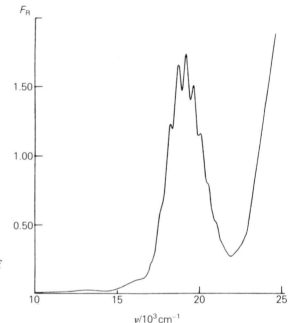

Fig. 6.20 Reflectance spectrum of K_2NiF_6. The ordinate is effectively absorbance. Redrawn from G.C. Allen and K.D. Warren, *Inorg.Chem.*, **8**, 753 (1969).

6.7.2   The information in an electronic spectrum

Transitions between two levels with long life-times are the most informative. Such an electronic transition has various possible changes in vibrational and rotational quantum numbers associated with it, so that the spectrum, however it is obtained, consists of a number of vibration bands, each with rotational fine structure, together forming an electronic system of bands. The selection rules governing the changes in vibrational and rotational quantum numbers depend on the nature of the electronic transition, and they can be ascertained by analyzing the pattern and structures of the bands. The fullest analysis can only be achieved if rotational structure is resolved. In this case rotation constants in upper and lower states can be calculated, as well as vibration frequencies, and a detailed picture of the structure of both states can be obtained.

Rotational structure is, however, rarely resolved except for molecules with only two or three atoms. This is due partly to the very high resolution required to resolve rotational detail for larger molecules, and partly because excitation often leads to states which are unstable. These result in short lifetimes for the upper states, and hence a loss of definition of the energy levels. The Heisenberg uncertainty principle tells us that $\Delta E \cdot \Delta t \approx h/2\pi$. If Δt is 10^{-10} s, the uncertainty ΔE in the energy level is $\sim 10^{-24}$ J per molecule, or 0.6 J mol^{-1}, some 0.05 cm^{-1}. This is quite enough to remove any resolvable rotational structure with the possible exception of sharp Q branches. Lines may also be broadened through the Doppler effect, and this broadening increases with the transition energy. Large molecules give spectra with a high density of lines, and line broadening can cause these to merge, so that rotational details are not discernable, even without dissociation or predissociation. Both the Doppler broadening and the density of lines in the spectrum can be dramatically reduced by nozzle-beam expansion, where a jet of gas, either pure sample or sample mixed with an inert gas diluent, is allowed to expand through a nozzle into an evacuated chamber at supersonic velocity. Most of the random (thermal) motion of the gas is thus converted into forward velocity, and the translational temperature (which governs the velocity measured perpendicular to the flow direction, and hence the Doppler broadening) can drop to within a degree or so of absolute zero. At the same time energy is lost from rotational degrees of freedom, by collisions during the expansion, and samples often end up almost entirely in the lowest few rotational levels, so that only a few transitions are observed in the spectrum.

For upper states that are directly-dissociated (*i.e.* with repulsive potentials) lifetimes may be as short as 10^{-14} s, giving energy uncertainties of hundreds of cm^{-1}. Under these circumstances there is no vibrational structure, and the electronic transition gives an unresolved band.

For solids and liquids, electronic absorption bands are usually broad and essentially featureless. This may be due to the overlapping of many vibration sub-bands, including hot bands corresponding to species in which low-frequency vibrations are excited. In such cases cooling the sample may lead to a simplification of the spectrum (Fig. 6.21). Another factor is that the lifetime of the excited state is short because of a high probability of re-emission (fluorescence), which broadens the levels.

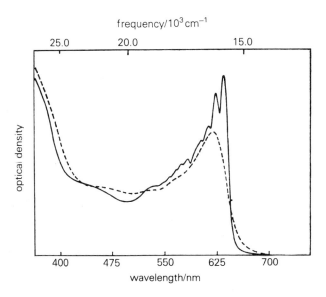

Fig. 6.21 Electronic absorption spectra of $[Fe(C_5H_5)_2]^+$ $[BF_4]^-$ in LiCl solution (---) at 300 K and (—) at 77 K. Note the clear vibrational structure visible in the low temperature spectrum. Redrawn from Y.S. John, D.N. Hendrickson and H.B. Gray, *J.Am.Chem.Soc.*, **93**, 3603 (1971).

The selection rules that govern the intensities of valence electronic transitions are derived from the quantum mechanical equations that are used to formalize and understand the processes involved. They depend on many factors, and it is not possible to give a full account of them here; you will find details in Refs [17] and [18]. It is important, however, that you understand some aspects of the rules if you are to make sense of the electronic spectra of inorganic species.

The rules can be divided into those concerned with electron spin and those related to orbital properties. The spin selection rules are very simple in principle:

1   only one electron is involved in any transition;
2   there must be no net change of spin – i.e. $\Delta S = 0$.

The first rule governs some processes that might not involve an overall change of spin, but it is conveniently considered with the spin rule. It is a very stringent rule; it is very rarely necessary to consider transitions involving more than one electron in interpreting the electronic spectra of inorganic species. The second rule is also stringent. Transitions that involve a change in overall spin are usually very weak indeed for species involving light elements. For instance, all the transitions in the visible spectrum of $[Mn(OH_2)_6]^{2+}$, a high spin d^5 complex, are spin forbidden, and the bands are extremely weak ($\varepsilon \simeq 0.04\ 1\ cm^{-1}\ mol^{-1}$), hence the pale color of maganese(II) salts. For complexes of heavy transition elements, however, spin–orbit coupling may lead to a relaxation of this rule, which only holds as long as the total spin quantum number S has physical meaning, and so spin-forbidden bands may be found to have reasonable intensity.

The orbital selection rule may be stated in different ways. It applies most strictly to species with a center of inversion. For them it may be stated in the following terms:

3a   Transitions that are allowed must involve an overall change in orbital angular momentum of one unit – i.e. $\Delta L = \pm 1$. Transitions within the same sub-level, i.e. $d–d$ or $p–p$ transitions, are forbidden.
3b   Transitions from g to g and from u to u states are forbidden; only transitions from u to g or from g to u states are allowed.

This rule is known as the Laporte selection rule. It too can be relaxed by various mechanisms. It is less stringent than the spin selection rule, partly because the mechanisms for getting round it are more effective. Consider, for instance, a regularly octahedral complex like $[Co(OH_2)_6]^{2+}$. The electronic bands in the visible spectrum are derived from $d–d$ transitions. These are Laporte-forbidden: $\Delta L = 0$. Moreover, this rule cannot be mitigated by mixing the d orbitals with p orbitals, because in a regular octahedron the d orbitals are g and the p orbitals are u. In tetrahedral $[CoCl_4]^{2-}$ there is no center of inversion; this means that mixing between d and p orbitals is possible, and the bands in the visible spectrum are much stronger than for $[Co(OH_2)_6]^{2+}$ (Fig. 6.22). However, even in a regularly octahedral complex there are ways of weakening the effectiveness of the Laporte selection rule. A $g–g$ transition may be forbidden if considered as a purely electronic process, but it may become

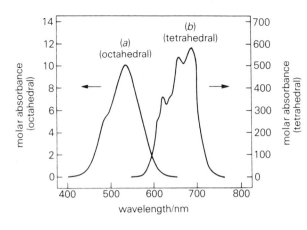

Fig. 6.22 Visible spectra of (a) $[Co(H_2O)_6]^{2+}$ and (b) $[CoCl_4]^{2-}$. Note that the intensity scales differ by a factor of 50, the tetrahedral complex giving a much more intense band. Redrawn from F.A. Cotton and G. Wilkinson, *Advanced Inorganic Chemistry*, 3rd edn, Wiley Interscience, New York (1972), p. 881.

allowed if it is coupled to a vibrational process of u symmetry – if, for instance, a u vibrational mode is excited in the upper state. This sort of process is described as vibronic. There are also other mechanisms for intensifying bands due to formally forbidden transitions, which we do not discuss here. And charge transfer transitions, which involve transfer of an electron from metal to ligand or ligand to metal (Section 6.8.5), are fully allowed, and bands may be very intense indeed, as for example in MnO_4^-.

Thus there is a very wide range of intensities observed for electronic bands; the weakest are those that are both spin and orbitally forbidden, the strongest both spin and orbitally allowed. Some typical intensities are given in Fig. 6.23.

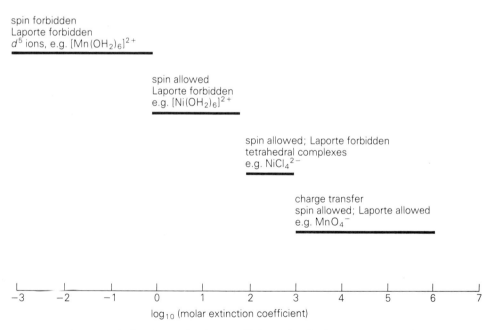

Fig. 6.23 Some typical molar extinction coefficients for d–d and charge-transfer transitions.

Another feature of electronic transitions is that the $\Delta v = \pm 1$ selection rule for pure vibrational transitions no longer applies. An electronic band can therefore show a progression of peaks due to $\Delta v = 0, 1, 2, 3\ldots$ bands for one or more vibrations (see

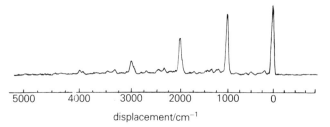

Fig. 6.24 Fluorescence spectrum of benzene, showing a progression in the symmetric skeletal mode of the ground electronic state. The weaker peaks are due to sets of similar progressions given by molecules with some other vibration excited as well. The spectrum was obtained by excitation via a single vibrational level of the upper electronic state, using light from an arc lamp and monochromator rather than a laser. Redrawn from C.S. Parmenter and M.W. Schuyler, *J.Chem.Phys.*, **52**, 5366 (1970).

Fig. 6.24). This gives us a much more detailed picture of the potential well than the single $\Delta v = 1$ pure vibrational transition. Analysis of progressions leading to dissocia tion (at which point the progression breaks off to be replaced by continuous absorption) can give us accurate values of dissociation energies for simple molecules. In absorption the progression relates to the vibration levels of the upper state. In emission or laser-induced fluorescence experiments the vibrational potential and dissociation limit(s) of the ground-state can be investigated. Thus selective excitation to the $v' = 0$ level of an electronically-excited state may be followed by emission accompanying transitions to the $v'' = 0, 1, 2, 3, 4 \ldots$ levels of the ground state. Even if we are not sure of the nature of the upper state involved, the energy differences in the progression give clear-cut information about the ground state. By changing the excitation frequency we can probe the upper-state vibrational levels, but it is often difficult to identify the $v' = 0$ level, which may have low probability of excitation or emission.

6.7.4 Calculation of valence excitation energies

In most cases, at least one of the states involved in valence excitation transitions has two unpaired electrons, and it is notoriously difficult to handle such states adequately. It is not surprising, then, that there are many unsatisfactory ways of calculating valence electron excitation energies, but none that give convincing results except over a very narrow group of compounds. For inorganic systems, especially those involving transition metals, there are special techniques which are used to predict $(d-d)$ transition energies, for example, but even those with adjustable parameters cannot always give unambiguous interpretations of the observed spectra. An old but superb treatment of this area is to be found in Ref. [19].

6.8 Electronic energy levels and transitions in transition metal complexes

The wide range of colours of transition metal complexes is among their most striking properties. This is one sign of the richness of the electronic spectra of compounds containing transition metal centers. These electronic spectra are full of information about the electronic structures of the species responsible for them; such spectra have been most extensively studied and analyzed, and there is no room in this book for a detailed account of their interpretation. Moreover, much of the work has been concerned with the analysis of the spectra and the interpretation of the parameters obtained in terms of theory; though this is an essential part of understanding of the chemistry of transition metals, it is less directly related to determining the structures of their compounds. We therefore give an indication of what causes bands in the electronic spectra of transition metal compounds, of how these bands can be analyzed, and of what sort of structural information can be obtained from them. If you want to go through the details of assignment and analysis, there are many excellent treatments that will help you, varying in depth from the brief but extremely clear account in Ref. [20] to long and complicated books on the subject, for example, Ref. [21].

<h3>6.8.1   Metal, ligand and metal–ligand bonding levels</h3>

Electronic transitions in transition metal compounds, as in all other species, take place between occupied and empty orbitals. The orbitals in these compounds can be considered to lie predominantly on the metal (or metals in a polynuclear species) or on the ligands, or to be involved in metal–ligand interactions and so be shared between metal and ligand. In the simplest electrostatic picture of a complex such as $[Ni(OH_2)_6]^{2+}$, for instance, the lone pairs of the oxygen atoms in the water molecules are regarded as in pure ligand orbitals and the eight d electrons of the nickel are held in pure d orbitals. Thus such a complex can be described as a d^8 complex. If the bonding between nickel and the water ligands is regarded as having some covalent character, then the lone pairs of the water molecules are in bonding levels, and two of the metal d orbitals take on some anti-bonding character. For a fairly simple mononuclear species this classification can be represented schematically as in Fig. 6.25. Remember, though, that even such simple ligands as O^{2-} may have non-bonding valence-shell orbitals.

In terms of this sort of classification, electronic transitions may be of five types: (a) ligand-based transitions

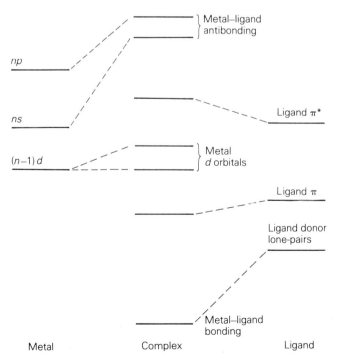

Fig. 6.25 Energy level diagram showing levels in a transition metal complex derived from metal, ligand and metal–ligand bonding orbitals.

(b) metal-based transitions

(c) metal-to-ligand transitions

(d) ligand-to-metal transitions

(e) intervalence transitions.

These transitions will be considered in turn.

6.8.2 Ligand–ligand transitions

Some ligands have low-lying empty orbitals, usually weakly antibonding in character, and thus possess characteristic electronic absorption spectra of their own. This is normally the case for organic ligands containing π-systems, such as pyridine and more complicated aromatic amines. As the metal–ligand bonding is likely to involve the very orbitals that are the lower levels in such transitions, the transition energies are greatly affected by the metal center, generally shifting to higher energy compared with those in the free ligand. More subtle effects occur when we compare one complex with another, so ligand–ligand transition energies will vary with the nature and charge of the central metal atom.

6.8.3 Metal-metal transitions (d–d bands)

Transitions between orbitals that are largely or completely localized on a metal center are usually described as d–d (or, for elements with partly filled f sub-shells, f–f) transitions. The bands associated with them are rarely very strong, and are sometimes very weak because of the electronic selection rules, but they have been studied extensively, and offer very detailed information about the electronic structures of transition metal derivatives. But it can be a complicated matter to extract useful information from the observed spectra. For a metal complex with d electrons, the analysis usually starts with the free ion. Knowing the number of d electrons, it is possible to work out the number of free ion terms by analyzing the microstates; this process is described in Ref. [20]. Once this has been done the terms must be put in order of energy. Hund's Rules will usually give the ground term correctly, but they are less reliable for the higher energy states. The energies of the terms depend on inter-electronic interactions. These can be described by a series of complicated integrals, but the integrals have been expressed in terms of just three parameters, A, B and C, called Racah Parameters. Only B and C come into the differences in energy between terms, so the process of ordering the terms can be made a lot simpler using Racah parameters.

Once the terms for the free ion have been worked out and put in order of energy, it is necessary to see how the environment of the ligands will affect them. The effect depends on the symmetry of the coordination environment. For a metal in an octahedral array of six identical ligands, the d orbitals split into a lower set of three with symmetry t_{2g} and an upper set of two with symmetry e_g. Environments of lower symmetry lead to different splitting patterns (see Fig. 6.26). The effects of environments of different symmetry upon the terms for different ions can be worked out using group theory. For example, for the very simple case of an ion with a single d electron, such as Ti^{3+}, the free ion term is ^2D. In an octahedral field this splits into a lower state $^2T_{2g}$ and an upper state ^2E_g, just as the d orbitals split into t_{2g} and e_g sets. The energy difference between the two terms is Δ, and so on this simple basis the electronic spectrum should consist of a single band whose energy gives Δ directly. The energy of the system can conveniently be represented by a diagram in which energy is

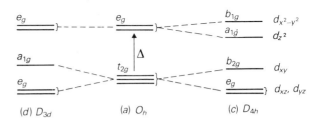

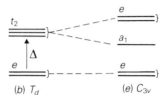

Fig. 6.26 One-electron energy levels associated with metal d orbitals in complexes of various symmetries: (a) octahedral; (b) tetrahedral; (c) tetragonal distortion of an octahedron; (d) trigonal distortion of an octahedron; (e) trigonal distortion of a tetrahedron.

plotted against Δ (Fig. 6.27). The energies of the $^2T_{2g}$ and ^2E_g states depend only on Δ, and the transition energy can be obtained easily from the diagram. For species with more than one d electron, things are much more complicated. To start with, there are more energy levels: for a d^3 system, for instance, the ground free ion term is usually ^4F, but this splits in an octahedral field into three levels – $^4A_{2g}$, $^4T_{2g}$ and $^4T_{1g}$, just as the f orbitals split into a_{2u}, t_{2u} and t_{1u} sets in an octahedral field. The number of states into which a term splits in a particular field is determined by group theory, but that is only the start of the problem; determining the relative energies is difficult. As with the free ion, the inter-electronic effects depend on a number of complicated integrals; these can be expressed in terms of Racah parameters, but B and C do not have the same values as they have for the free ion. There are two limiting cases to consider: the weak field case and the strong field case. In the weak field case the field of the ligands is weak compared with the inter-electron interactions, so that the quantum numbers L and S remain valid. In the strong field case the inter-electronic effects are treated as a perturbation. The results of very detailed calculations can be represented as plots of the variation of energy, E, with Δ. In these diagrams, known as Tanabe–Sugano diagrams, E/B is plotted against Δ/B (see Example 6.2). The units make the forms of the diagrams effectively independent of B, and since C is assumed to be about $4B$ they can be used to include variations in C. Tanabe–Sugano diagrams can be used to assign spectra empirically; they lead to the identification of allowed transitions, and very often the fit of the observed transition energies gives values of Δ/B that are unambiguous.

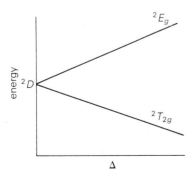

Fig. 6.27 Energy levels for a d^1 octahedral ion as a function of the ligand-field splitting, Δ.

SECTION 6.8: ELECTRONIC ENERGY LEVELS AND TRANSITIONS

Example 6.2

Assign the first three bands in the electronic spectrum of MnF_2, shown below, using the Tanabe–Sugano diagram for a d^5 metal ion in an octahedral field. Work out approximate values for Δ and B.

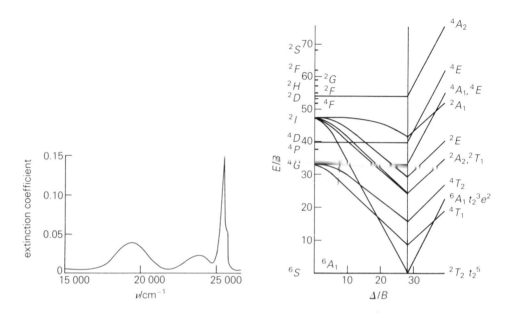

Note two things about the diagram. First, the ground state starts off as $^6A_{1g}$ for small values of Δ/B, but when Δ/B reaches about 28 the $^2T_{2g}$ state becomes the ground state. This is why there is a vertical line in the diagram at Δ/B 28. From this point the energy of the $^6A_{1g}$ state rises sharply as Δ/B increases. Second, not all states are represented by lines on the diagram. There are states derived from the free-atom terms ^4P, ^4D, etc., and while the energies for the free ions are marked on the vertical axis the energies of the states derived from them are not plotted.

There are no spin-allowed transitions for a d^5 ion. Fluoride is a ligand associated with small values of Δ and high-spin complexes, so our assignments must be based on the left-hand part of the diagram, with Δ/B less than 28. The transition energies are in the ratio 19:24:25. Transitions from a sextet state to a quartet state will be less strongly forbidden than transitions to doublet states, so we assume that the first three transitions are to the $^4T_{1g}$, the $^4T_{2g}$ and the $^4A_{1g}$ and ^4E_g levels (the last pair having the same energy on the diagram). Fitting the observed transition energy ratios to the diagram gives us a value for Δ/B of about 6 and a value of E/B of about 28 for the first transition, which is then assigned to $^4T_{1g} \leftarrow {}^6A_g$. We then know that E/B is about 28 but we have measured E as 19 000 cm^{-1}. It follows that B is about 680 cm^{-1}, and so Δ is about 4000 cm^{-1}.

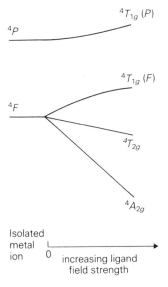

^4P

$^4T_{1g}\ (P)$

$^4T_{1g}\ (F)$

^4F

$^4T_{2g}$

$^4A_{2g}$

Isolated metal ion

0 increasing ligand field strength

Fig. 6.28 Orgel diagram for a d^3 system in an octahedral environment.

Similar plots, but with fixed B and C, are called Orgel diagrams. The Orgel diagram for a d^3 system is given in Fig. 6.28. The lowest energy state for the free atom or ion is usually ^4F with another quartet state, ^4P, to higher energy. In an octahedral environment the ^4P state remains single, having symmetry $^4T_{1g}$, but the ^4F state gives rise to three levels, whose energies diverge as the ligand field strength increases. All the levels involved are still g (symmetric with respect to the inversion center, since we are dealing with strictly octahedral complexes), so all transitions are formally orbitally forbidden ($\Delta L = 0$). There are only three possible spin-allowed ($\Delta S = 0$) excitations, from the lowest level, $^4A_{2g}$, to the three higher quartet states. In many Orgel and Tanabe–Sugano diagrams the lines are curved, because of interactions between states of the same symmetry.

Analysis of d–d spectra of transition metal complexes can give information about three parameters of structural importance: the symmetry of the species; the magnitude of Δ; and the magnitude of the Racah parameter B. Identifying the symmetry and coordination number of a complex is obviously important in determining its structure. The number of methods available for studying complicated species in solution is limited, and electronic spectroscopy is useful in this context. The electronic spectrum of nickel cytochrome c, for example, is interpreted as showing that the nickel atom is six-coordinated, rather than four coordinated. A similar analysis of the electronic spectrum of [Cu^{II}myoglobin], based on the frequency patterns from model compounds, indicated that the copper atom was five-coordinated. In oxyhemocyanin, too, the electronic spectrum, taken with the g-values from the ESR spectrum, shows that the copper(II) ions were in slightly distorted tetrahedral environments, and so the ready electron transfer to give tetrahedral Cu^I could be understood. The determination of Δ can also provide useful structural information, by showing what the ligand atoms are; in cases such as NCS$^-$ ligands may be bound to metals through either nitrogen or sulfur. Finally, the determination of B has led to the formulation of the nephelauxetic series of ligands, in which the position of a ligand is related to its power to expand the d orbitals of the metal.

Jahn–Teller distortion

It is found experimentally, and can be shown theoretically, that any complex with a degenerate (*e* or *t*) set of orbitals which is occupied unsymmetrically (*i.e.* neither entirely filled nor half-filled) distorts its ligand environment spontaneously in such a way as to remove the degeneracy [see Section 6.6.5]. As an example, an octahedral d^7 ion with a filled t_{2g} (lower) set of *d* orbitals and a single electron in the doubly-degenerate e_g upper set would distort in such a way that the e_g levels split, such as by an elongation along the *z*-axis, giving a tetragonal ligand arrangement as in Fig. 6.26(*c*). The single electron then occupies the lower orbital, d_{z^2}, and the overall energy of the system is reduced. This clearly allows a new transition, in which the odd electron moves from the d_{z^2} to the empty $d_{x^2-y^2}$ level, and should also increase the number of bands associated with excitation of an electron from the t_{2g} to the e_g set of levels (both sets are now split, so several extra bands will appear). Again, the effects of Jahn–Teller distortion must be taken into account in analyzing d–d band patterns.

Upper states can also be perturbed by Jahn–Teller distortions. In $[\mathrm{Ti(OH_2)_6}]^{3+}$, for instance, though the lower $^2T_{2g}$ state is not perturbed the upper ^2E_g state is split, leading in principle to the splitting of the d–d band into two, though in practice only one broad band with a shoulder is usually observed.

Metal–ligand and ligand–metal (charge-transfer) bands

A transition that involves the transfer of an electron from metal to ligand or vice versa is called a charge-transfer transition. One involving transfer of an electron from the metal, in the lower energy state, to a ligand, in the upper state, is known as an oxidative charge-transfer transition, because the metal is oxidized in the process. Similarly, one involving transfer from ligand to metal is called a reductive charge-transfer transition. These transitions are particularly important for second- and third-row transition metal complexes, and the associated bands are often extremely intense, so that they can dominate a whole spectrum and obscure the relatively weak d–d bands. Some possible

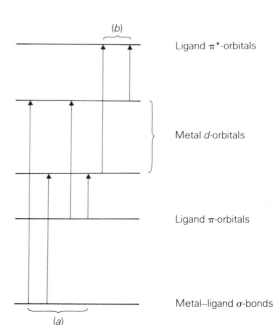

(*b*)

Ligand π^*-orbitals

Metal *d*-orbitals

Ligand π-orbitals

Fig. 6.29 Some possible ligand-to-metal charge transfer transitions (*a*) and metal-to-ligand charge transfer transitions (*b*). Note that the metal-ligand σ-bond orbitals are derived largely from ligand-based orbitals.

Metal–ligand σ-bonds

(*a*)

charge-transfer transitions are shown in Fig. 6.29. A well-known example of a reductive charge-transfer band is in the spectrum of the permanganate ion, $[MnO_4]^-$. Here we may formally write the ground-state as $[Mn^{7+}(O^{2-})_4]$, in which case the spin and orbitally allowed transition involves transferring an electron from one O^{2-} ligand to the metal, giving $[Mn^{6+}(O^{2-})_3(O^-)]$ as the upper state. Oxidative charge-transfer processes are observed in the spectra of complexes of low oxidation state metals with organic amines, such as $[TiCl_3(bipy)]$. Here the lower state is $[Ti^{3+}(Cl^-)_3(bipy)]$ and the excited state is $[Ti^{4+}(Cl^-)_3(bipy^-)]$.

6.8.6 Intervalence transitions

There is a large and growing number of compounds containing two or more transition metal atoms in different oxidation states. They have a range of interesting properties. For example, the striking color of Prussian Blue is an indication that electronic spectra may have unusual features. Intervalence compounds have been classified according to the ease of electron transfer between the distinct sites. In Class I compounds, electron transfer between the centers is slow, and the sites behave as if they were effectively independent. In Class II species the centers perturb each other but the sites remain distinct, while in Class III species the sites are indistinguishable and electron delocalization is effectively complete. There are other sub-classes, but these three represent the major divisions. There are some fascinating Class II chain species, such as $[Pt^{II}en_4][Pt^{IV}Cl_2en_4]Cl_4 \cdot 4H_2O$, where en is ethylenediamine, in which electronic transitions involve whole chains of metal sites. Binuclear species containing metal ions linked by ligands capable of acting as frameworks for electron transfer have been widely studied, and it has been shown that there should be an intervalence charge transfer (IVCT) transition at an energy which is simply related to the energetic barrier for thermal electron transfer. These IVCT bands appear in the near infrared. Species of Class III, such as $[(H_3N)_5RuNC\text{—}CNRu(NH_3)_5]^{5+}$, which can be regarded as a $[Ru(2.5)Ru(2.5)]$ dimer, show sharp IVCT bands that are not affected by solvent, whereas Class II species such as $[(H_3N)_5Ru(pyrazine)Ru(NH_3)_5]^{5+}$ give broad IVCT bands whose energy is sensitive to solvent. More details are given in Refs [22] and [23].

6.9 Assigning bands of transition metal complexes

The discussion above has indicated the most important observations that help in assigning the electronic spectra of transition metal complexes. The first criteria are the number, positions and intensities of bands; these may be compared with calculated parameters or semi-empirical predictions, and with the spectra of related species for which assignments are firmly established. In some cases, vibrational progressions may be observed, and careful analysis of the nature of the vibrations involved can give us an idea of both the symmetry and any changes in structure associated with the transition, just as it can for a band in a valence photoelectron spectrum (Section 6.6.2).

The effect of cooling the sample may be informative. In general, it is found that an allowed transition does not change significantly in intensity as the temperature is lowered, though the apparent energy may shift as the contribution from hot bands arising from vibrationally excited states is reduced. A forbidden band that appears by virtue of vibronic coupling will become weaker as the temperature falls, because the

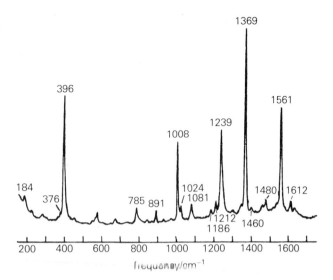

Fig. 6.30 Resonance Raman spectrum showing enhancement of ligand vibrations on excitation in a $\pi \rightarrow \pi^*$ ligand band. The spectrum is of a tetratolylporphyrin (TTP) complex TTPCr$\equiv$N, which has a characteristic Cr$\equiv$N stretch at 1017 cm^{-1} in its IR spectrum. This mode is not enhanced in the resonance Raman spectrum shown here. The five most intense bands are due to ligand modes. Redrawn from C. Campochiaro, J.A. Hofmann and D.F. Bocian, *Inorg.Chem.*, **24**, 449 (1985).

effect of the coupling is greater for more highly-excited vibrational states, which are less populated at lower temperatures.

For an oriented sample, such as a single crystal, study of the polarization of the absorption can show how the symmetry of the transition is related to that of the structure. Such information is not obtainable for samples in solution, but for optically active species the use of circularly polarized light can give a circular dichroism spectrum (see Section 6.10).

One technique which can be of some help in deciding on the nature of an electronic transition is resonance Raman spectroscopy (see Section 5.17). As we have seen, excitation of a Raman spectrum within the absorption profile of an electronic transition can lead to selective excitation of a progression in one or more molecular vibrations. If the nature of such a vibration can be established it may throw some light on the nature of the electronic band. Thus, if the electronic transition is accompanied by a significant change in the metal–ligand bonding it is likely that the most symmetrical metal–ligand stretching mode will be enhanced in intensity by the resonance Raman effect, as is observed for the MnO_4^- and CrO_4^{2-} ions (see Fig. 5.52). This observation indicates that the absorption band within which the excitation occurred is of the metal-to-ligand or the ligand-to-metal charge-transfer type. On the other hand, excitation within a band concerned mainly with the ligand might result in enhancement of the intensity of internal ligand vibrations of the resonance Raman spectrum (Fig. 6.30). Excitation within a d–d band may lead to very little resonance Raman enhancement, and indeed for some species with inversion centers it has been found that the Raman bands are weaker than in the normal Raman spectrum when excited by light in the absorption profile of a d–d band.

6.10 Circular dichroism

6.10.1 Optical rotatory dispersion and circular dichroism

A molecule or ion with no improper rotation axis is chiral, and will show optical activity. A chiral species rotates the plane of polarization of plane polarized light that is passed through it; the direction of rotation depends on which enantiomer is present,

Example 6.3

In hemocyanin two copper ions bind one molecule of O_2. The mode of attachment of O_2 and the coordination geometry round copper are both important in understanding the process of oxygenation. Each copper atom is also bound to three nitrogen atoms, part of the imidazole rings of histidines, and to a shared oxygen atom, part of another peptide. There are electronic bands at $ca.$ 345 and $ca.$ 570 nm; the CD spectrum suggests that two transitions of slightly different energies are responsible for the latter band. There are Raman bands at about 745 and about 280 cm^{-1} in the spectrum of the oxyprotein. The band at 745 cm^{-1} is markedly intensified by excitation in the envelope of the electronic band at 570 nm, but the Raman band at 280 cm^{-1} is not; it is enhanced by excitation at around 345 nm. The frequency of the Raman band at 745 cm^{-1} is sensitive to isotopic substitution of oxygen but not of copper, while that at 280 cm^{-1} is sensitive to isotopic substitution at copper but not at oxygen. Neither is affected by treatment with D_2O. Experiments with $^{16}O^{18}O$ show that the two oxygen atoms in the coordinated O_2 are equivalent (see Example 5.3). The O—O stretching frequencies in O_2 and in O_2^{2-} are $ca.$ 1100 and $ca.$ 800 cm^{-1} respectively. What can you deduce from these data about the structure of the complex?

....

The band at 745 cm^{-1} clearly represents the O—O stretching mode of coordinated O_2; its frequency shows that the coordinated molecule is more like O_2^{2-} than like O_2^-, and so that the anti-bonding π-levels of bound O_2 are effectively filled. The sensitivity of the band at about 280 cm^{-1} to the mass of copper but not of oxygen shows that it is due to modes involving copper and the imidazole ligands but does not involve significant Cu—O stretching; it also shows that the environment round copper cannot approximate to a regular tetrahedron, for if it were regularly tetrahedral the copper would not move significantly in the vibration.

If the O_2 ligand is treated as peroxide-like, then the Cu—O—O—Cu system will be non-planar, like H_2O_2, as shown in the figure below. In this structure the π-bonding and π-anti-bonding levels of O_2 are each split into a and b components. Each of the anti-bonding orbitals can interact with d orbitals on the copper atoms, as depicted below. The electronic band near 570 cm^{-1} would then be assigned to ligand–metal charge transfer (LMCT) from the a and b levels of bound O_2, derived from the π-anti-bonding levels of free O_2, thus accounting for the two transitions of slightly different frequency under the band envelope. The electronic band near 280 cm^{-1} is assigned to LMCT from the imidazole ligands of the protein.

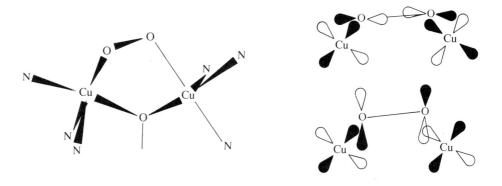

for the rotation is equal and opposite for the two enantiomers. Rotation of the plane of polarization of plane polarized light is therefore a useful way of studying the enantiomers of chiral species. A spectrum in which the magnitude of rotation for a particular enantiomer is plotted against the wavelength of the light is called an Optical Rotatory Dispersion (ORD) spectrum.

A chiral species is also circularly dichroic – that is, the absorption coefficients for right- and left-circularly polarized light are different. A plot of the difference between the molar absorption coefficients for right- and left-circularly polarized light against wavelength is called a circular dichroism (CD) spectrum. The sign of the differential absorption depends on which enantiomer is present, so the CD spectrum can provide information about the configuration of an enantiomer. The rotation depends on the scalar product of the electric dipole and magnetic dipole transition moments. Transitions that are both electronically and magnetically allowed can be seen in a CD spectrum, as in a normal electronic spectrum. In addition, however, transitions that are electronically forbidden but magnetic dipole allowed may be well defined in CD spectra. Thus CD spectra can help to reveal details of electronic spectra and to determine assignments. More details can be found in Refs [21] and [24].

6.10.2   Magnetic circular dichroism

The components of an electronically degenerate state will be split in a magnetic field by the Zeeman effect, if the ground and/or excited states possess non-zero angular momentum m_j; there is differential absorption of right- and left-circularly polarized light by the components of different m_j. A plot of the differential absorption for right- and left-circularly polarized light for a sample in a magnetic field is called a magnetic circular dichroism (MCD) spectrum. The species studied does not have to be chiral: the differential absorption comes from the effect of the magnetic field on degeneracy. Therefore the MCD spectrum gives information about the generacy of the ground and/or excited state of the species studied. If the excited state is degenerate but the ground state is not, the MCD spectrum will have the form known as an **A** term, with components with positive and negative differential absorption; the shape does not depend on temperature. If the ground state is degenerate but the excited state is not, the Boltzmann distribution will determine the intensities of the right- and left-circularly polarized absorptions from the Zeeman levels of the ground state. The resulting spectrum, described as a **C** term, will be all positive or all negative, and will be temperature-dependent. If there is a field-induced mixing of the various states, a **B** term may be observed. This looks like a **C** term but does not depend on temperature [21].

MCD spectra provide information about the degeneracy of states and help with the assignment of electronic spectra. For instance, the electronic spectrum of S_2N_2 [25] contains a single peak at 37 900 cm^{-1}. The MCD spectrum contains a strong peak centered near 39 000 cm^{-1} with a negative **B** term, and a strong second peak near 43 000 cm^{-1} with a positive **B** term. These observations show that the first absorption band of S_2N_2 at 37 900 cm^{-1} consists of two electronic transitions.

6.11   Detection of complexes and measurement of concentrations

Finally we return to the extreme importance of visible and UV absorption bands of transition metal complexes in detection and measurement of concentrations. The breadth of such bands means that those of different species often overlap, but it is generally possible to select a wavelength at which one compound may be monitored during the course of a reaction, for instance. The ease of measurement and relative cheapness of small visible/UV spectrometers makes their use very popular. High ordinate accuracy is obtainable because of the stability of source and detector, and concentrations can be inferred directly from the intensity of absorption if the Beer− Lambert Law is obeyed. This states that the absorbance, A (or the optical density, E) is directly proportional to the product of pathlength and concentration.

$$A = \varepsilon cl \tag{6.3}$$

The molar extinction coefficient ε has units of l cm^{-1} mol^{-1} if c is in mol l^{-1} and l is in cm, so that A is a pure number. A is related to the intensity of light I that is transmitted compared with the initial intensity I_0 by

$$A = \log_{10}(I_0/I). \tag{6.4}$$

The various types of absorption we have described above generally have fairly characteristic extinction coefficients. For this reason it is usual to report an electronic absorption spectrum as a list of peak positions and extinction coefficients, or as a list of the extinction coefficients at various wavelengths.

Another measure of intensity is the oscillator strength, f, given by

$$f = 4.135 \times 10^{-9} \times \text{integrated intensity.} \tag{6.5}$$

These experimental measures of intensity can be related to the wavefunctions for the upper and lower states using an integral known as the transition moment, D_{if}, which is of the form

$$D_{if} = [\int \psi_i{}^* M \psi_f \, d\tau]^2 \tag{6.6}$$

where ψ_i and ψ_f are the wavefunctions for the initial and the final states and M is the dipole moment operator, which can be obtained from symmetry tables. The transition moment is related to the oscillator strength by the expression

$$f = \frac{8\pi^2 mc}{3he^2} \nu s_f D_{if} \tag{6.7}$$

where s_f is the degeneracy number of the upper state, m is the mass of the electron and ν is the transition frequency.

A study of the changes in absorption spectrum of a system in the course of a reaction can give information about the mechanism, as well as about the nature of the product(s), the rate of the reaction and how the rate changes with concentration, with temperature and with the presence of other potential reagents. A useful feature of such dynamic spectra is the isosbestic point, which is a point where the total absorbance does not change with time, although the absorbances at points on either side do change as the concentrations of reagents and products evolve. The isosbestic point is in fact the point where the molar extinction coefficients of reagent and product are equal. A simple isosbestic point only arises if two species whose concentrations are changing as the reaction proceeds absorb significantly in the region of the spectrum concerned. If other absorbing products or intermediates occur there will be no isosbestic point. If the absorption spectra of reagents and products happen not to cross there can be no isosbestic point, but the same effect can be seen in the first-derivative spectra.

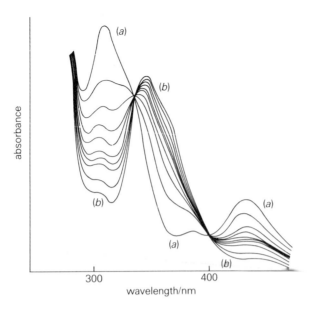

Fig. 6.31 Electronic spectra of solutions of $[Au_{11}(PMe_2Ph)_{10}]^{3+}$ (*a*) following addition of Cl^-, showing progressive transformation to $[Au_{13}(PMe_2Ph)_{10}Cl_2]^{3+}$ (*b*). The isosbestic points near 335 and 395 nm show that any intermediates have no absorption in this region. Redrawn from K.P. Hall and D.M.P. Mingos, *Prog.Inorg Chem.*, **32**, 237 (1984).

The example shown in Fig. 6.31 is remarkable, as it apparently shows the transformation of an Au_{11} cluster into an Au_{13} cluster without any intermediate or by-product. This is of course not possible, and we should rather conclude that any intermediate, formed by the initial attack of Cl^- on $[Au_{11}(PMe_2Ph)_{10}]^{3+}$, must have no significant absorption in the spectral range studied, or that its concentration is never so great that it contributes significantly to the spectrum. Another example is shown in Fig. 6.32. Here several isosbestic points are seen, showing that the absorption spectra of reagent and product (at equal concentrations) cross at several points. Only two species are involved in this example, which involves progressive electrochemical oxidation of an Ru^{II} complex to the Ru^{III} analog.

Fig. 6.32 Electronic spectra of a solution of $[Ru(CO)Cl(bipy)_2]^+$ showing the isosbestic points that develop as it is progressively oxidized electrochemically to the Ru^{III} species with the same ligands. The arrows show the direction of change in the spectrum as the reaction proceeds.

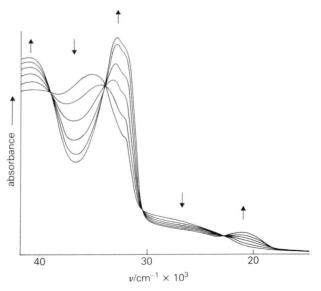

6.12 | **Spectra of compounds of elements with partly-filled _f_ subshells (lanthanides and actinides)**

The 4_f_ electrons in the lanthanides and their compounds and the 5_f_ electrons in the actinides and their compounds are well screened from external effects by the filled shells of higher principal quantum number. This has several important consequences, the first being that the _f_–_f_ bands of such species are characteristic of the element concerned, and hardly affected by even the immediate ligands. Of course, species with different numbers of _f_ electrons will have different states and hence different transitions, but all compounds of a particular ion will be expected to have very similar spectra. The lanthanide elements in particular are almost always found in the form of complexes of M^{3+} ions, and each element has its own characteristic color. The actinides are more variable in their oxidation states, and this gives them more varied colors, but again the spectra of compounds derived from a given oxidation state are likely to be very similar.

The second consequence of this insulation of the _f_ electrons from external influences is that the _f_–_f_ bands are sharp, much narrower than typical _d_–_d_ bands. They are also characteristically weak, the effects of deviations from perfect inversion symmetry (which is often found for complexes of lanthanide elements) being insufficient to relax the $g \rightarrow u$ selection rule. All _f_ levels are of _u_ symmetry in a centrosymmetric environment, so _f_–_f_ transitions are forbidden. The colors of lanthanide compounds are therefore pale; the actinides have rather stronger colors, because of the effects of spin–orbit coupling, which is greater for heavier atoms. Charge transfer transitions are less important for the _f_-block elements than for _d_-block elements, mainly because they are less likely to form complexes with ligands having suitable donor orbitals. Much less work has been published on spectra of these species than on those of _d_-block metal complexes.

Problems

6.1 The azide ion gives two peaks when the nitrogen $1s$ electrons are ionized by Al K_α radiation (1487 eV) the peaks correspond to electron kinetic energies of 1079 and 1084 eV. What are the velocities of the electrons ionized with these kinetic energies? What were their binding energies (in eV)? Which of the two peaks would you expect to be the more intense? Give your reasoning.

6.2 On irradiation of a molecule with He(I) radiation (wavelength 58.4 nm) an electron is ejected with kinetic energy 5.8 eV. What was the binding energy of that electron?

6.3 In the electronic spectrum of $[Cr(NH_3)_6]^{3+}$ there is a very intense absorption band at about $60\,000$ cm^{-1}, two moderately intense bands (ε_{max} _ca._ $100\,l\,cm^{-1}\,mol^{-1}$) near $20\,000$ and $30\,000$ cm^{-1} and a very weak band near $15\,000$ cm^{-1}. How can you account for these observations?

6.4 What would you expect to be the order of the energies of charge-transfer bands for the ions VO_4^{3-}, NbO_4^{3-} and TaO_4^{3-}? Give reasons for your answer.

6.5 Assign the hydrogen and nitrogen atomic orbitals in ammonia to their symmetry species, and indicate how they may combine to give molecular orbitals.

6.6 The He(I) photoelectron spectrum of CO is shown in Fig. 6.33. Assign the bands and comment on the nature of the HOMO.

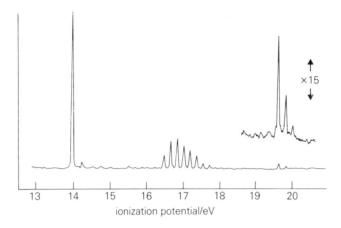

Fig. 6.33 Photoelectron spectrum of CO excited by He(I) radiation. Redrawn from Ref. [13].

6.7 Comment on the bands shown in the He(I)-excited photoelectron spectrum of CS_2 (Fig. 6.34), and suggest assignments for the first two bands observed.

6.8 By considering the symmetry of atomic orbitals, derive the symmetry species of the molecular orbitals expected for SF_4 (a) without inclusion of sulfur d orbitals and (b) including sulfur d orbitals.

6.9 A transition metal complex has an extinction coefficient of $140 \, l \, cm^{-1} \, mol^{-1}$ at a wavelength of 300 nm. What is the concentration of a solution with an absorbance of 1.7 in a cell 2 cm long?

6.10 A 10^{-3} molar solution of a transition metal complex has an absorbance of 1.8 in a 10 cm cell at a particular wavelength. What is the corresponding extinction coefficient?

6.11 How many d–d bands would you expect for a d^1 ion (such as Ti^{3+}) in an environment with (a) octahedral symmetry, O_h; (b) tetragonal symmetry, D_{4h}, with elongation along the z-axis; (c) tetrahedral symmetry, T_d; (d) trigonal symmetry, C_{3v}, derived from T_d by elongation along one C_3 axis?

6.12 The splitting of the ^5D ground state of a d^4 metal ion in an octahedral field and with tetragonal distortion (elongation along the z-axis) is shown in Fig. 6.35, with term labels. How many spin-allowed bands would you expect to see in the electronic spectrum of a high-spin d^4 metal complex with this tetragonal distortion? Compare qualitatively the magnitudes of the expected splittings of the ^5E_g and $^5T_{2g}$ levels in an elongated tetragonal field.

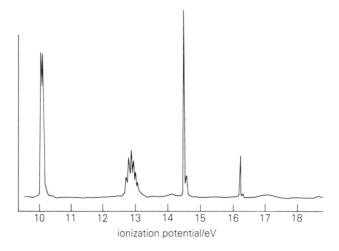

Fig. 6.34 He(I) photoelectron spectrum of CS_2. Redrawn from Ref. [13].

CHAPTER 6: ELECTRONIC AND PHOTOELECTRON SPECTROSCOPY

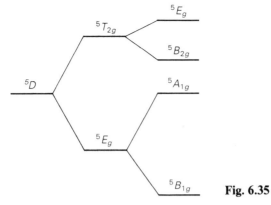

Fig. 6.35

6.13 The energy of the lowest energy charge-transfer band decreases across the series VO_4^{3-}, CrO_4^{2-}, MnO_4^- but increases across the series $[V(CO)_6]^-$, $[Cr(CO)_6]$, $[Mn(CO)_6]^+$. How do you account for this?

6.14 The electronic spectrum of octahedral VCl_6^{4-} in $AlCl_3$ at 500 K is shown in Fig. 6.36. Assign the bands to spin-allowed transitions using the Tanabe–Sugano diagram, Fig. 6.37. Use the values of the transition energies (8000, 13 100 and 20 300 cm^{-1}) to determine approximate values for B and Δ.

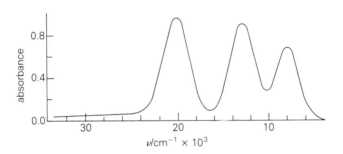

Fig. 6.36 Electronic spectrum of VCl_6^{4-}. Redrawn from Ref. [21].

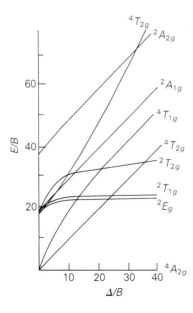

Fig. 6.37 Tanabe–Sugano diagram for octahedral d^3 complexes.

6.15 The ligand-field splitting parameter Δ is normally larger for octahedral complexes than for tetrahedral complexes of the same metal ion in the same oxidation state with the same ligand atoms. How do you account for the observation that the energy for the first d–d transition $(^6A_1 \rightarrow {}^4T_2)$ is around 11 000 cm^{-1} for $Fe^{III}O_6$ chromophores but around 22 000 cm^{-1} for $Fe^{III}O_4$ chromophores?

6.16 Parts of the visible spectra of the anions $OsCl_6^{2-}$, $OsBr_6^{2-}$ and OsI_6^{2-} are shown in Fig. 6.38. The corresponding regions of the spectra of complexes $[Os(py)_4X_2]$ are almost indistinguishable from one another. How can these observations be explained?

6.17 Using information provided in Table 6.1, explain what you would expect to see in the XPS spectrum of $[Fe(C_5H_5)_2(CO)_2]$, obtained using Al K_α radiation.

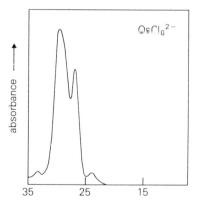
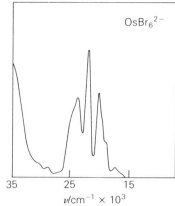
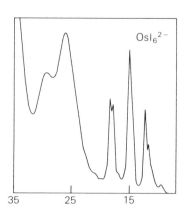

$\nu/\text{cm}^{-1} \times 10^3$

Fig. 6.38

Further reading

General — Refs [16] and [17].

X-ray photoelectron spectroscopy — Refs [1]–[3].

Valence-shell photoelectron spectroscopy — Refs [2] and [13]–[16].

Valence-shell excitation spectroscopy — Refs [19] and [21]. Also:

Molecular orbitals and symmetry — Refs [5]–[9].

Molecular orbital calculation methods — Refs [10]–[12].

References

1 K. Siegbahn, C. Nordling, G. Johansson, J. Hedmass, P.F. Héden, K. Hamrin, U. Gelius, T. Bergmark, L.O. Werme, R. Manne and Y. Baer, *ESCA Applied to Free Molecules*, North-Holland Publishing Co., Amsterdam (1971).
2 C.R. Brundle and A.D. Baker (ed.), *Electron Spectroscopy: Theory, Techniques and Applications*, Academic Press, London (1977), Vols I and II.
3 D.A. Shirley (ed.), *Electron Spectroscopy*, North-Holland Publishing Co., Amsterdam (1972).
4 W.B. Hughes and B.A. Baldwin, *Inorg. Chem.*, **13**, 1531 (1974).
5 D.S. Urch, *Orbitals and Symmetry*, Penguin Books (1970).
6 J.M. Hollas, *Symmetry in Molecules*, Chapman and Hall, London (1972).
7 F.A. Cotton, *Chemical Applications of Group Theory*, Interscience, London (1964).

8 J.N. Murrell, S.F.A. Kettle and J.M. Tedder, *Valence Theory*, Wiley, New York and London (1965).

9 H.H. Jaffé and M. Orchin, *Symmetry in Chemistry*, Wiley, New York (1965).

10 C.A. Coulson and R. McWeeney, *Coulson's Valence*, Oxford University Press, Oxford (1979).

11 J.A. Pople and D.L. Beveridge, *Approximate Molecular Orbital Theory*, McGraw-Hill, New York (1970).

12 L.C. Snyder and H. Basch, *Molecular Wave Functions and Properties*, Wiley-Interscience, New York (1972).

13 D.W. Turner, C. Baker, A.D. Baker and C.R. Brundle, *Molecular Photoelectron Spectroscopy*, Wiley, London (1970).

14 A.D. Baker and D. Betteridge, *Photoelectron Spectroscopy – Chemical and Analytical Aspects*, Pergamon, Oxford (1972).

15 J.H.D. Eland, *Photoelectron Spectroscopy*, Butterworths, London (1974).

16 G. Cooper, J.C. Green, M.C. Payne, B. Dobson and I.H. Hillier, *J.Am.Chem.Soc.*, **109**, 3836 (1987).

17 G. Herzberg, *Electronic Spectra of Polyatomic Molecules*, Van Nostrand, New York (1966).

18 J.M. Hollas, *High Resolution Spectroscopy*, Butterworths, London (1982).

19 C.J. Ballhausen, *Introduction to Ligand Field Theory*, McGraw-Hill, New York (1962).

20 P.F. Shriver, P.W. Atkins and C.H. Langford, *Inorganic Chemistry*, Oxford University Press, Oxford (1990).

21 A.B.P. Lever, *Inorganic Electronic Spectroscopy*, 2nd edn, Elsevier, Amsterdam (1984).

22 R.J.H. Clark, *Chem.Soc.Rev.*, **13**, 219 (1984).

23 C. Creutz, *Prog.Inorg.Chem.* **30**, 1 (1983).

24 R.D. Gillard, *Physical Methods in Advanced Inorganic Chemistry* (eds H.A.O. Hill and P. Day), Interscience, London (1968), p. 167.

25 H.-P. Klein, R.T. Oakley and J. Michl, *Inorg.Chem.*, **25**, 3194 (1986).

7 Mössbauer Spectroscopy

7.1 Introduction

Mössbauer spectroscopy differs from the other spectroscopic methods described in this book in a number of ways. The transitions are within the nucleus, and so involve emission and absorption of gamma-rays (the technique is sometimes called nuclear gamma resonance spectroscopy), and the associated problems of working with ionizing radiation. The only suitable sources of radiation are excited nuclei of the same isotope formed in the course of radioactive decay. There is no direct way of tuning the emitted quanta to the energy needed to excite the absorber. The necessary tuning, which is essential with monochromatic sources, is achieved using the Doppler effect by introducing controlled relative motion of source and absorber. This helps to explain why the energy differences in Mössbauer spectroscopy are recorded in the unfamiliar units of mm s^{-1}. These energy differences are minute in relation to the magnitudes of the emitted quanta; the emitted quanta are of the order of keV, and the units of mm s^{-1} are about 10^{-12} times smaller than this.

7.2 Principles

When a gaseous atom or molecule emits a quantum of radiation of energy E_v, the emitted quantum always has momentum E_v/c, where c is the velocity of light. By conservation of momentum, the emitter must recoil with momentum P_A which is equal and opposite, so that

$$P_A = Mv_R = -E_v/c \tag{7.1}$$

where M is the mass of the emitter and v_R is the recoil velocity. There is an associated recoil energy of the emitter, E_R, given by

$$E_R = Mv_R^2/2 = P_A^2/2M = E_v^2/2Mc^2 \tag{7.2}$$

For radiation in the near-ultraviolet region or at lower energy, and with emitters of normal atomic or molecular masses, v_R is so small that it can be neglected, and so can the recoil energy E_R. With gamma radiation this is not true. The energy E_γ of a quantum of gamma radiation is so large that the recoil velocity of a normal atom or molecule becomes significant. This leads to a very important consequence. The energy of the gamma quantum emitted comes from the decay of the nucleus from its excited state to its ground state. This is the only source of energy for the whole process of emission, and so it must also provide the recoil energy E_R. If we call the nuclear transition energy E_t, this means that

$$E_\gamma = E_t - E_R \tag{7.3}$$

This is illustrated in Fig. 7.1. The energy of the emitted quantum is therefore less than the transition energy by a significant amount. Furthermore, the thermal velocity of the emitter leads to Doppler broadening of the emitted radiation. In much the same way,

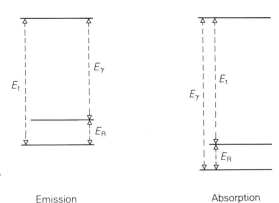

Fig. 7.1. Relationship between E_t, E_γ and E_R for emission and absorption (not to scale).

Emission

Absorption

there is significant recoil energy when an atom or molecule absorbs a quantum of gamma radiation. If a gamma quantum with energy E_γ is to excite a nucleus with transition energy E_t, then E_γ must be greater than E_t by an amount equal to the recoil energy, as shown in Fig. 7.1.

$$E_\gamma = E_t + E_R \tag{7.4}$$

So molecules in the gas or liquid phases that emit gamma quanta cannot reabsorb them, for the quanta and the transition energies do not match.

What Mössbauer discovered was how to eliminate the problem of the recoil energy. If emitter and absorber are part of a rigid solid lattice, the lattice vibrations are quantized and the recoil energy can only be transferred to the lattice in such amounts. If the recoil energy is smaller than these quanta, the whole lattice and not just the atomic unit recoils on emission or absorption. The effective value of M then becomes not the mass of one atomic or molecular unit, but that of the whole lattice, which is hundreds of thousands of times greater. Then the recoil velocity v_R becomes negligible, and so does the recoil energy. The Doppler broadening of the lines through the kinetic energies of source and absorber disappears, so that the energy of the emitted quantum (E_γ) becomes sharp and equal to the transition energy (E_t).

In practice, this is not always true, even for solids, because lattice vibrational modes may be excited during the absorption process. The extent to which this happens depends on the lattice forces, but if E_γ is greater than about 150 keV, nearly all events excite the lattice vibrations, and so E_γ equals E_t for only a small proportion of the incident radiation. It is the absorption or emission of radiation where E_γ is equal to E_t that gives rise to a Mössbauer spectrum. The proportion of the total radiation that is absorbed in this way is called the 'recoil-free fraction'; this fraction varies with temperature, and with some species is always low even at very low temperatures. However, study of the magnitude of the recoil-free fraction as a function of temperature can provide valuable information about molecular motions in solids.

7.3

Conditions for Mössbauer spectroscopy

To a fairly close approximation, the energy E_t of a nuclear transition depends only on the nucleus involved. There are well-defined excited states for many nuclei, but only some are suitable for Mössbauer spectroscopy. Several conditions have to be satisfied, and some of the most important are as follows:

1   The energy of the nuclear transition must be large enough to give useful gamma radiation, but not large enough to produce a recoil whose energy is much greater than the lattice vibration quanta. This means that E_γ must lie between 10 and 150 keV.

2   A substantial proportion of the excited nuclei must decay with emission of gamma photons. Other decay processes are of course possible, such as the emission of conversion electrons, but these must not dominate.

3   The lifetime in the excited state must be long enough to avoid introducing too much uncertainty in E_t through the Uncertainty Principle; it must be short enough to give intense lines of reasonable breadth, for lines that are extremely narrow are not useful. In practice, lifetimes of between 1 and 100 ns are suitable.

4   The excited state of the emitter must have a precursor which is long-lived and resasonably easy to handle.

5   The ground-state isotope must be stable; it should be present in fairly high natural abundance or enrichment should be easy.

6   The cross-section for absorption should be high.

As an example, consider the transition used in most of the work with ^{57}Fe Mössbauer spectroscopy. The precursor is ^{57}Co, which decays to an excited isotope of ^{57}Fe. Some of this decays directly to the ground state with emission of high energy (136.32 keV) gamma radiation. Some of it decays to a lower excited state, which itself decays with a lifetime of 99.3 ns to the ground state with emission of gamma photons of 14.41 keV.

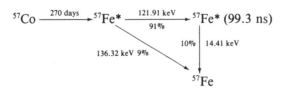

In the 14.41 keV transition, the energy of the photons emitted satisfies condition (1) and the lifetime of the excited state satisfies condition (3). The percentage of nuclei decaying by this route is not large, so condition (2) is not met very satisfactorily, but the large cross-section (condition 6) partly makes up for this. The properties of the precursor satisfy condition (4); ^{57}Co is obtained fairly readily by cyclotron radiation of iron [^{57}Fe$(d,n)^{57}$Co]. The ground-state of ^{57}Fe is stable, and though its natural abundance is only around 2%, enrichment is possible, but frequently unnecessary (which is just as well, since the process is expensive). A lot of elements with atomic numbers between about 25 and 96 have isotopes satisfying some or all of the necessary conditions. Lighter elements are not suitable because the energies of the nuclear transitions are too high. Tables of isotopes suitable for Mössbauer spectroscopy are given in standard texts.[1]–[3] By no means all have proved important to the chemist, though many have been extremely useful in solid-state physics. In this chapter the examples involve ^{57}Fe, ^{119}Sn, ^{121}Sb, ^{125}Te, ^{129}I, ^{129}Xe and ^{197}Au, and some of the relevant properties of these nuclei are given in Table 7.1. Of course, ^{129}I is an artificial species, a very long-lived fission product available in macroscopic amounts.

Mössbauer spectroscopy is usually studied using a standard substance as emitter and the material being investigated as absorber. Suitable source compounds have to be chosen with care. They must be well-defined and reproducible materials, and should give sharp lines with as little structure as possible; for ^{129}Xe, for instance, the

Table 7.1 Nuclear data for some nuclei used in Mössbauer spectroscopy.

Isotope	Γ_γ/mm s^{-1}	E_γ/keV	I_g	I_e	a/%	$t_{\frac{1}{2}}$/ns	σ_0/cm$^2 \times 10^{-18}$	Sign of $\Delta R/R$
^{57}Fe	0.192	14.412	1/2	3/2	2.17	99.3	2.57	$-$ve
^{119}Sn	0.626	23.875	1/2	3/2	8.58	18.3	1.40	$+$ve
^{121}Sb	2.1	37.15	5/2	7/2	57.25	3.5	0.21	$+$ve
^{125}Te	5.02	35.48	1/2	3/2	6.99	1.535	0.28	$+$ve
^{129}I	0.59	27.72	7/2	5/2	—	16.8	0.38	$+$ve
^{129}Xe	6.85	39.58	1/2	3/2	26.44	1.01	0.24	$+$ve
^{197}Au	1.87	77.34	3/2	1/2	100	1.892	0.041	$+$ve

Γ_γ = natural line width, E_γ = energy of γ-ray transition, I_g, I_e = ground and excited state nuclear spin quantum numbers, a = natural abundance of isotope, σ_0 = resonant absorption non-section, ΔR, R, see p. 284.

tetrahedral anion IO_4^- is often used. Some Mössbauer experiments have used the opposite arrangement: the excited nucleus is in the material being studied, and the standard is used as absorber rather than emitter. The work with halides of xenon, described later in this chapter, illustrates this arrangement.

7.4   Parameters from Mössbauer spectra

We have seen how it is possible, using solid absorbers and emitters, for gamma photons with energies between 10 and 150 keV to be emitted from, say, a lump of iron containing excited ^{57}Fe nuclei and then absorbed by other ^{57}Fe nuclei in a different lump. This does not seem very helpful so far. The nuclear energy transition E_t is determined to a close approximation by the nature of the nucleus involved. But if the lifetimes are suitable, the emitted photons have very precisely-defined energies. At this level of precision, E_t depends very slightly on the electronic environment of the nucleus, and on magnetic fields and electric field gradients at the nucleus too. These influences are discussed in more detail below. For the moment, though, we are still not much better off. We now know that E_t depends on the environment of the nucleus concerned as well as on what that nucleus is. For instance, the gamma photons emitted by our lump of metallic iron would be absorbed by ^{57}Fe nuclei in another lump of metallic iron, but not by ^{57}Fe nuclei in iron(II) sulfate. If we could measure the difference in E_t for ^{57}Fe in the two materials we should have a way of studying the environments of nuclei in solids. But we can only measure these variations in E_t if we can control the energies of the emitted quanta. We can do so by making use of the Doppler effect: source and absorber are moved relative to each other with relative velocity which is controlled and accurately measured (Fig. 7.2). This relative motion modulates the fre-

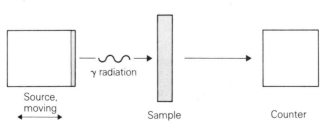

Fig. 7.2. Schematic diagram of a Mössbauer spectrometer.

quency of the photons received by the absorber, and when the energy of the modulated photons, which we can call $h\nu'$, is equal to E_t in the absorber, the radiation will be absorbed. So what we record is a plot of radiation absorbed against velocity (Fig. 7.3).

Luckily the energy differences involved correspond to relative velocities (of the order of mm s^{-1}) that can be obtained, controlled and measured precisely, so the experimental problems associated with using the Doppler effect in this way can be solved. There are more experimental problems associated with the counting of gamma photons and the accurate correlation of counting with velocity. With some nuclei, all measurements have to be made at low temperatures; with others, measurements over a range of temperatures, perhaps down to 1 K, are important, and this makes the equipment even more complicated. So does the recording of the spectrum of a sample in a magnetic field, which is also sometimes necessary. This normally involves the kind of powerful magnetic fields generated by superconducting magnets. Details of Mössbauer spectrometers and of the design tricks used to solve these problems are given in the reference books cited [1] – [3]. Here all you need to understand is that the problems can be solved.

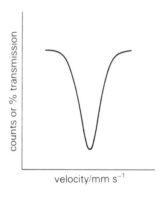

Fig. 7.3.   Representation of a Mössbauer spectrum.

7.4.1   Isomer shifts

How is E_t affected by the environment? First of all, there is the interaction between the nucleus and the electrons round it. This arises because the size of the nucleus is different in the ground state and the excited state. If the change in radius in going from the excited state to the ground state is ΔR and the nucleus is spherical in the ground state with radius R, the change in electrostatic energy on decay is given by

$$\delta = (\varepsilon_0/5)\,(Ze^2R^2)\,(\Delta R/R)\,[\,|\psi_s(\text{abs})|^2 - |\psi_s(\text{source})|^2\,] \tag{7.5}$$

where Z is the atomic number, e is the charge of an electron, ε_0 is the permittivity of free space, and $|\psi_s|$ represents the s-electron wavefunctions at the nuclei in the source and absorber.

There are several important things about this expression. First, it shows that the energy-change δ, usually called the isomer shift, the chemical isomer shift or the center shift, depends on the electron density at the nucleus. Since s-electron wave-functions have their maxima at the nucleus, where p, d and f wave-functions are zero, the isomer shift is most sensitive to changes in s-electron orbital occupancy. But changing the occupancy of p and d orbitals has some effect, because p and d electrons influence the s-electron density at the nucleus through screening.

Secondly, some nuclei are bigger in the excited state than in the ground state; others are bigger in the ground state than in the excited state. In other words, $\Delta R/R$ can be positive or negative, depending on the nucleus in question. So increasing the s-electron density at the nucleus leads to a positive isomer shift where $\Delta R/R$ is positive, and to a negative shift where it is negative.

All this shows that the isomer shift should be related to the oxidation state of the element concerned. Consider ^{119}Sn. The Sn atom has the valence electron configuration $5s^2 5p^2$. For ^{119}Sn, $\Delta R/R$ is positive. The Sn^{2+} ion has the formal electron configuration $5s^2$; the s-electron density at the nucleus is greater than in the tin atom, because the screening effect of the two $5p$ electrons has gone, and so the isomer shift of Sn^{2+} is positive relative to atomic tin. In Sn^{4+}, on the other hand, the two $5s$ electrons have been removed, reducing the s-electron density at the nucleus, and so the isomer shift of Sn^{4+} is negative. In Sn^{IV} compounds such as $SnCl_4$ or $SnCl_2Me_2$, the Sn shifts depend on the substituents and on the coordination number of tin. Empirical relationships between isomer shifts and substituents have been derived, and empirical parameters called partial isomer shifts or partial center shifts worked out for particular substituents, so that the overall shift for the Sn nucleus in a compound SnWXYZ is given by the sum of the partial shifts for W, X, Y and Z. Some values for ^{119}Sn isomer shifts are given in Table 7.2.

^{119}Sn isomer shifts have been useful in characterizing organo-derivatives of Sn^{II}. Compounds such as diphenyltin, which appear from their formulae to contain Sn^{II}, are quite common, but their formulae are often misleading. Many of them are polymeric, and contain tetravalent tin, which can be represented formally as Sn^{IV}. This is born out by their Mössbauer isomer shifts. It is usually supposed that Sn^{II} gives shifts greater than 2.1 mm s^{-1} and Sn^{IV} gives shifts smaller than 2.1 mm s^{-1} (both relative to SnO_2). Though the precise value of the change-over point is arguable, shifts below 2 mm s^{-1} are clearly associated with Sn^{IV} and shifts above 2.5 mm s^{-1} are equally clearly associated with Sn^{II}. The isomer shift for $(SnPh_2)_n$ is 1.56 mm s^{-1}, very much in the Sn^{IV} range (note the value for $SnPh_4$ in Table 7.2). In marked contrast, $Sn(C_5H_5)_2$ is an essentially monomeric derivative of Sn^{II} in the solid state, and its ^{119}Sn isomer shift is 3.74 mm s^{-1}, confirming that the compound contains Sn^{II}. The unusual compound $Sn[CH(SiMe_3)_2]_2$ is particularly interesting in this respect. [4] In the gas phase it is monomeric, with the sharp C—Sn—C angles characteristic of molecular Sn^{II} compounds. In the crystal, however, it is dimeric. As a dimer, it might be expected to resemble a tin-substituted alkene; after all, ethene is formally a dimer of methylene. But the geometry of the dimer of $Sn[CH(SiMe_3)_2]_2$ is peculiar. The bonds round each tin atom are not coplanar. The Sn—Sn distance is 2.76 Å, about the same as a normal Sn—Sn single bond, and the solid is diamagnetic. It is hard to know how to represent the electronic structure of the dimer, and 'bent donor–acceptor' Sn—Sn interactions have been suggested to explain the structure observed. The isomer shift is unusual, too: the value of 2.16 mm s^{-1} is on the dividing line between Sn^{II} and Sn^{IV}. It would be interesting to measure the isomer shift for the matrix-isolated compound, which might well behave more like a normal derivative of Sn^{II}.

Table 7.2 ^{119}Sn isomer shifts in some compounds of tin.

Compound	δ/mm s^{-1}
SnF$_4$	-0.47
Na$_2$SnF$_6$	-0.48
SnCl$_4$	0.85
SnBr$_4$	1.15
SnI$_4$	1.55
SnMe$_4$	1.21
SnPh$_4$	1.22
SnMe$_3$CF$_3$	1.31
Sn(C$_6$F$_5$)$_4$	1.04
Ge(SnPh$_3$)$_4$	1.13
SnH$_4$	1.27
SnO(black)	2.71
NaSn(OH)$_3$	2.60
SnS	3.16
SnSO$_4$	3.90
SnF$_2$(orthorhombic)	3.20
SnCl$_2$	4.07
SnBr$_2$	3.93
SnI$_2$	3.85
SnCl$_2$(pyridine)$_2$	3.02

Data from Ref. [1]; shifts relative to SnO$_2$.

It is a bit harder to use ^{57}Fe isomer shifts to characterize oxidation states of iron. There are more states: iron is found in states from 0 to VI, and they may differ from one another by one unit only. The electrons are taken from or returned to orbitals which are mainly d in character, and so the effect on the s-electron density at the nucleus is less direct than with tin. The spin states (which for a given oxidation state depend on the ligands) also affect the isomer shift. Some indication of the range of isomer shifts for various types of iron complexes is given in Fig. 7.4. Despite the fact that the various ranges overlap, some useful results have been obtained. The porphyrin complexes of iron are of great biological importance, and their oxidation and reduction plays an essential part in the chemistry of life. A lot is known about porphyrins of Fe^{II} and Fe^{III}. It is possible to reduce Fe^{II} porphyrins, and to oxidize Fe^{III} porphyrins, but the nature of the products is not obvious. Electrons can be taken from or added to the π-orbitals of the ligands as well as the metal. Mössbauer spectroscopy can be used to help to define the oxidation-states of the iron centers in these species. [5] Remember that for ^{57}Fe, $\Delta R/R$ is negative, so an increase in s-electron density at the Fe nucleus should be associated with a negative change in isomer shift.

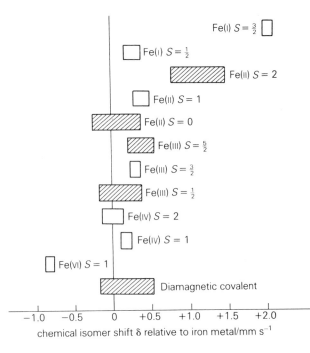

Fig. 7.4. Approximate representation of the ranges of chemical isomer shifts found in iron compounds. The most common configurations are cross-hatched. Redrawn from Ref. [1], where more detail is given.

Let us take reduction first. The tetraphenylporphyrinate complex of Fe^{II}, Fe(TPP), can be reduced under rigorously anaerobic conditions by sodium naphthalide in THF, giving a purple crystalline solid, [Na(THF)$_3$] [Fe(TPP)]. The magnetic moment is close to the value expected for low-spin Fe^I; in the crystal, the Fe—N distances are marginally longer than those in Fe(TPP). In confirmation that the iron center could be regarded as being in the +I state, the isomer shift has been found to be 0.65 mm s^{-1}. This should be compared with values of 0.52 mm s^{-1} for Fe^{II}(TPP) and around 0.45 mm s^{-1} for Fe^{III}-TPP derivatives. It therefore seems likely that the extra electron added to Fe(TPP) on reduction is mainly on the iron atom rather than on the porphyrin part of the complex. In contrast to these results, the isomer shift of a second reduction-product of Fe(TPP) is unexpectedly small. The anion [Fe(TPP)]$^{2-}$ can be produced by reduction with an excess of sodium naphthalide. Its crystal structure is like that of the monoanion, but with slightly longer Fe—N bonds, and the complex is diamagnetic. These results would be consistent with a formulation involving Fe^0, but the isomer shift is 0.48 mm s^{-1}, in the Fe^{II} range in this type of system. The Mössbauer parameters suggest that the two electrons added to Fe(TPP) are both on the porphyrin in this dianion, even though the single-electron reduction involves the metal rather than the ligand.

Now consider oxidation. The binuclear μ-peroxo-iron^{III}-porphyrin complexes can be cleaved homolytically by ligands such as 1-methylimidazole:

$$(\text{porph})\text{FeO—OFe(porph)} + 2(\text{1-MeIm}) \xrightarrow[\text{toluene}]{180\,\text{K}} 2\text{FeO(porph)}(\text{1-MeIm}) \quad (7.6)$$

The products are unstable above 240 K, but have been isolated as solids. The magnetic susceptibilities suggest the presence of two unpaired electrons per iron atom. No crystal structure has been obtained, but the isomer shift is 0.11 mm s^{-1}, fairly close to the value of 0.03 mm s^{-1} in Fe^{IV} hemoproteins. On this basis it is reasonable to call FeO(TPP) (1-MeIm) a complex of Fe^{IV}.

With some elements it is possible to use isomer shifts as the basis for detailed calculations of orbital occupancies and electron density distributions. With others, even simple and empirical correlations present problems, and a more detailed theoretical understanding is hard to ahieve. As has already been indicated, isomer shifts are sufficient with some elements to characterize oxidation states; with others, such as ^{197}Au, the ranges of observed isomer shifts for different oxidation states overlap, and the isomer shift by itself is not sufficient to determine the oxidation state of a gold center. So care is needed in interpreting a particular measurement.

Isomer shifts are always reported relative to a standard. As with chemical shifts in NMR spectroscopy, absolute values are not normally given. This means that it is important to notice what standard has been used in a particular paper. Nowadays, most people use the same reference standards; for instance, almost all ^{119}Sn shifts are reported relative to SnO$_2$, and ^{57}Fe shifts relative to iron metal (though the official standard is Na$_2$ [Fe(CN)$_5$NO]). But in some old papers different standards may have been used, and this point must be born in mind.

Electric quadrupole interactions

If a nucleus has spin greater than 1/2, it has a quadrupole moment, and its energy will be affected by any electric field gradient (efg) at the nucleus. In Mössbauer transitions there is a quadrupole moment associated with the upper or the lower state of the nucleus under study, and often with both. This means that Mössbauer emissions or absorptions will be split unless the efg at the nucleus is zero. The splitting observed will depend on the spin of the nucleus in ground and excited states and on the efg. For a nucleus like ^{119}Sn or ^{57}Fe, where the nuclear spin in the ground state, I_g, is 1/2, and the spin in the excited state, I_e, is 3/2, the energy-level patterns in the absence and in the presence of an efg are shown in Fig. 7.5(a). The ground state is unaffected, because it has no quadrupole moment; the upper state is split into two, with different m_I quantum numbers. The relevant selection rule for Mössbauer spectroscopy is $\Delta m_I = 0, \pm 1$. Therefore the effect of the efg on the spectrum is to split what was a single line into a doublet, with separation between the two components usually labeled Δ and called the quadrupole splitting. Its units are mm s^{-1}. It is related to the nuclear quadrupole moment, Q, the principal value of the efg tensor, and a parameter η, which is a measure of the departure of the efg from axial symmetry (Section 3.6). Sometimes the nuclear quadrupole coupling constant, e^2Qq, is quoted, measured in MHz. It is not possible to disentangle the full details of the efg tensor from this simple spectrum. To obtain all the necessary information either a single crystal must be used or the spectra must be taken with a sample in a magnetic field. But the value of Δ itself is often useful in solving chemical problems, and many of the examples that follow are based on this sort of simple analysis.

The case just described illustrates the simplest example of quadrupole splitting. Mössbauer transitions in which one of the nuclear states has $I = 1/2$ and the other has $I = 3/2$ are quite common, but those in which both the ground and the excited states have quadrupole moments are more complicated. Fig. 7.5(b) illustrates the pattern of splitting in an efg for a system in which I_g is 7/2 and I_e is 5/2, as with the Mössbauer

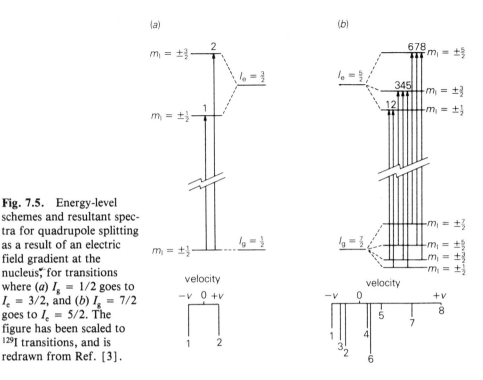

Fig. 7.5. Energy-level schemes and resultant spectra for quadrupole splitting as a result of an electric field gradient at the nucleus, for transitions where (a) $I_g = 1/2$ goes to $I_e = 3/2$, and (b) $I_g = 7/2$ goes to $I_e = 5/2$. The figure has been scaled to ^{129}I transitions, and is redrawn from Ref. [3].

transition for ^{129}I, with a line drawing of the spectrum. The spectra are more complicated to interpret, but more can be learnt from them. In particular, the quadrupole coupling constant can normally be determined: it is usually given in MHz.

The observation of quadrupole splitting in a spectrum is in itself significant, for it shows that there must be an efg at the nucleus in question. This precludes structures in which the environment is regularly octahedral or tetrahedral. In the ^{119}Sn spectrum of SnF_4, for instance, the resonance showed quadrupole splitting, and so the Sn atoms could not be in regular octahedral or regular tetrahedral environments, a conclusion later confirmed by X-ray crystallography.

It has been possible to explore the stereochemical activity of lone pairs of electrons in the compounds of heavy elements using Mössbauer spectroscopy. Everyone knows that lone pairs are usually stereochemically active in compounds of light elements, but what happens with heavy elements is less clear. Consider compounds containing Te^{IV} bound to six ligands. The tellurium atom has a lone pair of electrons. If this is stereochemically active, the atom will show what amounts to 7-coordination, with one position occupied by the lone pair; but if the arrangement of the six ligands round tellurium is regularly octahedral, the lone pair must be stereochemically inactive and might occupy the $5s$ orbital. All the techniques available — X-ray crystallography, vibrational spectroscopy, NQR — agree that ions of the form TeX_6^{2-} (where X is Cl, Br or I) are regularly octahedral; the lone pair is assumed to be in the $5s$ orbital. In keeping with this conclusion, the ^{125}Te Mössbauer spectra show no quadrupole splitting. Although the structures of $TeX_4Y_2^{2-}$ have not been determined crystallographically, their vibrational spectra suggest that the Y ligands are mutually *cis*: but whether they are *cis* or *trans*, the absence of quadrupole splitting is at first sight hard to explain. [6] It should be noted, however, that the unsymmetrical anions

$SnBr_2Cl_4^{2-}$ and $SnBr_4Cl_2^{2-}$ give [119]Sn Mössbauer spectra which show no quadrupole splitting, so unsymmetrical substitution in octahedral species does not always lead to an efg big enough to cause detectable quadrupole splitting. In marked contrast, the [129]I Mössbauer spectrum of $CsIF_6$, containing I^V, whose electronic structure is very like that of Te^{IV} in TeX_6^{2-}, shows a substantial quadrupole splitting (Fig. 7.6), implying that in $[IF_6]^-$ the lone pair is stereochemically active.[7] This is consistent with the structure of the isoelectronic XeF_6 in the gas phase; the structure of TeF_6^{2-} would be very interesting.

There is a parallel between the regularly octahedral geometry of TeX_6^{2-} and the ideal perovskite structure of $CsSnBr_3$. This remarkable compound is unlike other alkali metal tribromostannates, which are derivatives of tin(II). The others are white and their [119]Sn Mössbauer spectra show quadrupole splittings, suggesting that the Sn atoms are in distorted environments.[8] The Cs salt is black; the [119]Sn Mössbauer line is sharp and shows no quadrupole splitting, confirming that the Sn atoms are in cubic environments, so the lone pair at Sn must be in a 5s orbital. The color, geometry and other electrical properties have been explained by suggesting that 5s electron density is lost into a conduction band based on bromine orbitals. The salts of TeX_6^{2-} are also intensely colored, and their electronic structures may be explained in the same way.

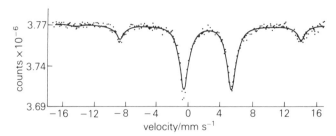

Fig. 7.6. [129]I Mössbauer spectrum of $CsIF_6$. The dots represent the actual observations and the continuous smooth curve has been fitted to them. Redrawn from Ref. [7].

The magnitude of the splitting can be useful in finger-printing. In [197]Au Mössbauer spectroscopy, the isomer shifts for Au^I and Au^{III} are not distinct enough to be characteristic on their own, but isomer shifts and quadrupole splittings taken together are usually enough to characterize both oxidation state and coordination number. In organo-tin esters, tin is found four-, five- or six-coordinated. Isomer shifts and quadrupole splittings have been used together to characterize the coordination of tin in compounds for which crystallographic data are not available. In some systems, care is needed in interpreting what is seen. Quadrupole splittings are not always observed where they would be expected. The [119]Sn Mössbauer spectra of tetraorgano-tin compounds, for example, often do not show quadrupole splittings even when the organic substituents are not the same. Other examples have been mentioned above.[9]

These ways of using quadrupole splittings to give information about structure are qualitatively based, though theoretical justification is often possible. More elaborate calculations make it possible to distinguish between *cis*- and *trans*-isomers in complexes MA_4B_2, and to make deductions about the relative π-donor and π-acceptor properties of ligands. If enough information can be obtained to determine the sign and magnitude of the efg, the full efg tensor and the quadrupole coupling constant (by studying single crystals), then the parameters obtained can be used to calculate orbital occupancies and to give details of electron distribution in complicated species. The

Example 7.1

The ligands 1,2-bis(dimethylarsino) tetrafluorocyclo butene, 1,2-bis (diphenylphosphino)-tetrafluorocyclobutene and 1,2-bis (diphenylphosphino)-hexafluorocyclopentane form complexes of the type:

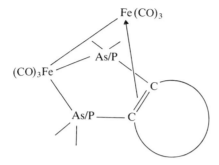

The Mössbauer spectra of the complexes at 80 K are illustrated below. Assign the lines in the spectra and measure as many Mössbauer parameters as you can.

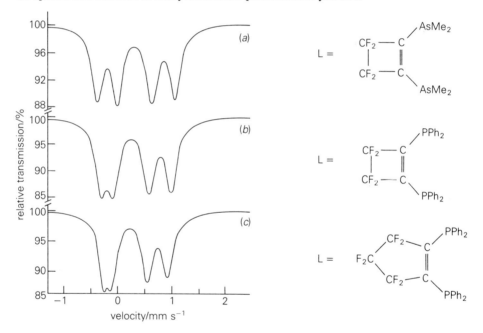

The spectra are redrawn from W.R. Cullen, D.A. Harbourne, W.V. Liengme and J.R. Sams, *Inorg. Chem.*, **8**, 95 (1969).

....

There are four lines in each spectrum, so each contains two quadrupole-split doublets. But in each case it is possible to pair the lines in three ways. For example, if we take the two high-velocity lines in the first spectrum together, and the two low (or negative)-velocity lines, we can measure δ 0.84, Δ 0.42 and δ – 0.20, Δ 0.39 mm s^{-1}. If we pair line 1 with 3 and 2 with 4 we obtain δ 0.53, Δ 1.05 and δ 0.12, Δ 1.02 mm s^{-1}.

Pairing lines 1 and 4, and 2 and 3, gives δ 0.32, Δ 1.44 and δ 0.28, Δ 0.64 mm s^{-1}. We must use our chemical understanding to choose between these options.

The iron atom bound to arsenic, which we will call atom A, is much closer to regular octahedral coordination than the other iron atom, B, and so should have the smaller quadrupole splitting. We can therefore rule out the first line assignment, because it yields two almost identical Δ values, and because there is a big difference between the two chemical shifts, with one of them falling outside the range expected for covalent iron(0) complexes (see Fig. 8.4). The second assignment implies that the two iron atoms have large and almost equal distortions from octahedral symmetry. We do not expect A to be so distorted. Moreover, with this assignment there is little change in chemical shift when the arsine ligand is replaced by phosphine, whereas we expect the shift for atom A to change. We also expect that changing from a cyclobutene to an equivalent cyclopentene should affect the quadrupole splitting for atom B, but should not affect the other quadrupole splitting, nor either chemical shift. These criteria are satisfied by the third assignment, which must therefore be presumed to be correct.

Using this assignment, we can measure the following parameters.

Compound	Atom	$\delta/mm\ s^{-1}$	$\Delta/mm\ s^{-1}$
(a)	A	0.28	0.64
	B	0.32	1.44
(b)	A	0.23	0.66
	B	0.32	1.30
(c)	A	0.22	0.65
	B	0.32	1.19

calculations are the same as those using similar information obtained by nuclear quadrupole resonance spectroscopy (Chapter 3).

7.4.3 Magnetic interactions

If a nucleus has spin greater than zero, it has a magnetic moment, and its energy will be affected by a magnetic field. If the nuclear magnetic moment is μ and the magnetic flux density is B, the energy levels of the nucleus are given by

$$E_m = -\mu B/I.m_I = -g\mu_N Bm_I \tag{7.7}$$

where μ_N is the nuclear magneton, $g = \mu/I\mu_N$, m_I is the magnetic quantum number of the nucleus, taking values I, $(I-1)$, $(I-2)$, ... $-(I-1)$, $-I$, and B is the magnetic field strength. What has happened is that the magnetic field has split the energy levels of the nucleus, which are degenerate in the absence of a magnetic field, into $(2I+1)$ different levels, separated by $\mu B/I$.

In a Mössbauer experiment, a nucleus is excited from its ground state g to an excited state e. The ground state has spin I_g, and the excited state has spin I_e, so that in a magnetic field the energy levels E_g and E_e are split. Transitions are allowed from one sub-level of the ground state to another of the excited state so long as $\Delta m_I = 0, \pm 1$. An example of the splittings and their effect on the spectrum of a nucleus with $I_g = 1/2$ and $I_e = 3/2$ (in this case ^{119}Sn) is given in Fig. 7.7. The diagram assumes that there is no efg at the nucleus. You can see that the magnetic field has split the single line spectrum that would be obtained without a magnetic field into a six-line pattern, which is symmetrical about the velocity at which the single line would have been observed.

Sometimes an external magnetic field is applied to a sample to help to sort out a spectrum, or to determine the sign of the efg and the magnitude of the asymmetry parameter η. For example, an interesting series of red complexes of iron with mixed chelating ligands such as 1,10-phenanthroline (phen) and oxalato (ox) or malonato

CHAPTER 7: MÖSSBAUER SPECTROSCOPY

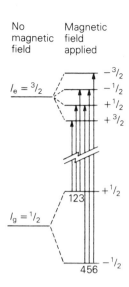

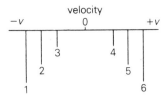

Fig. 7.7. Energy-level scheme and resultant spectrum for magnetic hyperfine splitting of a transition with $I_g = 1/2$ and $I_e = 3/2$. The relative splittings are scaled according to the magnetic moments of ^{119}Sn; the line intensity ratios of 3:2:1:1:2:3 are appropriate to a polycrystalline absorber. Redrawn from Ref. [3].

(mal) was reported in 1968. They had formulae like $[Fe(phen)_2(ox)].5H_2O$ or $[Fe(phen)_2(mal)].7H_2O$. The magnetic susceptibilities and other physical properties were taken as showing that these compounds contained two unpaired electrons per iron atom, and so should be described as 'spin-triplet' derivatives of Fe^{II}. In particular, in each case the ^{57}Fe Mössbauer spectra consisted of a simple narrow quadrupole doublet, with an isomer shift of about 0.3 mm s^{-1} and a quadrupole splitting of between 0.25 and 0.50 mm s^{-1}. Because only one Mössbauer resonance was observed, formulations of these compounds as mixed-valence species, containing both Fe^{II} and Fe^{III}, were excluded. However, when the Mössbauer spectra were recorded in a strong magnetic field and over a wide velocity-range, [10] the hyperfine pattern characteristic of Fe^{III} appeared, centered about a strong and complicated pattern of narrowly-spaced lines which was assigned to diamagnetic Fe^{II}. Mixed-valence complexes were then prepared from $[Fe^{II}(phen)_3]^{2+}$ and $[Fe^{III}(mal)_3]^{3-}$ or $[Fe^{III}(ox)_3]^{3-}$. These turned out to be red and appeared identical to the original red 'spin-triplet' complexes. In particular, they gave identical Mössbauer spectra. The Mössbauer spectrum of $[Fe(phen)_3]_3[Fe(mal)_3]_2.32H_2O$ is shown in Fig. 7.8. The spectra were recorded at 4.2 K, and with magnetic fields of different strengths. The magnetic hyperfine pattern becomes obvious as the strength of the magnetic field increases.

This example illustrates the important lesson that simple spectra may be deceptive; different species may still have the same isomer shifts and quadrupole splittings. In many compounds of iron and other transition metals, however, there are unpaired electrons which generate internal magnetic fields. These can lead to complicated effects, particularly where levels with different S-values are close together in energy. For such systems it may be necessary to record Mössbauer spectra over a range of temperatures (sometimes as low as 0.14 K) to disentangle what is going on. Allowing for the combined effects of magnetic and quadrupole interactions in a Mössbauer spectrum is in principle a complicated business, and will not be discussed further in this simple treatment. The magnetic field in most paramagnetic compounds usually relaxes so quickly, except at very low temperatures, that no magnetic splitting is observed. If compounds are ferromagnetic or antiferromagnetic, however, their spectra show magnetic splitting at room temperature.

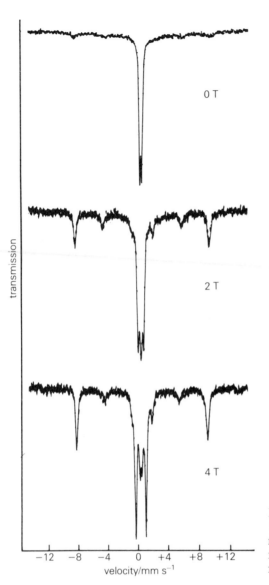

Fig. 7.8. ^{57}Fe Mössbauer spectra of [Fe(phen)$_3$]$_3$[Fe(mal)$_3$]$_2$. 32H$_2$O at 4.2 K recorded with the sample in magnetic fields of strength 0, 2 and 4 T. Redrawn from Ref. [10].

velocity/mm s^{-1} (axis: −12, −8, −4, 0, +4, +8, +12)

Labels: 0 T, 2 T, 4 T

transmission (vertical axis label)

7.4.4   Time- and temperature-dependent effects

Mössbauer spectra are determined by isomer shifts, by electric field gradients and by magnetic fields. All three may be affected if the compound studied changes with time. Suppose we are studying an iron complex that contains both Fe^{II} and Fe^{III} centers. The Mössbauer spectrum would be expected to show different resonances for the two types of Fe center, but this would only happen so long as any electron transfer between Fe^{II} and Fe^{III} was slow. If it were very fast, a single shift corresponding to the average of the two sites would be observed. In Mössbauer spectroscopy, 'fast' means fast in relation to the rates of decay of the excited nuclei, which are around 10^8 s^{-1}; where electron transfer is much slower than about 10^7 s^{-1}, separate lines will be observed. This shows that Mössbauer spectroscopy can be used to obtain information about rates of electron transfer that are of this order of magnitude. The basic acetates of iron provide a good example of this.[11] Basic iron acetate contains an oxygen atom surrounded by three iron atoms in a triangular planar arrangement, with acetate bridges round the edges of the Fe_3 triangle and water or other ligands completing the coordination shells of the metal atoms. The structure is quite common. Among many salts of this type, there is one which at least formally contains one Fe^{II} and two Fe^{III} atoms, of formula $[Fe^{II}Fe^{III}_2 Oac_6(H_2O)_3]$. It is interesting to ask whether the iron atoms in a single cluster can be distinguished experimentally, or whether they become equivalent through electron delocalization. In principle, this question might be answered by Mössbauer spectroscopy, which is often able to distinguish Fe^{II} from Fe^{III}. The ^{57}Fe Mössbauer spectrum of the mixed-valence species at room temperature shows a single slightly broadened line. This splits as the sample is cooled. At 17 K the pattern observed (Fig. 7.9) is interpreted as due to two overlapping doublets. These have different isomer shifts and quadrupole splittings, but the low-velocity components of each doublet overlap, so that only three lines are resolved, the one at lowest velocity being twice as

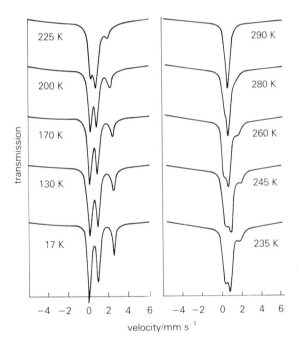

Fig. 7.9. Variable-temperature ^{57}Fe Mössbauer spectra of $[Fe(II)Fe(III)_2 Oac_6(H_2O)_3]$. The spectra are theoretical fits to the experimental data obtained using the relaxation model described in Ref. [11], from which the figure is redrawn.

strong as the other two. The spectra show that the iron atoms become equivalent on the Mössbauer timescale at 290 K, but are clearly distinct on that timescale at 17 K. The details of the changes in the spectra with temperature have been interpreted by supposing that electron transfer takes place within the Fe_3O units, and the activation energy for the process has been calculated to be 470 cm^{-1}. [11]

Rotation of anisotropic ligands can lead to changes in the efg at the central nucleus. This in its turn can lead to variations in the quadrupole splittings, and if the rate of rotation is of the right order of magnitude the Mössbauer spectra may be used to get information about this kind of process through analysis of temperature-dependent spectra. Variations in the recoil-free fraction with temperature have been used to find out about the forces holding molecules in lattices and molecular motions in solids. In the ^{129}I Mössbauer spectrum of ICN, for instance, there is no marked anisotropy in the recoil-free fraction; this observation led to the chemically significant conclusion that the solid did not contain chains of molecules linked by well-defined intermolecular interactions. If there had been such chains in the solid, binding should have been stronger along the chain axis and weaker perpendicular to it, and therefore markedly anisotropic. [12]

The magnetic hyperfine structure may also depend on temperature. This dependence may arise from relaxation phenomena or from changes in the population of magnetic states with temperature. For instance, the complex [{Cu(Mesalen)}$_2$Fe(acac)] (NO$_3$)$_2$, where MesalenH$_2$ is [N,N'-bis(methylsalicylidene)-ethylenediamine] shows interesting magnetic properties. [13] Measurements of the magnetic susceptibility suggest that the spins of the two Cu centers (each with $S = 1/2$) couple below 30 K in an antiferromagnetic way with the 5/2 spin of Fe^{III} to give a total spin of 3/2. At lower temperatures, the magnetic moment decreases because the upper of the two Kramer's doublets (see Section 3.3) becomes depopulated. The Mössbauer spectra at 300, 77 and 4.2 K show broad magnetic hyperfine patterns; between 4.2 and 1.3 K the patterns sharpen considerably, and at 1.3 K a broad asymmetric quadrupole doublet is observed ($\delta = 0.50$, $\Delta = 0.90$ mm s^{-1}). The sharpening of the pattern is connected with the changes in populations of the two Kramer's levels.

7.5 Difficulties

Mössbauer lines may sometimes be broad and ill-resolved. This is not a problem peculiar to Mössbauer spectroscopy, but its consequences can be awkward. Line positions can rarely be measured to much better than ±0.01 mm s^{-1}, and the precision is often less. The ranges of values for isomer shift and quadrupole splitting are not large compared with common linewidths, and overlapping lines may be hard to resolve. Indeed, it can be difficult to decide how many lines are covered by a single unresolved

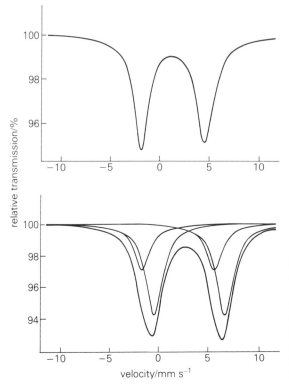

Fig. 7.10. ^{197}Au Mössbauer spectra of [Au$_4$(PPh$_3$)$_4$I$_2$] and [Au$_6$(dppp)$_4$]$^{2+}$. The upper spectrum can be fitted with two lines (linewidths 1.96, 2.34 mm s^{-1}). The lower spectrum has been fitted with four lines (linewidths 2.05 and 2.10 mm s^{-1}), as shown. Redrawn from Ref. [14].

CHAPTER 7: MÖSSBAUER SPECTROSCOPY

envelope. Curve-fitting methods are often used to help to determine this, and comparison between observed and calculated spectra gives the best available basis for making this kind of decision unless the effect of magnetic fields on the spectrum can be explored, but curve-fitting depends on assumptions about line-shapes and line-widths, and an element of uncertainty may well remain. Fig. 7.10 shows the Mössbauer spectrum [14] of $[Au_4(PPh_3)_4I_2]$ and $[Au_6(dppp)_4]^{2+}$, where dppp = $Ph_2P(CH_2)_3PPh_2$. The structure of $[Au_4(PPh_3)_4I_2]$ consists of a tetrahedron of gold atoms with iodine atoms bridging opposite edges [Fig. 7.11(a)]; the gold atoms are equivalent, and the spectrum appears as a single quadrupole doublet. The structure of $[Au_6(dppp)_4]^{2+}$ is more complicated. It contains four equivalent gold atoms at the corners of a tetrahedron, with a set of two other gold atoms (equivalent to each other but not to the first four) bridging opposite edges [Fig. 7.11(b)]. The observed spectrum (Fig. 7.10)

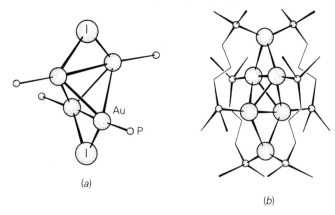

(a)

(b)

Fig. 7.11 Drawings of the structural units (a) $[Au_4(PPh_3)_4I_2]$ and (b) $[Au_6(dppp)_4]^{2+}$. Phenyl rings have been omitted from (a), for clarity. Redrawn from Ref. [14].

can be fitted by two overlapping quadrupole doublets, one twice as strong as the other. This example is fairly straightforward. A more complicated one is shown in Fig. 7.12. This shows the ^{57}Fe spectrum of polymeric vinylferrocene over a range of temperatures. The changes in linewidth are connected with molecular motion. It is suggested that the line-broadening arises from partly-resolved and overlapping doublets.

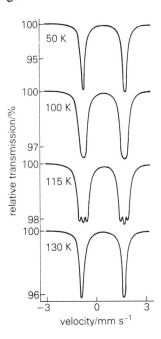

Fig. 7.12. ^{57}Fe Mössbauer spectrum of polyvinylferrocene with sample temperatures between 50 and 130 K. Redrawn from Ref. [15].

SECTION 7.5: DIFFICULTIES

The authors[15] have analyzed the spectrum at 115 K as due to three such doublets, but it is hard to decide how many may be hidden under such a poorly-resolved envelope.

7.6 Conversion electron Mössbauer spectroscopy (CEMS)

Mössbauer spectroscopy can be used to monitor mixtures containing elements such as iron in a range of chemical states and environments. It is therefore very suitable for studying processes like rusting and weathering. But as normally used it is a bulk technique. It can be adapted for studying surfaces, however, and one important new development involves the study of backscattered photons or electrons.[16] The excited state of ^{57}Fe that is responsible for emitting the 14.4 keV gamma photons normally used in ^{57}Fe Mössbauer spectroscopy also decays by other pathways, one of which involves the emission of conversion-electrons. The emitted electrons or radiation can be studied by detectors near the source, so that the spectra of surfaces can be recorded. For thin films, the equipment can be arranged so that transmitted and backscattered photons can be studied from the same sample, and a versatile experimental arrangement is shown in Fig. 7.13. The combination of electronic and photon spectra can even be used to give spectra as a function of depth, since the conversion electrons are attenuated by matter much faster than the gamma photons.

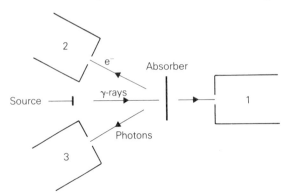

Fig. 7.13. Schematic drawing of possible geometries for Mössbauer experiments with detection of (1) transmitted gamma photons, (2) backscattered electrons, and (3) backscattered photons, X-rays and/or gamma photons. Redrawn from Ref. [16].

7.7 Structural deductions

In this section a few more examples are discussed. These make use of more than one of the aspects of Mössbauer spectroscopy that have been explained briefly above, and have been chosen to illustrate the sort of problems that Mössbauer spectroscopy can help to solve.

$I_2Br_2Cl_4$

The compound $I_2Br_2Cl_4$ was prepared by oxidizing NaI with $KBrO_3$ followed by Cl_2 gas. Its structure seems likely to be related to that of I_2Cl_6, made from NaI and Cl_2 without treatment by $KBrO_3$. I_2Cl_6 is known to be a planar molecule with a symmetrical Cl-bridged structure (7.I).

There are several possible structures for $I_2Br_2Cl_4$ even within this geometric

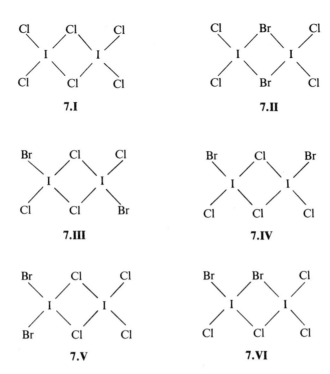

framework, however. The two Br atoms could both be bridging (7.II), or both terminal but on different I atoms (7.III and 7.IV), or both terminal but on the same I atom (7.V), or one bridging and one terminal (7.VI).

The question was answered by Mössbauer spectroscopy. [17] The parameters from the ^{129}I Mössbauer spectrum are given in Table 7.3. In the spectrum of I_2Cl_6 there was one set of lines, but in the spectrum of $I_2Br_2Cl_4$ there were two sets of lines due to non-equivalent iodine atoms. Since the sets were of equal intensity, it was most unlikely that one set was due to an impurity, and both sets were assigned to $I_2Cl_4Br_2$. This means that if the molecule has a single structure the two iodine atoms cannot be equivalent, and so structures 7.II, 7.III and 7.IV are ruled out. The isomer shift and quadrupole coupling constant for one of the iodine atoms are close to those found for I_2Cl_6 (see Table 7.3); this implies that one of the iodine atoms in $I_2Br_2Cl_4$ is in much the same environment as are the two in I_2Cl_6 – that is, bound to two terminal and two bridging Cl atoms. In structure 7.V, one of the iodine atoms satisfies this condition, but in structure 7.VI neither does, for both are bound to at least one bromine atom. Hence the correct structure must be 7.V.

Table 7.3 ^{129}I Mössbauer parameters for I_2Cl_6 and $I_2Br_2Cl_4$.

Compound		δ/mm s^{-1}	e^2qQ/MHz
I_2Cl_6		3.50 ± 0.10	$+3060 \pm 10$
$I_2Br_2Cl_4$	I_A	2.82 ± 0.02	$+2916 \pm 10$
	I_B	3.48 ± 0.02	$+3040 \pm 10$

Data from Ref. [17].

Detecting chlorides and bromides of xenon

Xenon fluorides and oxides are well known, but chlorides and bromides are not stable at room temperature and have proved hard to characterize. Mössbauer spectroscopy has been used to provide evidence that they are formed at low temperatures by making use of nuclear transformations. [18] Anions such as ICl_4^- and IBr_2^- are familiar chemical species. The radioactive isotope ^{129}I decays by β emission to give ^{129}Xe in an excited state; this then decays to the ground state, emitting 40 keV gamma-photons. The photons can be used in a Mössbauer spectrometer to excite ^{129}Xe in a standard absorber, and so it is possible to obtain the emission spectrum of ^{129}Xe formed in a chemical environment that was originally appropriate for iodine. Suppose that we start with $^{129}ICl_4^-$. As the iodine atoms emit β particles, they become xenon atoms. If during the timescale of the Mössbauer experiment the xenon atoms remain bound to the four chlorine atoms in the square-planar arrangement of the precursor ICl_4^-, the ^{129}Xe Mössbauer spectrum should show a characteristic isomer shift and quadrupole splitting. First, the absorption spectrum of ^{129}Xe in XeF_4 was recorded, using $^{179}IO_4^-$ as a source. The ^{129}I of course loses a β particle to become excited ^{129}Xe, which remains bound to the four oxygen ligands in a tetrahedral geometry (since XeO_4 is a characterized compound) and so emits a single line with no quadrupole splitting. The spectrum of XeF_4 is a quadrupole-split doublet, as expected for a square-planar molecule (Fig. 7.14). The emission spectrum of xenon from ICl_4^- was then recorded,

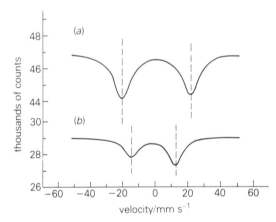

Fig. 7.14. Quadrupole splittings in the ^{129}Xe Mössbauer spectra of (a) XeF_4 and (b) $XeCl_4$. Redrawn from Ref. [18].

using XeO_4 as absorber, and this too is shown in Fig. 7.14. The spectrum is a well-defined doublet with much the same isomer shift as for XeF_4 and a rather smaller quadrupole splitting (see Table 7.4). This shows that $XeCl_4$ has a lifetime at least comparable with the timescale of the Mössbauer experiment at 4.2 K. Similar experiments using $K^{129}ICl_2$ or $K^{129}IBr_2$ as initial sources gave similar results, and the Mössbauer parameters for $XeCl_2$ and $XeBr_2$ are included in Table 7.4. When $Cs^{129}IBr_2$ was used as the starting-material, however, the resulting spectrum was a broad single line, implying that the environment of xenon in the lattice was different. Even at 4.2 K and on the timescale of the lifetime of $^{129}Xe^*$, $XeBr_2$ must be unstable in the $CsIBr_2$ lattice. What it decomposes to is not at all clear. The product may be atomic xenon or it may be some other XeBr complex. Whatever it is, the absence of quadrupole splitting implies that there is no significant efg at the Xe nucleus.

Table 7.4 Quadrupole splittings in the ^{129}Xe Mössbauer spectra of some xenon halides.

Compound	Δ/mm s^{-1}	Compound	Δ/mm s^{-1}
XeF$_4$	41.0	XeF$_2$	39.0
XeCl$_4$	25.6	XeCl$_2$	28.2
		XeBr$_2$	22.2

Data from Ref. [18].

Iron atoms and related species in matrices [19,20]

The ^{57}Fe Mössbauer spectra of dilute matrices of iron atoms in xenon show a sharp single line with an isomer shift of -0.75 mm s^{-1}. This is assigned to isolated iron atoms in an environment with cubic symmetry and hence no quadrupole splitting. At higher iron atom concentrations, an additional quadrupole-split resonance appeared ($\delta = -0.14$, $\Delta = 4.05$ mm s^{-1}); studies of the variation in relative intensities of the doublet and the single line with iron concentration have led to the assignment of the doublet to Fe$_2$ molecules. The reactions of these two species with less inert matrix materials have been studied by Mössbauer spectroscopy, and the dimer appears to be much more reactive than the isolated atoms. In N$_2$ matrices at 4.2 K, iron atoms at low concentrations give a single line with the same shift as found in xenon, but in N$_2$ the resonance shows quadrupole splitting ($\Delta = 2.70$ mm s^{-1}), presumably because the site symmetry in $\alpha - N_2$ is lower than cubic and so there is an efg at the iron nucleus. At higher concentrations of iron atoms the pattern of lines looks complicated, but it has been analyzed as due to the original doublet from iron atoms overlapping a second narrower doublet (Fig. 7.15). The parameters for the second doublet ($\delta = +0.71$, $\Delta = 1.17$ mm s^{-1}) are quite different from those for Fe$_2$ in xenon, and no resonances that could be assigned to Fe$_2$ were observed in the spectrum of the N$_2$ matrix. It seems that Fe$_2$ reacts with N$_2$ at 4.2 K whereas iron atoms do not. In methane matrices at 20 K, the line due to iron atoms again appears, this time with no quadrupole splitting, but here too there was no resonance that could be assigned to Fe$_2$; instead, new lines with isomer

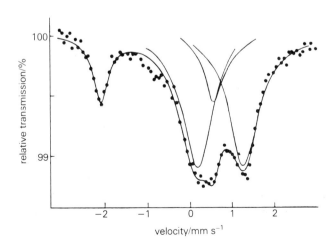

Fig. 7.15. ^{57}Fe Mössbauer spectrum of Fe in a matrix of solid N$_2$. The Fe concentration is 0.6%. The weaker doublet is assigned to Fe atoms, and the stronger doublet to a reaction-product. The dots represent the experimental data, and the smooth curves have been fitted to them. Redrawn from Ref. [19].

shifts of $+0.58$ and $+1.42$ mm s^{-1} were observed. Mössbauer resonances due to iron atoms but not to Fe_2 molecules were observed in matrices of NH_3 and C_2H_4, and in each of these systems new resonances were observed with isomer shifts greater than zero. Studies of the changes in relative intensities of lines with iron atom concentration can be used to deduce something about the nature of what has been formed, but without using other spectroscopic methods the conclusions are not well-defined.

The examples illustrate the applications of Mössbauer spectroscopy in inorganic chemistry. Many others will be found in standard texts [1] – [3] and in recent books. [21,22]

Biological systems

Iron is of course a vital chemical component of many biological systems, and so Mössbauer spectroscopy affords a way of studying the role of iron centers in biological processes. For example, it has been used to investigate model iron-containing compounds [23], to explore the intracellular distribution pattern of iron in *E. coli* by *in vivo* spectroscopy [24], and in studies of human hemoglobin [25]. Mössbauer spectroscopy has also been used to study the dynamic behavior of iron in iron-poor ferritin crystals [26].

Problems

7.1 The ^{57}Fe Mössbauer spectrum of $Fe_3(CO)_{12}$ at 4.2 K is shown in Fig. 7.16(*a*). How do you reconcile it with the molecular structure as determined by X-ray crystallography [Fig. 7.16(*b*)]? The figure is redrawn from F. Grandjean, G.J. Long, C.G. Benson and U. Russo, *Inorg. Chem.*, **27**, 1524 (1988).

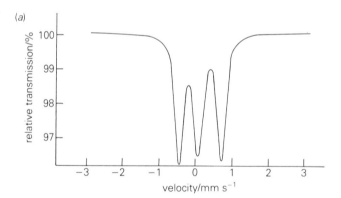

(*a*)

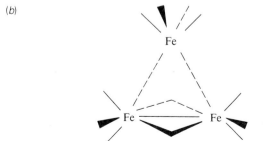

(*b*)

Fig. 7.16

CHAPTER 7: MÖSSBAUER SPECTROSCOPY

7.2 Pentamethylstibine reacts with oxine, with acetylacetone, and with anthranilic acid to give complexes of formula $Me_4SbC_9H_6NO$, $Me_4SbC_5H_7O_2$ and $Me_4SbC_7H_6NO_2$. Reaction with trichloroacetic acid or with phthalic acid gives complexes $Me_4Sb(Cl_3CCOO)_2H$ and $Me_4SbC_6H_4(COO)_2H$. The crystal structure of $Me_4SbC_9H_6NO$ shows that the antimony atom is in a distorted octahedral environment with planar bidentate oxinate. The ^{121}Sb Mössbauer parameters for these complexes are set out below. What can you deduce from them about the structures of the complexes?

Compound	δ/mm s^{-1}	Δ/mm s^{-1}
$Me_4SbC_9H_6NO$	-4.18	8.97
$Me_4SbC_5H_7O_2$	-4.51	8.12
$Me_4SbC_7H_6NO_2$	-5.06	-5.95
$Me_4Sb(Cl_3CCOO)_2H$	-5.40	0
$Me_4Sb(C_6H_4[COO]_2H)$	-5.43	0

7.3 The ^{57}Fe Mössbauer parameters for $[Fe(CN)_4(bpm)]^{2-}$, $[Fe(OH_2)_5(bpm)]^{2+}$ and $[(H_2O)_4Fe(bpm)Fe(CN)_4]$, where bpm is 2,2'-bipyrimidine, are set out below, with the spin states for the model compounds. Interpret the results for the binuclear species, comment on your conclusions and suggest other ways in which the electronic structure of this compound might be explored.

Compound	δ/mm s^{-1}	Δ/mm s^{-1}	Spin state
$K_2[Fe(CN)_4(bpm)]$	0.12	0.58	Low spin
$[Fe(OH_2)_4(bpm)]Cl_2$	1.12	2.57	High spin
$[(H_2O)_4Fe(bpm)Fe(CN)_4]$	$\begin{cases} 0.06 \\ 1.28 \end{cases}$	$\begin{cases} 0.72 \\ 2.42 \end{cases}$	

Measurements at 77 K; shifts relative to α-Fe.

7.4 How do you account for the following observations?
(a) The ^{119}Sn Mössbauer spectrum of black SnO shows a quadrupole splitting of 1.45 mm s^{-1}, whereas the spectrum of SnTe shows a sharp line with no quadrupole splitting.
(b) The ^{121}Sb Mössbauer spectrum of solid SbF_5 shows no resolved quadrupole splitting.
(c) The ^{57}Fe spectrum of $[(C_5H_5)_2Fe]_2SnCl_2$ contains a single resonance for the solid and for frozen solutions. The ^{119}Sn spectrum of the solid is also a single resonance, but there are two resonances in the spectrum of a frozen solution.

7.5 What could be learnt about the structures of the following using Mössbauer spectroscopy?
(a) CsAu (b) Me_3SnCN (c) I_4O_9

(d)

7.6 The ^{57}Fe Mössbauer spectrum of crystalline 1,1'-diisobutylbiferrocenium triiodide over a range of temperature shows two doublets at 77 K [isomer shifts 0.53, 0.54 mm s^{-1} relative to Fe foil; Δ 1.91, 0.59 mm s^{-1}], but only one doublet at 300 K [isomer shift 0.45 mm s^{-1}; Δ 1.09 mm s^{-1}]. What does this tell you about the electronic structure of the cation? The Mössbauer spectrum of the same material dispersed on poly(methylmethacrylate) shows two doublets at both 77 and 300 K. Comment on the significance of this observation.

7.7 The ^{57}Fe isomer shift for the complex anion $[(PhS)_2FeS_2MoS_2]^{2-}$ is 0.33 mm s^{-1}, relative to iron metal. What can you deduce about the oxidation states of iron and molybdenum?

7.8 Reaction between $SbCl_3$ and $Sb(C_5H_5)_3$ in dry hexane at 220 K for 12 hours gave a product showing two ^{121}Sb Mössbauer resonances. Their parameters are given below, with those for the two starting materials. Identify the products.

Compound	δ/mm s^{-1}	Δ/mm s^{-1}
Product 1	– 12.60	– 4.10
Product 2	– 14.00	– 5.50
$Sb(C_5H_5)_3$	– 11.17	– 2.67
$SbCl_3$	– 14.40	– 5.90

7.9 The ^{57}Fe Mössbauer spectra of $[Fe_3O(O_2CCH_3)_6 (3\text{-Me-py})_3]$.solv., where solv. is CH_3CN (*a*) and toluene (*b*) are shown in Fig. 7.17. Comment on these and deduce what you can about the electronic structure of the complex. What additional measurements could be made to help to understand the system better? The figure is redrawn from S.M. Oh, S.R. Wilson, D.N. Hendrickson, S.E. Woehler, R.J. Wittebort, D. Inniss and C.E. Strouse, *J. Am. Chem. Soc.*, **109**, 1073 (1987).

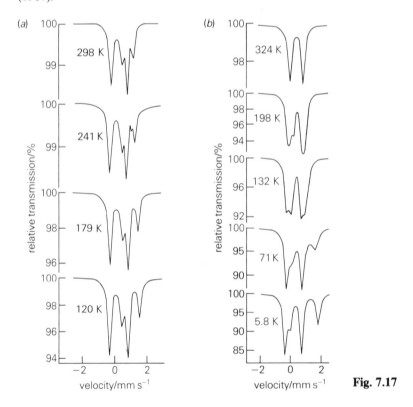

Fig. 7.17

7.10 oxidation of I_2 in $(CH_3CO)_2O$ in the presence of H_3PO_4 gives a material described as $I(PO_4)$. Its ^{129}I Mössbauer parameters are given below, with those of some model compounds. Deduce what you can about the nature of $I(PO_4)$.

Compound	δ/mm s^{-1}	e^2Qq/MHz
$I(PO_4)$	1.97	– 2600
$I(CH_3COO)_3$	3.60	+ 2400
I_2Cl_6	3.50	+ 3060
ICl	1.73	– 3131

7.11 The crystal structure of triphenylstannylpotassium-[18]aneO$_6$ (the crown ether [18]crown-6) shows that the units in the crystal are crown ether-potassium ion complexes and pyramidal triphenylstannyl anions. The Mössbauer spectrum of this solid consists of a quadrupole-split doublet (δ 2.17 mm s^{-1}, Δ 1.86 mm s^{-1}) that is just in the region normally associated with Sn^{II}. Solid triphenylstannylpotassium shows two Mössbauer resonances; one, a doublet (δ 2.16 mm s^{-1}; Δ 1.86 mm s^{-1}) corresponds to the doublet observed in the spectrum of the crown ether complex, the other, a singlet (δ 1.46 mm s^{-1}) is in the regional normally associated with Sn(IV). The singlet shows no resolvable quadrupole splitting. How do you interpret these observations?

Further Reading

Refs. [1]–[3] are full accounts of Mössbauer spectroscopy, and Refs. [21]–[22] are more recent books on the subject. Many applications of the method are described in chapters on Mössbauer Spectroscopy in the annual series, *Spectroscopic Properties of Inorganic and Organometallic Compounds*, Specialist Periodical Reports, Royal Society of Chemistry, London.

References

1   N.N. Greenwood and T.C. Gibb, *Mössbauer Spectroscopy*, Chapman and Hall, London (1971).
2   G.M. Bancroft, *Mössbauer Spectroscopy*, McGraw-Hill, London (1973).
3   T.C. Gibb, *Principles of Mössbauer Spectroscopy*, Chapman and Hall, London (1976).
4   J.D. Cotton, P.J. Davison, M.F. Lappert, J.D. Donaldson and J. Silver, *J. Chem. Soc., Dalton Trans.*, 2286 (1976).
5   C.A. Reed, *Adv. Chem. Ser.*, **201**, 333 (1982).
6   N.S. Dance, P. Dobud and C.H.W. Jones, *Can. J. Chem.*, **59**, 913 (1981).
7   S. Bukshpan, J. Soriano and J. Shamir, *Chem. Phys. Letters*, **4**, 241 (1969).
8   J.D. Donaldson, J. Silver, S. Hadjiminolis and S.D. Ross, *J. Chem. Soc., Dalton Trans.*, 1500 (1975).
9   J.J. Zuckerman, *Adv. Organomet. Chem.*, **9**, 21 (1970).
10 E. König, G. Ritter and H.A. Goodwin, *Inorg. Chem.*, **20**, 3677 (1981).
11 C.T. Dziobkowski, J.T. Wrobleski and D.B. Brown, *Inorg. Chem.*, **20**, 671 (1981).
12 M. Pasternak and T. Sonnino, *J. Chem. Phys.*, **48**, 2009 (1968).
13 I. Morgenstern-Baradau and H.H. Wickman, *J. Chem. Soc., Chem. Commun.*, 176 (1985).
14 J.W.A. van der Velden, J.J. Bour, J.J. Steggerda, P.T. Beurskens, M. Roseboom and J.H. Noordik, *Inorg. Chem.*, **21**, 4321 (1982).
15 F.J. Litterst, A. Lerf, D. Nuyken and N. Alcala, *Hyperfine Interactions*, **12**, 317 (1982).
16 M.J. Tricker, *Adv. Chem. Ser.*, **194**, 63 (1981).
17 M. Pasternak and T. Sonnino, *J. Chem. Phys.*, **48**, 1997, (1968).
18 G.J. Perlow and M.R. Perlow, *J. Chem. Phys.*, **41**, 1157 (1964); G.J. Perlow and H. Yoshida, *J. Chem. Phys.*, **49**, 1474 (1968).
19 P.H. Barrett and P.A. Montano, *J. Chem. Soc., Faraday Trans. II*, **73**, 378 (1977).
20 P.H. Barrett, M. Pasternak and R.G. Pearson, *J. Am. Chem. Soc.*, **101**, 222 (1979).
21 G.J. Long (ed.), *Mössbauer Spectroscopy Applied to Inorganic Chemistry*, Plenum Press, New York (1984) Vol. 1.
22 F.J. Berry and D.P.E. Dickson (ed.), *Mössbauer Spectroscopy in Perspective*, Cambridge University Press (1985).
23 M. Schappacher, L. Ricard, J. Fisher, R. Weiss, R. Monteil-Montoya, E. Bill and A.X. Trautwein, *Inorg. Chem.*, **28**, 4639 (1989); J. Adler, J. Ensling, P. Gutlich, E.L. Bominaar, J. Guillin and A.X. Trautwein, *Hyperfine Interactions*, **42**, 869 (1988).
24 B.F. Matzanke, G.L. Müller, E. Bill and A.X. Trautwein, *European J. Biochem.*, **183**, 371 (1989).

25 M.I. Oshtrakh, V.A. Semionkin, B.N. Burykin and V.I. Khleskov, *Mol. Phys.*, **66**, 531 (1989).

26 E.R. Bauminger, P.M. Harrison, I. Nowik and A. Treffry, *Hyperfine Interactions*, **42**, 873 (1988).

8 Diffraction Methods

8.1 Introduction

There can be no doubt that when a compound has to be identified there is nothing to beat a crystal structure determination. After all, the final outcome – a drawing of the molecule, with all the atoms nicely shaded and labeled (Fig. 8.1) – is convincing, and will silence all but the most stubborn critics! And it is all based on very simple physical principles. Just as light is diffracted at a pair of slits, and the resulting interference pattern depends on the wavelength of the light and the distance between the slits, so X-rays (or neutrons or electrons) are diffracted by a pair of atoms, and the interference pattern which is produced depends on the wavelength of the X-rays, and the distance between the atoms. In the optical experiment, the aim is usually to measure the wavelength of the light. In the X-ray experiment, the wavelength is known, and the experiment thus provides a direct measurement of distances between atoms. It is this directness of diffraction experiments which makes them so appealing.

Diffraction methods, particularly X-ray studies of crystalline materials, have become so easy and quick to apply, and relatively cheap, that we may be tempted to

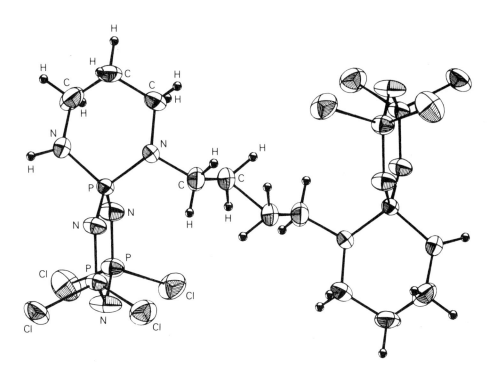

Fig. 8.1 Perspective view of the product of the reaction between $N_3P_3Cl_6$ and spermine $[NH_2(CH_2)_3NH(CH_2)_4NH(CH_2)_3NH_2]$. Taken, with permission, from J.-F. Labarre, G. Guerch, F. Sournies, R. Lahana, R. Enjalbert and J. Galy, *J. Mol. Struct.*, **116**, 75 (1984).

think that *without* a crystal structure, there must remain some doubt about identification of a new compound. But there are many compounds which, for one reason or another, do not give crystals suitable for crystallographic study. It would be unfortunate if large areas of chemistry, or even individual compounds, were ignored, simply because crystals could not be obtained easily, just as it would be a mistake, for example, to work only with those elements having spin 1/2 isotopes, convenient for NMR study. It is perfectly possible to characterize a compound unequivocally by chemical, spectroscopic and other non-diffraction methods. All these techniques are simply tools for the task of identifying molecules and understanding their properties. The tools must be chosen for the particular task in hand – not the other way round. Nevertheless, having said that, let us emphasize that when crystals can be obtained, determination of their structure is highly desirable, both to confirm the identity of the compound, and to obtain the geometric parameters, which may reveal a lot about the bonding between the atoms and the interactions between molecules, or parts of the same molecule. In exactly the same way, it is desirable to determine the structures of volatile compounds as gases, when they are free of intermolecular interactions, although gas-phase work is not as routine or straightforward as solid-phase work has become.

In a book such as this, it is not possible to go into great detail in discussing the theory of diffraction methods, or the experimental arrangements, or the techniques for refining structures. Such things are important, but provided the general principles are understood, the details can wait until they are needed – when a structure is to be determined. What is needed first is an understanding of *what* information can be obtained by the various diffraction methods, so that it will be possible for you to decide whether a diffraction study can help in some particular situation. We also aim to stress the limitations of the techniques, so that you can look at the mass of literature data with a critical eye, and recognize the reliable results, and the occasional errors.

8.2 Diffraction of electrons, neutrons and X-rays

In order to observe the diffraction and interference of light waves it is necessary to have two (or more) slits with a separation comparable with the wavelength of the light. It follows that if we wish to study diffraction patterns arising from crystals, we must use radiation with a wavelength of the same order of magnitude as the distances between atoms in the crystal, that is 10^{-10} m. In the spectrum of electromagnetic radiation, it is X-rays that have the appropriate wavelength, and after their discovery by Röntgen in 1895, it was not long before diffraction by single crystals was demonstrated.

Moving particles also have wavelengths, which depend on their mass and velocity. Only the lightest of particles have convenient wavelenghts, and in practice only electrons and neutrons are used in diffraction studies. Typically, electrons accelerated through 50 kV have a wavelength of about 5 pm (0.05 Å), which is comparable with interatomic distances, and convenient, for a variety of reasons. The advantage of using electrons is that it is easy to obtain an intense, monchromatic beam, and this is particularly important in studying gases, in which the density of molecules is low. Despite this low concentration of molecules, a data set in a typical electron diffraction experiment can be obtained in 20 seconds, whereas for an X-ray study of a crystalline solid, several hours or days are needed. The disadvantage of using electrons is that they are charged particles, and so do not penetrate solids (or liquids) very well, and so are not generally used for studies of condensed phases. The one major exception is the use of low energy electron diffraction for the specific purpose of looking at the surface layers of a crystal — a particularly important technique in surface chemistry, but beyond the scope of this book.

Thus electrons can give an intense beam, ideal for looking at gases, for which their low penetrating power does not matter. X-rays, of course, have high penetrating power, and are ideal for studying solids, but it is not possible to get a really intense monochromatic X-ray beam. Intense beams can be obtained from synchrotrons and storage rings [see Section 1.5.1], and it is in principle possible to observe X-ray diffraction by gases with such a source, but rarely could the cost be justified with respect to the importance of the experiment. On the other hand, studies of single crystals, powders and amorphous materials using synchrotron radiation are now of considerable importance. In particular, use is made of anomalous scattering, which occurs when the X-ray energy is tuned closed to an absorption edge of a particular element. In this situation it is possible to extract a data set which depends on the positions of atoms of that element alone, and this may be invaluable in the early stages of solving the structure, as well as in the subsequent refinements.

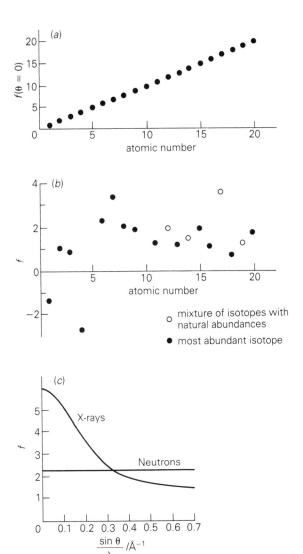

Fig. 8.2 Variation of scattering power with atomic number (a) for X-rays; and (b) for neutrons; and (c) variation with angle for X-ray and neutron scattering by carbon.

Both X-ray and electron diffraction suffer from the major disadvantage that the intensity of scattering depends on the atomic numbers of the scattering atoms [Fig. 8.2(a)]. It is therefore difficult to locate light atoms in the presence of heavy ones. Thus positions of hydrogen atoms are invariably less well defined than those of heavier ones, and if one or more really heavy atoms are present, even such atoms as carbon or oxygen may be hard to find. In such circumstances, neutron diffraction may be useful. Neutron scattering intensities have no simple relationship to atomic number, and are not the same for different isotopes of the same element, so heavy atoms may contribute no more to the total scattering pattern than light ones [Fig. 8.2(b)]. In particular, scattering by deuterium is as strong as that by typical heavy atoms, and neutron diffraction is therefore of great importance in studies of hydrogen bonding, or of hydrido ligands in transition metal complexes. Another important feature of neutron diffraction is that atomic scattering factors are constant, and do not decrease with diffraction angle, in marked contrast to X-ray or electron scattering factors [Fig. 8.2(c)]. Neutrons are highly penetrating, like X-rays, as they are small and uncharged, but flux densities for neutrons with suitable velocities are low, and experiments can only be performed on condensed phases, and not gases. Moreover, neutron sources are not widely available: usually a nuclear reactor is required, although there are sources designed specifically to provide neutron beams.

The principles of diffraction are more or less the same, no matter what is being diffracted. So in the following sections we will concentrate on the material being studied. In studies of gases, electron diffraction is normally used, and so we will refer to electron diffraction in the section on gases, although more or less the same information could be obtained, with greater expenditure of time and money, from X-ray or neutron diffraction experiments. In the same way, X-rays (or more rarely, neutrons) are used for solids, whether powders or single crystals. However, it is important to be aware that electron and neutron diffraction enable nuclear positions to be determined, whereas X-ray diffraction gives the positions of centers of electron density – which is not quite the same thing. The consequences of this difference will be discussed in the following sections.

We start by considering diffraction by gases, as the interpretation of the data is readily visualized, and so it provides a straightforward introduction to the principles of structure determination.

8.3   Diffraction by gases

8.3.1   Principles

The essential requirements for a gas-phase electron diffraction experiment are a monochromatic beam of electrons, interacting with the gas at only one point, and some device for recording the diffraction pattern (Fig. 8.3). The beam of electrons is generated by accelerating electrons emitted from a cathode (usually a hot tungsten wire) through an anode, and it is then focused by magnetic lenses. The gas to be studied flows through a fine nozzle, with a backing pressure of typically 10 Torr, and is then trapped on a cold surface. The whole apparatus must be pumped continuously to maintain a high vacuum, so that the electron beam is diffracted only at the point where it crosses the beam of molecules emerging from the nozzle. The diffraction pattern is

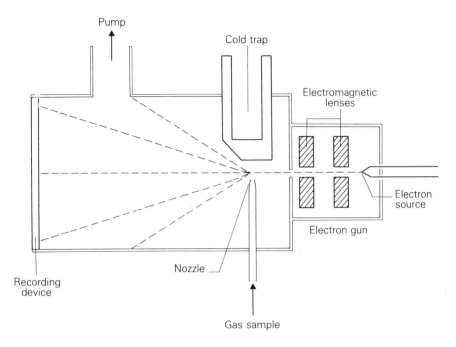

Fig. 8.3 Apparatus for the diffraction of electrons by gases.

usually recorded on a photographic plate, from which intensities can be measured later, but sometimes electron counters are used.

Electrons are scattered mainly by interaction with the electric fields of atomic nuclei. As these fields extend over quite a wide area, on an atomic scale, the proportion of electrons scattered is large (somewhat larger than for X-rays and much larger than for neutrons, at least for heavy atoms): even with the low density of molecules in a stream of gas, a data set may be collected in a few seconds or minutes. However, the intensity of scattered electrons falls off rapidly with scattering angle (roughly as the fourth power) and the range of intensities is thus too large to be recorded accurately on a photographic plate. A differential filter is therefore placed adjacent to the plate: this consists of a rotating sector, cut so that the width of its opening increases as the fourth power of the distance from its center.

As a gas consists of molecules in random orientations, its diffraction pattern consists of diffuse, concentric rings, and it can be described simply in terms of intensities which are a function of ring radius, or scattering angle. Analysis of data collected in an electron diffraction experiment first of all involves making allowances for experimental arrangements, such as the sector, which must be calibrated carefully. The total scattered intensity can then be expressed as a function of scattering angle, θ, and then of the variable, s, where $s = 4\pi \sin(\theta/2)/\lambda$, and λ is the electron wavelength. At this stage the scattering intensity is a steeply falling curve having two major components. The first is known as atomic scattering. An isolated atom causes diffraction of an electron beam, but no interference, and the total atomic scattering for a molecule is simply the sum of the contributions from each of the atoms [Eqn (8.1)].

$$I_{atomic}(s) = \sum_i F_i(s)^2/s^4 \tag{8.1}$$

The atomic scattering factors, $F_i(s)$, are functions both of electron wavelength (*i.e.* accelerating voltage) and of scattering angle and may be calculated or obtained from published tables. Thus for $TiCl_4$, the chlorine (four times) and titanium scattering factors are added, and after scaling they are subtracted from the total scattering. This leaves the molecular scattering, which contains the desired structural information.

The intensity of molecular scattering for a diatomic molecule can be expressed in a simplified form by Eqn (8.2), in which F_i and F_j

$$I_{mol}(s) = \frac{F_i(s)F_j(s)}{s^4} \cdot \frac{\sin(sr_{ij})}{sr_{ij}} \cdot \exp(-u_{ij}^2 s^2/2) \tag{8.2}$$

are the scattering factors for atoms i and j, r_{ij} is the distance between the atoms, and u_{ij} is their root mean square amplitude of vibration. The overall form is therefore that of a damped sine wave. The frequency of the oscillations is a simple function of the inter-atomic distance, and the rate of exponential decay enables us to determine the extent to which the atoms are moving relative to each other. The full expression describing molecular scattering also includes terms which take account of anharmonicity in the vibration (which causes modulation of the sine wave frequency) and of a phase shift on scattering. This last involves a cosine term, and is particularly important for the combination of a light atom and heavy atom, for which the amplitude of the sine wave pattern can reduce to zero, and then increase again. Full details of these effects, and descriptions of the theory and experimental arrangements for electron diffraction, are given in Refs [1] and [2].

For a polyatomic molecule the total molecular scattering is the sum of components for each pair of atoms in the molecule, of which there are 3 for a triatomic molecule, 6 for 4 atoms, 10 for 5 atoms, and 1225 for 50 atoms! Each component is described by an expression with the form of Eqn (8.2), with anharmonic and phase shift terms, and ideally 3-atom scattering should also be considered. The total pattern for a molecule of any size is therefore extremely complicated, and although computer analysis is possible, it is not easy to visualize the significance of a molecular scattering curve, such as that shown in Fig. 8.4. However, Fourier transformation of this curve gives the radial distribution curve, Fig. 8.5. This curve, for the phosphorus compound 8.I, shows clear peaks corresponding to the bonded-distances P—F, P = S and P—Br, and to the non-bonded distance Br···S. The F···F atom pair gives rise to the weak shoulder above the P—Br peak, while the two remaining peaks, Br···F and F···S, overlap.

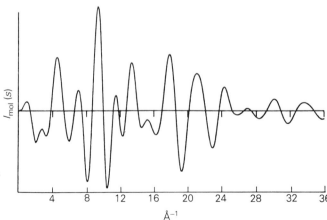

Fig. 8.4 Electron diffraction molecular scattering intensity, $s^4 I_{mol}(s)$, for $PBrF_2S$.

CHAPTER 8: DIFFRACTION METHODS

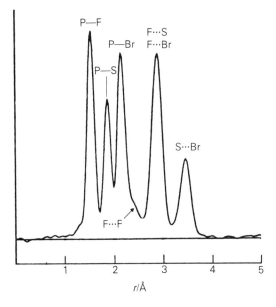

Fig. 8.5 Radial distribution curve, $P(r)/r$, for $PBrF_2S$. This curve was obtained by Fourier transformation of the molecular scattering intensity curve shown in Fig. 8.4.

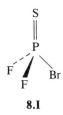

8.I

Each of the peaks has an approximately Gaussian shape, and their widths depend on the amplitudes of vibration for the appropriate atom pairs. The relative areas of peaks in the radial distribution curve are given by Eqn (8.3), in which Z_i and Z_j are the atomic numbers of atoms i and j, and n_{ij} is the number of times the distance r_{ij} occurs in the molecule

$$\text{Area} \propto \frac{n_{ij} Z_i Z_j}{r_{ij}}. \tag{8.3}$$

Thus in principle there are two pieces of information relating to each pair of atoms, without going into anharmonicity or other more complicated matters, and these items of information can be determined from electron diffraction data. Refinement of structures involves the construction of a mathematical model of the molecule, and this must be a reasonable approximation to the true structure if the parameters are to refine to correct values. The refinement process is invariably based on comparison of experimental and calculated molecular scattering intensities, but radial distribution curves are essential for human appreciation of the situation. In the following section the problems are discussed in terms of radial distribution curves only.

8.3.2 Interpretation of results

Using the intensity data shown in Fig. 8.4, the parameters defining the structure of $PBrF_2S$ (8.I) could be refined [3]. As the molecule has a plane of symmetry, only six parameters are required, and the three bond lengths (P—F, P = S and P—Br) are easily determined, as they give distinct peaks in the radial distribution curve (Fig. 8.5). The refined distances have estimated standard deviations of 0.003-0.004 Å, about 0.2% of the distances. The peaks due to non-bonded distances are slightly more confused in the radial distribution curve, but nevertheless there is enough information to enable the three angles to be determined, with uncertainties of 0.3 to 1.0°. (Any three of the angles FPF, BrPF, BrPS and FPS can be refined: the fourth is dependent on the other three.) Thus the whole three-dimensional structure is obtained, even though only one-dimensional data are available. In addition, amplitudes of vibration were obtained for the seven different types of atom pair − again, only seven, because of the symmetry. This is a fairly typical example of work on a molecule which is well suited for an electron diffraction study.

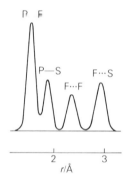

Fig. 8.6 Radial distribution curve for PF_3S. Redrawn from K. Karakida and K. Kuchitsu, *Inorg.Chim.Acta*, **16**, 29 (1976).

Increasing the symmetry decreases the number of peaks in the radial distribution curve, and also reduces the number of parameters required to describe the structure. Thus there are only four peaks in the radial distribution curve for PF_3S (Fig. 8.6), and only three parameters are needed. These are the P—F and P = S bond lengths, and one inter-bond angle. So this structure can be determined even more easily than that of $PBrF_2S$, but most molecules of this simplicity have been well studied, and electron diffraction is increasingly applied to more complex species.

Provided symmetry remains high, matters are still fairly straightforward. Fig. 8.7 shows the radial distribution curve for $Si(SiMe_3)_4$, which (making some minor assumptions) can be described by just seven geometrical parameters, even though it has 53 atoms, and hundreds of different inter-atomic distances. The first part of the curve has distinct peaks corresponding to the C—H, Si—C and Si—Si bonds, but the rest is a complicated shape, which depends on all the parameters. It is not possible to look at it and assign specific peaks to particular non-bonded distances. But it is by analysis of the whole curve (or more accurately, of the whole molecular scattering intensity curve) that the geometrical parameters, and some vibrational amplitudes, are determined.

In a similar way, it is possible to study quite large and asymmetric molecules, determining up to about 20 geometrical parameters. As a rule of thumb, it is possible to refine one parameter for every peak or shoulder visible in the radial distribution curve, but with good data a computer can distinguish features which may elude the human eye. In deciding whether an electron diffraction study of a compound should be

Example 8.1

The radial distribution curve for gaseous B_4Cl_4, determined by electron diffraction, is shown in Figure (a) below. The positions of the peaks are marked. What can be deduced from this information?

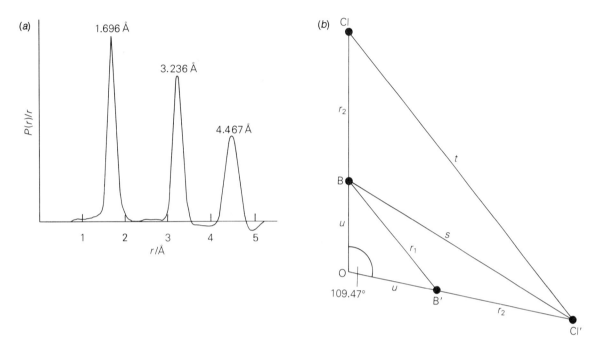

There are only three peaks in the radial distribution curve, and this indicates that the B_4Cl_4 molecule must have high symmetry. The only structure giving so few peaks, even allowing for overlap, is a tetrahedral B_4 cage, with each boron atom bound to one terminal chlorine atom. In such a structure there are four different inter-atomic distances, two for the bonded B—B and B—Cl atom pairs, and two for non-bonded B$\cdots$Cl and Cl$\cdots$Cl atom pairs. The first two of these peaks must overlap at 1.7 Å, and so this peak does not give us the individual bond lengths. These must be calculated from the non-bonded inter-atomic distances.

Consider the section through two boron and two chlorine atoms and the centroid of the B_4 tetrahedron [Fig. (b) above].

Here r_1 and r_2 are the B—B and B—Cl bond lengths, s and t are the B$\cdots$Cl and Cl$\cdots$Cl distances (3.236 and 4.467 Å respectively), and u is the distance of the boron atom from the centroid.

From triangle BOB' we have $r_1/2u = \sin(109.47/2)$, so $u = 0.6124\,r_1$. From triangle ClOCl', $t/[2(u + r_2)] = \sin(109.47/2)$, so $u + r_2 = 2.7355$.

From triangle BOCl, using the cosine rule, $s^2 = u^2 + (u + r_2)^2 - 2u(u + r_2)\cos(109.47)$, i.e. $3.236^2 = 0.6124^2\,r_1^2 + 2.7355^2 - 2 \times 0.6124\,r_1 \times 2.7355 \times (-0.3333)$. Thus $0.3750\,r_1^2 + 1.1167\,r_1 - 2.9887 = 0$. Solving for r_1 we find $r_1 = 1.7028$ or -4.3084. Clearly the first solution is the correct one.

This then gives $u = 1.0428$ and $r_2 = 1.6927$.

Thus the B—B distance is 1.703 Å and the B—Cl distance is 1.693 Å. These two peaks would overlap. The ratio of the intensities of the peaks is $6 \times 5 \times 5$:$4 \times 17 \times 5$, and so their weighted mean position is 1.696 Å.

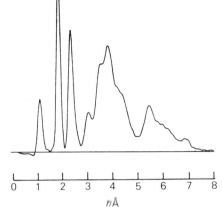

Fig. 8.7 Radial distribution curve for $Si(SiMe_3)_4$, obtained from gas-phase electron diffraction data. Redrawn from L.S. Bartell, F.B. Clippard and T.L. Boates, *Inorg. Chem.*, **9**, 2436 (1970).

attempted, it is always important to try to estimate how much information the experiment is likely to give, compared with what is required, given the probable symmetry of the molecule.

8.3.3   **Limitations**

As electron diffraction provides direct measurement of interatomic distances, it is an ideal method for the determination of the molecular structures of gases. Moreover, it is always better in principle to study structures in the gas phase, rather than in the crystalline state, as the molecules are free of external constraints and packing forces, which can distort the structure (particularly the conformation) or even change it completely. Certainly for any comparison with structures obtained by molecular orbital calculations, gas phase experimental data should, if possible, be used. But there are, of course, limits to the usefulness of electron diffraction, of which the most obvious is that gaseous samples are needed. The essential requirement is that the compound to be studied should have a vapor pressure of about 1 Torr or more at a temperature at which it is stable. Lower vapor pressures can occasionally be used, but the experiments are more difficult to perform. The temperature does not really matter, and such involatile species as alkali metal salts and some metal oxides have been studied at temperatures of 1000 to 2000 K or more. However, we should remember that raising the temperature of a sample increases amplitudes of vibration, and may change the relative populations of isomers or conformers.

The volatility requirement reduces the range of compounds suitable for electron diffraction study, but even if data can be collected, it may not be possible to determine a structure fully by this method. We have seen that the area of a peak in the radial distribution curve depends on the atomic numbers of the atoms involved. Thus contributions to the scattering from light atoms, particularly hydrogen, are small, and it is often impossible to determine their positions accurately. For example, the radial distribution curve for PF_2HS (8.II) looks very like that for PF_3S (Fig. 8.6), having peaks corresponding to P—F and P=S bonded and F$\cdots$F and F$\cdots$S non-bonded atom pairs. (The relative intensities of the peaks are somewhat different, as PF_2HS has only two P—F bonds instead of three.) The three remaining contributions, P—H,

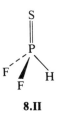

8.II

F$\cdots$H and S$\cdots$H, involve hydrogen and simply cannot be seen by eye, whereas the equivalent P—Br, F$\cdots$Br and S$\cdots$Br contributions for PBrF$_2$S are very prominent in Fig. 8.5. Consequently, the P—H distance and the angle HPS cannot be determined: if they are allowed to refine they go to unrealistic values, with enormous standard deviations. This problem cannot be overcome using electron diffraction data alone. The only solution is to refine structures using both diffraction data and either rotational constants (Section 4.6.3) or dipolar couplings, measured for solutions in liquid crystals (Section 2.13).

The second major difficulty encountered in electron diffraction arises from the overlap of peaks in the radial distribution curve. As the peaks have an approximately Gaussian shape, if two similar peaks are almost superimposed, their sum is also a Gaussian, and it is impossible to determine the positions and amplitudes of vibration for both components. One solution is to calculate the vibrational amplitudes from spectroscopic data, and then to freeze these at the calculated values in subsequent refinements. But even then, the two peak positions could be reversed, without changing the appearance of the curve at all. For example, Fig. 8.8 shows the radial distribution curve for perchloric acid (8.III), for which, assuming C_{3v} symmetry for the ClO$_4$ group, there are two O$\cdots$O distances almost the same. Even with the amplitudes of vibration fixed, there were two structures that could fit the data well, one with the O=Cl=O angles 112.8° and the O=Cl—O angles 105.8°, and the other with angles of

Fig. 8.8 Radial distribution curve for perchloric acid, ClO$_3$(OH), showing the positions of the two Cl—O bonded distances, and the two O$\cdots$O non-bonded distances. Redrawn from A.H. Clark, B. Beagley, D.W.J. Cruickshank and T.G. Hewitt, *J. Chem. Soc. (A)*, 1613 (1970).

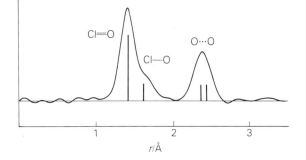

8.III

117.3 and 99.6°. These two, apparently very different, structures both gave non-bonded O···O distances of about 2.34 and 2.42 Å, which is why both could fit the experimental data. In fact, the first one produced a slightly better fit to the data, and was more reasonable, when compared with structures of related molecules, and so it was possible to select one solution with confidence. But it illustrates well how, even with a system involving only four major contributions to the radial distribution curve, there can be very serious problems caused by peak overlap. These problems can escalate rapidly as the number of atoms in the molecule increases, but the ease of determining a particular structure depends on how the various distances happen to fall in relation to each other. In silyl monothioacetate (8.IV) the silicon atom is bound to oxygen, and

<table>
<tr><td align="center">CH₃—C with =S, O—SiH₃

8.IV</td><td align="center">CH₃—C with =O, S—GeH₃

8.V</td></tr>
</table>

the main bond lengths in the molecule are 1.35 (C—O), 1.48 (C—C), 1.62 (C=S) and 1.72 Å (Si—O), all very close to each other, so that a single peak with a shoulder is observed in the radial distribution curve [4]. However, in germyl monothioacetate (8.V) the germanium atom is bound to sulfur, and that small change makes an enormous difference to the electron diffraction study [5]. Now the main bond lengths are 1.22 (C=O), 1.49 (C—C), 1.77 (C—S) and 2.23 Å (Ge—S), all well separated from each other, and easily determined. So peak overlap is a serious constraint on electron diffraction studies, but each molecule needs to be considered on its own merits. And it must be remembered that the overlap of two peaks does not necessarily mean that those two distances cannot be measured; the structure determined for the whole molecule depends on the analysis of all the data.

There is one further problem associated with electron diffraction, which arises from the fact that the atoms are not stationary. Any one electron 'sees' a molecule effectively frozen at one instant, but the total picture depends on millions of electrons, which 'see' molecules in different positions in the course of their vibrations. Each distance determined by electron diffraction is therefore averaged over the vibrations. Consider a molecule such as $HgCl_2$, undergoing a bending vibration, as shown in Fig. 8.9. In the average position, which is linear, the Cl···Cl distance is twice the Hg—Cl distance. But at all other positions, the Cl···Cl distance is less than twice the bond length, and so the average over the vibration is less than expected. This is known as a shrinkage effect, and if no correction is made for it (involving calculations which

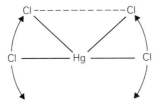

Fig. 8.9 Origin of the shrinkage effect. When a molecule such as $HgCl_2$ bends, the distance between the outer pair of atoms decreases. The Cl···Cl distance, averaged over the vibration, is thus less than twice the Hg—Cl distance.

CHAPTER 8: DIFFRACTION METHODS

require knowledge of the vibrational frequencies) one would conclude that the molecule was bent. For reasonably rigid molecules, such as CO_2, the effect is not very large, and apparent deviations of a few degrees from linearity are normal, but if the bending mode has a low frequency and large amplitude, the apparent angle may be as much as 40° away from the true average value. The effect is not limited to linear systems: large shrinkage effects can arise from any low-frequency modes, particularly torsions. It is always important to understand how reported interatomic distances have been averaged over molecular vibrations.

8.4   Diffraction by liquids

Studying structures of liquids is a slow and thankless task − you need to start when you are young if you want to give yourself time to get somewhere! For a start, there are very few techniques available. NMR studies can give valuable information about structures of simple compounds in solutions in liquid crystals (Section 2.13), but otherwise the diffraction methods and EXAFS (Section 8.9) are the only generally available techniques. It is not difficult to collect diffraction data. It can be done very easily using electrons, X-rays or neutrons. The hard part is knowing what to do with the data once they have been obtained.

The problem arises from the nature of the structures of liquids, which are not just super-dense gases. Certainly they do contain molecules in random positions, so that they give diffraction patterns consisting of diffuse rings, but there is also some preferred orientation of one molecule with respect to its neighbors. However, this ordering between molecules does not extend far, except perhaps very close to the freezing point, so that regular repetition of interatomic spacings, characteristic of crystalline solids, is not found. There are many theories of liquid structure, the most important of which are clearly and simply described in Ref. [6]. Perhaps the easiest model to visualize is that in which molecules have relationships to each other similar to those in solids, but with some sites unoccupied. The molecules are moving around rapidly, filling the gaps, and thereby creating others. Thus the shortest intermolecular distances are about the same as in the solid phase, but their multiplicities are relatively small. Longer and longer distances become more and more confused, and soon there is just a random distribution of distances, continuing indefinitely, such as is shown for a monoatomic liquid in Fig. 8.10.

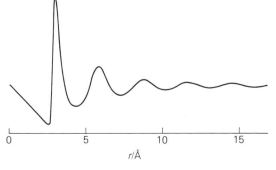

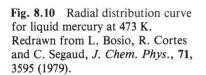

Fig. 8.10   Radial distribution curve for liquid mercury at 473 K. Redrawn from L. Bosio, R. Cortes and C. Segaud, *J. Chem. Phys.*, **71**, 3595 (1979).

Now consider what a radial distribution curve for a polyatomic molecule in the liquid phase or in solution would look like. It would have an intramolecular part, which would be exactly equivalent to the curve for the same molecule in the gas phase. Then there would be an intermolecular part, like Fig. 8.10, but more complicated, because there are many types of intermolecular contact to consider. The outcome is a curve such as Fig. 8.11. Unfortunately, we have arrived at an explanation of the figure, by starting with the answer! If we performed an experiment and arrived at a radial distribution curve, it would extremely difficult to analyze it, as the intramolecular and intermolecular parts may overlap. Useful information can be obtained for very small, symmetrical molecules, and sometimes, if the molecular structure is assumed (perhaps to be the same as in another phase), some information relating to the ordering of the molecules can be extracted. But as a general method of determining the structures of molecules in liquids, it is just not practicable. For species in solution, of course, the situation is even worse than for pure liquids, as the many solvent molecules all contribute to the total scattering pattern. EXAFS has a major advantage over diffraction methods in the study of liquids and solution, as it enables us to obtain curves representing the distribution of neighboring atoms round central atoms of a particular element. This simplification of the radial distribution curve can be invaluable, particularly in the study of metal-containing species. The technique is described in Section 8.9.

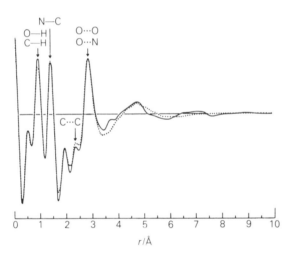

Fig. 8.11 Radial distribution curve for a 10:1 molar mixture of water and trimethylamine, a composition which gives a crystalline hydrate on cooling. The curve shows peaks corresponding to bonded and non-bonded intramolecular contacts, such as C—O, O—H, N—C and C···C, and intermolecular contacts, particularly between oxygen atoms of one water molecule and oxygen or nitrogen atoms of adjacent molecules. The solid line represents experimental data, and the dotted line shows values calculated from a theoretical model of the structure. Adapted, with permission, from C. Folzer, R.W. Hendricks and A.H. Narten, *J. Chem. Phys.*, **54**, 799 (1971).

So we are left with the situation that, by and large, we are ignorant of the structures of liquids and solutions. As the majority of chemical reactions occur in solution, and chemical reactivity is highly dependent on structure, this is unfortunate. But at present it is not possible to find out much about liquids. Instead we will look at solids, for which the situation is dramatically different.

8.5   Diffraction by single crystals

8.5.1   Principles

In discussing the diffraction of electrons by gases, we used the analogy of the double slit experiment in optics, which gives rise to an interference pattern. In a crystal there are regular arrays of atoms, all of which can diffract electrons, X-rays or neutrons, so that the analogous optical experiment is one involving a diffraction grating. A grating has a series of parallel lines, evenly spaced, and so there is a cooperative effect, with the interference patterns for each pair of lines adding together to give an extended series of maxima and minima. A two-dimensional grating gives a much more complicated pattern, from which it is possible to deduce the spacings between the two sets of rulings, and the angle between them. In the same way, the diffraction pattern produced by a crystal shows the effects of cooperative scattering by a regular array of centers, and the repeat distances can be determined. Of course, a crystal is more complicated than a grating, as it is a three-dimensional array, which can be placed in different orientations relative to the incident beam. It may also have a lot of atoms within the repeating unit, and it is the geometrical relationships between those atoms that we want to investigate.

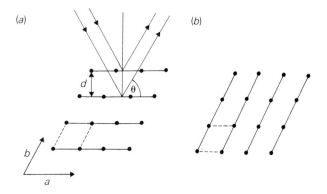

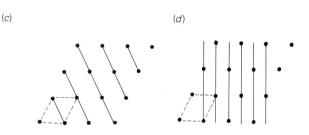

Fig. 8.12 Lines of atoms in a two-dimensional lattice. (*a*) reflection from (01) planes, (*b*) (10) planes, (*c*) (11) planes and (*d*) (21) planes. The unit cell is marked with dotted lines.

In practice, it is usual to think in terms of reflecting planes, rather than a three-dimensional diffraction grating. Consider the two-dimensional array of atoms in Fig. 8.12(*a*) to be one layer of a three-dimensional array. The incident beam is reflected, partly from the top row, and partly from the second row. For constructive interference to occur, the path difference must be a multiple of the wavelength, as expressed by Bragg's Law [Eqn (8.4)], where d is the distance between the planes.

$$n\lambda = 2d\sin\theta \tag{8.4}$$

But there are other planes of atoms which can also give reflections. In Fig. 8.12, which just represents atoms in a two-dimensional matrix, (b) shows the lines joining adjacent atoms in different rows. These lines would give the reflection with indices 1 and 0, with the unit cell (the repeating unit of the structure) defined as shown in the Figure, whereas the first one shown was the (01) reflection. In (c), the lines join atoms which are shifted one step in each direction and so give the (11) reflection, while (d) shows the (21) lines. Thus there is a whole set of lines, each described by two indices for the two-dimensional array. For a three-dimensional lattice, there are planes, each described by three constants known as the Miller indices, h, k and l, and each group of parallel planes gives reflections. The Bragg Law, defining the condition for constructive interference, can be written as $\lambda = 2d_{hkl}\sin\theta$.

Thus it can be seen that if a beam of X-rays (or neutrons or electrons) is directed at a crystal consisting of a regular array of atoms such as that in Fig. 8.12, a pattern of reflections can be recorded. The pattern will depend on the orientation of the crystal with respect to the incident beam, but by changing the orientation it should be possible to observe all the reflections in turn. However, with the exception of crystals of some elements, most crystals are not like Fig. 8.12 at all, as they contain many different atoms. Each atom type has the same repeating pattern, and the types all give rise to the same sets of reflections at the same angles. But because the patterns for the various types of atom do not exactly coincide, the diffracted beams are not in phase. It follows that the total intensity of a reflection depends on how far apart the various lattices are, or in other words, on the positions of the atoms in the cell. The square root of the total intensity of a reflection from a particular set of planes with Miller indices h, k and l is called a structure amplitude, $|F_{hkl}|$. It depends on the fractional coordinates of the atoms, x, y and z, and their scattering powers, f, and also on parameters describing the vibrational movement of the atoms in the lattice. Assuming that each atom i moves isotropically, with mean square amplitude of vibration U_i, we can define quantities A and B by Eqns (8.5) and (8.6).

$$A_{hkl} = \sum_i f_i \cos[2\pi(hx_i + ky_i + lz_i)][\exp(-8\pi^2 U_i\sin^2\theta/\lambda^2) \tag{8.5}$$

$$B_{hkl} = \sum_i f_i \sin[2\pi(hx_i + ky_i + lz_i)]\exp(-8\pi^2 U_i\sin^2\theta/\lambda^2) \tag{8.6}$$

Here 2θ is the diffraction angle as shown in Fig. 8.12, λ is the wavelength of the X-rays, and the summation is over all atoms in the unit cell. Then the amplitude (F) and phase (ϕ) of the structure factor are given by:

$$|F_{hkl}| = \sqrt{(A^2 + B^2)} \tag{8.7}$$

$$\phi_{hkl} = \tan^{-1}(B/A). \tag{8.8}$$

For anisotropic motions these expressions are more complicated.

So the object of a crystallographic experiment is to measure the intensities accurately for as many reflections as possible, and then to refine the coordinates and vibrational parameters for all the atoms, to fit the observed structure amplitudes. We cannot

CHAPTER 8: DIFFRACTION METHODS

measure the phases, and deducing them is a major problem in analyzing diffraction data, to which we will return at the end of this section.

The diffraction experiment, like many others, requires a source, a sample and a detector. Here we do not need to go into details which can be found in any book on crystallography (see the suggestions for further reading at the end of this chapter). The usual laboratory source consists of a metal anode (normally copper or molybdenum) which emits X-rays when irradiated with high-energy electrons. A monochromator selects radiation of the required wavelength (normally 1.542 Å for a copper radiation, or 0.711 Å for molybdenum), and the beam passes through a long, thin collimator tube, as X-rays cannot be focused. High intensity X-rays, with a wide range of wavelengths, can be obtained from a storage ring or synchrotron. Intense sources of thermal neutrons are not easy to come by: the neutrons produced in a nuclear reactor can be used, but after selection of energy and direction the resultant flux is inevitably low. A few sources built specifically to provide neutrons do exist.

The sample is a single crystal, typically about 0.3 mm in each direction for X-ray work, with a minimum of about 0.1 mm, but larger for neutron diffraction. Smaller crystals can be studied using an intense X-ray source, such as a storage ring. The crystal must be mounted on a goniometer head, so that its position and orientation can be changed. The recorder is either a camera, or a detector and counter, and various ingenious devices are used to ensure that each reflection is measured separately, with no overlap, and as accurately as possible. With a modern four-circle diffractometer the whole sequence of measuring reflections is entirely automatic, once the initial setting-up procedures have been completed.

Fig. 8.13 Different arrangements of a molecule A—B in a one-dimensional crystal. The vertical bars mark the unit cells, and the dotted lines indicate division of a cell into asymmetric units. The arrangements (a) and (c) have the same symmetry, as do (b) and (d).

We have seen that the reflection angles describing a diffraction pattern from a crystal enable the repeat distances to be determined, and that the intensities of reflections depend on the positions of atoms within the lattice. We also need to consider the effects of symmetry in the crystal. To illustrate the concepts involved, we will use a one-dimensional model, and a diatomic molecule A—B. Fig. 8.13 shows various ways in which the molecules could be arranged in a one-dimensional crystal, some more likely than others. In (a), the simplest arrangement, the repeating unit includes just one molecule. In (b), the repeat involves two molecules, and so we say that the unit cell is about twice as long as in (a). However, there is symmetry within the cell, and so the asymmetric unit only contains two atoms (the other two then being generated by symmetry), and solving the structure involves determining the coordinates of just the two atoms. The arrangement (c) has the same symmetry as (a) (there is no symmetry within the cell), but the cell is about three times as large, and six atoms need to be located, while (d) has the same symmetry as (b), but has four molecules in the unit cell, and two in the asymmetric unit. This demonstrates, in a very trivial way, some important concepts, such as the ideas of unit cells and asymmetric units, and also of space groups. Arrangements (a) and (c) belong to the same space group (no symmetry in the cell) while (b) and (d) belong to another (inversion symmetry in the cell, and therefore also between cells). In one dimension, the cell is described by just one parameter, its length, and there are only two possible space groups. In two dimensions, a cell is described by two lengths and an angle, as shown in Fig. 8.14(a). As it is possible to

(a)

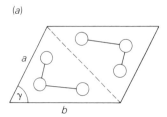

(b)

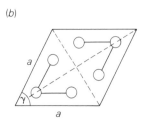

Fig. 8.14 Two-dimensional unit cell, described by two lengths, a and b, and the angle γ. In (a) the two molecules in the cell are related by an inversion center, and so there are two asymmetric units in the cell, separated by the dotted line. In (b) the two sides of the cell are of equal length, and the contents of the cell now have two planes of symmetry. There are four asymmetric units in the cell, each containing half a molecule. One atom of each molecule must lie on a mirror plane.

pack rectangles, triangles, squares or hexagons, but not pentagons, heptagons *etc.*, in two-dimensional space, there can be 2-fold, 3-fold, 4-fold or 6-fold rotational symmetry. Altogether there are 17 possible space groups, which is important if you are a wallpaper designer. In three dimensions there are 230 space groups [7], which is important if you are a crystallographer. (There are 4895 in four dimensions which doesn't seem to be important at all! In fact they are used in the description of magnetic structures.) A three-dimensional unit cell is described by three lengths and three angles. But a cell, and its contents, may also have symmetry. For example, Fig. 8.14(b) shows a two-dimensional cell with all its sides equal, and a cell could also be rectangular, or square. Table 8.1 lists the seven possible crystal systems for three dimensions, and the relationships between their axes and angles.

The symmetry properties of the crystal obviously have an important bearing on the pattern of reflections. A crystal with cubic symmetry, for example, will have identical intensities for reflections from the (100), ($\bar{1}$00), (010), (0$\bar{1}$0), (001) and (00$\bar{1}$) planes, while a general reflection will have 48 equivalents. So it is only necessary to measure

Table 8.1 Coordinate systems

Coordinate system	Axes and angles
Triclinic	$a \neq b \neq c, \ \alpha \neq \beta \neq \gamma$
Monoclinic	$a \neq b \neq c, \ \alpha = \gamma = 90° \neq \beta$
Orthorhombic	$a \neq b \neq c, \ \alpha = \beta = \gamma = 90°$
Tetragonal	$a = b \neq c, \ \alpha = \beta = \gamma = 90°$
Rhombohedral	$a = b = c, \ \alpha = \beta = \gamma \neq 90°$
Hexagonal	$a = b \neq c, \ \alpha = \beta = 90°, \ \gamma = 120°$
Cubic	$a = b = c, \ \alpha = \beta = \gamma = 90°$

intensities for relatively few reflections. If there is translational symmetry i.e. glide planes and/or screw axes (see Appendix), the intensities of some reflections may be zero, and from the distribution of these systematic absences it is possible to identify the space group, or at least to limit the number of possibilities. The determination of symmetry properties and measurement of cell parameters are carried out at an early stage of the structural study and are very important, as any errors are carried over into all subsequent calculations. In particular, correct identification of the space group is critical, as any mistake here can introduce errors in structural parameters which are many times greater than random errors normally found in the analysis. It has been suggested that as many as 10% of published structures are based on incorrectly assigned space groups. This is probably an over-pessimistic estimate, but it does emphasize the need for care, both in doing the work, and in assessing the published data.

Once the intensities of reflections have been measured and various corrections applied for absorption, extinction *etc.*, it should be possible to reconstruct a map of electron density in the cell (assuming that X-rays have been used). Unfortunately, there is a major stumbling block here. The structure factors are complex quantities corresponding to sums of wave motions, and therefore have both amplitude and phase [Eqns (8.7) and (8.8)]. But the detectors used (counters or photographic plates) measure intensities integrated over a period of time, and so all we can obtain are the moduli of the structure factors, $|F_{hkl}|$. The problem therefore is to deduce phases as well as amplitudes. Various methods are used to achieve this end and thereby solve the structure.

One method is to calculate Patterson maps, which represent all interatomic vectors in the cell. A Patterson function is therefore the three-dimensional equivalent of a radial distribution curve, which we encountered in connection with one-dimensional diffraction data, obtained for gases [Section 8.3.1]. For simple structures, it is possible to work from this to the arrangement of the atoms, particularly if one atom is much heavier than the others so that the vectors involving that atom stand out clearly. Thus the 'heavy-atom' method of solving structures is very useful, particularly for transition metal complexes, which often have one atom much heavier than the rest.

An alternative strategy is to use so-called direct methods [8], which are based on two fundamental physical principles. First, the electron density in the unit cell cannot be negative at any point, and so the large majority of possible sets of values for the phases of the various structure factors are not allowed. A random set of phases would give positive and negative regions in the electron density map with equal probability. Secondly, the electron density in the cell is not randomly distributed, but is mainly

Example 8.2

The mineral cooperite (PtS) has tetragonal crystals with $a = b = 3.47$ Å, $c = 6.10$ Å. The fractional coordinates of the atoms in the cell are:

Pt: $0, \frac{1}{2}, 0$ and $\frac{1}{2}, 0, \frac{1}{2}$.
S: $0, 0, \frac{1}{4}$ and $0, 0, \frac{3}{4}$.

By drawing appropriate projections, show the full environment of one platinum and one sulfur atom, and describe the coordination of each giving the site symmetry. Calculate the shortest Pt–S distance.

Projections down the a and c axes are shown in the figure.

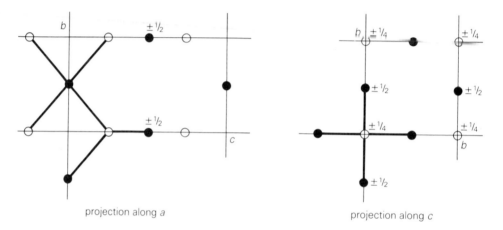

| projection along a | projection along c |

Sulfur atoms are marked as hollow circles and platinum atoms as solid circles. The atoms lie in the plane, unless they are marked otherwise: $\pm\frac{1}{2}$ means that atoms lie half a cell above and below the point indicated.

The projection down a shows that each platinum atom has four nearest neighbor sulfur atoms, in an arrangment that is nearly, but not exactly, square planar. Two S$\cdots$S distances are 3.47 Å, while the other two are half of c, i.e. 3.05 Å, and the site symmetry of these atoms is therefore D_{2h}. The sulfur atoms also have a coordination number of four. One atom is shown with its four nearest neighbors in each projection. One of these illustrates one of the mirror planes of an SPt$_4$ group, while the other includes a view along a two-fold axis. The sulfur atom has approximately tetrahedral coordination, but the tetrahedron is not regular, and the site symmetry is D_{2d}. The structure thus combines square planar and tetrahedrally coordinated atoms, in a very simple arrangement.

The shortest distance between Pt and S atoms is the bonded distance. It is given by

$$r(\text{Pt–S})^2 = (b/2)^2 + (c/4)^2$$

$$= 1.735^2 + 1.525^2$$

i.e. $r(\text{Pt–S}) = 2.310$ Å

concentrated in small areas which we identify as atoms. A consequence of these two principles is that for certain sets of reflections (usually groups of three) having particular combinations of Miller indices, there are theoretical probability relationships among their phases. Thus phases can be assigned to some reflections, usually the most intense, and then the positions of some or all of the heaviest atoms can be located. Of course, once some atoms have been located, whether by direct methods or using Patterson maps, difference maps show up areas of electron density arising from other atoms, and so the whole structure can be solved. The difficult part is nearly always in getting started. Once atoms are in more-or-less the right places, least squares refinements adjust coordinates and vibrational parameters. A final difference map should show no significant regions of electron density unaccounted for.

8.5.2 Interpretation of results

The intensities of reflections observed in an X-ray diffraction study of a single crystal depend on the distribution of electrons within the crystal. Provided the phases as well as the amplitudes of the structure factors can be deduced, a three-dimensional Fourier transformation can give a three-dimensional map of electron density within the unit cell. These maps can be presented as a series of cross-sections, or as projections onto a plane. Fig. 8.15 shows such a projection, dating from 1940, for platinum phthalocyanine (8.VI), which has all its atoms lying in one plane. The scattering is very much dominated by the platinum, and so there are some distortions in its vicinity, but nevertheless the positions of the carbon and nitrogen atoms are immediately obvious.

The aim of structure refinement is to model the experimental electron density distribution as accurately as possible. Once an atom has been placed in approximately the right place, up to nine parameters can be refined to describe it. There are three coordinates defining its position and six vibrational parameters (known as temperature factors) which together define the extent and direction of its movement, by describing an ellipsoid of motion. It is clear from Fig. 8.15 that all these parameters will be more accurately determined for the heavy platinum atom which has 74 electrons, than for the lighter carbon and nitrogen atoms, which only have six or seven. In this case it is not possible to refine hydrogen atom parameters at all, although there are suggestions of electron density around the outer carbon atoms. For some atoms, it is only possible to refine a single vibrational parameter; it is assumed that the movements of an atom are equal in all directions, as described by an isotropic temperature factor.

Fig. 8.16 shows a structure determined by the heavy-atom method [9], one of thousands reported each year. In this diagram most atoms are represented by ellipsoids so that the amplitudes and directions of motion, as well as their positions, can be seen. The movement of the oxygen atom O(2) is seen to be mainly perpendicular to the Rh(2)—C(2) bond, whereas the rhodium atoms and the carbon atoms move relatively little and almost isotropically. The hydrogen atoms of the allene group are shown as circles, as anisotropic temperature factors were not obtained for these, although their positions were refined. Other hydrogen atoms are not depicted at all, for clarity. Their positions were not refined, but they were included in the model, set at fixed distances from their adjacent carbon atoms.

Refinement of structures gives atomic coordinates (and vibrational parameters) with their estimated standard deviations. From these, interatomic distances and angles can be obtained, and it is instructive to look at a few of these, as this is a fairly typical

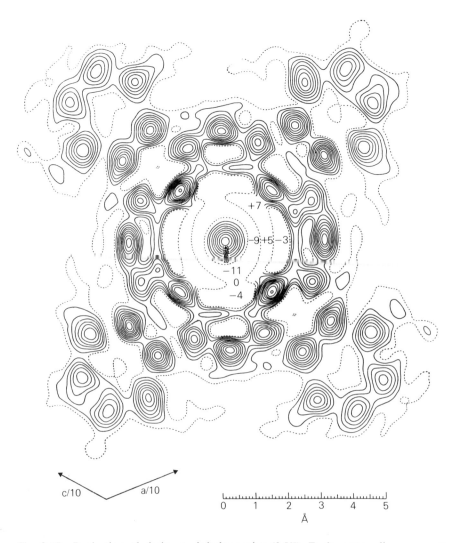

Fig. 8.15 Projection of platinum phthalocyanine (8.VI). Each contour line represents a density increment of one electron per Å^2 except in the central portion where the increment is 20 electrons per Å^2. The one-electron line is dotted. Taken, with permission, from J.M. Robertson and I. Woodward, *J. Chem. Soc.*, 36 (1940).

8.VI

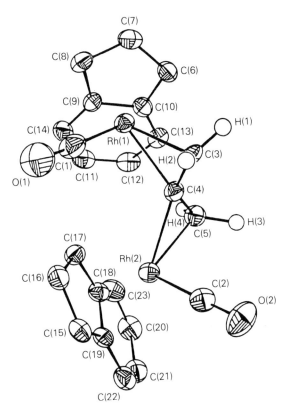

Fig. 8.16 Molecular structure, as determined by X-ray diffraction, of μ-allene-dicarbonyl bis(η^5-indenyl)dirhodium. The non-hydrogen atoms are represented by 50% probability ellipsoids. Some hydrogen atoms are shown as circles, and others are omitted. Redrawn from Ref. [9].

structure determination. The Rh—Rh distance is precisely determined at 3.711 ± 0.001 Å. The Rh—C distances have estimated standard deviations of 0.002 or 0.003 Å while the errors are 0.003 or 0.004 Å for C—C and C—O distances, but 0.03 to 0.05 Å for C—H distances. Thus the correlation between accuracy of determined parameters and the atomic numbers of the atoms involved is clear. Moreover, when X-rays are used, the C—H distances are not very useful, because we are measuring distances between centers of electron density, and for hydrogen bonded to another atom, this is very much less than the internuclear distance. Similarly, uncertainties are 0.1° for Rh—C—Rh and Rh—C—C angles, 0.2° for C—C—C angles, but about 2 to 3° for C—C—H angles.

Such clear-cut results are invaluable. Not only is the compound unequivocally identified, but some interesting parameters of chemical importance emerge. For example, in this case the coordinated allene molecule is shown to be substantially bent [the angle C(3)—C(4)—C(5) is 137.8 ± 0.2°], whereas free allene has a linear skeleton. The hydrogen atoms are also bent back, away from the rhodium atoms. Observation of such structural features leads to understanding of the mode of coordination of the ligand, which in turn points the way for further synthetic work. The only unsatisfactory feature is the uncertainty in locating the hydrogen atoms.

The structure shown in Fig. 8.17 was solved by direct methods. The ten silicon and phosphorus atoms were located immediately, and a difference map showed the presence of the 28 carbon atoms. When their positions had been refined, all but three of the hydrogen atoms could be located immediately and the rest were found later. Thus the structure was determined more or less automatically, without need for preconceived

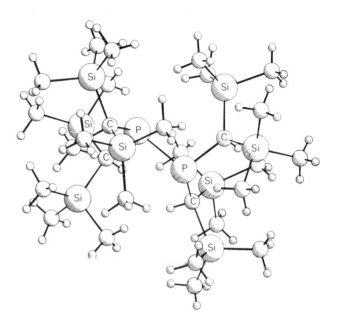

Fig. 8.17 Molecular structure, as determined by X-ray diffraction, of a tetra-alkylbiphosphine. The alkyl groups are —CH(SiMe$_3$)$_2$.

ideas about where the atoms should be or what bond connections there were. In this respect, three-dimensional data are much superior to the one-dimensional data obtained in studies of gases, as in the latter cases models of structures are needed to start the refinement process. However, the more that is known about the structure, the better direct methods work. Actually, some input is needed in the present case. We have assumed that the two atoms at the middle of the molecule are phosphorus and that the eight outer ones are silicon. As they have 15 and 14 electrons respectively, it would be possible to get them confused. Of course, it is obvious which is which, on chemical grounds, but that is not always so. For example, it is often impossible to distinguish carbonyl and nitrosyl ligands when both are present.

The structure shown in Fig. 8.17 is of interest, in that it shows the effects of the bulky substituents in several ways. The P—P, P—C and some of the Si—C bonds are unusually long, and angles at the phosphorus, silicon and inner carbon atoms are all substantially distorted from the normal arrangements. The estimated standard deviations are about 0.003 Å for P—C and Si—C bond lengths and 0.1 to 0.2° for C—Si—C angles. As there are no very heavy atoms, the hydrogen atom positions are quite well determined, with uncertainties of about 0.03 Å for C—H bond distances and 2° for Si—C—H angles.

These two examples which we have looked at in some detail show how routine crystallographic studies can give invaluable results, both in identifying unknown compounds and in explaining their properties. It is also possible to determine the absolute configuration of an optically active molecule, provided it crystallizes in a space group which has no glide or mirror planes or inversion centers. In general, the relative scattering factors of atoms are independent of the wavelength of the X-rays, but this is not true for wavelengths near to an absorption edge for a particular atom. This anomalous dispersion can be used to ascertain the absolute configuration of molecules. The process involves comparing observed and calculated intensities for a

Example 8.3

The complex $\{Os_3PtH_4(CO)_{10}[P(cyclo-C_6H_{11})_3]\}$ forms orthorhombic crystals, with cell dimensions a, 20.85 Å; b, 36.25 Å, and c, 18.35 Å. Its density is 2.550 Mg m^{-3}. How many molecules are there in the unit cell? Fractional coordinates of the metal atoms in the asymmetric unit are listed below. Calculate the bond distances between metal atoms, and deduce and comment on the shape of the metal core of the molecule.

	x/a	y/b	z/c		x/a	y/b	z/c
Os(1)	0.5918	0.1254	− 0.0205	Os(5)	0.4072	0.1278	0.7426
Os(2)	0.6686	0.0632	0.0432	Os(6)	0.3248	0.0691	0.8088
Os(3)	0.6785	0.0786	− 0.1098	Os(7)	0.3282	0.0763	0.6536
Pt(4)	0.7206	0.1363	− 0.0086	Pt(8)	0.2787	0.1390	0.7381

....

The volume of the unit cell is $20.85 \times 36.25 \times 18.35$ Å^3 = 1.3869×10^{-26} m^3. The molecular weight of the complex is 1330. The number of molecules per unit cell is given by (cell volume × density)/(molecular weight/Avogadro's number) = $(1.3869 \times 10^{-26} \times 2.55 \times 10^6)/(1330/6.02 \times 10^{23})$, i.e. 16.

As the crystals are orthorhombic, the fractional coordinates can be converted to Cartesian coordinates simply by multiplying by the appropriate cell dimensions to give

	$x/$Å	$y/$Å	$z/$Å		$x/$Å	$y/$Å	$z/$Å
Os(1)	12.339	4.546	− 0.376	Os(5)	8.490	4.633	13.627
Os(2)	13.940	2.291	0.793	Os(6)	6.772	2.505	14.841
Os(3)	14.147	2.849	− 2.015	Os(7)	6.843	2.766	11.994
Pt(4)	15.025	4.941	− 0.158	Pt(8)	5.811	5.039	13.544

The two groups of atoms are obviously well separated from one another. Distances between metal atoms within a group can then be calculated as follows:
$r[Os(1) - Os(2)] = \sqrt{[(13.940 - 12.339)^2 + (2.291 - 4.546)^2 + (0.793 + 0.376)^2]} = 3.003$ Å.
Similarly we obtain Os(1) – Os(2) 3.003; Os(1) – Os(3) 2.972; Os(2) – Os(3) 2.870; Os(1) – Pt(4) 2.724; Os(2) – Pt(4) 3.017; Os(3) – Pt(4) 2.932; Os(5) – Os(6) 2.992; Os(5) – Os(7) 2.977; Os(6) – Os(7) 2.860; Os(5) – Pt(8) 2.711; Os(6) – Pt(8) 3.005 and Os(7) – Pt(8) 2.938 Å.

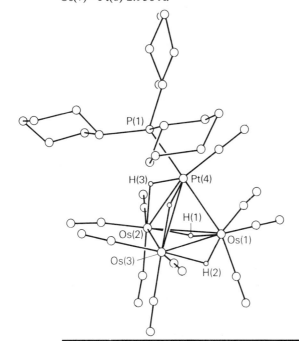

Within each group of four atoms the distances are roughly equal, so the metal core is a slightly distorted tetrahedron. The mean Os—Os bond length (2.945 Å) is slightly longer than the mean Os—Pt length (2.887 Å), as expected, but in each cluster one Os—Pt bond is distinctly shorter than the other two, while one Os—Os bond, between the other two osmium atoms, is also short. This is consistent with the hydrogen atoms bridging the four longest edges of the tetrahedron, as shown in the figure. This is adapted, with permission, from L.J. Farrugia, M. Green, D.R. Hankey, M. Murray, A.G. Orpen and F.G.A. Stone, *J. Chem. Soc., Dalton Trans.*, 177 (1985), which should be consulted for more information.

series of reflections F_{hkl} and $F_{\bar{h}\bar{k}\bar{l}}$. Alternatively, the whole set of intensities can be calculated, for both possible configurations, and the quality of fit of theory and experiment can be compared for the two cases.

Although crystallography is convenient and has become very straightforward and rapid in application, it does have limitations. In the next section we will consider some of these, as it is by finding out about the circumstances in which the method is less than ideal that we become aware of how useful it is. But even when a structure can be determined, we must realize that the structure is that of a crystalline solid and that most chemical reactions occur in solution or liquid phases. It is therefore important to look at the intermolecular contacts in the crystal, something completely ignored in drawing the pictures of molecules shown in Figs 8.16 and 8.17. Fig. 8.18 shows one layer of the structure of silyl acetate [10]. There are no strong interactions between layers, but within each layer there are chains of molecules, with fairly short contacts between a silicon atom in one molecule and an oxygen atom in the next. The effect of these interactions is quite substantial: bond lengths in the gas phase differ by up to 0.046 Å from those in the solid phase, and there are changes of up to 4.4° in bond angles, but in both phases there are short intramolecular Si···O contacts. Thus it is always important to be aware that the crystal structure is that and no more. Other phases may be quite different. The biphosphine shown in Fig. 8.17 dissociates in solution or in the gas phase, to give stable dialkylphosphenyl radicals, and the solid structure is then of no relevance at all.

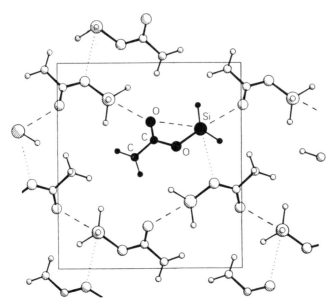

Fig. 8.18 Crystal structure of silyl acetate, $CH_3COOSiH_3$, showing one layer of molecules. The molecules are linked in chains, with fairly short Si···O interactions. These, and similar short intramolecular interactions, are shown as dashed lines. Some weaker contacts between chains are marked as dotted lines.

8.5.3 Limitations

Before embarking on a discussion of the limitations of single-crystal diffraction studies we should emphasize again the positive side. Although this book concentrates on the structural properties of molecules and polyatomic ions, crystallography involves the solid state in general. Thus it can be applied not only to molecules great and small, from diatomics to viruses, but to minerals, salts, polymers and metals. The range of applications is immense, but it is not universal, and we must therefore explore the limits.

The first and most obvious restriction of single-crystal diffraction studies is to compounds which form suitable single crystals. There are very many which do not, for various reasons. Some do not crystallize at all, but give glasses. Other form only very small crystals, while others form twinned crystals. Some are susceptible to radiation damage. Some only crystallize at low temperatures, which is just inconvenient, but if on cooling they undergo another phase change, and shatter, that may make crystallographic study impossible. Such compounds can only be studied as powders (Section 8.6) or glasses (which are the same as liquids so far as diffraction is concerned) or by non-diffraction methods. Perhaps worst of all, the few good crystals in a sample sometimes turn out to be an impurity.

If a compound does crystallize satisfactorily, there are still pitfalls and difficulties that may arise. X-ray data relate to electron density distribution in the crystal, and provided atoms of about the right atomic number are put into the correct positions, a reasonable fit to the data will be obtained. So if one starts off by making the wrong assumptions, it may be that they will not be challenged. It has been suggested that a reported structure for $ClF_6^+ CuF_4^-$ is really that of $[Cu(H_2O)_4]^{2+} SiF_6^{2-}$. The chemical difference is enormous, but the electron distributions are not very different: the ions ClF_6^+ and SiF_6^{2-} are actually isoelectronic and $[Cu(H_2O)_4]^{2+}$ has just one electron more than CuF_4^-.

A further potential source of trouble arises when there is nearly, but not quite, an element of symmetry in the crystal. Consider the arrangement of atoms in the two-dimensional unit cell shown in Fig. 8.19(a). The two molecules are related by an

(a)

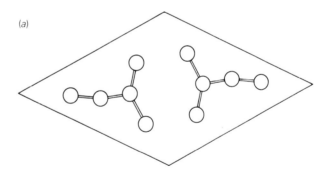

(b)

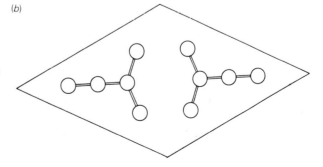

Fig. 8.19 (a) Molecules having no internal symmetry in a two-dimensional unit cell. The two molecules are related by a center of inversion. (b) If it is assumed that there are planes of symmetry in the cell, the positions of atoms must be changed.

SECTION 8.5: DIFFRACTION BY SINGLE CRYSTALS   357 inversion center, but in addition there is almost a plane of symmetry bisecting the molecules. It would be very easy to assign the wrong space group and to end up with parameters which are in error by substantial amounts. In particular, the three-atom chain would be forced to be exactly linear, whereas it ought to be significantly bent.

The reverse situation can also arise. If the unit cell really did have the extra symmetry, as shown in Fig. 8.19(b), and refinements were performed without enforcing this, then the atom coordinates would be free to move a little out of line. Such relaxation of symmetry requirements can only lead to a better fit to the data, and careful statistical testing may be necessary to see whether the changes are significant, and so to be sure that mistakes are not made. To err either way is serious. But it is not always easy to be sure what the truth is. Working from systematic absences of reflections does not always give an unambiguous answer about the space group. Crucial reflections may be weak if there is approximate symmetry, and spots may appear in places where they should not be seen, arising from multiple reflections or some radiation in the wrong wavelength from the X-ray source.

(a)

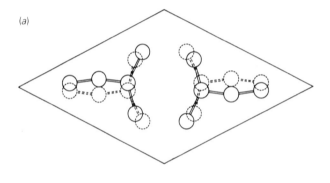

(b)

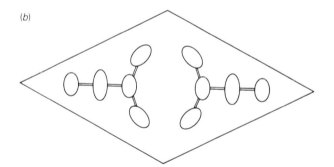

Fig. 8.20 (a) Disordered molecules in a two-dimensional unit cell. One set of positions is represented by solid lines, and the other by dashed lines. (b) If the disorder is ignored, the atoms will appear to be in average positions, with artificially large temperature factors.

A further problem that is often found in analysis of crystallographic diffraction data is that of disorder. Suppose that the molecules shown in Fig. 8.19 have the less symmetric structure (a), but that some of these lie in one position in the cell, and others in the mirror image position, as shown in Fig. 8.20(a). If the overall distribution of the molecules is random, the apparent symmetry will be the same as in Fig. 8.19(b). Refinement with this symmetry imposed would give the results shown in Fig. 8.20(b). Provided the 'half-atoms' are not too far away from each other they appear as a single atom, with an enhanced thermal parameter along the vector joining them. Clearly, the

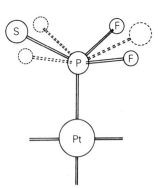

Fig. 8.21 Disorder in a platinum complex with a —PF$_2$S ligand. One set of positions for the fluorine and sulfur atoms is represented by solid lines, and the other by dashed lines. The two phosphorus positions are so close that there appears to be one atom location only.

geometrical parameters describing the molecule in Fig. 8.20(*b*) are not those of the real individual molecules. Yet it can easily happen that disorder is not noticed and so errors can creep in.

In other circumstances disorder can be observed, but inadequately modelled. For example, the —PF$_2$S ligands in a platinum complex were found to be disordered, as shown in Fig. 8.21. It was easy to see six half-atoms, and to refine their positions. But the resulting bond lengths and angles seemed strange. The explanation was that the phosphorus atom must also be disordered, but the two sites were close and not clearly resolved. When this atom was also split between two sites, the geometry became much more reasonable. But what about the platinum atom and every other atom in the molecule? They must also be disordered to some extent. The trouble is that it is not possible to be sure how big the effects are, as they cannot be distinguished from vibrational effects. So even when the type of disorder is recognized, it can give rise to errors in positions of atoms which do not appear to be involved.

In each of the examples just considered, the two possible sites were equally populated, but that is not necessarily the case. In studying the structure of the iridium complex 8.VII it was found that there were two molecules in the asymmetric unit [11].

$$\begin{array}{c}\text{PEt}_3 \\ | \quad \text{CO} \\ \end{array}$$

Cl —— Ir —— P
Cl
PEt$_3$
O
Cl
Cl

8.VII

One of these was perfectly ordered, but the other showed two different forms of disorder. The conformation of the PCl$_2$ group was disordered, with rotation about the Ir—P bond, so that there were three chlorine positions, with occupancies of 0.8, 0.8 and 0.4. Then it was found that there was also disorder involving the carbonyl group and the chlorine atom *trans* to it, with some of the molecules having one arrangement and the rest having their positions interchanged. When such problems arise, it is very difficult indeed to obtain accurate parameters. In this particular case, that did not matter much, as there was another, ordered molecule present. But in other circumstances it may mean that the structural study cannot give reliable geometrical information.

8.6 Diffraction by powders

When single crystals are being studied, it is essential to determine the exact orientation of the crystal axes with respect to the beam of X-rays or neutrons, so that data collection, indexing *etc.* can be carried out properly. In a sample of a powder there are many very small crystals, randomly orientated, and each crystallite can diffract in the normal way. Any one reflection has a particular Bragg angle, and the whole assembly of crystals will give cones of reflection, which are observed as circles. As each reflection is now described simply by one angle, and its intensity, there is inevitably loss of information compared with what can be obtained from single crystal data. Reflections may be superimposed, or at least unresolved, and it is usually very difficult indeed to index a powder pattern. It must be done by trial and error, unless the unit cell dimensions are known beforehand. Indeed, a series of possible cells may thus be found, and these must be evaluated critically. However, once indexing has been done, the reflections can be used to obtain accurate cell dimensions, which can be of importance in connection with analysis of single crystal data.

As the pattern of lines arising from a powder depends on the cell dimensions and symmetry, and line intensities depend on the atom positions, powder diffraction photographs provide excellent fingerprints for crystalline substances. Files of patterns for thousands of elements and compounds exist and unknown materials can often be identified by simple comparison. Even components of mixtures can be recognized, as the patterns are fairly simple. Powder methods are used widely, as for example in identifying minerals. If it is thought to be possible that a material in an old building contains asbestos, a powder diffraction photograph will show what is present very easily and quickly, and so it can be seen whether there is a potential health hazard. Powder data can also be very useful in identifying solid products of reactions. It is too easy to say, for example, that 'a solid product, presumed to be ammonium chloride, was formed'. With a simple experiment, the presumption can be replaced by an observation.

So far we have considered the powder patterns, but said little about the intensities. Yet those intensities give information about atom positions within the unit cell, just as they do for single crystal data. Of course, it is very difficult to solve a complicated structure using just powder data, but if a structure is partly known, or if it is fairly simple, then useful results can be obtained. Fig. 8.22 shows powder data for neutron scattering by a low-temperature phase of ammonium nitrate. This can only be obtained as a powder, as several phase changes that occur on cooling are associated with

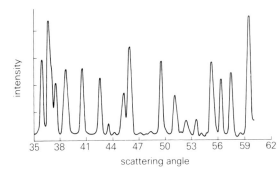

Fig. 8.22 Part of the neutron scattering profile for a low temperature phase of ammonium nitrate powder. The curve is a calculated profile, which fits the experimental data. Adapted, with permission, from C.S. Choi and H.J. Prask, *Acta Crystallogr., Sect. B*, **39**, 414 (1983).

substantial expansions and contractions, which destroy the crystals. The full three-dimensional structure was obtained by fitting the intensity profile data, and eventually gave N—O bondlengths and O—N—O angles with estimated standard deviations of 0.004 to 0.009 Å and 0.2 to 0.5°, respectively. These errors are just two to three times larger than would be expected in a normal single crystal study, and so the results are quite respectable. In general, it is better and easier to work with single crystals, if suitable ones are available. But powder neutron diffraction studies are regularly reported, as large samples are needed for neutron work, and it is not always possible to prepare large enough single crystals. Powder X-ray diffraction studies leading to full structure determinations are less frequently reported, but they can be important, for example in studies of structure changes associated with phase changes. In general, however, determinations of new structures using powder data alone are rare: refinements, based on known or partially known structures, are more common.

8.7 Neutron diffraction

Although both X-ray and neutron diffraction are good techniques for the study of structures of crystalline substances, the vast majority of published crystal structures are based on X-ray data. The reason for this is that X-ray sources are readily available in the laboratory, whereas neutron sources are not. Neutrons having suitable velocities (and thus wavelengths) are usually obtained by moderating the velocities of fast neutrons formed in nuclear reactors, although purpose-built sources with higher fluxes are now available in a few places. The flux of a more-or-less monochromatic, collimated beam of neutrons is inevitably small, and large crystals (typically 1.5 to 5 mm in each dimension) and long counting times are needed. The experiments are therefore relatively time-consuming, and expensive. (It has been estimated that each neutron counted costs about one cent!) But nevertheless, neutron diffraction is important, because the information it provides is not exactly the same as that obtained by X-ray diffraction.

X-rays are scattered primarily by electrons, and the intensity of scattering by an atom depends on the square of its atomic number. In contrast, neutrons are scattered directly by atomic nuclei. The scattering for light atoms, such as hydrogen or carbon, is not very different from that for heavy atoms, such as second or third row transition metals (Fig. 8.2), and so it is possible to locate light atoms accurately, even when heavy ones are present. [The disadvantage of this is that structure solving is made more difficult. The compound $N[Si(CH_3)_3]_3$ has essentially 13 atoms to be found in an X-ray study, but 40 in a neutron study.] Moreover, neutron diffraction gives the positions of atomic nuclei, whereas X-ray diffraction gives the centers of electron density. For atoms with many electrons, these are more or less the same thing, but for light atoms, particularly hydrogen, they are quite different, as electrons are displaced relative to the nucleus into covalent bonds. So neutron diffraction must be used if hydrogen atom positions are required.

Many carboxylic acids exist as hydrogen-bonded dimers in the solid phase, and this is also true of 1,1′-ferrocenedicarboxylic acid (8.VIII), which has four hydrogen bonds joining each pair of molecules. A neutron diffraction study [12] at low temperature gave all the atom positions accurately, and all bond lengths, including those involving hydrogen, were determined with uncertainties of only 0.002 to 0.003 Å; estimated standard deviations of angles were 0.1 to 0.2°. All the O—H···O linkages were found to be very nearly linear, and there was an inverse correlation between the lengths of the

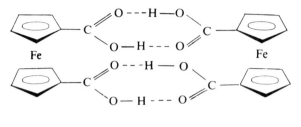

8.VIII

O—H bonds and those of the associated H···O hydrogen bonds. Thus the nature of the links between molecules could be clearly established. Similarly, in the neutron diffraction study of powdered ammonium nitrate discussed in Section 8.6, the hydrogen bonding involving ammonium hydrogen atoms and nitrate oxygen atoms was characterized. In that case the deuteriated compound, ND_4NO_3, was used. In neutron diffraction work deuterium is often preferred to hydrogen, which gives rise to a high level of inelastic scattering, and thereby produces a diffuse background.

The value of neutron diffraction in studying hydrogen bonding lies in its ability to give reliable estimates of the positions of the nuclei involved. X-ray studies of carboxylic acids often locate the hydrogen atoms, but the parameters obtained are of no great value. In other circumstances, X-ray diffraction simply cannot give any useful information at all about hydrogen atoms. For example, many transition-metal cluster compounds are known, and often these have hydrogen atoms bonded to the metals, and it is important to know how these groups of atoms are connected. Fig. 8.23 shows

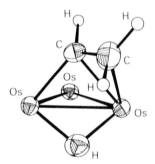

Fig. 8.23 Structure of decacarbonyl-μ-hydrido-μ-vinyl-*triangulo*-triosmium, omitting the carbonyl groups, as determined by a combined X-ray and neutron diffraction study. Redrawn from A.G. Orpen, D. Pippard, G.M. Sheldrick and K.D. Rouse, *Acta Crystallogr., Sect. B,* **34,** 2466 (1978).

the arrangement of the central core of atoms in such a complex. There are three osmium atoms, a bridging hydrogen atom, and a bridging vinyl group, which is formally σ-bonded to one metal atom, and π-bonded to another. There are also (not shown) ten carbonyl ligands, four bound to one osmium atom, and three on each of the others. This structure was determined using both X-ray and neutron diffraction data, so that information about the light atoms was obtained, but without the expense of collecting a very large neutron data set. The combination of data from two techniques in this way is often valuable. Single crystal X-ray data can also be combined with powder neutron data, to give results unobtainable by either method alone.

Finally, we should note that neutrons have spin, and so have magnetic moments. There is therefore an additional interaction with atoms having one or more unpaired electron spin, and magnetic scattering can be observed. If the magnetic moments of the neutrons are randomly orientated, as they usually are, this magnetic scattering appears as a diffuse background. But if a polarized beam of neutrons is used, and the magnetic moments in the crystal also tend to be aligned with each other, then the diffraction experiment can yield information about the magnetic structure of the crystal. This technique has been of particular value in studying ferromagnetic and antiferromagnetic materials. The principles and their applications are discussed in Ref. [13].

8.8   Study of electron density distributions

In the last section we emphasized that neutron and X-ray diffraction do not yield the same information. Neutron diffraction gives nuclear positions, whereas X-ray diffraction locates the centers of electron density. The difference between the two is therefore a measure of the extent to which the electron distribution in a crystal differs from that predicted using a model with spherical distributions for each atom, and each atom also being exactly neutral. This model is clearly false, as atoms in molecules normally carry some charge, and valence shell electrons are not spherically distributed, as they are concentrated into covalent bonds or lone pairs, orbitals which can be markedly directional, and so the difference between the spherical-atom model and the true distribution can be mapped.

Fig. 8.24 is an example of such a difference map. The positions of all the atoms in cyanuric acid were found by neutron diffraction, and the electron density distribution

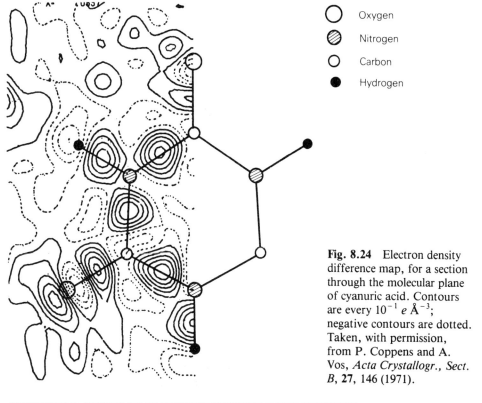

○ Oxygen
◍ Nitrogen
○ Carbon
● Hydrogen

Fig. 8.24 Electron density difference map, for a section through the molecular plane of cyanuric acid. Contours are every $10^{-1} e\ \text{Å}^{-3}$; negative contours are dotted. Taken, with permission, from P. Coppens and A. Vos, *Acta Crystallogr., Sect. B*, **27**, 146 (1971).

corresponding to these positions, with a spherical atom model, was calculated. The difference between this and the distribution calculated from X-ray data was then derived. A section through this, passing through all the atoms of the planar molecule, is shown. The map shows clear maxima in all bonds, C—O, C—N and N—H, and so the concentration of electrons between atoms in covalent bonds is clearly demonstrated. There are corresponding areas of reduced electron density in the vicinity of the nuclei. However, there is one other major region of high electron density, just outside one oxygen atom, and this must correspond to electrons in a non-bonding orbital located on this atom.

Such results are interesting and important, and it is possible to compare experimental results with those calculated theoretically. But it is inconvenient to have to collect high quality X-ray and neutron diffraction data sets. Moreover, roughly the same information can be obtained by using X-ray data alone, as the information relating to the distribution of the valence shell electrons is contained mainly in the low-angle reflections. The positions of nuclei can therefore be found using wide-angle data, and the deviation from spherical symmetry around neutral atoms (known as the deformation density) can then be found using the low-angle data. The non-spherical deformation density around each atom can be described as the linear combination of a number of localized functions, each of which is refined. By this method, known as multi-polar refinement, random noise in electron density maps can be filtered out, so that it is possible to make meaningful comparisons between experimental and theoretical deformation densities.

Fig. 8.25 shows a deformation density map for a section through dimolybdenum tetra-acetate (8.IX), passing through one molybdenum atom and the four oxygens bonded to it, and it can be seen that the map is dominated by electron density variations around the metal atom, and any deviation from spherical symmetry is masked. So in this case the net charges on molybdenum, carbon and oxygen were refined. The molybdenum atoms were found to have a net charge of $+1.16 \pm 0.08$, the oxygen atoms had charges of -0.29 ± 0.04, while the carbon atoms were neutral. Using these

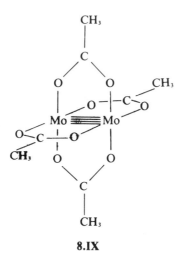

8.IX

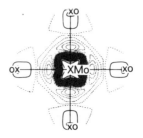

Fig. 8.25 Experimental deformation density map for dimolybdenum tetra-acetate, assuming all atoms to be neutral. The map is for a section through a plane formed by one Mo atom and four O atoms bonded to it. Negative contours are dotted. Taken, with permission, from K. Hino, Y. Saito and M. Benard, *Acta Crystallogr., Sect. B*, **37**, 2164 (1981).

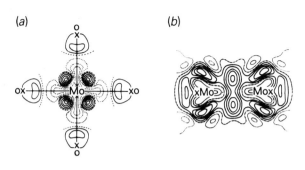

(a)   (b)

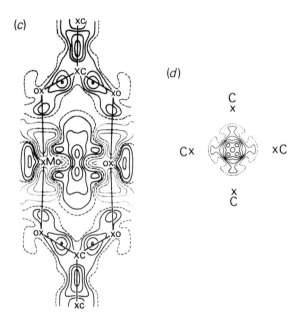

(c)

Fig. 8.26 Experimental deformation density maps for dimolybdenum tetra-acetate, with refined charges on all atoms. (*a*) Section through a plane formed by one Mo atom and four O atoms bonded to it. (*b*) Section through the Mo≡Mo bond and bisecting the two planes of the adjacent acetate groups. (*c*) Section through the Mo≡Mo bond and two acetate groups. (*d*) Cross-section through the middle of the Mo≡Mo bond. Negative contours are dotted. Taken, with permission, from K. Hino, Y. Saito and M. Benard, *Acta Crystallgr., Sect. B*, **37**, 2164 (1981).

(d)

charges, a new deformation density map, corresponding to that illustrated in Fig. 8.25, was obtained. The new map, shown in Fig. 8.26(*a*), shows that there are regions of electron density close to the molybdenum atoms in the MoO_4 planes, but lying between the ligands. These regions are also shown in the section through the Mo—Mo bond, but passing between the ligands [Fig. 8.26(*b*)]. This map also shows substantial electron density between the molybdenum atoms, but on either side of the bond. Another section through the Mo—Mo bond, this time including the acetate ligands [Fig. 8.26(*c*)], also shows electron density around the bond, so that overall there is an annulus of charge surrounding the bond. A cross-section through the middle of the bond [Fig. 8.26(*d*)] shows this quite clearly. Fig. 8.26(*c*) also indicates that there is some electron density on the molecular axis, outside the molybdenum atoms, and also, as expected, in the C—C and C—O bonds. The results are clearly consistent with ideas about bonding in the complex, which is one of the few examples of a compound having a quadruple bond.

Deformation density studies are important, as they give information obtainable by no other method. But great care must be taken to get useful results. High quality data are needed, obtained at low temperature, to minimize effects of vibrations. There is an excellent account of the techniques involved, with many examples, in Ref. [14].

8.9 Extended X-ray absorption fine structure

It has been known since the early days of the study of X-rays that when they are absorbed by atoms, there are oscillations of the absorption coefficient on the high energy side of the absorption edge. Only during the 1970's was the origin of this phenomenon explained, so that it could be put to good use. When an X-ray photon with sufficient energy is absorbed by an atom, a core electron is ejected, and the outcome can be

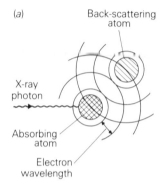

(a)

Back-scattering
atom

X-ray
photon

Absorbing
atom

Electron
wavelength

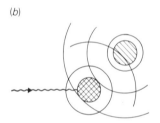

(b)

Fig. 8.27 Origin of EXAFS. The circles represent the photoelectron wave originating at the absorbing atom and back-scattered from neighboring atoms. In (a) the waves are in phase at the absorbing atom, but in (b) the photoelectron has a longer wavelength, and destructive interference occurs at the absorbing atom.

represented as an outgoing spherical wave, originating at the absorbing atom (Fig. 8.27). If the atom has neighbors, back-scattering from each of them gives an incoming electron wave, and so causes interference. The outcome is that there are variations in the absorption coefficient, which depend on the spatial relationships of the central atom and its neighbors. It is therefore possible to deduce the distances from the absorbing atom to its various neighbors, and it matters little whether the site of interest is in a small molecule or a very large one, or in a crystalline or amorphous solid, liquid, solution or even gas. The technique has proved to be most valuable in probing specific sites in complex molecules, particularly metal centers in metallo-proteins.

The X-ray absorption spectrum may also show detailed structure *below* the absorption edge. This X-ray absorption near edge structure (XANES) arises from excitation of core electrons ($1s$ in the case of a K edge) to high level vacant orbitals. Valuable information can be obtained by analysis of this part of the spectrum, but it does not provide geometrical information, and is not discussed any further here.

The experiment simply involves measuring the absorption spectrum in the vicinity of the absorption edge for the chosen element. This, of course, is easier said than done.

A high intensity, tunable X-ray source is required, and in practice almost all experiments use synchrotron radiation, which gives useful spectra for solids, liquids or concentrated solutions in a few minutes. The K edges of elements down to about phosphorus or sulfur can be studied by this method, and K or L edges of all heavier elements are also accessible. For lighter elements, down to carbon, a laser-produced plasma can give suitable soft X-rays, or equivalent information for light atoms can be obtained by electron energy loss spectroscopy (see Section 5.3.4).

The EXAFS intensity oscillations are described by Eqn (8.9), which gives the relative modulation, χ, of the absorption coefficient, μ, of the atom as a function of

$$\chi(\varkappa) = \frac{1}{\varkappa} \sum_i \frac{f_i(\varkappa)}{r_i^2} e^{-\sigma_i^2 \varkappa^2} e^{-2\mu r_i} \sin[2\varkappa r_i + \alpha_i(\varkappa)] \qquad (8.9)$$

the variable $\varkappa$. This is defined as $\sqrt{ZmE}/\hbar$, where m and E are the mass and energy of the ejected photoelectron, and $\varkappa$ has units of length^{-1}. The summation is over all atoms in the vicinity of the absorber, and so the total intensity enables us to determine the numbers of neighbors at a particular distance from the absorbing atom, although such numbers have uncertainties of 20% or more. For each back-scattering atom there are four terms. The amplitude term $f_i(\varkappa)$ is characteristic of the atom, and depends primarily on its atomic number. There are then two exponential terms, one accounting for the relative motions of absorbing and back-scattering atoms in terms of their mean square amplitude of vibration, σ_i^2, and the other relating to the decrease in amplitude due to inelastic scattering of the photoelectron. Finally there is a sine wave term, with frequency dependent on the distance, r_i, between absorbing and scattering atoms, and on a phase shift experienced as the electron travels from the absorber to the scatterer and back. Knowledge of these phase terms is essential if the distance is to be determined, and they may be evaluated theoretically or by studying a similar compound of known structure. However, they are not strictly transferable from compound to compound, and this makes it very difficult to derive reliable structure information for any compound which does not closely resemble one or more with known structures. There are also sine waves arising from multiple scattering, and for these the frequency depends on the total distance along the multiple-scattering pathway. These terms damp out quickly, as do those arising from neighbors not bound directly to the absorbing atom, and so contribute mainly to the low-energy end of the spectrum. The number of possible multiple-scattering pathways can be very large, but failure to consider their contribution to the total pattern may lead to serious errors in the analysis of the data. The relative magnitudes of the scattering associated with various pathways in $[Co(CO)_4]^-$ are illustrated in Ref. [15], which also shows that refined distances may change by 0.2 Å or more when the multiple-scattering terms are introduced.

The form of Eqn (8.9) is analogous to that used to describe diffraction of electrons by gases [Eqn (8.2)], with the variable $\varkappa$ in one equivalent to s in the other, and the principles of analysis are therefore much the same. The total pattern is the sum of a lot of damped sine waves, superimposed on the absorption edge, as shown in Fig. 8.28. After removal of the background, the oscillating pattern [Fig. 8.29(a)] is very similar to a molecular scattering curve such as is observed for a gas (Fig. 8.4). The major new difficulties are the uncertainty about the phase terms, which affect the refined distances, and ignorance of the origin of the curve, a parameter which must therefore also be refined. Fourier transformation gives a radial distribution curve [Fig. 8.29(b)], which

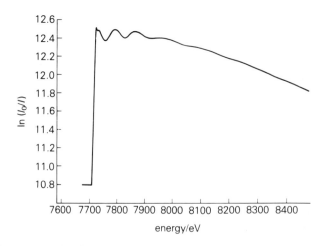

Fig. 8.28 X-ray absorption spectrum (Co K edge) of the cobalt complex 8.X. Redrawn from Ref. [15].

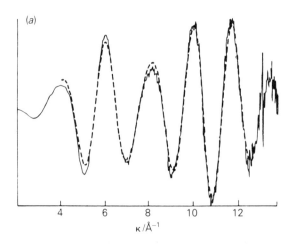

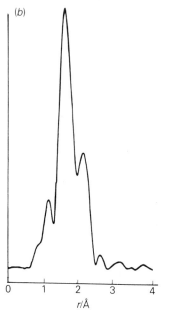

Fig. 8.29 (*a*) EXAFS data $x^3\chi(x)$ for the complex 8.X. The continuous line represents unfiltered data, and the dashed line filtered data. (*b*) Radial distribution curve obtained by Fourier transformation of the unfiltered EXAFS data (*a*). Redrawn from Ref. [15].

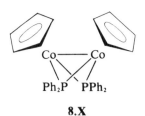

8.X

shows the distribution of the atoms surrounding the central, absorbing atom. However, because the phase terms are part of the sine wave arguments, the positions of peaks in the radial distribution curve do not correspond to the true interatomic distances.

Nevertheless, refinement of a structural model to fit the observed intensity data gives distances which can be reliable to within 0.01 Å. In the study [16] of the cobalt complex 8.X, the Co—C, Co—P and Co—Co distances obtained were 2.034 ± 0.004, 2.169 ± 0.019 and 2.572 ± 0.009 Å, respectively, in good agreement with values of 2.046 ± 0.020, 2.16 ± 0.01 and 2.56 ± 0.01 Å, found by single crystal X-ray crystallography. In this case, little new information has been obtained, but in the same paper parameters are also presented for the cation formed by oxidation of the neutral complex, a product for which crystallographic data are not available. Thus it can be seen that EXAFS may give accurate and reliable results, providing data can be collected first for related compounds with known structures. Note that only distances to neighboring atoms can be measured, and that it is not possible to determine the angles between the various bonds to the central, absorbing atom.

EXAFS only gives one-dimensional data, except when measured for single crystals, and in this respect it is like diffraction of electrons or X-rays by gases or liquids. However, it can be applied equally well to powders, liquids, solutions, polymeric materials or surfaces, and as sensitivity is quite high, the species being studied need not be present in high concentrations. The main attraction of the technique is that it provides specific information about the environment of atoms of a particular element. The total radial distribution curve for the cobalt compound 8.X would be extremely complicated, with many overlapping peaks, but the curve obtained from cobalt EXAFS contains contributions only from atom pairs involving cobalt, and so is relatively simple to interpret. This characteristic of the technique has made it of unique value in the study of metal-containing biological systems, or of active sites of homogeneous and heterogeneous catalysts. For example, the environments of the iron atoms in haemoglobin and deoxyhaemoglobin have been studied, and anti-tumor platinum drugs have been observed in whole cells, without the need to separate out particular components of the cell contents. In some cases, more than one element is well suited for EXAFS studies, and so even more information can be obtained. For example, molybdenum and iron EXAFS data have been used in attempts to elucidate the structure of the nitrogen fixation enzyme, nitrogenase.

Thus EXAFS spectroscopy is an invaluable technique, having great versatility, sensitivity and specificity, but unfortunately, at the present time, it is not as accurate as other structural methods. Despite inherent problems with the method, it has many important areas of application, for which other techniques are inadequate. For fuller details, see Refs [17]–[19].

SECTION 8.9: EXTENDED X-RAY ABSORPTION FINE STRUCTURE

Example 8.4

Describe the radial distribution curves that you would expect to obtain by Fourier transformation of Co EXAFS for $[Co(CO)_4]^-$, $[Co_3(CO)_9CH]$ and $[Co_4(CO)_{10}(PPh)_2]$, which have the following structures.

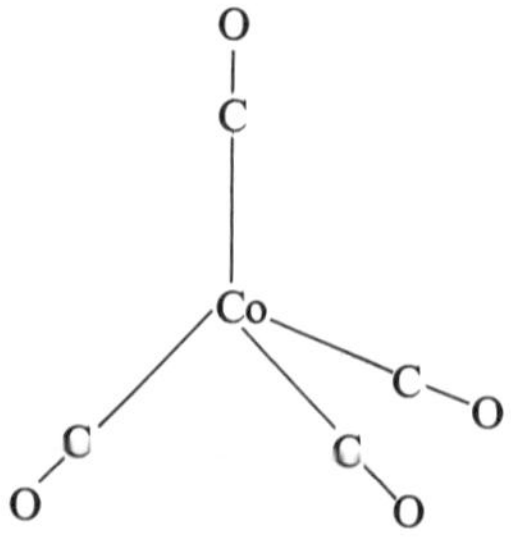

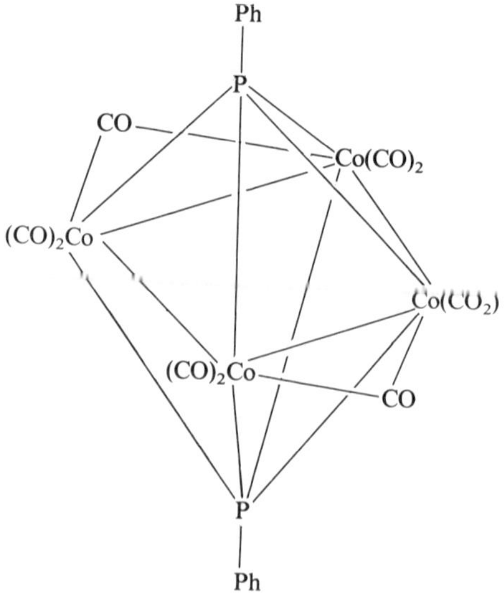

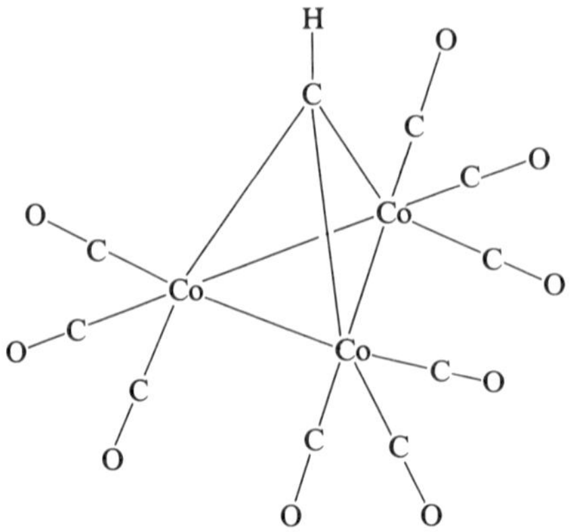

The tetrahedral $[Co(CO)_4]^-$ ion will give just two peaks in the Co EXAFS radial distribution curve, corresponding to the bonded Co—C and non-bonded Co···O atom pairs. The experimental (solid line) and theoretical (dotted line) curves are shown as the first section of the figure below, with the peaks at 1.77 and 2.92 Å.

For $[Co_3(CO)_9CH]$ the situation is more complicated, as there are many more atoms, and some peaks in the curve overlap, while others are at greater distances than can easily be observed by this method. The major components of the curve arise from Co—C (carbonyl) and Co—C (methinyl), overlapping near 1.8 Å (the distances are 1.81 and 1.86 Å respectively), Co—Co at 2.46 Å, and Co···O at 2.94 A. The next weak peak, at 3.37 Å, can be attributed to non-bonded Co···C atom pairs, involving a cobalt atom and two of the carbonyl groups on each of its neighboring cobalt atoms. That is all that can be seen on the radial distribution curve [(b) below], which only goes to 4 Å, but at longer distances there would also be another non-bonded Co···C peak, involving the two remaining carbonyl groups, and two long-range Co···O peaks.

Although the radial distribution curve for $[Co_4(CO)_{10}(PPh)_2]$ is in principle even more complex, there are still only four distinct peaks below 4 Å [curve (c) below]. We expect to observe Co—C, for both terminal and bridging carbonyl ligands, Co—P, and two different (bridged and non-bridged) bonded Co—Co atom pairs. The shortest non-bonded distances will arise from Co···O, with both terminal and bridging carbonyl groups, from the diagonal Co···Co pairs, and from Co···C, involving carbonyl ligands on neighboring cobalt atoms.

It was found that the first peak is due entirely to the Co—C bonded pairs, the second to Co—P and bridged Co—Co, the third to the unbridged Co—Co and Co···O, and the fourth to Co···C and diagonal Co···Co. All nine distances were in fact refined, with most standard deviations in the range 0.01 to 0.04 Å, but up to 0.22 Å for Co···O(bridge).

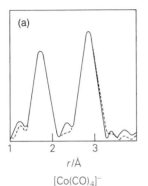

(a)

$r/\text{Å}$

$[\text{Co(CO)}_4]^-$

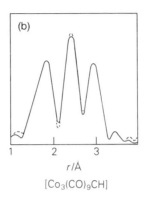

(b)

$r/\text{Å}$

$[\text{Co}_3(\text{CO})_9\text{CH}]$

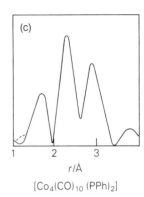

(c)

$r/\text{Å}$

$[\text{Co}_4(\text{CO})_{10}(\text{PPh})_2]$

The figure is adapted with permission, from Ref. [15]. Copyright (1987) American Chemical Society.

Problems

8.1 By considering radial distribution curves for the compounds GeF_4, TeF_4 and XeF_4, show how it is possible to distinguish tetrahedral, trigonal bipyramidal and square planar structures by electron diffraction. List the positions and relative areas of the peaks in each case, assuming all bond lengths to be 1.80 Å, and that the structure of TeF_4 does not deviate from the regular trigonal bipyramidal arrangement. How would the radial distribution curve for TeF_4 be changed if (a) the axial bonds were longer than the equatorial bonds, (b) the angle between the equatorial Te—F bonds was greater than 120° and (c) the axial Te—F bonds were bent away from the lone pair of electrons?

8.2 Describe the radial distribution curves you would expect to obtain for electron diffraction by ClF_5 and PF_5, showing how the different symmetries are reflected in the distribution of peaks.

8.3 Which of the following molecules would be good for study in the gas phase by electron diffraction? Discuss the problems likely to be encountered in each case.

(a) Ge_2Cl_6 (b) SF_5H (c) $\text{Cr(C}_6\text{H}_6)_2$ (d) $\text{FP}\overset{\displaystyle \text{O—CH}_2}{\underset{\displaystyle \text{O—CH}_2}{<}}$

(e) SbH_3 (f) OsO_4 (g) $\text{Co}_2(\text{CO})_8$

8.4 What methods would you use to determine each of the following? [The best techniques do not necessarily involve diffraction.]
   (a) The Si—H bond length in SiH_3Cl
   (b) The Si—H bond length in $\text{SiBu}^\text{t}_3\text{H}$
   (c) Whether the CCS fragment in $\text{CF}_3\text{C}{\equiv}\text{SF}_3$ is linear or bent
   (d) The arrangement of the carbonyl ligands in $\text{Fe}_3(\text{CO})_{12}$
   (e) The positions of the hydrogen atoms in $\text{Os}_4(\text{CO})_{12}\text{H}_4$
   (f) The positions of the hydrogen atoms in $\text{Fe(CO)}_4\text{H}_2$

8.5   How would you determine the C≡C distances in $SiH_3C≡CH$ in the (a) gas, (b) liquid or solution and (c) solid phase? How would you determine the C—Br distance in $SiH_3C≡CBr$ in the same three phases?

8.6   There are two isomers of disulfur difluoride, $S=SF_2$ and FSSF. Explain how you could distinguish the isomers and determine the structure of each by electron diffraction. What other methods could be used to determine their structures?

8.7   What would you expect to see in the radial distribution curve for iron(II) phthalocyanine, 8.XI, obtained in an iron EXAFS study? How would that curve be modified if the iron atom was displaced from the plane of the ligand? In what important way would the curve obtained by EXAFS differ from that obtained by electron diffraction?

8.XI

8.8   Crystals of the copper protein plastocyanin (anhydrous molecular weight = 10 500) are orthorhombic with $a = 29.6$, $b = 46.7$, $c = 58.1$ Å; there are four molecules in the unit cell. The observed density is 0.928 Mg m^{-3}. Estimate the number of water molecules per protein molecule in the crystals.

8.9   A projection of a portion of one unit cell of crystalline aluminum(III) chloride is shown in Fig. 8.30. The a (6.03 Å) and b (10.44 Å) axes are shown; the c-axis is 17.04 Å. No atoms with other fractional z-coordinates make bonding contacts to the atoms shown. Extend the diagram along a and b to show at least two adjacent unit cells. Determine the length of the shortest Al—Cl distance and describe the coordination of the aluminum and chlorine atoms. Is the structure best described as discrete molecules, infinite chains, infinite sheets or an infinite framework?

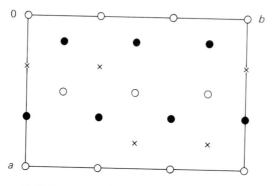

O   Cl atoms at $z = 0$

×   Al atoms at $z = 0.0833$

●   Cl atoms at $z = 0.1667$

Fig. 8.30

8.10 Crystals of a compound with empirical formula BCl_2 are orthorhombic with a 11.90, b 6.28 and c 7.69 Å. Assuming a density of 1.870 Mg m^{-3} calculate the number of molecules per unit cell. Atomic coordinates for half a molecule are as follows:

	x/Å	y/Å	z/Å
B	0.76	0.43	0.00
Cl(1)	0.78	2.16	0.00
Cl(2)	2.24	− 0.46	0.00

The other atoms are related to these by inversion through the origin. Sketch the molecule and calculate the bond lengths. What is the apparent molecular point group? Describe the radial distribution curve for gaseous electron diffraction that you would expect on the basis of this structure. In fact, the radial distribution curve has only three peaks, centered at 1.74, 3.02 and 4.06 Å. Deduce the structure and point group of this compound in the gas phase.

8.11 Dibenzene chromium crystallizes in the cubic space group $Pa3$, and the chromium atom must occupy a site of symmetry $\bar{3}$ (S_6). Assuming an ordered structure, describe the distortion of an idealized molecule with D_{6h} symmetry that would be consistent with these observations.

8.12 A complex containing four rhodium atoms forms orthorhombic crystals with a 21.337, b 41.943 and c 31.815 Å. The fractional coordinates of two metal atoms are − 0.0577, 0.0262, 0.1411 and 0.1391, − 0.0731, 0.1752, and the other two atoms of the same molecule are related by a two-fold rotation, and have coordinates given by $\frac{1}{4} - x, y, \frac{1}{4} - z$. Are there any metal–metal bonds in the complex?

8.13 Anhydrous copper(II) chloride gives monoclinic crystals with a = 6.85, b = 3.30, c = 11.49 Å, β = 90.0° ($\alpha = \gamma = 90°$). It may be conveniently considered in the space group $F2/m$, (point group C_{2h}), which requires that every atom at fractional coordinates 1: x, y, z has identical atoms at 2: $-x, -y, -z$; 3: $x, -y, z$; and 4: $-x, y, -z$. What is the nature of the symmetry operations giving positions 2, 3 and 4? In addition, the unit cell is face centered, i.e. the environments of the points: $\frac{1}{2} + x, \frac{1}{2} + y, z; \frac{1}{2} + x, y, \frac{1}{2} + z;$ and $x, \frac{1}{2} + y, \frac{1}{2} + z$ are identical to those of x, y, z. In the structure, a copper atom is located at 0,0,0 and a chlorine atom at 0.125, $\frac{1}{2}$, 0.125. Give the coordinates of all atoms in a unit cell, showing that there are four copper and eight chlorine atoms per cell.

Sketch the structure as seen in projection down the b-axis. Indicate probable chemical bonds in your drawing and estimate their lengths. Describe the coordination of the copper atoms and relate it to the electronic configuration of copper (II)

8.14 Assuming the most symmetrical structure that is chemically reasonable, deduce the point group for an isolated molecule of benzene chromium tricarbonyl. How many symmetry-independent Cr—C, C—O, C—C and C—H bond lengths will there be?

Crystals of benzene chromium tricarbonyl are monoclinic with a = 6.17, b = 11.07, c = 6.57 Å, β = 101.5°, and the density is 1.650 Mg m^{-3}. Calculate the unit cell volume and the number of molecules per unit cell.

The point group of the crystal lattice is C_{2h}. This implies that in each unit cell there may be either 4 molecules with crystallographic symmetry C_1, or 2 molecules with symmetry C_i, C_2 or C_s, or 1 molecule with symmetry C_{2h}. What is the crystallographic symmetry of the molecule, and how many independent Cr—C, C—O, C—C and C—H bond lengths will there be?

8.15 TlMe$_2$I crystallizes in a tetragonal space group with a 4.29 and c 14.01 Å. Its measured density is 4.6 Mg m^{-3}. Show that this is consistent with the presence of two formula units per unit cell.

In this space group, atoms occurring only twice per cell must have D_{4h} symmetry and lie on one of the sets of positions

$$0 \quad 0 \quad 0 \quad \text{and} \quad \tfrac{1}{2} \ \tfrac{1}{2} \ \tfrac{1}{2} \quad \text{or} \quad 0 \ 0 \ \tfrac{1}{2} \quad \text{and} \quad \tfrac{1}{2} \ \tfrac{1}{2} \ 0$$

Treating methyl groups as spheres of electron density, use this information to deduce the structure of TlMe$_2$I and calculate the shortest distance between Tl and I atoms.

8.16 Fig. 8.31 shows the structures of $NMe_2(SiH_3)$, $NMe_2(SiClH_2)$ and $NMe_2(SiCl_2H)$ as determined by X-ray crystallography. Comment on the observed structures.

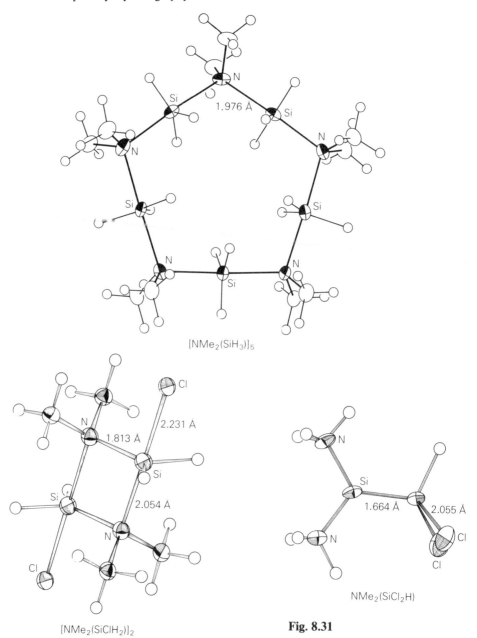

$[NMe_2(SiH_3)]_5$

$[NMe_2(SiClH_2)]_2$

$NMe_2(SiCl_2H)$

Fig. 8.31

8.17 The pyridyl-naphthyridine ligand can link to a metal–metal bonded system by all three nitrogen atoms, as in structure 8.XII. How many such ligands could bond in this way to a pair

8.XII

of octahedral metal atoms? Would it then be possible to coordinate more ligands acting as bidentate chelates, and if so, how would they restrict the possible arrangement of tridentate ligands?

The structure of such a complex has been determined by X-ray diffraction [A.T. Baker, W.R. Tikkanen, W.C. Kaska and P.C. Ford, *Inorg. Chem.*, **23**, 3254 (1984).] Would it be possible to determine the arrangement of ligands by any other method?

8.18 In neutron diffraction experiments vanadium atoms cannot be located directly because vanadium has almost zero scattering power. In a study of liquid VCl_4, a radial distribution curve was obtained. This consisted of a distinct peak at 3.45 Å, a much broader one centered around 3.80 Å and unresolved intensity at greater distances. Assuming that VCl_4 has a regular tetrahedral shape, account for the observations, and calculate the V–Cl bond length.

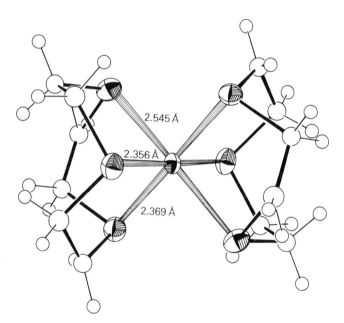

2.545 Å

2.356 Å

2.369 Å

Fig. 8.32

8.19 Fig. 8.32 shows the structure of the $[Pd(1,4,7\text{-trithiacyclononane})_2]^{3+}$ ion as determined by X-ray crystallography. The palladium atom lies on an inversion center, and Pd–S bond lengths are shown. Discuss the coordination at palladium in the light of its electronic structure.

8.20 The Cartesian coordinates of the atoms of an ammonium ion in crystalline ammonium oxalate monohydrate, as determined by X-ray diffraction and neutron diffraction, are as follows. The standard deviations apply to each of the coordinates. Calculate the N–H bond lengths, with standard deviations, and comment on the differences between values given by the two techniques.

| Atom | X-ray | | | | Neutron | | | |
	x	y	z	e.s.d	x	y	z	e.s.d
N	3.1071	2.3352	1.6188	0.0012	3.1079	2.3362	1.6158	0.009
H(1)	3.817	2.743	1.102	0.050	3.8166	2.7610	1.0027	0.0014
H(2)	3.415	1.567	2.064	0.045	3.4832	1.5207	2.1236	0.0014
H(3)	2.338	2.072	1.095	0.069	2.3093	2.0311	1.0350	0.0017
H(4)	2.852	3.031	2.208	0.052	2.8131	3.0621	2.2886	0.0017

8.21 A compound with composition $CoHg(CNS)_4$ gives tetragonal crystals with a 11.109 and c 4.379 Å. The space group is $I\bar{4}$, implying that there is an S_4 axis through the origin parallel to z, and that the environment of the point $\frac{1}{2}, \frac{1}{2}, \frac{1}{2}$ is identical to that of 0,0,0. There is a mercury atom at the origin, and a cobalt atom at $\frac{1}{2}, 0, \frac{1}{4}$. Fig. 8.33 shows a section through the three-dimensional difference electron density map phased on the cobalt and mercury positions.

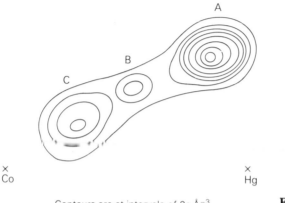

Contours are at intervals of $2e$ Å^{-3} **Fig. 8.33**

Identify the peaks A, B, and C whose maxima all lie on the given plane. Calculate the Hg$\cdots$Co distance and, using that to provide a scale for the diagram, estimate bond lengths and angles in the fragment shown.

Further reading

For a simple, general introduction: J Wormald, *Diffraction Methods*, Clarendon Press, Oxford (1973).

On electron diffraction by gases: Refs [1] and [2].

On diffraction by liquids: D.W.L. Hukins, *X-ray Diffraction by Disordered and Ordered Systems*, Pergamon Press, Oxford and New York (1981). (This book also covers liquid crystals, fibers, polymers *etc.*, which are not discussed in this chapter.)

On X-ray crystallography: J.D. Dunitz, *X-ray Analysis and the Structure of Organic Molecules*, Cornell University Press, Ithaca (1979); P. Luger, *Modern X-ray Analysis on Single Crystals*, de Gruyter, Berlin (1980); M.F.C. Ladd and R.A. Palmer, *Structure Determination by X-ray Crystallography*, Plenum, New York (1985); J.P. Glusker and K.N. Trueblood, *Crystal Structure Analysis*, 2nd edn, Oxford University Press, Oxford (1985).

For an excellent paper, encouraging us to evaluate published crystal structures critically, see P.G. Jones, *Chem. Soc. Rev.*, **13**, 157 (1984).

On practical details of X-ray crystallography: G.H. Stout and L.H. Jensen, *X-ray Structure Determination*, Macmillan, New York (1968).

On EXAFS spectroscopy: Refs [17]–[19], and also A. Bianconi, L. Incoccia and S. Stiptich, *EXAFS and Near Edge Structure*, Springer-Verlag, Berlin (1983). For a good simple description see S.P. Cramer and K.O. Hodgson, *Prog. Inorg. Chem.*, **25**, 1 (1979).

References

1 H.M. Seip, in *Molecular Structure by Diffraction Methods*, Specialist Periodical Report, The Chemical Society, London (1973), Vol. 1, Chap. 1.
2 I. Hargittai and M. Hargittai (eds), *Stereochemical Applications of Gas-phase Electron Diffraction*, VCH Publishers, New York (1988).
3 L. Acha, E.R. Cromie and D.W.H. Rankin, *J. Mol. Struct.*, **73**, 111 (1981).

4 M.J. Barrow, E.A.V. Ebsworth, C.M. Huntley and D.W.H. Rankin, *J. Chem. Soc., Dalton Trans.*, 1131 (1982).

5 E.A.V. Ebsworth, C.M. Huntley and D.W.H. Rankin, *J. Chem. Soc., Dalton Trans.*, 835 (1983).

6 D. Tabor, *Gases, Liquids and Solids*, 2nd edn, Cambridge University Press, Cambridge (1979).

7 Theo Hahn (ed.), *International Tables for Crystallography*, D. Reidel Publishing Company, Dordrecht, Netherlands (1983), Vol. A.

8 M.F.C. Ladd and R.A. Palmer, *Theory and Practice of Direct Methods in Crystallography*, Plenum Press, New York (1980); C. Giacovazzo, *Direct Methods in Crystallography*, Academic Press, New York (1980).

9 Y.N. Al-Obaidi, P.K. Baker, M. Green, N.D. White and G.E. Taylor, *J. Chem. Soc., Dalton Trans.*, 2321 (1981).

10 M.J. Barrow, S. Cradock, E.A.V. Ebsworth and D.W.H. Rankin, *J. Chem. Soc., Dalton Trans.*, 1988 (1981).

11 E.A.V. Ebsworth, R.O. Gould, N.T. McManus, N.J. Pilkington and D.W.H. Rankin, *J. Chem. Soc., Dalton Trans.*, 2561 (1984).

12 F. Takusagawa and T.F. Koetzle, *Acta Cryst.*, **B35**, 2888 (1979).

13 G.E. Bacon, The Determination of Magnetic Structures by Neutron Diffraction, in *Advances in Structure Research by Diffraction Methods* (eds R. Bill and R. Mason), Vieweg/Interscience (1966), Vol. 2.

14 P. Coppens, in *Chemical Crystallography, International Review of Science, Physical Chemistry Series 2*, Butterworth, London (1975), Vol. 11.

15 N. Binsted, S.L. Cook, J. Evans, G.N. Greaves and R.J. Price, *J. Am. Chem. Soc.*, **109**, 3669 (1987).

16 B.-K. Teo, P. Eisenberger and B.M. Kincaid, *J. Am. Chem. Soc.*, **100**, 1735 (1978).

17 B.-K. Teo and D.C. Joy (eds), *EXAFS Spectroscopy*, Plenum Press, New York (1981).

18 H. Winick and S. Doniach (eds), *Synchrotron Radiation Research*, Plenum Press, New York (1981).

19 D.C. Koningsberger and R. Prins, *X-ray Absorption: Principles, Applications, Techniques of EXAFS, SEXAFS and XANES*, Wiley, New York (1987).

9 Mass Spectrometry

9.1 Experimental arrangements

A mass spectrometer is an instrument in which ions are produced from a sample, separated according to their mass-to-charge ratios, and then recorded, in terms of intensity (number of ions). Thus a spectrometer is in principle very simple, consisting of a source of ions, an analyzer and a detector. Of course, in practice there are many complications, and a high-resolution mass spectrometer is a sophisticated and expensive device. In this book we can only consider the principles of operation very briefly; for full details, standard texts should be consulted [1], [2]. Refs [3] and [4] have good, short accounts of mass spectrometry, as well as many examples of applications of the technique in inorganic chemistry.

There are now many different types of ion source available, but for a long time the only widely-used method was electron impact. This required a supply of sample in the gaseous state (a serious limitation for very many compounds) at about 10^{-6} Torr pressure, interacting with a collimated beam of electrons. What happens to the sample molecule M depends on the electron energy, but at energies commonly used (*ca.* 70 eV) an electron may be stripped off, leaving the radical cation $[M\cdot]^+$ (usually called the molecular ion), which may then break down into a mixture of neutral and positively charged fragments. Alternatively, the sample molecule may capture an electron, with or without fragmentation, giving negatively charged ions. In a variation of this technique, chemical ionization, the ionization of the sample is achieved by using a primary source of ions (obtained by electron impact) instead of direct interaction with electrons.

In both the methods described above, vaporization of the sample is a prerequisite, and so study of involatile or thermally fragile compounds is impossible. To get round this major limitation, several desorption methods have been developed, enabling ions to be obtained directly from condensed (solid or liquid) phases [5]. The earliest such method to be widely available was field desorption, in which the ions were desorbed by a very strong electric field (10^9-10^{10} V m^{-1}), produced by applying a high potential to a sharp edge or a fine wire. In this, as in other techniques, the aim is to transfer energy to the sample, causing transfer of molecules, or even pre-existing ions, to the gas phase. Other sources of energy can be a beam of ions (secondary ion mass spectrometry, often referred to as SIMS), a laser or high energy neutral atoms, typically argon (fast atom bombardment, FAB). This is an area in which developments are taking place rapidly, but FAB has caught on quickly, is commercially available, and is applicable to diverse systems, such as proteins, alkali metal halide clusters, and organometallic complexes. We have easily obtained spectra for mixed-metal complexes such as 9.I by this method, having failed completely to get anything useful using electron impact sources. One characteristic of the desorption methods is that they give high concentrations of the molecular ion (without fragmentation occurring), and so are particularly useful for identification of unknowns or for molecular weight measurement, and even for analysis of mixtures. As large molecules can now be studied, implementation of these

378

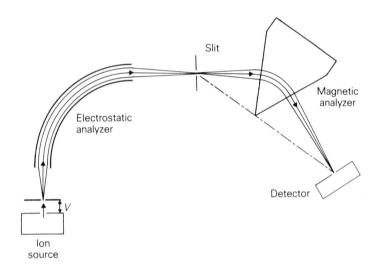

9.1

new methods has taken place at the same time as the mass range of spectrometers has increased, and upper mass limits of 9000 a.m.u. or more are routinely achievable.

Thermospray mass spectrometry involves direct injection of an electrolyte solution through a heated capillary into a vacuum. The solvent then evaporates rapidly from the tiny droplets, leaving single positive and negative ions. No further energy need be supplied to generate ions, so again very little fragmentation is observed, and even di-cations such as $[Me_3N(CH_2)_3NMe_3]^{2+}$ can be detected [6].

Slit

Magnetic analyzer

Electrostatic analyzer

Detector

V

Ion source

Fig. 9.1 Double-focusing mass spectrometer. Ions emerging from the source a accelerated by the potential and then deflected and focu ed by an electrostatic analy After passing through a slit they are deflected and refoc ed by a magnetic analyzer, and detected after passing a third slit.

In most high resolution spectrometers the ions are analyzed, first being accelerated by application of a potential of some thousands of volts, and then passing through two analyzers, one electrostatic, and one magnetic (Fig. 9.1). An electrostatic or magnetic field perpendicular to the motions of the ions causes them to move in circles, the radii of which depend on their charge-to-mass ratio, and also have a focusing effect. Thus ions emerging from the source slit are focused by the electrostatic analyzer at a second slit placed between the two analyzers. They are then refocused by the magnetic analyzer, and are detected as they pass through a collector slit. The radius r of the path taken by ions with mass-to-charge ratio m/e accelerated through a voltage V in a magnetic field B is given by Eqn (9.1)

$$r^2 = \frac{2Vm}{B^2 e}.$$ (9.1)

Variation of the magnetic field or the accelerating voltage thus brings the various ions in turn to the detector, where the ion current (proportional to the number of ions) is measured. The spectrum is simply a plot of ion current against mass-to-charge ratio. In reverse geometry spectrometers the magnetic analyzer precedes the electrostatic one (see Section 9.4). In some other spectrometers (often used for studying negative ions) the ions are all subjected to the same potential, and so accelerate to different speeds depending on their masses, and their flight times are measured. In quadrupole spectrometers [7], which are usually used as small, cheap, low resolution and fast scanning instruments, direct and radio-frequency voltages are applied to four parallel rods. The theory of what happens is complicated, but the outcome is the same: separation of ions of different masses. It should be noted that most of the ions have the same charge − that of a single electron − but that doubly charged and even triply charged ions do occur.

Another form of mass spectrometry is ion cyclotron resonance, which is particularly important in the study of ion−molecule reactions. In this spectrometer the ions move in a circular path under the influence of a magnetic field, with an angular frequency which depends on their mass-to-charge ratio. When this frequency is in resonance with an alternating electric radio-frequency field they absorb energy, and are thus detected.

As a mass spectrum can be obtained quickly and easily, and the spectra are characteristic of the compounds for which they are obtained, it is natural that combination of mass spectrometry with chromatographic methods has become a standard technique for the analysis of mixtures. This technique is used particularly in organic chemistry, but not exclusively so. It is perhaps most often applied to the routine analysis of different mixtures of the same few compounds, as for example in analyzing mixtures of hydrocarbons when studying the catalytic properties of transition metal complexes. But mass spectrometry can also be applied quickly to the identification of unknown compounds, as there are several large data bases, containing information about the mass spectra of many thousands of compounds [8].

9.2 Molecular ions

As we have already seen, a mass spectrum is plotted in the form of ion current as a function of mass-to-charge ratio. The majority of ions have single positive charges, and so appear close to integral mass values, but it is quite normal to observe small numbers of doubly, or even triply, charged ions, particularly for metal complexes, and these can give peaks between the integral values. Negative ions can also be formed, and study of these can give much useful information, particularly relating to bond dissociation energies.

The simplest process occurring in a mass spectrometer involves interaction of an electron with a molecule, resulting in the loss of a further electron from the molecule, leaving a radical cation. This ion, which has effectively the same mass as the parent molecule, is called the molecular ion. Usually it is the ion with highest mass in the spectrum, although species with higher masses can occasionally be formed by subsequent ion−molecule reactions. Thus it is often possible to determine the molecular weight of a compound simply by looking for the highest mass peak in its mass spectrum. For example, manganese carbonyl gives a parent ion at m/e 390 atomic mass units (a.m.u.), and as the masses of a manganese atom and a carbonyl group are 55 and 28, respectively, it does not take long to deduce that the formula of manganese carbonyl is $Mn_2(CO)_{10}$. Determining molecular weights to this accuracy can be invaluable in proving the identity of a compound. The iridium complex 9.II was not easy to characterize fully. Nuclear magnetic resonance and vibrational spectroscopy gave a lot of information, but it was difficult to show that the chlorine atoms were present. When a mass spectrum was obtained, this became easy. The highest mass peaks could only be accounted for if there were four chlorine atoms in the molecule.

$$
\begin{array}{c}
PEt_3 \\
| \quad Cl \\
Cl - Ir - PCl_2 \\
OC \quad | \\
PEt_3
\end{array}
$$

9.II

However, the molecular ion does not necessarily give the strongest peak in the spectrum, and it may not be visible at all. With electron impact sources, the relative intensity of the molecular ion peak can be increased by reducing the energy of the electron beam, thus decreasing the amount of energy available for fragmentation. By reducing the energy still further it is possible to measure the appearance potential for the ion, and the ionization potential of the molecule. Thus some information found by this method (Section 9.5) can relate to that found by photoelectron spectroscopy (Chapter 6).

With ionizing methods other than electron impact, the molecular ions are much more prominent (particularly so with laser desorption), and these methods provide very satisfactory ways of measuring molecular weights. However, it must be remembered that desorption methods produce ions direct from the condensed phase, and that the structures of molecules in solids and liquids may be quite different from those in the gas phase. When ionization is achieved by fast atom bombardment (FAB), fragment ions are not prominent, but the most intense peaks do not necessarily correspond to molecular ions. In such a source the sample is mixed intimately with a mulling agent, usually glycerol. The positive ions obtained from neutral species are formed by protonation, and thus appear with a mass one greater than the molecular weight, while anions are often found with a mass one less than the molecular weight, having been formed by deprotonation. One special advantage of a FAB source is that it can be used to study both positive and negative ions ejected from ionic compounds. In this case the ions may be seen without protonation or deprotonation, but the spectrum may be complicated by ion clusters. Thus the spectrum of a sodium salt Na^+X^- may show peaks due to species such as $[Na_2X]^+$. The extreme example is CsI, which gives a series of ions $[Cs_nI_{n-1}]^+$, with n going up to at least 60. This happens to provide an extremely convenient way of calibrating spectra.

Simply determining the mass of an ion to the nearest integer may not be good enough. A carbonyl complex of iron gives a molecular ion at 504 a.m.u. As iron has a mass of 56, that total mass can be fitted equally well by formulae $Fe(CO)_{16}$ (admittedly somewhat unlikely!), $Fe_2(CO)_{14}$, $Fe_3(CO)_{12}$, $Fe_4(CO)_{10}$, $Fe_5(CO)_8$ and so on. But with high resolution instruments it is possible to determine the masses of ions to within a few parts per million, and as atomic masses are not exact integers, it is usually possible to distinguish between the various options. In our example, the exact masses for $Fe_3(CO)_{12}$ and $Fe_4(CO)_{10}$, are 503.7438 and 503.6889 a.m.u., respectively – and so it is quite easy to decide which formula is correct. It is always a good idea to measure the mass of the parent ion in a spectrum accurately, as a check that the proposed formula is correct. When the ester CH_3COSPF_2 was prepared, the molecular ion was found, as expected, at 144 a.m.u. But there are other conceivable products with the same molecular weight, and final confirmation was provided by measuring the mass as 143.9613, compared with the expected value of 143.9610. In modern mass spectrometers determination of precise masses is a routine operation.

At some mass numbers it may be possible to resolve several peaks due to different ions. In the spectra of boron hydrides, for example, it is often difficult to sort out what is going on, because boron has isotopes ^{10}B and ^{11}B, and of course ^{10}BH and ^{11}B are both of mass 11. But not quite. The two masses are actually 11.0209 and 11.0093, and so with high resolution instruments the various ions contributing to one composite peak can be separately identified.

Discussion of boron also reminds us that many elements do not have just one isotope, and if such elements are present in a compound, there will be not just one molecular ion, but a whole series. The pattern, which depends on the relative abundances of the constituent isotopes, can be distinctive and diagnostic, provided the group of

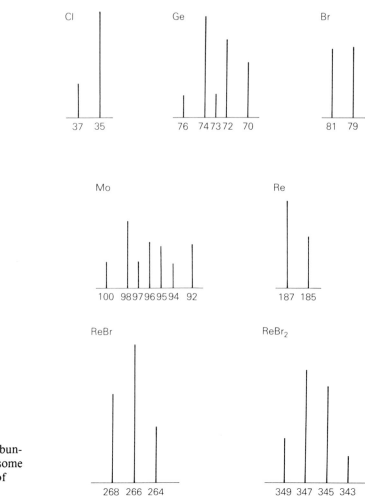

Fig. 9.2 Isotope abundance patterns for some atoms and groups of atoms.

CHAPTER 9: MASS SPECTROMETRY

peaks does not overlap another such group. Fig. 9.2 shows isotope patterns for several elements, and also for some of the elements in combination. In the latter case, the relative probabilities of finding the various combinations of isotopes must all be calculated by multiplying the abundances of the constituent isotopes. For example, rhenium has isotopes ^{185}Re and ^{187}Re, with abundances 37% and 63%, respectively, while bromine exists as ^{79}Br (51%) and ^{81}Br (49%). Thus the ion [ReBr]$^+$ can exist in four isotopic forms. ^{185}Re^{79}Br has a mass of 264 a.m.u., and its abundance is $37 \times 0.51\% = 19\%$. Similarly, the abundances of ^{185}Re^{81}Br (mass 266), ^{187}Re^{79}Br (also mass 266) and ^{187}Re^{81}Br (mass 268) are 18, 32 and 31%, respectively, giving the overall pattern shown in Fig. 9.2. Having worked out this distribution, we can easily calculate the pattern for [ReBr$_2$]$^+$, and so on. Once the patterns of possible combinations of atoms are calculated and known, it is very easy to spot them in a spectrum. Problems will occur when several such patterns overlap, and this is particularly severe when hydrogen atoms are present, so that several patterns, shifted relative to each other by one mass unit, may be superimposed. Isotope abundances and masses are given in Table 9.1, and extensive compilations of accurate masses for assemblies of atoms have been prepared [9], so that it is often possible to determine the constitution of an ion from its mass alone, without any need to know anything at all about what elements are present in the sample.

Table 9.1 Isotopic abundances of naturally occurring isotopes

Element	Z	Mass	%	Mass	%	Mass	%	Mass	%	Mass	%	Mass	%	Mass	%	Mass	%	Mass	%	Mass	%
H	1	1	100.0																
He	2	4	100.0																
Li	3	6	7.4	7	92.6														
Be	4	9	100.0																
B	5	10	19.6	11	80.4														
C	6	12	98.9	13	1.1														
N	7	14	99.6	15	0.4														
O	8	16	99.8	18	0.2														
F	9	19	100.0																
Ne	10	20	90.9	21	0.3	22	8.8												
Na	11	23	100.0																
Mg	12	24	78.7	25	10.1	26	11.2												
Al	13	27	100.0																
Si	14	28	92.2	29	4.7	30	3.1												
P	15	31	100.0																
S	16	32	95.0	33	0.8	34	4.2												
Cl	17	35	75.5	37	24.5														
Ar	18	36	0.3	38	0.1	40	99.6												
K	19	39	93.1	41	6.9														
Ca	20	40	97.0	42	0.6	43	0.1	44	2.1	48	0.2								
Sc	21	45	100.0																
Ti	22	46	7.9	47	7.3	48	73.9	49	5.5	50	5.3								
V	23	50	0.2	51	99.8														
Cr	24	50	4.3	52	83.8	53	9.6	54	2.4										
Mn	25	55	100.0																
Fe	26	54	5.8	56	91.7	57	2.2	58	0.3										
Co	27	59	100.0																
Ni	28	58	67.9	60	26.2	61	1.2	62	3.7	64	1.1								
Cu	29	63	69.1	65	30.9														
Zn	30	64	48.9	66	27.8	67	4.1	68	18.6	70	0.6								
Ga	31	69	60.4	71	39.6														
Ge	32	70	20.5	72	27.4	73	7.8	74	36.5	76	7.8								
As	33	75	100.0																
Se	34	74	0.9	76	9.0	77	7.6	78	23.5	80	49.8	82	9.2						
Br	35	79	50.5	81	49.5														
Kr	36	78	0.4	80	2.3	82	11.6	83	11.6	84	56.9	86	17.4						
Rb	37	85	72.2	87	27.8														

Table 9.1 *Contd*

Element	Z	Mass	%	Mass	%	Mass	%	Mass	%	Mass	%	Mass	%	Mass	%	Mass	%	Mass	%	Mass	%
Sr	38	84	0.6	86	9.9	87	7.0	88	82.6												
Y	39	89	100.0																		
Zr	40	90	51.5	91	11.2	92	17.1	94	17.4	96	2.8										
Nb	41	93	100.0																		
Mo	42	92	15.9	94	9.0	95	15.7	96	16.5	97	9.5	98	23.8	100	9.6						
Ru	44	96	5.5	98	1.9	99	12.7	100	12.6	101	17.1	102	31.6	104	18.6						
Rh	45	103	100.0																		
Pd	46	102	1.0	104	11.0	105	22.2	106	27.3	108	26.7	110	11.8								
Ag	47	107	51.8	109	48.2																
Cd	48	106	1.2	108	0.9	110	12.4	111	12.8	112	24.1	113	12.3	114	28.9	116	7.6				
In	49	113	4.3	115	95.7																
Sn	50	112	1.0	114	0.7	115	0.4	116	14.3	117	7.6	118	24.0	119	8.6	120	32.9	122	4.8	124	5.9
Sb	51	121	57.3	123	42.7																
Te	52	122	2.5	123	0.9	124	4.6	125	7.0	126	18.7	128	31.8	130	34.5						
I	53	127	100.0																		
Xe	54	128	1.9	129	26.4	130	4.1	131	21.2	132	26.9	134	10.4	136	8.9						
Cs	55	133	100.0																		
Ba	56	130	0.1	134	2.4	135	6.6	136	7.0	137	11.3	138	71.7								
La	57	139	99.9																		
Ce	58	136	0.2	138	0.3	140	88.5	142	11.1												
Pr	59	141	100.0																		
Nd	60	142	27.1	143	12.2	144	23.9	145	8.3	146	17.2	148	5.7	150	5.6						
Sm	62	144	3.1	147	15.0	148	11.2	149	13.8	150	7.4	152	76.7	154	22.7						
Eu	63	151	47.8	153	52.2																
Gd	64	152	0.2	154	2.2	155	14.7	156	20.5	157	15.7	158	24.9	160	21.9						
Tb	65	159	100.0																		
Dy	66	160	2.3	161	18.9	162	25.5	163	25.0	164	28.2										
Ho	67	165	100.0																		
Er	68	162	0.1	164	1.6	166	33.4	167	22.9	168	27.1	170	14.9								
Tm	69	169	100.0																		
Yb	70	168	0.1	170	3.0	171	14.3	172	21.8	173	16.1	174	31.8	176	12.7						
Lu	71	175	97.4	176	2.6																
Hf	72	174	0.2	176	5.2	177	18.5	178	27.1	179	13.8	180	35.2								
Ta	73	181	100.0																		
W	74	180	0.1	182	26.4	183	14.4	184	30.6	186	28.4										
Re	75	185	37.1	187	62.9																
Os	76	186	1.6	187	1.6	188	13.3	189	16.1	190	26.4	192	41.0								
Ir	77	191	37.3	193	62.7																
Pt	78	192	0.8	194	32.9	195	33.8	196	25.3	198	7.2										
Au	79	197	100.0																		
Hg	80	196	0.1	198	10.0	199	16.8	200	23.1	202	13.2	201	29.8	204	6.9						
Tl	81	203	29.5	205	70.5																
Pb	82	204	1.5	206	23.6	207	22.6	208	52.3												
Bi	83	209	100.0																		
U	92	235	0.7	238	99.3																

9.3   Fragmentation

When a molecule is ionized in a mass spectrometer, excess energy may be imparted to the ion. This is particularly true for ions formed from gas phase molecules in an electron impact source. The ion may just be formed in some excited state, but it is probable that it will fragment, giving one neutral part and a new ion. Further fragmentation can then occur, so that the spectrum which is recorded contains many ions, the abundances of which depend on a number of factors, including their stabilities or lifetimes, and the stabilities of their precursors. By careful study it is possible to determine the whole pattern of breakdown, and to measure parameters such as bond dissociation energies, but in this section we will concern ourselves mainly with the interpretation of the simple mass spectrum.

Even for a simple triatomic molecule, ABC, there are four possible initial fragmentation steps. The bond A—B can break, giving either [A]$^+$ and BC, or A and [BC]$^+$, or the B—C bond can break, giving either [AB]$^+$ and C, or AB and [C]$^+$. More than one of these processes may occur, and so even in the simplest of cases the situation is potentially complicated. The ions [AB]$^+$ and [BC]$^+$ can break down further, and again there are two possibilities in each case. Thus [AB]$^+$ could give [A]$^+$ and B, or A and [B]$^+$. So the spectrum of ABC could include some or all of the ions [ABC]$^+$, [AB]$^+$, [BC]$^+$, [A]$^+$, [B]$^+$ and [C]$^+$. The only one missing is [AC]$^+$, and its absence is quite strong evidence that our compound is ABC, and not BAC or ACB. However, rearrangement reactions do occur, and it is not possible to be absolutely sure that a group of atoms appearing in a fragment ion were necessarily bonded to each other in the parent molecule. Moreover, ion−molecule reactions occasionally occur, so we may even see ions such as [ABCAB]$^+$ − but fairly rarely. In the mass spectrum of Re$_2$Cl$_2$(CO)$_8$ [10] there is a series of peaks corresponding to the molecular ion, and ions showing successive loss of all eight of the carbonyl groups. The fact that the Re$_2$Cl$_2$ unit holds together is very strong evidence that these atoms are bound to each other,

9.III

9.IV

and the structure (9.III) does in fact have the two chlorine atoms bridging between the rhenium atoms. It is generally true that metal carbonyls lose all the carbonyl groups easily in a mass spectrometer, and this provides a convenient way of counting them. It was this technique that provided the first identification of the cluster complex Ru$_6$(CO)$_{17}$C, an unexpected structure [Ru$_6$(CO)$_{18}$ seemed at the time to be much more likely] subsequently confirmed by X-ray crystallography. The method even applies to such large complexes as 9.IV, which loses all ten carbonyl groups before the extensive central part fragments at all.

As with molecular ions, identification of fragment ions may be aided by measurement of accurate masses, or by observing isotope patterns. Fig. 9.3 shows part of the mass spectrum of ReBr(CO)$_5$. This is not difficult to analyze by any means, but at a glance it is possible to recognize the triplet patterns (shown in Fig. 9.2) of ReBr units, and the doublets arising from ions containing rhenium, but not bromine.

Thus by considering the ions observed in a mass spectrum, it is often possible to deduce the structure of the molecule involved. In the spectrum of the thio-ester 9.V one would expect to find ions such as [SnMe$_3$O]$^+$, but not [SnMe$_3$S]$^+$, whereas the

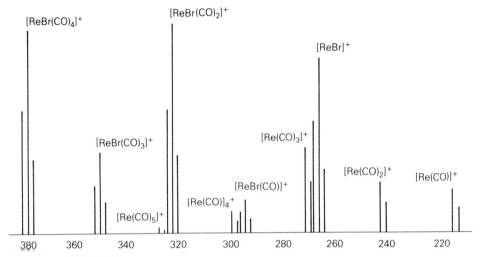

Fig. 9.3 Representation of part of a mass spectrum of ReBr(CO)$_5$. In addition to the peaks shown, there will be weak ones arising from ions containing ^{13}C, ^{17}O or ^{18}O.

9.V

9.VI

reverse would be true for the isomer 9.VI. Of course, the possibility of rearrangement must always be taken into account, but it is usually possible, when the whole pattern of ions is considered, to reach the correct conclusions.

9.4 Ion reactions

A mass spectrum includes peaks from many different ions, and it is clear that complicated series of reactions takes place within the mass spectrometer. Normally, the information obtained from a spectrum concerns the identity of the products, and it is not possible to say much else about the reactions themselves, except perhaps by making intelligent guesses about the ancestry of particular ions. However, it is possible to obtain direct information about ion reactions, the simplest way being by observing in the spectrum peaks associated with metastable ions. Some ions have such short lifetimes that they dissociate while moving through the spectrometer, with the result that one ion (of mass m_1) is accelerated after the initial ionization, but a different ion (mass m_2) (called a daughter ion) passes through the magnetic analyzer. The resulting peak comes neither at m_1 or m_2 in the spectrum, but at mass m^*, given by:

$$m^* = m_2^2/m_1. \tag{9.2}$$

The ions usually observed are those formed during the 10^{-5} seconds or so that they spend between the electrostatic and magnetic analyzers, and they give spectral peaks which are quite broad compared with those normally seen. An example is shown in Fig. 9.4. Interpretation of peaks due to metastable ions usually involves calculating

CHAPTER 9: MASS SPECTROMETRY

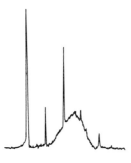

Fig. 9.4 Part of a mass spectrum showing four normal ions and one peak arising from a metastable ion.

masses of daughter ions for all reasonable parent ions, until a sensible solution is found. Alternatively, a table of metastable transitions can be used [11].

In a study of the mass spectrum of $P(OPF_2)_3$ [12], five weak peaks attributable to metastable ions were observed, and all could easily be assigned in terms of parent and daughter ions which were prominent in the spectrum. For example, the peak at 143.3 a.m.u. was attributed to the ion $[P(OPF_2)_3]^+$ (mass 286) losing the neutral fragment PF_2O to give $[P(OPF_2)_2]^+$ (mass 201). In three of the other cases, the neutral molecule formed was PF_3 or PF_3O, so migration of a fluorine atom must have taken place. In the remaining case, the ion formed was $[P_2F_3O]^+$, which must again have involved some rearrangement, which this time is revealed by the charged rather than by the neutral fragment. When all the information from these metastable ions is put together, it is possible to make a plan of the fragmentation pathway, as shown in Fig. 9.5. The picture is not complete, but all that is not known is the origin of the small ions $[PF_2O]^+$, $[PF_2]^+$ and $[PFO]^+$ (there may be more than one source for each, of course), and of $[P_2F_4O]^+$. This last ion is more interesting, as it is most likely to have the structure $[O(PF_2)_2]^+$, and its formation must therefore have involved migration of a complete PF_2 group.

Sequences of ion reactions can be studied in more detail by a variety of fairly sophisticated methods, which have developed rapidly in recent years. When a metastable

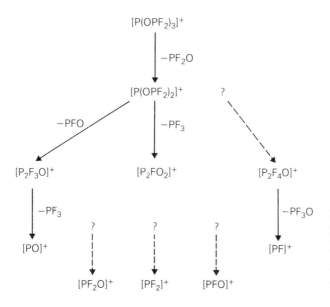

Fig. 9.5 Fragmentation pathway for $P(OPF_2)_3$. Reactions marked with arrows have been shown to occur by observation of appropriate metastable ions.

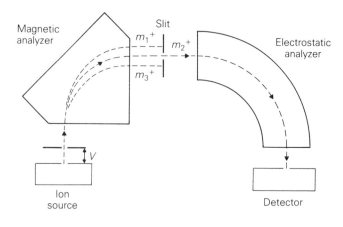

Fig. 9.6 Arrangement of reverse geometry mass spectrometer, used for recording MIKES spectra. Ions having a chosen mass (here m_2) are selected, and daughter ions formed by their decomposition are separated by the electrostatic analyzer.

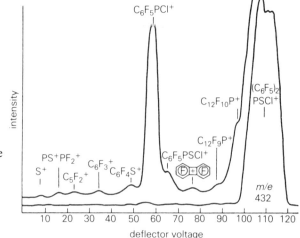

Fig. 9.7 MIKES spectrum of the molecular ion of $PClS(C_6F_5)_2$, showing its major breakdown product to be $[PCl(C_6F_5)]^+$. Redrawn from T.R.B. Jones, J.M. Miller and M. Fild, *Org. Mass Spectrom.*, **12**, 317 (1977).

ion decomposes, its daughter ion only takes a proportion (m_2/m_1) of its kinetic energy – less than is taken by ions formed in the source. It follows that in a conventional spectrometer with electrostatic and magnetic analyzers, if the magnetic analyzer is tuned to a particular ion, and the accelerating voltage is then swept upwards, all precursors of that ion will in turn be observed, in what is known as an ion kinetic energy spectrum. More usefully, the positions of the magnetic and electrostatic analyzers can be reversed (Fig. 9.6). The magnetic field can then be adjusted so that only ions having a chosen mass pass the second slit. Decay of metastable ions then produces daughter ions which have different kinetic energies, and scanning the electrostatic analyzer voltage gives a spectrum showing all the daughter ions of a particular ion. (These reverse geometry spectrometers suffer from the major disadvantage that normal metastable transitions are not observed.) An example of such a mass-analyzed ion kinetic energy (MIKES) spectrum is shown in Fig. 9.7. This spectrum shows that the molecular ion of the phosphine sulphide (9.VII) decomposes primarily by loss of C_6F_5S, and it is therefore deduced that an Arbusov rearrangement to the form 9.VIII occurs before decomposition. In further refinements and variations of the method, it has been possible to detect consecutive processes and to show, for example, that an ion formed from a particular

9.VII   9.VIII

precursor then goes on to give identified daughter ions. Thus by careful study the whole pattern of fragmentation can be analyzed.

Another development has been in the field of tandem mass spectrometry, or multi-stage mass spectrometry (MS/MS). In this a sample is ionized, and the ions are separated in a mass spectrometer. A particular ion is selected, and energy is transferred to it by a collision process, whereupon it fragments. The product ions are then analyzed in a second mass spectrometer. There are two major applications of this. First, it is a very powerful analytical tool for mixtures, working on picogram quantities, and it is possible to distinguish even between isomers, as fragmentation patterns of otherwise indistinguishable molecular ions can be quite different. Second, it is a means of studying the decomposition of ions, in a selective manner. Fig. 9.8 shows MS/MS spectra

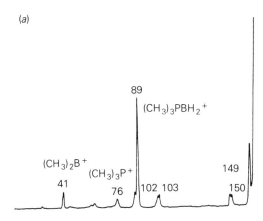

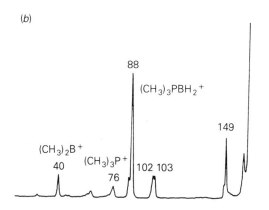

Fig. 9.8 MS/MS spectra for the ions $[(Me_3P)_2BH_2]^+$ (a) with ^{11}B and (b) with ^{10}B. Redrawn from G.L. Glish, P.J. Todd, K.L. Busch and R.G. Cooks, *Int. J. Mass Spectrom. Ion Phys.*, **56**, 177 (1984).

obtained for ions of mass 164 and 165 obtained from $[(Me_3P)_2BH_2]^+Br^-$. The two parent cations differ only in having different isotopes of boron. So peaks having the same masses in the two MS/MS spectra must be from ions containing no boron, while those with a mass difference of one must represent boron-containing species. The empirical formulae of daughter ions can thus be determined.

9.5   Thermodynamic data

It is obvious that in an experiment which involves supplying energy to a molecule or ion to cause it to fragment, we ought to be able to control the amount of energy available, and deduce something about the energetics of the reactions involved. In the first place, with an electron impact source, it is possible to vary the energy of the electrons and to note the lowest energy at which the molecular ion is formed. In fact there is no sudden onset of ionization, and in practice it is necessary to obtain an ionization efficiency curve, from which the ionization potential of the molecule can easily be derived. On increasing the electron beam energy further, fragment ions appear, and for each of these an appearance potential may be obtained in a similar manner. Fig. 9.9 shows an ionization efficiency curve for O_2. In this case the appearance potential is quite clear, and the ion current rises linearly as the electron beam increases. The gradient changes at several higher electron energies, and the points of inflection correspond to the onset of ionization from different molecular orbitals of O_2. Thus in this very simple case the mass spectrum gives information which could otherwise be obtained by ultraviolet photoelectron spectroscopy [Section 6.6.2]. For more complicated molecules, combined studies of mass spectra and photoelectron spectra can give detailed information about ion decomposition as a function of excess energy.

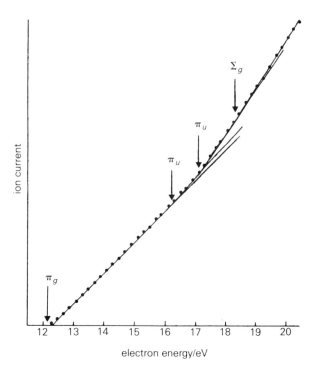

Fig. 9.9 Ionization efficiency curve for O_2^+. The ion current is zero for electron energies below 12.3 eV, and there are further points of inflection in the curve at 16.2, 17.0 and 18.2 eV. These correspond to the onset of ionization from the π_u level to give quartet and doublet ion states, and from the σ_g orbital to give a quartet state. The curve should be compared with the photoelectron spectrum of O_2, shown in Fig. 6.15. Redrawn from B. Cantone, V. Emma and F. Grasso, *Adv. Mass Spectrom.*, **4**, 599 (1968).

If an ion $[AB]^+$ decomposes to give $[A]^+$ and B, then the appearance potential of $[A]^+$ is the sum of the ionization potential of A and the bond dissociation energy of AB. If the ion is formed in an excited state, we obtain only an upper limit for the dissociation energy: for positive ions this is rarely a serious problem. Of course, if the ionization potential of B is lower than that of A, we will get A and $[B]^+$ as products instead. Thus if the ionic fragment has a known I.P., bond dissociation energies may be derived. Or if the dissociation energy can be guessed, the I.P. can be deduced. Thus, for example, a series of compounds of the type $[Fe(\eta^5\text{-}C_5H_5)(CO)L(MX_3)]$ (L = CO or PPh_3, M = Si or Sn, X = Me or Ph), such as 9.IX, were studied [13], and the appearance potentials of the ions $[MX_3]^+$ were measured. The $MX_3 \cdot$ radicals are well-known species, and their ionization potentials had already been reported, and so the dissociation energies of the Fe—Si and Fe—Sn bonds could be derived. The results are summarized in Table 9.2.

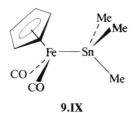

9.IX

Table 9.2 Dissociation of complexes $[Fe(\eta^5\text{-}C_5H_5)(CO)L(MX_3)]$

	$[MX_3]^+$ appearance potential/eV	$MX_3 \cdot$ ionization potential/eV	Fe—M dissociation energy /eV	/kJ mol^{-1}
$[Fe(C_5H_5)(CO)_2(SiMe_3)]$	9.22	7.25	1.97	190
$[Fe(C_5H_5)(CO)_2(SnMe_3)]$	9.12	6.81	2.31	223
$[Fe(C_5H_5)(CO)_2(SnPh_3)]$	8.87	6.29	2.58	249
$[Fe(C_5H_5)(CO)(PPh_3)(SiMe_3)]$	9.48	7.25	2.23	215

In dealing with negative ions great care is needed, as products may have considerable amounts of translational or vibrational energy, or even be in excited electronic states. For example, in the formation of $[SnCl_3]^-$ from $SnCl_4$, a chlorine atom is produced, and it is not known whether it is in the $^2P_{3/2}$ or $^2P_{1/2}$ state. So even after allowing for the translational energy of the products, there is still this degree of uncertainty in the measured electron affinity for $SnCl_3$. Nevertheless, a great deal of thermodynamic data can be, and has been, obtained by mass spectrometry – data which otherwise can be very difficult indeed to acquire.

Problems

N.B. Isotopic masses and abundances are listed Table 9.1.

9.1 A rhodium carbonyl complex consists of a cluster of rhodium atoms, surrounded by carbonyl ligands, and includes another atom at the center of the cluster. The mass spectrum includes peaks at 1078, 1050 and decreasing every 28 a.m.u. down to 448 a.m.u. Identify the complex.

9.2 Mass spectra of Fe^{III} trisacetylacetonate show prominent peaks at m/e values of 607, 508, 452, 409, 353 and 254. Suggest an explanation for these observations.

Example 9.1

The compound $CH_2=PH$ is produced by thermal decomposition of $ClCH_2PH_2$. Estimate the heat of formation of $CH_2=PH$ from the following information: ionization potential of $CH_2=PH = 10.3 \pm 0.2$ eV; appearance potential of $[CH_2PH]^+$ from $ClCH_2PH_2 = 11.0 \pm 0.2$ eV (see figure below); $\Delta H_f°(HCl) = -92$ kJ mol^{-1}; $\Delta H_f°(ClCH_2PH_2) = -44 \pm 4$ kJ mol^{-1}

Use standard heats of atomization for the elements involved (H: 218, C: 717, P: 315 kJ mol^{-1}) and standard single bond energies (C—H: 413, P—H: 321 kJ mol^{-1}) to estimate the bond energy of the C=P bond in $CH_2=PH$.

(a)
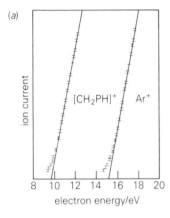

(b)
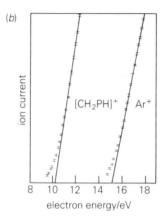

(a) Ionization efficiency curve for $CH_2=PH$, with Ar calibration [IP(Ar) = 15.76 eV]. (b) Measurement of appearance potential of $[CH_2PH]^+$ from $ClCH_2PH_2$, with Ar calibration. Redrawn from J.R. Chow, R.A. Beaudet and H. Goldwhite, *J. Phys. Chem.*, **93**, 421 (1989).

The appearance potential tells us that the reaction $ClCH_2PH_2 \rightarrow [CH_2PH]^+ + e^- + HCl$ is endothermic by 11.0 ± 0.2 eV, or 1061 ± 20 kJ mol^{-1}. The heat of formation of the ion plus electron is therefore:

$$\Delta H_f([CH_2PH]^+ + e^-) = 1061 - 44 + 92 = 1109 \pm 20 \text{ kJ mol}^{-1}.$$

If we subtract the observed ionization potential, 10.3 ± 0.2 eV or 994 ± 20 kJ mol^{-1}, we find the heat of formation of $CH_2=PH$.

$$\Delta H_f(CH_2=PH) = (1109 \pm 20) - (994 \pm 20) = 115 \pm 30 \text{ kJ mol}^{-1}.$$

[We have assumed that the measurement errors of ± 0.2 eV on the two potentials are random and uncorrelated. As both measurements are made in the same way and using the same apparatus and calibrant this probably almost certainly overestimates the standard error of the difference.]

The bond energy terms may be estimated using the cycle:

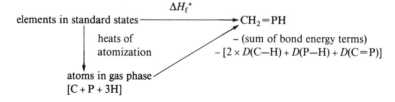

Thus $\Delta H_f°(CH_2=PH) = \Sigma(\text{heat of atomization}) - (\text{bond energy terms})$, i.e. $115 \pm 30 = 1686 - [1147 + D(C=P)]$, so $D(C=P) = 539 - 115 = 424 \pm 30$ kJ mol^{-1}.

This may be compared with the standard single bond energy, $D(C—P) = 264$ kJ mol^{-1}, and the C=S double bond energy, $D(C=S) = 578$ kJ mol^{-1}.

9.3 Fast atom bombardment mass spectroscopy of solid $[\{V(\eta^5\text{-}C_5Me_5)\}_4(\mu_2\text{-}O)_6]$ generates ions with m/e values of 840 (M$^+$), 706, 689, 571, 554, 435, 419, 403 and 300. Suggest fragmentation steps that would lead to the observed ions.

9.4 A compound containing a cluster of three heavy metal atoms, thought to contain Rh, Os or both, with one cyclopentadienyl ligand (Cp) and some carbonyls, shows ten groups of peaks showing the same metal isotope pattern, each group separated from its neighbors by 28 mass units. If the highest mass peak in the highest group is at m/e 804 suggest a formula for the compound.

9.5 Calculate the isotope patterns expected for $[Re(CO)_3]^+$ and $[ReBr]^+$, ignoring contributions from ions containing ^{13}C, ^{17}O or ^{18}O. By measuring the heights of peaks between 260 and 280 a.m.u. in Fig. 9.3, estimate the relative contributions of these ions to the mass spectrum of $ReBr(CO)_5$.

9.6 The peaks due to the molecular ion of $[Ce(C_5H_4Me)_3]$ include one at m/e 377 as expected for ^{140}Ce, the most abundant isotope of cerium, and another with 18% of the intensity at m/e 378. Suggest an assignment of the peak at m/e 378.

9.7 The three major isotopes of lead are ^{208}Pb, ^{207}Pb and ^{206}Pb, with abundances of 52%, 23% and 24%, respectively, and the abundances of ^{35}Cl and ^{37}Cl are 76% and 24%, respectively. Calculate and draw the isotope distribution patterns for (a) Pb_2, (b) PbCl and (c) $PbCl_2$.

9.8 The isotopes ^{69}Ga and ^{71}Ga are 60% and 40% abundant, respectively. In a mass spectrum there is a group of ions at 69, 70, 71, 72, 73 and 74 a.m.u. with relative intensities 6:19:28:26:16:11. Calculate the relative numbers of the ions $[GaH_3]^+$, $[GaH_2]^+$, $[GaH]^+$ and $[Ga]^+$ contributing to the spectrum.

9.9 In the mass spectrum of a carbaborane there is a peak at 24 a.m.u. which at high resolution is seen to consist of two components, one at 24.03859, and the other at 24.01713 a.m.u. Write down all the ions containing boron, carbon (^{12}C) and/or hydrogen which have a mass of 24 a.m.u., and from the exact masses determine which two contribute to the observed peak. (The masses of ^1H, ^{10}B, ^{11}B and ^{12}C are 1.007825, 10.012939, 11.009305 and 12.0000 respectively).

9.10 A compound isolated from a reaction of CpFeCp′ (where Cp′ is C_5H_4NHCHO) gave a molecular ion with an exact mass 211.0080 a.m.u. Suggest a formulation for this product. [Exact masses ^1H 1.0078, ^{12}C 12.0000, ^{14}N 14.0031, ^{16}O 15.9949, ^{56}Fe 55.9349].

9.11 In the electron impact mass spectrum of $[Os(CO)(C_2H_4)(mesitylene)]$ a prominent fragment ion appears at 28 mass units below the parent ion M$^+$. How could you show whether this peak was due to loss of CO or of C_2H_4?

9.12 The mass spectrum of $[Yb(C_5H_5)_2(acac)]$ (abbreviated to $[YbCp_2L]$) shows peaks ascribed to the ions $[YbCp_2L]^+$, $[YbCpL]^+$ and $[YbL]^+$. Where would you expect to see metastable ion peaks if the two fragment ions were formed by successive loss of Cp ligands in simple processes without rearrangements? (Use ^{174}Yb as the central atom.)

9.13 In the mass spectrum of PMe$_3$, a metastable transition is observed at 22.4 a.m.u. The metastable ion is known to be the ion formed simply by loss of H· from the molecular ion. What are the daughter ion and the neutral fragment? Comment on the nature of the decomposition reaction.

9.14 The appearance potential of the ion $[SiH_3]^+$ in the mass spectrum of SiH$_4$ is 12.10(5) eV, and in the mass spectrum of SiH$_3$ it is 8.14 eV. Deduce the first bond dissociation enthalpy for SiH$_4$.

9.15 The appearance potentials for the $[Mn_2(CO)_{10}]^+$ and $[Mn_2]^+$ ions in the mass spectrum of $Mn_2(CO)_{10}$ are 8.6 and 18.7 eV, respectively, while for $Re_2(CO)_{10}$ the two equivalent appearance potentials are 8.3 and 29.0 eV. Calculate the mean bond dissociation energies for the Mn—C and Re—C bonds in these compounds. For $MnRe(CO)_{10}$ the appearance potentials for $[MnRe(CO)_{10}]^+$, $[MnRe(CO)_5]^+$ and $[MnRe]^+$ are 8.2, 12.1 and 26.0 eV, respectively. What can you deduce from this extra information?

9.16 Reaction of Fe^+ ions in the gas phase with iodobenzene leads to the ion $[FeC_6H_5]^+$ and iodine atoms. Given heats of formation (all kJ mol^{-1}) for Fe^+, 1176.7; C_6H_5I, 164.8; I, 106.7 and the C_6H_5 radical, 330.5, deduce a limit for the Fe-phenyl bond strength in the $FeC_6H_5^+$ ion.

Further reading

Refs [1] and [2] are good accounts of mass spectrometry, while Refs [3] and [4] include shorter accounts of the method, as well as giving many examples of its use. Useful older accounts of mass spectrometry include R.W. Kiser, *Introduction to Mass Spectrometry and its Applications*, Prentice-Hall, Englewood Cliffs, New Jersey (1965); J. Roboz, *Introduction to Mass Spectrometry; Instruments and Techniques*, Interscience, New York (1968).

The Specialist Periodical Reports on *Mass Spectrometry*, published every two years by the Royal Society of Chemistry, London, provide accounts of recent developments in the field, and include a chapter reviewing applications to inorganic and organometallic chemistry.

For information on some specialist topics, the following may be useful:

R.G. Cooks, J.H. Beynon, R.M. Caprioli and G.R. Lester, *Metastable Ions*, Elsevier, Amsterdam (1973).

C.E. Melton, *Principles of Mass Spectrometry and Negative Ions*, Marcel Dekker, New York (1970).

F.W. McLafferty (ed.), *Tandem Mass Spectrometry*, Wiley-Interscience, New York (1983).

H. Hartmann and K.-P. Wanczek, *Ion Cyclotron Resonance Spectrometry*, Springer-Verlag, Berlin (1978), Vol. 1 and (1982), Vol. 2.

References

1 D.H. Williams and I. Howe, *Principles of Organic Mass Spectrometry*, McGraw-Hill, Maidenhead, U.K. (1972).
2 J.R. Chapman, *Practical Organic Mass Spectrometry*, Wiley, New York (1985).
3 J. Charalambous (ed.), *Mass Spectrometry of Metal Compounds*, Butterworths, London (1975).
4 M.R. Litzow and T.R. Spalding, *Mass Spectrometry of Inorganic and Organometallic Compounds*, Elsevier, Amsterdam (1973).
5 K.L. Busch and R.G. Cooks, *Mass Spectrometry of Large, Fragile and Involatile Molecules*, *Science*, **218**, 247 (1982).
6 G. Schmelzeisen-Redeker, U. Geissmann and F.-W. Röllgen, *Angew. Chem., Int. Ed. Engl.*, **23**, 892 (1984).
7 P.H. Dawson (ed.), *Quadrupole Mass Spectrometry and its Applications*, Elsevier, Amsterdam (1976).
8 S.R. Heller and G.W.A. Milne, *EPA/NIH Mass Spectral Data Base*, National Standard Reference Data System, National Bureau of Standards (U.S.); E. Stenhagen, S. Abrahamsson and F.W. McLafferty, *Registry of Mass Spectral Data*, Wiley, New York (1974); *Index of Mass Spectra*, Atomic Weapons Research Establishment, Aldermaston, Berkshire, England.
9 J.H. Beynon and A.E. Williams, *Mass and Abundance Tables for Use in Mass Spectrometry*, Elsevier, Amsterdam (1963).
10 K. Edgar, B.F.G. Johnson, J. Lewis, I.G. Williams and J.M. Wilson, *J.Chem.Soc. (A)*, 379 (1967).
11 J.H. Beynon, R.A. Saunders and A.E. Williams, *Tables of Metastable Transitions for Use in Mass Spectrometry*, Elsevier, Amsterdam (1965).
12 E.A.V. Ebsworth, G.M. Hunter and D.W.H. Rankin, *J. Chem. Soc., Dalton Trans.*, 1983 (1983).
13 T.R. Spalding, *J. Organomet. Chem.*, **149**, 371 (1978).

10 Case Histories

10.1 Introduction

Some 85% of this book has been devoted to the description of experimental techniques for identifying inorganic compounds, and for investigating their molecular and electronic structures. The only satisfactory way to present this material involves describing the techniques one by one, and we have emphasized the applications of these methods to chemical problems, rather than concentrating on the theoretical aspects. In doing this, we may well have given a wrong impression. We have used many examples to illustrate the ways in which each particular type of experiment can give useful chemical and structural information, and we have, of course, tended to choose examples which give clear-cut results. But when we come to apply these techniques we often find that no one method can tell us all we need to know about a compound. Or we compare results obtained in different ways, and discover apparent conflicts. On other occasions we think that we understand a structure fully, but then realize that what exists in the solid state is very different from what is found in solution or in the gas phase.

This chapter is an attempt to restore the balance. We aim to encourage wide-ranging investigations of inorganic compounds, using whatever methods may be relevant to the particular systems being studied, and to emphasize that structural techniques are often complementary, and are much more powerful when used in combination rather than alone. It is not possible to describe in a rigorous way how this should be done. All we can do is to provide some case histories – accounts of what has been achieved – and hope that the general principles of the broad approach to structural questions can thus be demonstrated. In the first example we show how study of xenon hexafluoride, by many people, over 20 years, has given understanding of a compound which has unusual behavior in gas, liquid and crystalline phases. This is followed by an account of some experiments which demonstrate the formation of the cations Xe_2^+, $[RCN-XeF]^+$ and related species. The next section is concerned with some tetrahydroborates, and in particular beryllium bis(tetrahydroborate). This compound has been studied in great detail over some 40 years, but there is still uncertainty about the gas-phase structure.

These examples involve compounds which have been investigated in exceptional detail, but not all systems need such depth of study. So most of the remaining sections of this chapter describe the methods used to characterize particular compounds, or to answer specific structural questions. We have tried to select a range of compounds studied by widely differing means. Some are large molecules, and some are small. Some can be identified by a single experiment, while others require careful detective work, using clues from diverse sources. Some are easy to characterize, and others present problems which are difficult to solve – or remain unsolved. And some show unusual behavior, which repays careful study.

These case histories are not meant to be reviews, and so we have not given comprehensive lists of references. However, if you are interested in following up any of the cases, the references given will quickly provide a route into the relevant literature.

Many of the problems at the end of the chapter are intended to provoke thought and discussion about possible approaches to further investigation of these systems. We have not provided answers to any of these questions, because at this stage we want to encourage debate and argument, leading to new ideas and new experiments.

10.2   Xenon hexafluoride

The preparation of the first compounds of xenon in 1962 marked a major turning-point in the development of inorganic chemistry. Until then the rare gases of Group 18 were known as inert gases, and few people thought to question their inertness. After all, did they not have the stable octet of electrons? Yet this thinking ignored the fact that expansion of the octet had been known since long before the discovery of electrons! Familiar compounds such as PF_5 and SF_6 have 10 and 12 electrons in their valence shells. Indeed, iodine compounds such as IF_5 and IF_2^- were known, so why should not the isoelectronic XeF_5^+ and XeF_2 also exist?

The flurry of activity in the 1960's produced many xenon compounds of great structural interest. The simple fluorides, XeF_2 and XeF_4, are widely used as examples in the teaching of electron pair repulsion theory; their linear and square planar configurations arise from shells with 10 and 12 electrons, with three and two lone pairs of electrons, respectively. In contrast, XeF_6 is structurally far from being simple. Much effort has gone into investigation of its behavior in gas, liquid and solid phases, and the story is undoubtedly still incomplete.

Xenon hexafluoride is quite volatile, and so it is not surprising that the first structural studies were concerned with the gas phase. On the basis of electron pair repulsion theory, it might be expected that XeF_6 should not have regular octahedral (O_h) symmetry, as the xenon atom has fourteen valence shell electrons, including one lone pair. However, the ion $TeCl_6^{2-}$ and other related species were believed to have regular structures, and the lone pair of electrons was described as 'stereochemically inert'. Simple molecular orbital ideas led to the suggestion that the xenon compound should behave similarly, and have a regular octahedral structure, with three sets of 'three-center, four-electron' bonds, and the lone pair in a pure s orbital. Electron diffraction studies of XeF_6 were carried out in three laboratories, and the most comprehensive work is described in two papers by Gavin and Bartell [1, 2] All the studies agreed that the molecule shows distortion from O_h symmetry. The radial distribution curve (Fig. 10.1) shows quite unequivocally that most of the fluorine–fluorine distances between 2.2 and 3.5 Å (the *cis* distances in the octahedral molecule) are shortened, while a few are lengthened. Moreover, all the peaks except the one at 3.7 Å are much wider than those for other hexafluorides, suggesting that (*a*) there are several different Xe—F bonded distances, (*b*) the fluorine atoms are unusually mobile, and (*c*) movements of opposed Xe—F bonds are correlated, so that the peak corresponding to the longest (*trans*) F· · ·F distance is not unduly broad.

The electron diffraction data can be fitted reasonably well by several static models; those giving the closest fit can be represented as distortions of octahedral symmetry by a lone pair of electrons. This pair may occupy either a face of the octahedron, giving C_{3v} symmetry, or an edge, giving C_{2v} symmetry, as shown in Fig.

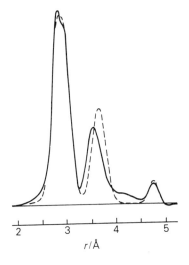

Fig. 10.1 Radial distribution curve for electron diffraction by gaseous XeF_6. The continuous line represents the experimental curve, and the dashed line is a synthetic curve calculated for a model with O_h symmetry. Redrawn from Ref. [1].

10.2. The latter structure is marginally preferred, but neither fits the data quite as well as would normally be expected. And in each case the deviation from regular octahedral symmetry is less than would be expected on the basis of electron pair repulsion theory, although that theory predicts correctly that the bonds adjacent to the unoccupied site should be longer than the others. So it seems that there is no satisfactory simple explanation of the observations, and we must look at rather more complicated possibilities.

One interesting suggestion [3] was that XeF_6 has a low-lying electronic triplet state, significantly populated at room temperature, which would be deformed by the Jahn–Teller effect. In the resulting D_{3d} molecular symmetry the $^3T_{1u}$ electronic state of O_h would separate into $^3A_{2u}$ and ^3E_u states. This idea was supported by studies [4] of vibrational spectra, both of the gas phase and of samples isolated in a matrix of solid argon. Octahedral XeF_6 would have only two IR-active modes (one stretch and one deformation) and three Raman-active modes (two stretches and a deformation). The observed spectra show at least three or four bands in the Xe—F stretching region in the IR, and three in the Raman. This was interpreted in terms of the three electronic isomers, which were also said to interconvert remarkably slowly (over periods of several minutes). However, the numbers of observed stretching modes can also be explained in terms of a static structure of C_{3v}, C_{2v} or other symmetry, and so the vibrational data do not provide strong support for the electronic isomer model. Two other experiments finally provided firm evidence that this model must be incorrect. First, deflection experiments with an inhomogeneous magnetic field showed that if XeF_6 contains a paramagnetic component, that component cannot be more than 1% of the total. Then a study [5] of the electronic spectrum in the region

Fig. 10.2 Representations of deformation from O_h symmetry consistent with electron diffraction data for gaseous XeF_6. The lone pair of electrons is shown to be repelling the fluorine atoms, to give structures of C_{3v} and C_{2v} symmetry. Redrawn from Ref. [2].

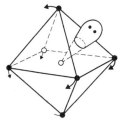 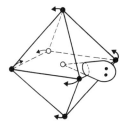

SECTION 10.2: XENON HEXAFLUORIDE 397

between 50 and 170 eV, using a synchrotron source, showed that the gas phase molecules must all be in the same electronic state, with close to O_h symmetry.

That leaves us with a single isomer, and some distortion from a regular octahedral structure, but we know from molecular beam experiments that there is no significant permanent electric or magnetic dipole moment. There is therefore no static, deformed structure. The best explanation of all these observations is that the XeF_6 molecules are rapidly changing between eight equivalent C_{3v} structures, such as is shown in Fig. 10.2, via twelve equivalent transitional structures of C_{2v} symmetry (there are eight faces and twelve edges of an octahedron for the lone pair to occupy). This idea was first proposed to account for the details of the electron diffraction data [2], and has been applied in an extensive analysis of all the available gas-phase data [6]. It seems probable that there is a potential minimum for the C_{3v} form, but that the energies of C_{2v} and O_h forms are only slightly higher. The molecule therefore spends most of its time in asymmetric configurations. But the simplest way to visualize what is happening is to consider first a molecule with O_h symmetry. When the molecule deforms during the t_{1u} bending vibration, there is mixing of the ground electronic state with the low-lying T_{1u} state. This mixing stabilizes the deformed structure, so that the energy required for bending becomes very small, or even negative, in which case the equilibrium structure is no longer of octahedral symmetry. This effect, known as the second-order Jahn–Teller effect, accounts for the observed distortion. But the energies of the different structures are so similar that the molecule very rapidly pseudorotates, changing between the eight equivalent C_{3v} configurations.

Although a xenon hexafluoride molecule with C_{3v} symmetry would have two chemically distinct types of fluorine atom, the rapid pseudorotation would make them equivalent in the ^{19}F NMR spectrum of a solution at room temperature. The spectrum does have just one resonance, which is broad and shows no sign of coupling to ^{129}Xe (26% abundant, spin $\frac{1}{2}$) or ^{131}Xe (21% abundant, spin $\frac{3}{2}$). The ^{129}Xe spectrum is also a single broad line, so it appears that intermolecular exchange of fluorine atoms occurs.

However, on cooling a remarkable change occurs in the spectra. At very low temperatures the ^{19}F spectrum [7] [Fig. 10.3(a)] shows a pattern of at least seven lines, while the ^{129}Xe spectrum [8] [Fig. 10.3(b)] has at least eleven lines. The intensity distribution in the latter group indicates that what are seen are the central peaks of a 25-line multiplet, and that each xenon nucleus is therefore coupling to 24 equivalent fluorines. The ^{19}F spectrum is consistent with this. If 'Xe_4F_{24}', molecules exist, there will be some with four, three, two, one and no ^{129}Xe nuclei, and superposition of the ^{19}F spectra of all these isotopic species would give a total of nine resonances: the relative intensities of the central seven of these agree with what is observed.

We therefore have Xe_4F_{24} units, in which all xenons are equivalent, all fluorines are equivalent, and all XeF couplings are equivalent. This is impossible on symmetry grounds, unless some rapid scrambling process is occurring. The nature of this process (which is still rapid even at 125 K) is not clear, but for the coupling to be retained the exchange must be intramolecular. Thus the fluorines must be continually migrating between all the different positions In an Xe_4F_{24} unit – surely the most extreme example of a fluxional molecule? There have been several suggestions about possible structures for the tetramer. One idea was that xenon atoms are bonded together in a tetrahedral core, with terminal and bridging fluorine atoms, as shown in Fig. 10.4. It is easy to visualize how the fluorines could scramble, via the bridge positions, but the idea of four xenon atoms being bonded to each other is hard to accept. However, if we

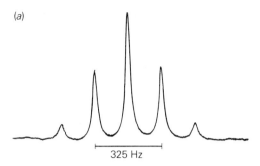

(a)

325 Hz

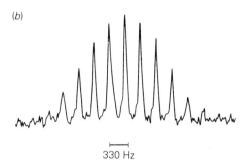

(b)

330 Hz

Fig. 10.3 NMR spectra for Xe_4F_{24}: (a) ^{19}F spectrum of a solution in $O(SF_5)_2$ at 155 K and (b) ^{129}Xe spectrum of a solution in $SClFO_2/CCl_2F_2$ at 128 K. Spectrum (a) is redrawn from Ref. [7], and spectrum (b) from Ref. [8].

simply remove the direct Xe—Xe bonds from this structure, we are left with a fluorine-bridged form which still can account for the fluxional behavior: and this cyclic bridged tetramer is very similar to what is found in crystalline xenon hexafluoride. The solution simply shows association, which is a necessary step on the way from gaseous monomeric molecules to the crystalline state.

In fact it is not as simple as that: it seems that with xenon hexafluoride nothing is simple! There are at least four crystalline states, all of which have been studied by X-ray crystallography. The one formed on freezing liquid XeF_6 is monoclinic, with 8

Fig. 10.4 Suggested structure for Xe_4F_{24} in solution. The similarity of this structure to that shown in Fig. 10.5(a) should be noted. Redrawn from Ref. [7].

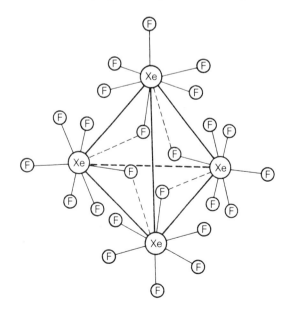

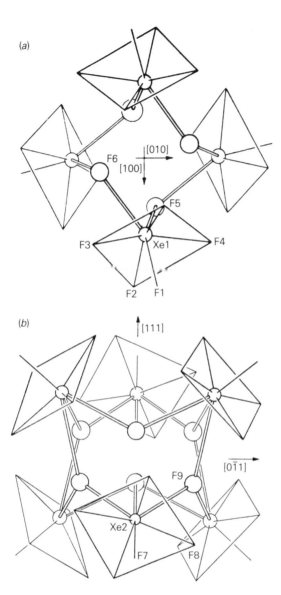

(a)

(b)

Fig. 10.5 Views of the polymeric xenon hexafluoride clusters found in the cubic phase, redrawn from Ref. [9]. Xenon atoms are represented by small circles and bridging fluorine atoms by large circles. In (a) the tetramer is viewed with the $\bar{4}$ axis parallel to [001], and in (b) the hexamer is shown with a 3-fold axis parallel to [111].

molecules per unit cell. On further cooling this changes to an orthorhombic form with 16 molecules in each unit cell, and then to another monoclinic phase, this time with 64 molecules per unit cell. In all these phases there are tetrahedra of four XeF_6 molecules. Each xenon atom has five fluorine atoms occupying positions of a square pyramidal structure, while the sixth fluorine atom is further away from the xenon, and bridges to another XeF_6 unit. There are no free ions, but the XeF_5 units with five short bonds have a structure closely resembling that of the free XeF_5^+ ion.

The fourth phase is formed when XeF_6 is sublimed, slightly below room temperature. It is cubic, and contains 144 molecules in the unit cell [9]. Two-thirds of the molecules are in Xe_4F_{24} units with four bridging atoms. These units can be regarded as tetrahedral or as eight-membered rings [Fig. 10.5(a)], and are essentially the same as those found in the other three crystalline phases. The remainder of the molecules are in hexameric units, which can be regarded as octahedra or as 12-membered rings [Fig. 10.5(b)]. These units also have square pyramidal XeF_5 components, but in this case the remaining fluorines (which are randomly distributed among the possible sites) are triply bridging. Not only are the bridging atoms disordered, but the orientations of the polymeric units are also disordered, and there are right- and left-handed enantiomorphs of the hexamer. The description [9] of the structure, and of the packing of the almost spherical polymers, makes fascinating reading.

So in gas, liquid and solid phases, xenon hexafluoride demonstrates its unwillingness to sit still. It is an extreme example, but it does show very clearly how important it is to consider structures of molecules in all phases, and to take into account *all* the observations that are available, from spectroscopic, diffraction and other techniques.

10.3 The cation Xe_2^+

Simple binuclear cations of the rare gases have been known in the gas phase for a long time, but until recently no compound containing such a species had been character-

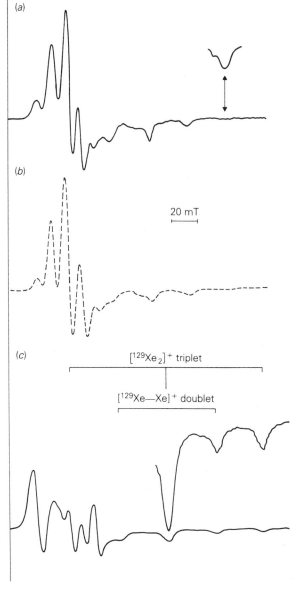

Fig. 10.6 ESR spectra (*a*) of the green product formed from xenon with natural isotopic abundance, (*b*) the spectrum calculated for Xe_2^+ with the same isotopic composition and the following parameters: $g_\perp$, 2.304; $g_\parallel$, 1.855. $A_\perp$ (^{129}Xe), 22, (^{131}Xe), 6.52 mT; $A_\parallel$, (^{129}Xe), 60, (^{131}Xe), 17.8 mT. Spectrum (*c*) was obtained from xenon containing 80% ^{129}Xe; the patterns from the $g_\perp$ part of the spectrum are marked. Parts of spectra obtained with increased recorder gain are shown as insets. Redrawn from Refs [10] and [11].

20 mT

[$^{129}Xe_2$]$^+$ triplet

[$^{129}Xe—Xe$]$^+$ doublet

Table 10.1 Isotopic Species of Xe_2^+

Species	Natural abundance	ESR multiplicity ($g_\perp$)
*Xe–*Xe	27.5%	Singlet
^{129}Xe–*Xe	27.7%	Doublet
^{131}Xe–*Xe	22.2%	Quartet
^{129}Xe–^{129}Xe	7.0%	Triplet
^{131}Xe–^{131}Xe	4.5%	Septet
^{129}Xe–^{131}Xe	11.2%	Doublet of quartets

*Xe represents all non-spinning isotopes of xenon.

ized. In view of the stability of its other compounds, xenon is the most likely member of the group to form such compounds; the reaction between $[XeF]^+[Sb_2F_{11}]^-$ and a reducing agent such as moisture gives a green product which is stable under a pressure of xenon at room temperature and has been shown to contain Xe_2^+ by an elegant combination of ESR and resonance Raman spectroscopy.

The green species gives a well-defined ESR spectrum when frozen in a matrix of SbF_5 at 4.5 K [10] The resonance is characteristic of an axially symmetric system, as Xe_2^+ would be, with $g_\perp$ 2.304 and $g_\parallel$ of 1.885. There are well-defined hyperfine splittings on both parts of the spectrum, which can be fitted by calculating the spectrum expected for Xe_2^+ using the natural abundances of the spin-active isotopes ^{129}Xe (spin $\frac{1}{2}$) and ^{131}Xe (spin $\frac{3}{2}$), optimizing the A-values to give the best match of calculated and observed spectra; the fit is shown in Fig. 10.6. The ESR spectrum of the same species prepared from Xe containing 80% ^{129}Xe was also recorded [11]. While the region associated with $g_\parallel$ was second-order and has not yet been sorted out, the region associated with $g_\perp$ shows clearly that whatever was responsible for the resonance contained two equivalent Xe atoms. Naturally-occurring xenon contains substantial proportions of both ^{129}Xe and ^{131}Xe, as well as non-spinning isotopes. The statistical distribution of these in Xe_2^+ leads to six different possible spin-systems and six different first-order splitting patterns which would all be superimposed in a first-order ESR spectrum (see Table 10.1). The part of the spectrum associated with $g_\perp$ should consist of a singlet, a doublet and a quartet, all with comparable total intensities, on top of the other weaker multiplets. With 80% ^{129}Xe, the triplet due to $[^{129}$Xe–^{129}Xe$]^+$ should make up two-thirds of the total intensity, most of the rest arising from the doublet due to species containing one ^{129}Xe atom bound to some other isotope with spin zero. In the observed spectrum, the low-field component of the $g_\perp$ part of the resonance is overlapped by the second-order $g_\parallel$ peaks, but four of the five lines can be seen clearly.

The UV/visible spectrum of the green species in SbF_5 (Fig. 10.7) contains new and prominent bands at 335 and 710 nm. These are important in connection with the Raman spectrum. The Raman spectrum of a similar solution, excited by radiation of wavelength 530.8 nm, was complicated [Fig. 10.8(a)], but besides the bands due to the solvent and to $[XeF]^+[Sb_2F_{11}]^-$ there was a new polarized band at 123 cm^{-1}. The frequency of this band was unaffected if the green solution was prepared from $[XeF]^+[Sb_2F_{11}]^-$ and $H_2^{18}O$, so whatever is responsible for it does not contain oxygen. If ^{136}Xe is used in the preparation instead of natural xenon, however, the band drops in frequency by 2.4 ± 0.9 cm^{-1}; the shift calculated for Xe_2^+ is 2.3 cm^{-1}.

The wavelength of 530.8 nm used to excite this spectrum lies well away from the envelope of the electronic band whose maximum is at 710 nm in the UV/visible spectrum of the green solution. It was possible to show that this electronic band and

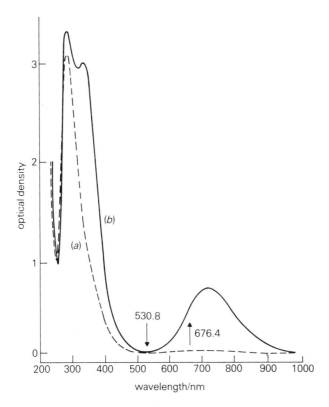

Fig. 10.7 UV/visible spectrum of a solution of $[XeF]^+[Sb_2F_{11}]^-$ in SbF_5, (*a*) before exposure to moisture, when the solution is yellow, and (*b*) after exposure to moisture, when the solution turned green. The positions of the lines used to excite the Raman spectra are also marked. Redrawn from Refs [10] and [11].

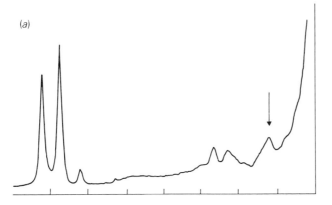

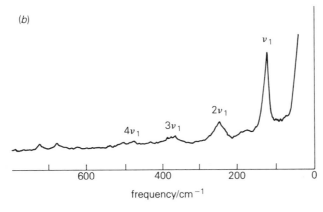

Fig. 10.8 (*a*) Raman spectrum excited at 530.8 nm of the green solution obtained when a solution of $[XeF]^+$ $[Sb_2F_{11}]^-$ in SbF_5 is exposed to moisture and held under a pressure of xenon. The band marked with an arrow is attributed to the product, Xe_2^+. (*b*) Raman spectrum of the same solution, excited at 676.4 nm. Redrawn from Ref. [12].

the Raman band at 123 cm^{-1} were due to the same species by exciting the Raman spectrum with radiation of 676.4 nm. This lies well within the envelope of the electronic band (see Fig. 10.7). In the Raman spectrum obtained [12], the intensity of the band at 123 cm^{-1} was markedly enhanced, and at least four components of the kind of progression characteristic of resonance Raman spectra [Section 5.17] could be observed [Fig. 10.8(b)]. All these observations provide very convincing evidence that the green species is Xe_2^+.

10.4 Complexes of xenon and krypton fluoro-cations with nitriles

It has been known for some years that the reaction of XeF_2 with fluoride ion acceptors such as AsF_5 gives ionic species XeF^+, and similar species are thought to be formed

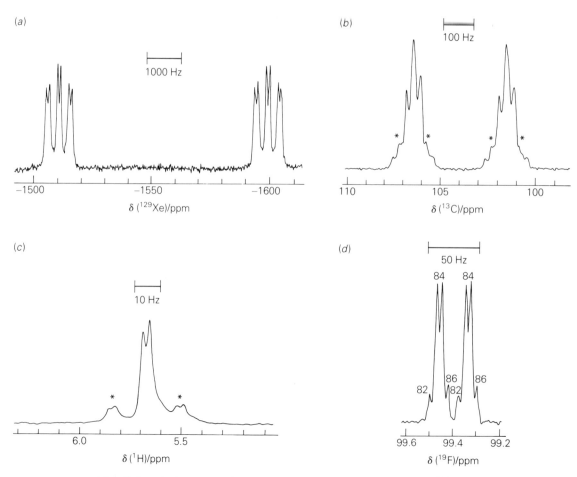

Fig. 10.9 NMR spectra of [HCN—XeF]$^+$ and [HCN—KrF]$^+$. (a) ^{129}Xe spectrum and (b) ^{13}C spectrum of a sample of [HCN—XeF]$^+$ labeled with ^{13}C and dissolved in liquid HF. The ^{129}Xe spectrum shows couplings to ^{19}F, ^{14}N and ^{13}C, while the ^{13}C spectrum exhibits splittings due to ^{1}H and ^{14}N, with satellites (marked with asterisks) due to ^{129}Xe. The ^{1}H spectrum (c) of an unlabeled sample dissolved in BrF_5 also contains ^{129}Xe satellites, and shows a small, four-bond, coupling to ^{19}F. (d) The ^{19}F spectrum of ^{15}N-labeled [HCN—KrF]$^+$ in BrF_5 shows doublet splittings due to ^{15}N and ^{1}H, while krypton isotopic shifts confirm that these nuclei are all linked in one chemical species. The spectra are taken, with permission, from Refs [13] and [14].

from KrF_2, though the much lower stability of krypton compounds makes it harder to be sure. It has been found that the XeF^+ ion reacts with nitriles, RCN, where R may be H, alkyl or perfluoroalkyl, to give simple ionic species $[RCN-XeF]^+$, while reaction of KrF_2 with complexes of RCN with AsF_5 gives $[RCN-KrF]^+$ for R = H or fluoroalkyl. The HCN-KrF complex cation appears to explode on warming above 223 K, even in solution in liquid HF. All these species clearly contain bonds between rare gas atoms and nitrogen, which is a significant extension of the range of chemical behavior of the rare gases.

The reactions of $XeF^+AsF_6^-$ with HCN in liquid HF or BrF_5 solvent were studied by multinuclear NMR and Raman spectroscopy [13]. The ^1H, ^{19}F, ^{13}C, ^{14}N and ^{129}Xe spectra [Fig. 10.9(a)–(c)] showed the expected signals and couplings for the simple species $[HCNXeF]^+$, and confirm that all five atoms are bonded to one another. The Raman spectrum of the solid isolated on removal of solvent contained lines assigned to the anion AsF_6^- and others compatible with the linear cation. Some of the lines, including the Xe—F stretch at 564 and 569 cm^{-1}, showed a doublet structure, probably because of factor group splitting, while others were probably hidden by the strong lines due to the anion. On the whole, though, the Raman evidence seems to confirm the strong evidence from the NMR spectra that a complex has been formed. A weak band at 368 cm^{-1} was tentatively assigned to the Xe—N stretching mode.

The krypton complex cation $[HCN-KrF]^+$ was formed [14] in the reaction of KrF_2 with $[HCNH]^+AsF_6^-$ in liquid HF or BrF_5 at 213 K. The Raman spectrum was obtained from the solid left on evaporating the solvent from a solution in HF, below 220 K; it included lines at 560 cm^{-1}, assigned to the KrF—F stretch, and 2158 cm^{-1}, assigned to the C—N stretch. The NMR spectra [Fig. 10.9(d)] were studied in liquid BrF_5 solvent. The ^{19}F spectrum of the normal species showed a doublet due to coupling to the single proton, while ^{15}N- or ^{13}C-substituted samples showed an additional doublet coupling in each case. The presence of H, C, N and F in a single species was confirmed by study of the ^1H and ^{15}N spectra.

A very unusual aspect of the ^{19}F spectra was the appearance of additional signals close to the basic doublet due to three of the five naturally occurring zero-spin isotopes of krypton [Fig. 10.9(d)]. The secondary isotopic chemical shift differences are large enough (about 0.0138 ppm per atomic mass unit, or 5 Hz per 2-unit mass difference between the major isotopes at 236 MHz) to give distinct signals for ^{82}Kr, ^{84}Kr and ^{86}Kr, which have the largest natural abundances (11.6%, 56.9% and 17.4%).

Similar reactions with perfluorinated nitriles [15] gave both xenon and krypton complexes, the latter decomposing above about 213 K. A related compound,

10.I

trifluoro-s-triazine, gave crystals of $[C_3F_3N_3.XeF]^+AsF_6^-$ (10.I). The ^{19}F NMR spectrum included resonances attributable to the two types of ring fluorine, and also a signal assigned to a unique fluorine on xenon, with ^{129}Xe satellites and a triplet coupling to two equivalent fluorine nuclei on the ring. The Raman spectrum of the isolated crystals of this compound showed bands assigned to the Xe—F stretch at 553 and 544 cm^{-1}, very close to those in the $[HCNXeF]^+$ ion, at 564 and 559 cm^{-1}. It is interesting that the Xe—F and Kr—F stretching frequencies are apparently so close in these cations, despite the smaller mass of krypton, reflecting the weakness of the Kr—F bonds. As in the other complexes, the Xe—N stretch seems to be very weak, and has been tentatively assigned at 313 cm^{-1}.

10.5   Tetrahydroborates

The tetrahydroborates are a remarkable family of compounds, most of which have unusual chemical and structural characteristics. The alkali metal tetrahydroborates (also known as borohydrides) are very ordinary, well-behaved, ionic compounds. The BH_4^- ion is isoelectronic with CH_4 and NH_4^+, and is about as interesting. It is a convenient source of hydride ion, and as the alkali metal salts are soluble in water, they are valuable reducing agents. But so far as structural matters are concerned, they are not very exciting.

In contrast, almost all other tetrahydroborates are more or less covalent. They are extremely reactive, igniting spontaneously in air, even at very low temperatures; they are often volatile [$U(BH_4)_4$ is among the most volatile of uranium compounds]; and they exhibit an astonishing variety of structural forms. In all cases, the BH_4 groups are linked to a central (usually metal) atom by hydrogen bridges. Usually there are two such bridges, as in the case of $Ti(BH_4)(\eta^5-C_5H_5)_2$ (10.II), the structure of which has been determined by X-ray crystallography [16]. Triple bridges are also found quite frequently, in such compounds as $U(BH_4)(\eta^5-C_5H_5)_3$ (10.III) and $Zr(BH_4)_4$. Singly-

$$Cp = \eta^5-C_5H_5$$

10.II   **10.III**

bridged tetrahydroborates are much rarer. One well-characterized example is $Cu(BH_4)(PMePh_2)_3$ part of which is shown in Fig. 10.10. This structure was determined by analysis of X-ray and neutron diffraction data [17], a procedure which is often used for these compounds, as the hydrogen atoms contribute little to the X-ray diffraction intensities. The structure shown in Fig. 10.11, determined by X-ray diffraction, is even more remarkable than those we have already described. Of the four hydrogen atoms bonded to each boron atom, one is in a terminal position, two are bridged to cobalt atoms, while the fourth one is in a triple-bridging position, linked to boron and to both cobalt atoms [18].

The different modes of bonding of tetrahydroborates give rise to distinct and characteristic patterns of vibrational frequencies [19]. A doubly-bridged group with

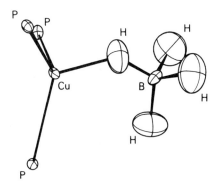

Fig. 10.10 P$_3$CuBH$_4$ core of the complex Cu(BH$_4$)(PMePh$_2$)$_3$ showing 50% probability ellipsoids. Redrawn from Ref. [17].

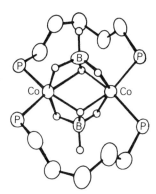

Fig. 10.11 Part of the structure of {Co(BH$_4$)[Ph$_2$P(CH$_2$)$_5$PPh$_2$]}$_2$·0.5C$_6$H$_6$. The phenyl groups and all hydrogen atoms bonded to carbon are omitted. Redrawn from Ref. [18].

C_{2v} local symmetry has four stretching modes, two of the terminal bonds (typically between 2400 and 2600 cm^{-1}) and two of the bridges (1950 to 2150 cm^{-1}). A triply-bridged group with C_{3v} local symmetry has one terminal and two bridge stretching modes, at 2450-2600 and 2100-2200 cm^{-1}, respectively, while a singly bridged group, also with C_{3v} symmetry, has two stretches of terminal bonds and one of the bridge, at 2300-2450 and *ca.* 2000 cm^{-1}. The deformations of the groups are also characteristic, and so it is usually possible to ascertain the number of bridges in the BH$_4$ groups in a new compound, simply by recording vibrational spectra.

In contrast to this, some other spectroscopic techniques are superficially much less helpful. The fact that different structural forms occur in otherwise similar molecules suggests that the energy differences between isomers are small. If, therefore, triply-bridged forms are transition states between doubly-bridged states, or vice versa, we should expect the barrier to scrambling of the hydrogens to be small. This is found to be so. In almost every case, the ^{1}H NMR spectrum of a tetrahydroborate shows only one type of boron hydride, usually with coupling to ^{11}B, indicating that fast exchange is occurring. This behavior persists down to the lowest temperatures at which spectra are obtainable (typically 125 K), so the rearrangements must be extremely facile.

The exceptions to the rule of fast exchange are rare. One such example is the doubly-bridged anion, [Mo(BH$_4$)(CO)$_4$]$^-$, for which the fluxional process is sufficiently slow at 193 K for two chemically distinct boron hydride resonances to be observed [20] (Fig. 10.12). Similarly the iridium complex 10.IV was shown to be non-fluxional at room temperature [21], and several other similar systems have now been discovered. But perhaps most remarkable of all is the osmium complex 10.V,

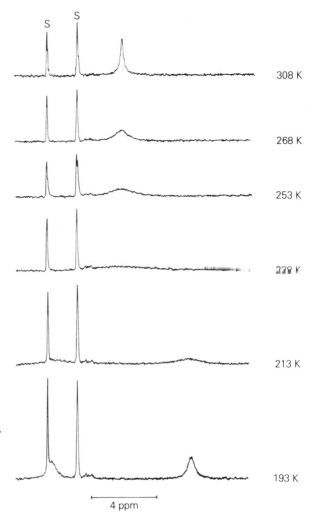

Fig. 10.12 ^{1}H NMR spectra, with ^{11}B decoupling, of $[Mo(BH_4)(CO)_4]^-$. The peaks marked S are due to the solvent. At the lowest temperatures, two resonances of equal area are seen: on warming, these broaden, coalesce, and then sharpen. Adapted, with permission, from Ref. [20]. Copyright (1977) American Chemical Society.

308 K

268 K

253 K

233 K

213 K

193 K

4 ppm

which on warming to 360 K shows exchange between bridging hydrogens and the metal hydrides, while leaving the terminal boron hydrides unaffected [22].

In general, the exchange processes occurring in these systems are fast on the NMR timescale, but slow on the timescales of rotational or vibrational spectroscopy, or of diffraction methods. So it is from NMR experiments that we can hope to gain ideas about the mechanism of exchange. For an example we will consider $Al(BH_4)_3$, as the nuclear spin of aluminum (^{27}Al, 100% abundance, spin 5/2) gives some important extra information. The compound has doubly-bridged tetrahydroborate groups: this is clear both from its vibrational spectra, and from analysis of gas-phase electron

10.IV

10.V

R = cyclo$-C_5H_9$

diffraction data [23]. The aluminum and the three boron atoms lie in a plane, and the three AlH_2B planes are probably perpendicular to this, so that the overall symmetry is D_{3h}, although slight twisting, giving D_3 symmetry, cannot be completely ruled out.

The hydrogen atoms are therefore of two chemically distinct types, but as usual with these compounds only one multiplet is seen in the proton NMR spectrum, at all temperatures studied [24]. The observed pattern has many overlapping lines, and appears as one broad resonance [Fig. 10.13(a)], so that the interpretation is not clear without the help of decoupling experiments. With ^{27}Al decoupling the spectrum is a 1:1:1:1 quartet, due to coupling with ^{11}B (80%, spin 3/2), with weaker lines attributed

(a)

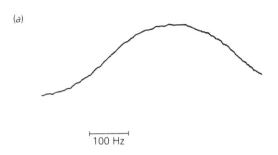

100 Hz

(b)

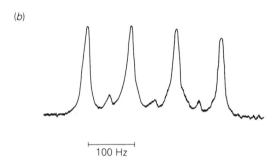

100 Hz

(c)

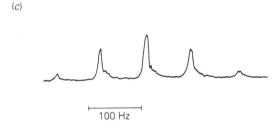

100 Hz

(d)

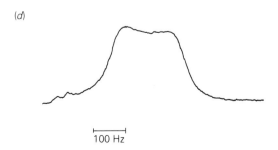

100 Hz

Fig. 10.13 NMR spectra of $Al(BH_4)_3$: (a) ^1H spectrum, (b) ^1H spectrum with ^{27}Al decoupling, (c) ^{11}B spectrum and (d) ^1H spectrum with ^{11}B decoupling. Redrawn from Ref. [24].

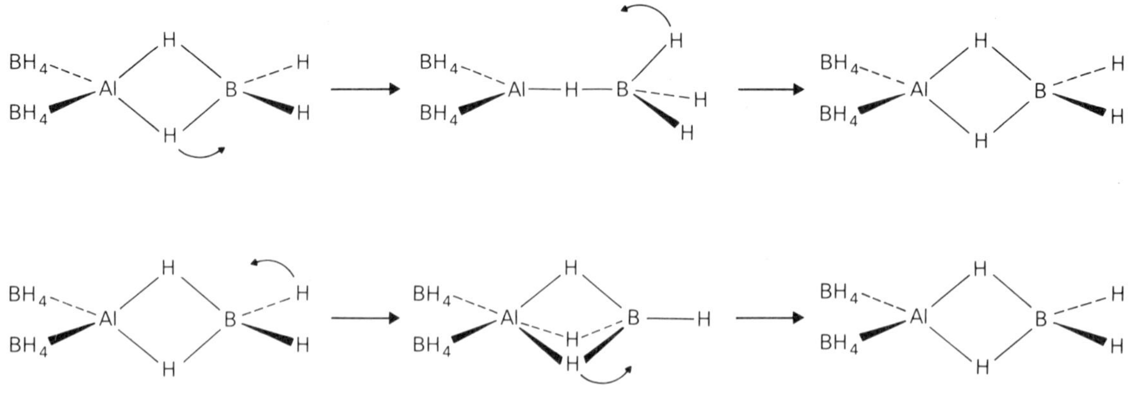

Fig. 10.14 Possible mechanisms for exchange of terminal and bridge hydrogen atoms in Al(BH₄)₃. For clarity, only one BH₄ group is shown in detail.

to hydrogens bound to ^{10}B (20%, spin 3) [Fig. 10.13(b)]. Observation of BH couplings shows that the hydrogen atoms stay bonded to the same boron atom. This is confirmed by observing the ^{11}B spectrum, which shows a 1:4:6:4:1 quintet [Fig. 10.13(c)], with each boron atom therefore remaining bonded to the same four hydrogen atoms. The ^{1}H spectrum with ^{11}B decoupling is a broad, flat-topped resonance [Fig. 10.13(d)], which is composed of six broad lines of equal intensity. The exchange process therefore involves retention of AlH coupling, indicating that the same twelve hydrogens stay attached to the aluminum, and all are equally coupled to it. So each tetrahydroborate unit remains intact, and does not exchange hydrogen atoms with other such groups, in the same or in different molecules, and the aluminum retains its three rapidly tumbling BH₄ units. How this happens is not known, but it seems probable that singly- or triply-bridged intermediates are involved, as shown in Fig. 10.14.

One interesting point about the quite sophisticated NMR experiments described above is that they were first reported in 1955, when NMR was still in its infancy [24]. They gave an understanding of the structure and dynamics of Al(BH₄)₃, and our ideas about this molecule have not changed greatly since then. Further studies have shown that on warming solutions to *ca.* 360 K the ^{1}H NMR spectrum becomes a well-resolved multiplet, showing coupling to ^{11}B and ^{10}B only, and that this new form of spectrum persists on cooling to room temperature, only reverting to the original form at *ca.* 230 K [25]. This behavior was said to indicate that a second structural form of Al(BH₄)₃ exists, but it is also possible that at high temperatures a small amount of something catalyzing intermolecular exchange of BH₄ groups is formed. If the material is redistilled it behaves as it did before warming.

In contrast to this, our ideas about beryllium bis(tetrahydroborate) have changed several times, and the present position is perhaps more confused than ever. The first structural study, by electron diffraction, indicated that the heavy atoms had a linear B—Be—B configuration, with three bridging hydrogen atoms arranged round each Be—B bond. Overall symmetry was D_{3d} (structure 10.VIa) or D_{3h}. However, the infra-red spectrum was interpreted in terms of a doubly-bridged structure (10.VIb) and the electron diffraction data were shown to be consistent with this idea. Then another electron diffraction study was published [26]. It is clear from the radial distribution curve [Fig. 10.15(a)] that there is no peak at 3.6 Å, which is where the

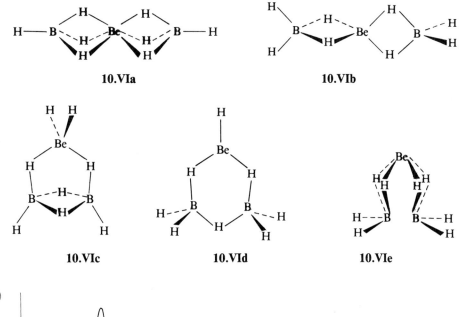

10.VIa

10.VIb

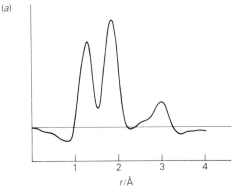

10.VIc

10.VId

10.VIe

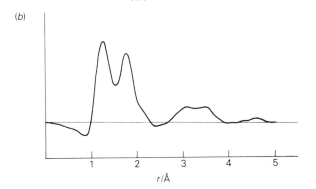

(a)

(b)

Fig. 10.15 Radial distribution curves for gaseous $Be(BH_4)_2$, obtained in separate experiments. Curve (a) is redrawn from Ref. [26], and curve (b) from Ref. [27]. Note particularly the absence of a peak near 3.6 Å in curve (a).

$B \cdots B$ distance in a linear structure must be, as the B—Be distance is close to 1.8 Å. The curve could be fitted only by assuming a triangular structure (10.VIc), which has all four bridging hydrogen atoms at the same distance from the beryllium atom. Publication of this structure was followed by a spate of reports favoring triangular structures. Analysis of vibrational data led to structure 10.VId, with three bridging hydrogen atoms and terminal BH_2 groups, and this was supported by mass spectroscopic evidence. Then the same authors suggested structure 10.VIe instead, with no terminal Be—H bonds. Non-linear structures were also supported by measurement of a dipole moment $(2.1 \pm 0.5$ D) and by electric deflection experiments. There seemed to be no doubt – beryllium borohydride was triangular. Then suddenly, a new electron diffraction study was reported [27] – linear again! The new radial distribution curve [Fig. 10.15(b)] is markedly different from its predecessor. The relative intensities of the first two peaks have changed dramatically, and a peak due to a long $B\cdots B$ distance at 3.6 Å is clearly visible. So whatever else is happening, it is now plain that the substances studied in the two electron diffraction experiments were not the same. The new data were fitted best by structure 10.VIa, with D_{3d} symmetry, but other linear structures having C_{3v} symmetry could not be ruled out.

At about this time crystalline $Be(BH_4)_2$ was studied by X-ray diffraction, and a polymeric structure (Fig. 10.16) was found [28]. This involves some BH_4 groups which are bound only to one beryllium atom, and some which are bound to two, so

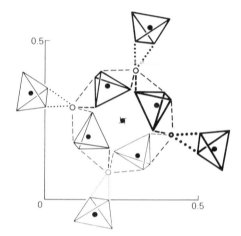

Fig. 10.16 Representation of the structure of crystalline $Be(BH_4)_2$. Half of the BH_4 groups (shown as tetrahedra) are linked by two hydrogen bridges to beryllium atoms (shown as circles). The remaining-groups and the beryllium atoms are joined in helical polymers, each BH_4 being joined to two Be atoms by two pairs of hydrogen bridges. The figure shows one turn of a helix around a 4_1 axis. Redrawn from Ref. [28].

that the beryllium atoms are six-coordinate. Could it be that on initial evaporation of this crystalline solid, one form of gaseous $Be(BH_4)_2$ was formed, which then rearranged to a second form? New vibrational data based on both Raman and infra-red spectra [29] were interpreted in terms of a mixture of two forms, only one of which persisted in matrix-isolated samples. This form was believed to have triple-bridged groups, with stretching frequencies of ca. 2645 cm^{-1} (terminal), 2255 and 2175 cm^{-1} (bridging), as in structure 10.VIa, but with the Be atom displaced, giving C_{3v} symmetry. Then one more very interesting electron diffraction experiment was performed [30]. One sample of $Be(BH_4)_2$ was split into two. One part was crystallized, and the other was rapidly frozen, to give an amorphous solid. Diffraction patterns from the gases evaporating from these two samples were recorded – and were different. Unfortunately, the results of the analyses of the two patterns have not been reported. So we are now certain that two different gases can be obtained, but we still cannot be sure that both consist entirely of $Be(BH_4)_2$ molecules.

The final pieces of experimental information come from a study of NMR spectra of gaseous $Be(BH_4)_2$ [31]. Such experiments are not easy, but are of much greater value than studies of solution spectra, as we do not need to consider whether the structure in solution is the same as that of the gas. The ^1H spectrum [Fig. 10.17(a)] shows just one multiplet, with coupling to ^{11}B and ^{10}B, while the ^{11}B spectrum [Fig. 10.17(b)] is a quintet, with coupling to four hydrogens. As all hydrogens have the same chemical shift there can only be one chemical species present, and the BH_4 groups must undergo rapid internal rearrangement of hydrogen atoms, just as in other tetrahydroborates. Moreover, the retention of couplings between one boron and four

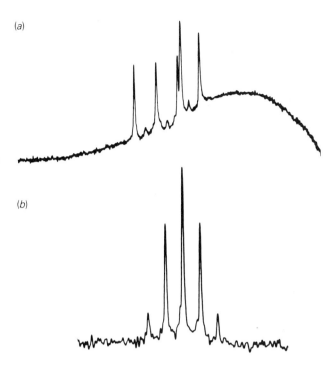

(a)

(b)

Fig. 10.17 (a) ^{1}H and (b) ^{11}B NMR spectra of gaseous $Be(BH_4)_2$. The ^{11}B spectrum shows coupling to four hydrogens, while the hydrogens show coupling to ^{11}B and to ^{10}B. This extra peak in the ^{1}H spectrum is due to residual protons in deuteriated $SiMe_4$. Adapted, with permission, from Ref. [31]. Copyright (1978) American Chemical Society.

hydrogen nuclei excludes the possibility of exchange of hydrogens between groups, via structures such as 10.VIc or 10.VId. We can therefore say quite unequivocally that the compound, even after being in the gas phase for several hours, consists of two separate BH_4 groups bound to a Be atom, probably with a linear heavy-atom arrangement. The question of the number of bridging hydrogens is still unanswered, but it may be that there is an equilibrium between forms such as 10.VIa, 10.VIb and the intermediate form 10.VIf. The identity, mode and circumstances of formation, and the structure of the second species, which may be a triangular isomer or some other unsuspected substance, remain obscure.

10.VIf

10.6 Sodium orthonitrate, Na_3NO_4

Nitrates of the formula MNO_3, containing the trigonal planar NO_3^- ion, are of course very familiar; they contrast with the equally familiar phosphates M_3PO_4, containing the tetrahedral PO_4^{3-} ion. Compounds of the formula M_3NO_4 were first prepared many years ago, but nothing was known of their structures until 1977, when a microcrystalline material of formula Na_3NO_4 was obtained by fusing an equimolar mixture of $NaNO_3$ and Na_2O in a sealed crucible at 570 K for seven days. Its Raman spectrum [32] showed no bands attributable to NO_3^- ions (ruling out a possible formulation as a double salt of NO_3^- and O^{2-} ions) and an assignment of the spectrum (Table 10.2) in terms of the expected vibrations of a tetrahedral anion

Table 10.2 Raman bands of Na_3NO_4

Mode[2]	Observed frequency/cm^{-1}	Frequency in NF_4^+/cm^{-1}
ν_1 (a1)	843	843
ν_2 (e)	540	437
ν_3 (t_2)	1024, 1012, 1007, 988, 977	1160, 1150
ν_4 (t_2)	669, 651	604

[2]Symmetry labels as for T_d symmetry.

NO_4^{3-} was proposed. The splittings of the bands assigned to ν_3 and ν_4 were attributed to the effects of low site symmetry in the solid (no more than D_{2d}). It was pointed out that the spectrum was very similar to that of the isoelectronic NF_4^+ ion, as expected.

Attempts to grow crystals suitable for X-ray diffraction studies were hampered by the instability of the material; it decomposes on heating (probably by attacking the crucible), and on exposure to H_2O or CO_2 in air. Crystals were finally grown by controlled annealing in an argon atmosphere at 688 K over 240 days (!), with slow heating and cooling [33]. The crystals (~0.1 mm in size) proved to contain the expected NO_4^{3-} anions, on sites with no crystallographic symmetry. Despite this the ions were tetrahedral within experimental error, with bond lengths of 1.384 Å, compared with the expected 1.50 Å for a single bond. Analysis of the vibration frequencies gave a force constant f(NO) of nearly 700 N m^{-1}, which is larger than that found for the corresponding PO_4^{3-} ion (~560 N m^{-1}), but less than that in NF_4^+ (~1000 N m^{-1}). The splittings observed in the Raman spectrum can be interpreted [34] in terms of eight anions in a centrosymmetric array (factor group D_{2h}), each having no required symmetry elements, but many fewer components than expected are actually seen.

10.7 Some cyclopentadienyl compounds

There are many cyclopentadienyl (Cp) derivatives, of metals and of non-metals, including main group and transition elements, actinides and lanthanides. Some, like the alkali metal derivatives, are essentially ionic. Most d-block transition metals form covalent η^5 π-bonded complexes, MCp_2, while the lanthanides and actinides typically form MCp_3, again with π-bonded η^5 rings. There is a third common mode of bonding, in which the ring is bound by a single σ bond, and there are also intermediate bonding arrangements. With such diverse possibilities, these compounds will inevitably raise structural questions, and we deal with some of these in this section. First, we consider the conformation of the rings in ferrocene, and then look at the electronic structure of per-methylated rhenocene. The bonding in beryllocene is the subject of the third subsection – is it η^1 or η^5, or both, or neither? Finally we turn to one of the many cyclopentadienyls which exhibit fluxional behavior.

10.7.1 Ferrocene: eclipsed or staggered?

The story of the preparation and identification of ferrocene in 1951 is an instructive one. During attempts to prepare the hydrocarbon fulvalene, cyclopentadienyl magnesium bromide was treated with iron(III) chloride [35]. Reaction was expected in accord with Eqn (10.1).

$$6C_5H_5MgBr + 2FeCl_3 \rightarrow 3C_5H_5{-}C_5H_5 + 2Fe + 3MgBr_2 + 3MgCl_2 \qquad (10.1)$$

The reaction yielded an orange solid, soluble in organic solvents. It had a molecular weight of 186.5, and was shown by analysis to have the formula $C_{10}H_{10}Fe$. It was recognized at once that a stable organo-iron compound had been prepared for the first time, and this accidental discovery was the first step in the development of modern organometallic chemistry.

Originally it was suggested that the new compound was $Fe(\eta^1\text{-}C_5H_5)_2$ (although that nomenclature was not used at the time), but further studies [36] of the infrared and ultraviolet spectra, and observations of diamagnetism and a very small or zero dipole moment, led to the proposal that the compound had two parallel rings, with the iron atom sandwiched between them, and equidistant from all ten carbon atoms. It was recognized that in such a structure the rings could eclipse one another (10.VIIa), with the molecule having D_{5h} symmetry, or that they could be staggered (10.VIIb), with D_{5d} symmetry. An intermediate structure with D_5 symmetry is also possible.

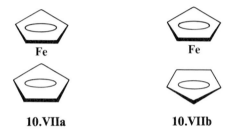

10.VIIa   **10.VIIb**

As bis(cyclopentadienyl)iron, also called ferrocene, could easily be obtained in crystalline form, it was soon studied by X-ray diffraction, and the sandwich structure was confirmed. In the first study using three-dimensional data [37] a monoclinic unit cell was found, and the space group was $P2_1/a$. With a cell volume of 409 Å^3 and density of 1.51 Mg m^{-3} there must be two molecules per unit cell, and on symmetry grounds the two molecules are required to be centrosymmetric. This rules out D_{5h} symmetry for the molecules, but allows D_{5d}. Of course, there cannot be perfect five-fold symmetry in a crystal, and the refined carbon atom coordinates deviated substantially from ideal positions. But uncertainties were large, as the diffraction intensities were measured visually from films, and the final conclusion was therefore that ferrocene molecules favored the staggered conformation.

Ferrocene is fairly volatile, so it is not surprising that it was soon studied in the gas phase by electron diffraction. Early investigations confirmed that the sandwich structure was retained in the vapor phase, but gave no clear information about the conformation. A later study [38] was more successful. Distinguishing eclipsed and staggered forms depends on analysis of the part of the radial distribution curve relating to the inter-ring carbon–carbon distances. In the eclipsed conformation there are five equal C$\cdots$C distances of one type, ten of a second (longer) type, and ten of the longest type. In the staggered conformation the distances are slightly different, but there are ten of each of the shorter two types, and five of the longest. The differences are therefore not very great, but comparison of the experimental curve with theoretical ones for the two types (Fig. 10.18) shows clearly that the eclipsed (D_{5h}) form is favored, although deviations of a few degrees from the perfectly eclipsed form cannot be ruled out. Complete analysis of these data enabled the energy difference between the forms (i.e. the barrier to rotation) to be estimated as 4 kJ mol^{-1}.

By this time it had become clear that the original X-ray study was inadequate, as neutron diffraction work had shown that there was disorder within the crystal, and

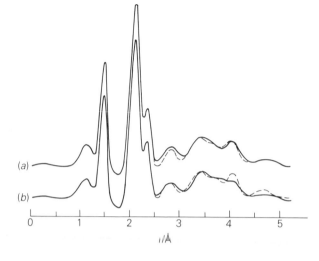

Fig. 10.18 Radial distribution curve for electron diffraction by gaseous ferrocene. The solid lines represent theoretical models for (*a*) eclipsed and (*b*) staggered models, and the dashed lines in the region above 2.5 Å represent the experimental data. Redrawn from Ref. [38].

that conclusions about molecular symmetry could not be drawn from arguments about crystal symmetry. It was also known that ruthenocene, $Ru(\eta^5\text{-}C_5H_5)_2$, had an almost eclipsed structure in the crystalline phase. Two sets of new data for ferrocene were collected with a diffractometer, one with the crystal at room temperature, and one with it at 173 K [39]. With the new, higher quality data, it was possible to obtain reliable information about the thermal parameters for the carbon atoms. These showed that the major axes of the thermal ellipsoids lay tangential to the rings (Fig. 10.19), and that cooling the crystal did not reduce this component of the apparent

Fig. 10.19 Projections of the vibrational ellipsoids for the carbon atoms in ferrocene, onto the mean plane of the ring, (*a*) at room temperature and (*b*) at 173 K. Redrawn from Ref. [39].

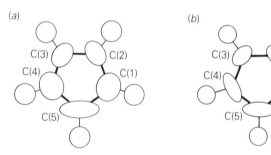

movement. However, the ellipsoids for the five atoms in a ring were not equally eccentric, suggesting that the effect could not be accounted for simply in terms of rotational disorder. The most satisfactory model involved disorder as shown in Fig. 10.20, with each ring represented by ten carbon atoms of half weight. The positions

Fig. 10.20 Result of refinement of disordered ring positions in ferrocene. The pentagons show positions of carbon atoms in the disordered structure of one ring. and the positions found assuming no disorder are shown as labeled dots. Redrawn from Ref. [39].

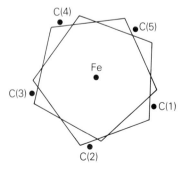

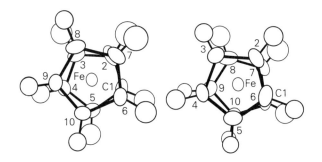

Fig. 10.21 Views of the two molecules of ferrocene in the asymmetric unit of the triclinic phase, at 148 K. The rings are twisted about 9° from the fully eclipsed conformation. Redrawn from Ref. [40].

which combine to give the full-weight C(5) atom are seen to be much further separated from each other than those giving C(2) and C(3).

On cooling ferrocene below 164 K, a triclinic phase is formed, and in this form there is no disorder. An X-ray diffraction study [40] shows that the molecules now have D_5 symmetry, with the rings twisted about 9° away from the eclipsed orientation, as shown in Fig. 10.21. The result is then very close to what exists in the high temperature modification. A molecule in this form has one ring as in Fig. 10.20, while the second is inverted through the metal atom. The outcome is a molecule with D_5 symmetry, having a conformation which is very similar to that which had finally been deduced to exist in the disordered low temperature modification. In yet another phase, an orthorhombic form studied at 98 K [41], the rings are perfectly eclipsed, and the molecules have D_{5h} symmetry. In the end, therefore, all the diffraction results are consistent. In the crystalline phases the rings are twisted at the most a few degrees from the eclipsed conformation, while in the gas they are again eclipsed, or very nearly so.

10.7.2 Decamethylrhenocene

Although there are very many transition metal complexes of formula $M(C_5R_5)_2$ known, few containing second- or third-row metals in the d^5 configuration have been cliaracterized. The geometry and electronic structure of such species are of considerable interest, so when it was discovered that $[Re(C_5Me_5)_2H]$, formed from rhenium metal and pentamethyl-cyclopentadiene, gave a deep purple solid on photolysis, the nature and structure of the product were seen as potentially important [42]. The product was shown by analysis to be $[Re(C_5Me_5)_2]$; its molecular weight was found by mass spectrometry and from vapor pressure measurements to correspond to that of the monomer. Since the compound was stable at room temperature in the absence of air moisture, its crystal structure could be determined by conventional methods. The compound was found to have a sandwich structure with parallel and eclipsed rings, with methyl groups arranged with two hydrogens towards the metal and one away. The infrared and Raman spectra of both solid and matrix-isolated species showed the expected ligand-based bands at frequencies around 2900, 1450 and 1030 cm^{-1}, and low frequency features assigned to skeletal modes; there was no sign of resonance enhancement with change in the frequency of the exciting line in the Raman spectrum. In the ultraviolet and visible spectra of matrix-isolated samples [Fig. 10.22(a)] several well-defined progressions were observed, and progressions could be detected (though they were less well defined) in the spectra of samples dissolved in toluene. The compound also showed laser-induced fluorescence when irradiated in argon matrices.

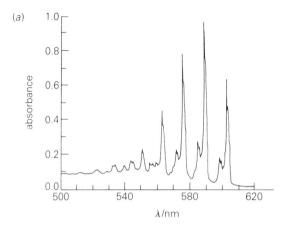

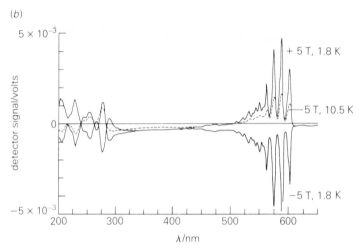

Fig. 10.22 (*a*) Visible absorption spectrum of [Re(C$_5$Me$_5$)$_2$] in an argon matrix at 20 K. (*b*) Magnetic circular dichroism of [Re(C$_5$Me$_5$)$_2$] in argon matrices at 1.8 K, recorded at + 5 and – 5 T magnetic flux density (solid lines), and at 10.5 K and + 5 T (broken line). The spectra are adapted with permission, from Ref. [42]. Copyright (1988) American Chemical Society.

Once the molecular geometry had been determined, the interesting question about this compound concerned its electronic structure. If we suppose, as is generally accepted, that the electronic structure of such species can be defined satisfactorily by the ligand field model, the metal *d*-orbitals in such a parallel sandwich compound split under the operations of the point group D_{5h} into $e_2(xy, x^2 - y^2)$, $a_1(z^2)$ and $e_1(xy, yz)$ in order of decreasing tightness of binding. Depending on the energy gaps and other factors, there are three possible electron configurations [Fig. 10.23(*a*)] giving rise to the possible terms $^2A_1$, $^2E_2$ and $^6A_1$. The $^2E_2$ and $^2A_1$ terms would be split into higher and lower components by spin-orbit coupling [Fig. 10.23(*b*) and (*c*)]. Establishing the ground state for [Re(C$_5$Me$_5$)$_2$] turned out to be a complicated matter, resolved by an elegant combination of magnetic circular dichroism, magnetic susceptibility measurements, ultraviolet photoelectron spectroscopy and ESR spectroscopy.

The MCD spectrum of [Re(C$_5$Me$_5$)$_2$] in an argon matrix at 1.8 K [Fig. 10.22(*b*)] showed a structured peak at 600 nm with positions of the maxima in very good agreement with those found in the absorption spectrum [Fig. 10.22(*a*)]. The inverse

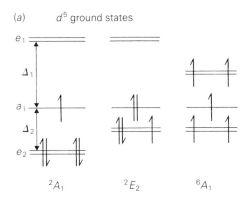

(a) d^5 ground states

$^2A_1$ $^2E_2$ $^6A_1$

(b)

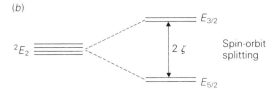

$^2E_2$ $E_{3/2}$

2ζ Spin-orbit splitting

$E_{5/2}$

(c)

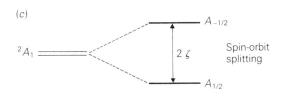

$^2A_1$ $A_{-1/2}$

2ζ Spin-orbit splitting

$A_{1/2}$

Fig. 10.23 (*a*) The three possible electronic configurations and terms that can be adopted by d^5 sandwich complexes as their ground states. The effects of spin-orbit coupling on the $^2E_2$ and $^2A_1$ terms are shown in (*b*) and (*c*).

temperature dependence of the MCD spectrum showed that the spectrum arose from the C term of a paramagnetic molecule (see Section 6.10.2). This tells us that ground state is degenerate (i.e. in this case $^2E_2$) and the excited state non-degenerate. Moreover, a method has been developed that allows the estimation of $g_\parallel$ either from the ratio of integrated differential absorbance to integrated absorbance, or from the shape of the magnetization curve. The first gave $g_\parallel$ as 5.3 ± 1.4 at 10.9 K and the second gave $g_\parallel$ as 5.07 ± 0.19. Since for a molecule with the $^2A_1$ ground state a value for $g_\parallel$ of about 2 would be expected, and since $[V(\eta^6\text{-toluene})_2]$, which has the ground state $^2A_1$, gave g-values close to 2, it is clear that $[Re(C_5Me_5)_2]$ has the ground state $^2E_2$, or $E_{5/2}$ if spin orbit coupling is allowed for, and not $^2A_1$. It was not possible to determine by how much $g_\perp$ differed from zero (the value expected for molecules with true axial symmetry) by these methods. ESR experiments gave no signal down to 77 K for samples in frozen solutions, but a solution in toluene gave a sharp sextet centered at very low field, the sextet arising from coupling to the two isotopes of Re with $I = 5/2$. The value of $g_\perp$ determined from the ESR spectrum was 5.081 ± 0.003, which agrees very well indeed with that determined from the MCD spectrum. No corresponding signal for $g_\perp$ was detected up to a field limit of 1.9 T, implying that $g_\perp < 0.34$. However, no ESR signal would be observed if $g_\perp$ were zero.

All the data obtained in matrices at low temperatures and for condensed phase samples were consistent with the $E_{5/2}$ ground state, but the UPS spectrum of gaseous samples, recorded with both He(I) and He(II) radiation, could not be fitted satisfactorily on the basis of the $E_{5/2}$ ground state. The fit was slightly better on the basis of the $A_{1/2}$ configuration, but the most satisfactory fit was obtained for a mixture, with

80% of the $A_{1/2}$ state and 20% $E_{5/2}$. Ligand field parameters for [Re(C$_5$Me$_5$)$_2$]$^+$ were determined from the electronic spectra and by iterative calculations using analogies with related species, and suggest that for [Re(C$_5$Me$_5$)$_2$] the $A_{1/2}$ state is about 0.125 eV above the $E_{5/2}$ state. However, for the neutral species it is possible that electron repulsion is less than for the cation because the orbitals of the neutral species are more extended, and so the ground state in the gas phase might be $A_{1/2}$. Measurements of magnetic susceptibility for the crystal at 4.3 K gave μ 1.59-1.74 μ_B, close to the value expected for the $A_{1/2}$ ground state (μ 1.73 μ_B), and so this seems to be the ground state in the crystal. The magnetic moment measured in solutions at 298 K is 1.95 μ_B, suggesting that there is an equilibrium, between $A_{1/2}$ and $E_{5/2}$ states. However, electronic absorption spectra, recorded for solutions over a range of temperatures, indicate that there is only one electronic state occupied, and that this is the same one that is observed in low-temperature matrices and frozen solutions.

It is therefore concluded that in argon and N$_2$ matrices and in frozen toluene the ground state is $E_{5/2}$, but that in the gas phase there is an equilibrium between $E_{5/2}$ and $A_{1/2}$ states. In solution, there may be an equilibrium, but the evidence is not decisive.

10.7.3   How asymmetric is beryllocene?

In the preceding two sections we have considered the geometric and electronic structures of two sandwich compounds, in which the metal atoms lie between two parallel rings. They are described as η^5 π-bonded complexes. In the solid phase, there is no clear distinction between this type of compound and ionic metal cyclopentadienyls. The cyclopentadienyl groups have local five-fold symmetry, and characteristically simple vibrational and NMR spectra. However, there is a third mode of bonding – the η^1 or σ-bonded metal-ring link found in HgCp$_2$ and R$_3$SiCp. Such compounds are usually fluxional, the metal atom and ring moving with respect to each other so that first one, then another, carbon atom is bonded to the metal. This process may leave all ring carbon and hydrogen atoms apparently equivalent on the NMR timescale, but decreasing the rate by lowering the temperature may allow the motion to be frozen out, so that distinct ^{1}H and ^{13}C signals are seen. As the timescale of vibrational motions is much shorter than that for NMR spectroscopy, species containing rings with low symmetry have characteristic bands including in particular one C–H bond stretching mode which is at low frequency (<3000 cm^{-1}) because of the saturated character of the carbon atom involved.

The beryllium compound BeCp$_2$, usually called beryllocene, proved to be one of the strangest of the simple metal cyclopentadienyls, and the question of its structure generated a long-running controversy. A number of possible structures were suggested and at times there seemed to be real discrepancies between the results of different experiments. Quantum mechanical calculations turned out not to be very useful, as the energies found for very different structures were remarkably similar. The discrepancies now seem to have been resolved, and the structure is no longer in doubt; but it is still an unusual compound.

After beryllocene was first prepared in 1959, early work showed its IR spectrum to be more complex than those of the symmetrical sandwich molecules such as ferrocene, and it was also found to have a large dipole moment in solution, 2.24 D in cyclohexane. It was thus clear from the start that the molecular symmetry was lower than that of ferrocene, though it was not clear whether this involved two dissimilar

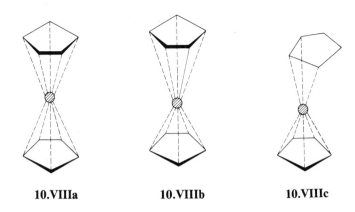

10.VIIIa 10.VIIIb 10.VIIIc

rings or two identical but asymmetric rings. An early X-ray study appeared to show a unit cell structure formally identical to that of ferrocene, implying a centrosymmetric molecule [10.VIIIa], which is incompatible with the observation of a dipole moment in solution. Thus there was a possibility that the molecular structure in the crystalline phase was not the same as that in gaseous and solution phases, but it was also possible that the crystallographic results could be due to disorder in the crystal.

The compound is volatile, and the first electron diffraction study suggested a structure that was widely accepted for some years. In this, the two rings were parallel but the beryllium atom was closer to one than to the other with distances of 1.48 and 1.98 Å to the ring centers, giving overall C_{5v} symmetry for either a staggered or an eclipsed arrangement of the rings [10.VIIIb]. An IR study concluded that the spectra in solid and solution were compatible with this structure.

The ^{1}H NMR spectrum [43], however, showed only a single peak due to C_5H_5 protons, and it was necessary to assume that the beryllium atom moved rapidly from the position nearer one ring to a corresponding position near the other to account for the apparent equivalence of the two rings. This equivalence persisted down to the lowest temperature studied, 138 K. The ^{9}Be NMR spectrum was also reported to show a single peak, with no sign of coupling to protons, probably because of fast relaxation by the ^{9}Be quadrupole ($I = 3/2$), but this observation sheds no light on the position of the beryllium atom.

An X-ray study at low temperature, 153 K, showed that the structure was, in fact, disordered, with the two rings approximately parallel but not coaxial. The beryllium atom was effectively η^5 coordinated by one ring, while the other ring had slipped sideways so that the beryllium atom was under one carbon atom [10.VIIIc]. The most recent crystallographic work [44] confirms this arrangement [Fig. 10.24(a) and (b)], and the structure at room temperature [45] can be described in similar terms, except for relative rotations of the rings about their five-fold axes [Fig. 10.24(c)].

In the light of all these structural results, it was decided to re-examine the electron diffraction data. At first it appeared that they were incompatible with the X-ray structure, but a later study with new diffraction data and more systematic treatment of background subtraction [46], showed that the slipped structure fitted the data as well as (but no better than) the originally suggested C_{5v} form. The rings were parallel, 1.47 and 1.86 Å from the beryllium atom, and one ring was displaced 0.8(1) Å to one side.

Further studies of the infrared [47] and Raman [48] spectra have confirmed that the structure is essentially the same in solid, liquid, solution and vapor phases. Highly

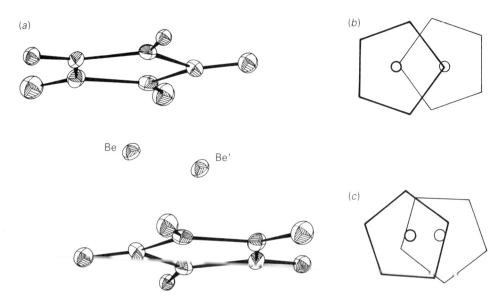

Fig. 10.24 (a) Lateral view of the beryllocene molecule in the crystalline phase at low temperature, showing the two equivalent sites for the beryllium atom. Adapted, with permission, from Ref. [44]. Top views of the low temperature (b) and room temperature (c) structures show the change in the orientation of the rings with temperature.

symmetrical ferrocene-like structures can be excluded, and it is concluded that the slipped structure probably involves the asymmetrically-coordinated ring binding through more than one carbon atom, as there is no C—H stretching band below 3000 cm^{-1}, as would be expected for a σ-bonded, η^1 ring. More recently, a study of the ^{13}C NMR spectrum [49] has shown that there is partially resolved coupling to ^{9}Be. The coupling is independent of magnetic field, and is the same for solutions in diethyl ether and cyclohexane, but collapses as the sample is cooled. These observations lead to the conclusion that relaxation of the ^{9}Be nuclei is caused by molecular inversion (i.e. rapid exchange of the beryllium atoms between the equivalent sites) rather than by tumbling of the whole molecule. Lineshape analysis leads to a value of 5.2 kJ mol^{-1} for the exchange process, and an exchange rate of between 10^9 and 10^{11} Hz at room temperature. However, that is not the end of the story. Solutions of beryllocene in benzene and toluene give paramagnetic products [50], and it seems likely that there is some interaction between the π electrons of the aromatic compounds and suitable acceptor orbitals on beryllium.

10.7.4   **Cyclopentadienyldifluorophosphine**

Reaction of $PClF_2$ or $PBrF_2$ with the potassium [51] or thallium [52] salts of cyclopentadiene, MC_5H_5, gives a volatile compound whose vapor density and mass spectrum are consistent with the formula $PF_2C_5H_5$. The IR spectrum shows the bands expected for a compound containing a PF_2 group and a σ-bonded (η^1) C_5H_5 ring, rather than a 'half-sandwich' (η^5) π-bonded ring. There are three structures (10.IXa-c) fitting these criteria.

The ^{1}H NMR spectrum at low temperatures [Fig. 10.25(a)] has three groups of resonances, with relative intensities 2:2:1. The two higher frequency groups lie in a region associated with hydrogen atoms bound to unsaturated carbons, while the low

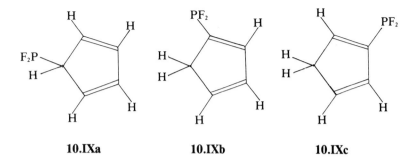

10.IXa   **10.IXb**   **10.IXc**

frequency group (which is a triplet of doublets, coupling to one phosphorus and two fluorine nuclei) arises from a hydrogen bonded to a saturated carbon atom. The compound is thus identified as isomer 10.IXa. However, on warming the resonances broaden and then coalesce to give a single group of resonances [Fig. 10.25(*b*)] with a chemical shift equal to the average of those for the five contributing protons. These changes suggest that there is a fluxional process, in which the PF_2 group can migrate round the ring, making the five hydrogens equivalent.

This interpretation is confirmed by observation of the ^{19}F NMR spectrum, which is also temperature-dependent. At 215 K it consists of a wide doublet ($^1J_{PF}$ 1172 Hz) of doublets ($^2J_{FH}$ 11.5 Hz) [Fig. 10.26(*a*)], so showing coupling only to the single proton on the carbon carrying the PF_2 group. On warming to 300 K, the pattern [Fig. 10.26(*b*)] changes to a doublet of sextets, with the fluorine now coupling equally to all

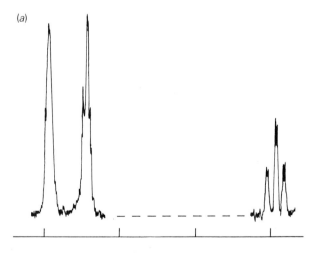

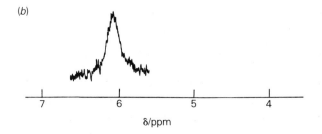

Fig. 10.25 100 MHz ^1H spectra of $PF_2C_5H_5$ (*a*) at 215 K and (*b*) at 350 K.

SECTION 10.7: SOME CYCLOPENTADIENYL COMPOUNDS   **423**

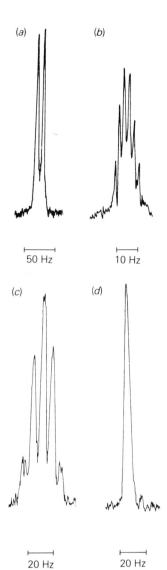

Fig. 10.26 ^{19}F and ^{31}P NMR spectra of $PF_2C_5H_5$. The ^{19}F spectrum is split into a wide doublet by ^{31}P, and only one half is shown: the ^{31}P spectrum is similarly split into a wide triplet by two fluorine nuclei, and details of the center part of this are shown. (a) ^{19}F spectrum at 215 K, (b) ^{19}F spectrum at 300 K, (c) ^{31}P spectrum at 180 K and (d) ^{31}P spectrum at 295 K.

five ring protons. The new coupling constant, J_{FH}, is 2.5 Hz, which is again the average of the contributing couplings, one of 11.5 Hz and four of approximately zero. The alterations of spectra are typical of those for fluxional systems. At low temperatures the effective lifetime of the species is long enough for the different hydrogen sites to be distinguished, and separate chemical shifts or couplings to be observed, but at high temperatures the lifetime is reduced, and everything is averaged.

Similar effects are also observed in the ^{31}P NMR spectra, but at first sight the spectra are confusing. The low-temperature spectrum [Fig. 10.26(c)] seems to be basically a triplet of quintets, while in the high-temperature spectrum [Fig. 10.26(d)] the quintets are replaced by singlets. So apparently the phosphorus nucleus couples equally to four hydrogens at low temperature, but to none when the fluxional process is fast. Of course this cannot be true. In fact, the quintet structure is a triplet of triplets, with $^3J_{PH}$ and $^4J_{PH}$ being almost equal, at 7-8 Hz. There is also a small doublet coupling, due to $^2J_{PH}$, of 2 Hz, and this can be clearly seen in the outermost lines of the quintet, as well as in the ^1H spectrum in Fig. 10.25(a). So on warming we should see sextets with an average value for J_{PH} of *ca.* 6 Hz, but we see only singlets. Why? The answer is that the three-bond and four-bond PH couplings are of opposite sign, so that the average is not 6 Hz, but *ca.* 0.5 Hz, too small to be resolved.

10.8 Some clusters of gold and other metals

The chemistry of cluster compounds has been a subject of intense activity in recent years. Some such compounds have been regarded as models of metal surfaces, and have indeed shown potential as catalysts. Others are of great theoretical interest, and a lot is now known about their electronic structures. Cluster compounds of gold are particularly interesting. Clusters with from two to thirteen atoms of gold have been characterized. The upper limit corresponds to a system having a central gold atom surrounded by twelve others, giving a centered icosahedral core, with phosphine and halogen ligands round the outside.

Some clusters containing nine gold atoms are described in the first subsection below, and the transformation of a cluster of eleven atoms into one containing an Au_{13} core is described in Section 6.8.8. Not that the chemistry ends with thirteen atoms: in Section 10.8.2 below we present evidence for the existence of clusters containing 55 metal atoms, of gold or of other metals. In the last part of this section we describe compounds in which nitrogen or carbon atoms are held at the centers of groups of five or six gold atoms, and pose the question: are these clusters with different atoms in the middle, or do they consist of gold atoms covalently bound to the central atoms, but not to one another?

10.8.1 Some Au₉ clusters

In this section we shall consider some simple clusters with just nine gold atoms, $[Au_9(PAr_3)_8]$ $[BF_4]_3$ (Ar = Ph, p-C_6H_4Me and p-C_6H_4OMe). The ^{31}P NMR spectra of these compounds in solution all show only singlets [53], suggesting either that the structures have a high degree of symmetry, so that all eight phosphine sites are symmetry-related, or that a fluxional process occurs, making the phosphines equivalent on the NMR timescale. The electronic spectra of the three complexes in solution are very similar [54], and this indicates that the structures of the metal atom cores in solution are probably the same. It is generally found in gold cluster chemistry that electronic spectroscopy provides an excellent way of distinguishing different forms: changing ligands has very little effect, but changing the core structure or the number of atoms in the cluster has very marked effects.

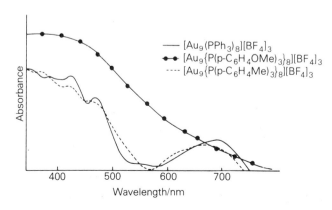

Fig. 10.27 Diffuse reflectance spectra for some gold cluster compounds in the solid phase. One spectrum is very different from the other two, suggesting that the metal cores of the compounds can have two different structures. Redrawn from Ref. [54].

— $[Au_9(PPh_3)_8][BF_4]_3$
•—• $[Au_9\{P(p-C_6H_4OMe)_3\}_8][BF_4]_3$
---- $[Au_9\{P(p-C_6H_4Me)_3\}_8][BF_4]_3$

It is therefore remarkable that diffuse reflectance spectra of the solids obtained when the three compounds are crystallized are not all the same (Fig. 10.27) [55]. The spectrum of the complex having $P(p\text{-}C_6H_4OMe)_3$ ligands, which forms golden-brown crystals, is markedly different from those of the other complexes, which form green crystals. It seems, therefore, that one or more of the clusters changes structural form on crystallization, and that two isomers of these compounds can thus exist. This is confirmed by the observation that the nitrate $[Au_9\{P(p\text{-}C_6H_4OMe)_3\}_8]$ $[NO_3]_3$ crystallizes simultaneously in two forms, one golden-brown and one green [54].

The structures of the two isomers are important, and at least one example of each has now been studied by X-ray crystallography. One of the green crystals, $[Au_9\{P(p\text{-}C_6H_4Me)_3\}_8]$ $[BF_4]_3$, has a metal core as shown in Fig. 10.28(a), with one central gold atom, while the other eight can be regarded as occupying all but four of the apices of an icosahedron [56]. The structure of a cation in $[Au_9\{P(p\text{-}C_6H_4OMe)_3\}_8]$ $[BF_4]_3$ (a golden brown form) is shown in Fig. 10.28(b) [57]. The arrangement of the gold atoms is strikingly different from that in the other form. Again, there is a central atom, but this time the remaining atoms are all equivalent, giving a crown, with very nearly perfect D_{4d} symmetry. Clearly, therefore, these gold clusters demonstrate how rearrangements of the cores of metal atoms can occur, giving major changes in structure, and therefore of structure-related properties.

It is instructive to consider how far one could characterize these solid gold compounds *without* using X-ray crystallography. Mössbauer spectra are not helpful. Those for compounds with the structures shown in Fig. 10.28 can be interpreted in terms of a single quadrupole pair of resonances, and resonances due to the central atom and the two types of peripheral atom cannot be distinguished. X-ray photo-electron studies of gold $4f$ electron binding energies are not much better. There is

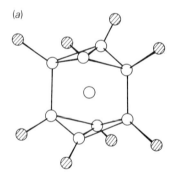

(a)

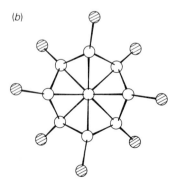

(b)

Fig. 10.28 Structures of gold clusters $[Au_9(PAr_3)_8]$ $[BF_4]_3$, (a) Ar = $p\text{-}C_6H_4Me$ and (b) Ar = $p\text{-}C_6H_4OMe$. The gold atoms are represented by plain circles and phosphorus atoms by shaded circles. Other atoms are omitted.

some evidence that it is possible to identify peaks due to central and peripheral atoms, but again, further resolution of the latter group is definitely impossible. Perhaps the most useful method so far applied is solid state NMR. The ^{31}P spectrum, with ^{1}H decoupling, of $[Au_9(PPh_3)_8] [NO_3]_3$ has two resonances, consistent with the structure of Fig. 10.28(a). However, without a crystal structure of this particular compound, rather than of one that is just related to it, we cannot be sure that the observed inequivalence does not arise from two molecules or parts of molecules unrelated by symmetry in the unit cell.

We are left with X-ray crystallography as the only satisfactory method of studying these compounds as solids. If they do not crystallize, we may be unable to characterize them properly. And if structures of solids are obtained we must always remember that the structures in solution may be completely different.

10.8.2   Some enormous clusters

Very large metal clusters are difficult to characterize, but many have now been reported. A few very stable species have been prepared that appear to contain 55 metal atoms as a core, surrounded by a sheath of ligands. The reason for the special stability of this type of cluster is presumably that it is a closed shell, in terms of the twelve-fold packing of spheres; we would expect 'magic numbers' of 13, 55, 147 . . . metal atoms to behave in this way. The 55-atom cluster has a central core of 13 metal atoms, and a surface layer of 42 atoms forming a cuboctahedron. Such clusters have been prepared for metals such as gold, rhodium, ruthenium and platinum, by simply reducing phosphine-protected metal complexes in benzene solution. The apparent generality of the preparation, the high yields (up to 80% in some cases) and the ease of formation from mono-metallic species are all in favor of some special stability of the clusters.

The gold species, formulated as $[Au_{55}(PPh_3)_{12}Cl_6]$, was the first to be reported [58]. It is prepared by reducing $[AuCl(PPh_3)]$ in benzene solution with gaseous diborane, removing volatile material and washing away the triphenylphosphine-borane adduct formed as a by-product. It is very soluble in some organic solvents, but decomposes slowly in solution on exposure to air or on heating, depositing gold mirrors. In the solid state it is stable to air. The IR spectrum shows mainly bands attributed to the triphenylphosphine ligand, with an Au—Cl stretching band assigned at 280 cm^{-1}, rather lower than in $[AuCl(PPh_3)]$ (330 cm^{-1}). The ^{31}P-$\{^{1}$H$\}$ NMR spectrum contains a single line. The formulation is based on the empirical formula from elemental analysis, and the molecular weight from studies of sedimentation in a centrifuge. No macroscopic crystals were formed, so an X-ray study was not possible.

The Mössbauer spectrum (Fig. 10.29) was most helpful. The complex, asymmetric peak was analyzed in terms of four distinct gold atom environments – the 13 core atoms, whose isomer shift is very similar to that in metallic gold, 24 peripheral atoms bearing no ligands, 12 peripheral atoms coordinated to triphenylphosphine ligands and 6 peripheral atoms coordinated to chlorine. The coordinated atoms appear as doublets due to quadrupole coupling, but the uncoordinated atoms give single peaks. The relative intensities and Mössbauer parameters of the coordinated gold atoms are in the expected ranges for phosphine and chloride ligands respectively. Nevertherless, there is still some doubt about the structure of the gold cluster. Studies of mass spectra

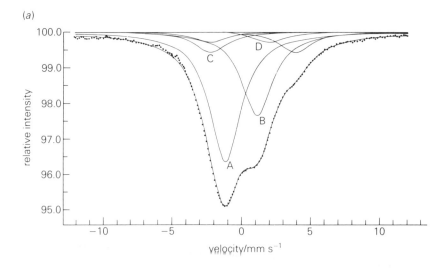

(a)

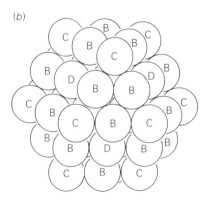

(b)

Fig. 10.29 (a) Mössbauer spectrum and (b) arrangement of gold atoms for $[Au_{55}(PPh_3)_{12}Cl_6]$. The atoms, and the corresponding peaks in the spectrum, are labeled A for the central core, B for uncoordinated peripheral atoms, C for atoms coordinated by triphenylphosphine and D for atoms coordinated by chlorine. Adapted, with permission, from Ref. [58].

suggest that the sample is inhomogeneous and that the preparation is not exactly reproducible. An alternative model based on vertex-sharing icosahedra has been proposed.

Two distinct Rh_{55} clusters [59, 60], each with 12 phosphine and 20 chloride ligands, have been studied. It was hoped that ^{103}Rh coupling to the ^{31}P ligand atoms might be observed in the NMR spectra. However, no sign of coupling to rhodium was seen in the $^{31}P\text{-}\{^1H\}$ spectrum, and no ^{103}Rh spectrum at all could be observed, even at low temperature. It was concluded that the metal core and the ligand sheath are relatively free to move, so that no unique coupling constants manifest themselves. The timescale of the relative motion seems to be of the order of 10^{-7} seconds or less.

A Pt_{55} cluster with 12 tris(t-butyl)arsine ligands and 20 chloride ligands has also been reported [60], as well as a Ru_{55} cluster with 12 tris(tert-butyl)phosphines and 20 chlorines.

Very interesting reactions occur when the compounds are electro-transported to a platinum electrode in methylene chloride solution [61]. It appears that each cluster loses the ligand sheath and the outermost layer of metal atoms, leaving only the central 13-atom core. These then coagulate into 13-member clumps of 13-atom cores, precipitating as finely divided metal with a distinctive X-ray powder pattern, which contains reflections characteristic of the packing planes within and between the cores.

Part at least of the solid seems to have undergone a further coagulation of the clumps into larger units, which retain the internal six-fold symmetry, and exhibit clear hexagonal faces when observed in a scanning electron microscope. The powder patterns are quite distinct from those of the bulk metals, which are mostly cubic. The metals revert to the normal forms on heating.

10.8.3   More gold clusters – or not?

Several interesting compounds reported recently may be viewed *either* as Au_5 or Au_6 clusters with a first-row atom (carbon or nitrogen) at the center, *or* as carbon *or* nitrogen five- or six-coordinated by two-coordinate gold. The first formulation can be regarded as analogous to those of atom-centered cluster carbonyl complexes of transition metals [e.g. $Rh_6C(CO)_{16}$], whereas the alternative view is more revolutionary, implying possible violation of the octet rule even for first row atoms.

As an example [62], $[AuCl\{P(C_6H_5)_3\}]$ reacts with $C[B(OCH_3)_2]_4$ in the presence of CsF in very polar solvents to give $[C\{AuP(C_6H_5)_3\}_6]^{2+}[B(OCH_3)F_3]_2^-$, isolated as

(a)

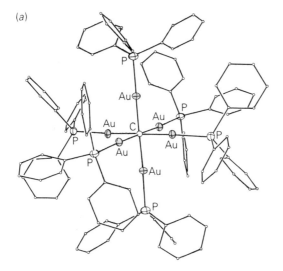

(b)

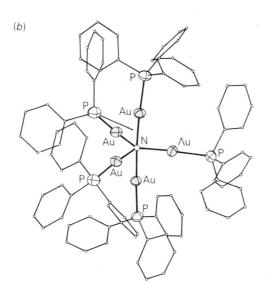

Fig. 10.30 The structures of the cations (*a*) $[C\{AuP(C_6H_5)_3\}_6]^{2+}$ and (*b*) $[N\{AuP(C_6H_5)_3\}_5]^{2+}$ as determined by X-ray crystallography. No Au–Au bonds have been drawn in these diagrams. Adapted, with permission, from Refs [62] and [64].

a diamagnetic solid soluble in CCl_2H_2 but only very sparingly soluble in non-polar solvents. FAB mass spectroscopy shows the expected M^{2+} ion, and also the singly-charged M^+ ion with m/e 2767.8. The structure as determined by X-ray diffraction [Fig. 10.30(a)] shows the cation to have a slightly distorted octahedron of $AuP(C_6H_5)_3$ units surrounding the central carbon atom. The Au—C distances are all in the range 2.12–2.13 Å, and the Au—P bonds appear normal for gold(I) at 2.27 Å, while the PAuC angles are all close to 180°, as expected for simple two-coordinate Au^I. The Mössbauer spectrum is also consistent with this picture, including just one doublet, whose isomer shift and quadrupole splitting parameters are consistent with the presence of Au^I. On the other hand, the Au$\cdots$Au distances along the edges of the octahedron average 3.00 Å, (ranging from 2.91 to 3.09 Å), and are thus similar to distances found in non-centered Au_n polyhedral clusters.

Fairly simple molecular orbital considerations show that an atom in the center of an Au_6 cluster should strongly stabilize it, lowering the energy of the lowest a_{1g} orbital by interaction with the s orbital of the central atom. A similar but less significant energy saving is achieved by interaction of a t_{1u} cluster orbital with the p orbitals of the included atom. This stabilization is maximal for $C(AuL)_6$ clusters with an overall charge of $2+$, as observed. Detailed computations [63] suggest, however, that the result is to reduce the Au$\cdots$Au overlap drastically, lending support to the non-cluster picture.

A five-coordinate nitrogen atom features in the related $[N\{AuP(C_6H_5)_3\}_5]^{2+}$ dication [64], the structure of which is shown in Fig. 10.30(b). As with the carbon cation, the individual linear LAuN linkages are consistent with the description of the compound as containing normal, linear, two-coordinate gold (I), but the trigonal bipyramidal arrangement at nitrogen is most unusual. The gold Mössbauer spectrum (Fig. 10.31) shows an overlapping pair of doublets consistent with Au^I in two sites (axial and equatorial in the trigonal bipyramid). But again, the Au$\cdots$Au distances are in the range 2.88–3.08 Å, and it is hard to exclude Au—Au bonding interactions. If these are significant, then we should regard the compound as a cluster, as shown on the cover of this book. Apparent equivalence of all five phosphorus atoms in the ^{31}P NMR spectrum suggests the ion may be fluxional in solution.

The carbon cation iso-electronic with the nitrogen dication, $[C\{AuP(C_6H_5)_3\}_5]^+$, has been prepared, and shown to contain a trigonal bipyramidal carbon atom, and there are even reports of a nitrogen-containing tri-cation, $[N\{AuP(C_6H_5)_3\}_6]^{3+}$, which presumably is based on an octahedral nitrogen atom.

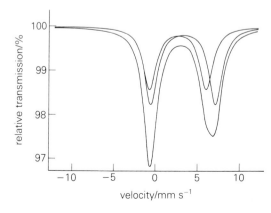

Fig. 10.31 The ^{197}Au Mössbauer spectrum of crystalline $[N\{AuP(C_6H_5)_3\}_5]^{2+}(BF_4^-)_2.(C_4H_8O)_2$ at 4 K, showing how the observed peaks have been resolved into two doublets. Adapted, with permission, from Ref. [64].

CHAPTER 10: CASE HISTORIES

Iron–molybdenum–sulfur cubane-type clusters

At present there is a lot of interest in the electron-transfer processes that occur in the working of nitrogenase. This interest has led to studies of oxidation and reduction in clusters containing iron, molybdenum and sulfur. The complexes are hard to work with, and since a series of electron-transfer steps is involved it is rare to be able to isolate and study directly all the members of a chain of reaction-products. The work normally uses crystallography, electrochemistry, electronic spectroscopy and any other techniques that may help to indicate what species have been formed, how the structures change with oxidation or reduction, and how the electrons are distributed within the clusters.

To illustrate the range and present limitations of studies like these, consider clusters of the form $[Fe_6Mo_2S_8(SR)_9]^{n-}$, which will be written as Q^{n-} [65]. These are based on two cubane-type subclusters Fe_3MoS_4 linked through the Mo atoms by a triple SR bridge (Fig. 10.32). The complexes $[NBu_4^n]_3[Fe_6Mo_2S_8(SR)_9]$, containing Q^{3-}, have been shown by cyclic voltammetry and differential pulse polarography to undergo two (and in some cases four) successive one-electron reduction steps, and one one-electron oxidation. These process can be represented like this:

$$Q^{2-} \Leftrightarrow Q^{3-} \underset{\text{rev}}{\Leftrightarrow} Q^{4-} \underset{\text{rev}}{\Leftrightarrow} Q^{5-} \Leftrightarrow Q^{6-} \Leftrightarrow Q^{7-}$$

Note that at least two of the steps are electrochemically reversible.

Salts containing the anions Q^{3-} and Q^{5-} have been isolated and studied by X-ray crystallography, and some geometrical parameters for the species where R = Ph are given in Table 10.3; the parameters for the latter salt are not as precise as those for the former, and since the cations were not the same (NEt_4 for Q^{5-} and NBu_4^n for Q^{3-}), there may be some associated variations in structure, but it seems generally true that reduction leads to a slight increase in the size of the cluster. The electronic spectra of

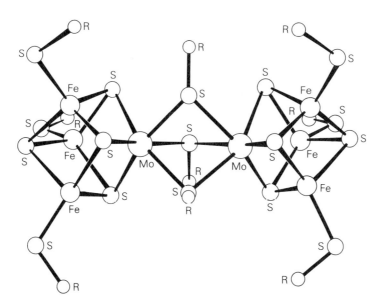

Fig. 10.32 Structure of the central part of the cluster $[Fe_6Mo_2S_8(SPh)_9]^{3-}$. Redrawn from G. Christou, C.D. Garner, R.M. Miller, C.E. Johnson and J. Rush, *J.Chem.Soc., Dalton Trans.*, 2363 (1980).

Table 10.3 Selected structural parameters for Q^{3-} and Q^{5-}

Distance	$Q^{3-}/\text{Å}$	$Q^{5-}/\text{Å}$
Mo$\cdots$Mo	3.813 ± 0.002	3.685 ± 0.003
Mo–S$_{bridge}$	2.62 ± 0.02	2.58 ± 0.02
Mo–S$_{cubane}$	2.37 ± 0.01	2.34 ± 0.02
Mo$\cdots$Fe	2.76 ± 0.01	2.71 ± 0.02
Fe–S	2.28 ± 0.02	2.25 ± 0.03
Fe-SPh	2.30 ± 0.01	2.25 ± 0.03

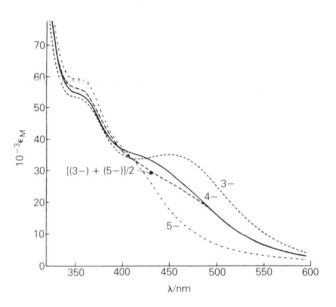

Fig. 10.33 Electronic absorption spectra of $[Fe_6Mo_2S_8(SPh)_9]^{n-}$ in dimethylformamide solution. Spectra for ions with charges 3 –, 4 – and 5 – are shown, with the calculated mean spectrum of 3 – and 5 – species. Redrawn from Ref. [65].

Q^{3-}, Q^{4-} and Q^{5-} were recorded using a device known as an OTTLE (Optically Transparent Thin Layer Electrode). This makes it possible to reduce species in a spectroscopic cell under controlled electrochemical conditions, to characterize new oxidation-states and to check the reversibility of the processes. The results for the species with R = Ph are shown in Fig. 10.33. On reduction, the peaks and shoulders are shifted to higher energies.

There is also an important point concerning the electronic structure of Q^{4-}. Since the two Fe_3MoS_4 subclusters are equivalent in Q^{3-} the charges must be balanced by associating one negative charge with the bridge and one with each subcluster. In Q^{4-}, the subclusters are inequivalent if the extra electron is localized on one of them. They only become equivalent if the extra electron is delocalized, or there is very fast electronic exchange between the two subclusters or between separate anions. Addition of a further electron to give Q^{5-} would once more lead to equivalent subclusters, now both reduced. These ideas are represented below.

The interesting point concerns what happens with Q^{4-}. If the two cubane-type subclusters behave independently on the timescale of the technique used, then the spectrum of Q^{4-} will be the same as those of Q^{3-} and Q^{5-} superimposed. Looking at the electronic spectra obtained (Fig. 10.33), the match between that of Q^{4-} and the superposition of Q^{3-} and Q^{5-} is far from perfect, but the differences could probably be accounted for in terms of the perturbation of one cluster by the other in Q^{4-}. Thus the clusters in Q^{4-} may well be effectively independent on the timescale of their electronic spectra ($\sim 10^{-12}$ s).

The Mössbauer spectra give further information about the electronic structures of these species. The spectra of solutions of Q^{3-}, Q^{4-} and Q^{5-} with R = Ph at 80 K are given in Fig. 10.34. The spectrum of Q^{3-} consists of a broad doublet, though it has been suggested that the spectrum is best fitted by superimposing three or four similar doublets with slightly different shifts and quadrupole splittings, arising from different but similar iron sites in the crystalline complex. On reduction to Q^{4-}, an additional doublet with a similar shift but bigger quadrupole splitting appears, and on further reduction to Q^{5-} the new and wider doublet becomes relatively stronger (Fig. 10.34): the isomer shifts increase somewhat with successive reduction. Working from empirical relationships between isomer shifts and the oxidation states of Fe sites in Fe_4S_4 species, these measurements have been used to give rough local charges on the metal atoms in Q^{3-}, Q^{4-} and Q^{5-}, but the conclusions are debatable. It appears by comparing the spectra of Q^{4-} with a superposition of those from Q^{3-} and Q^{5-} that in Q^{4-} the subclusters remain inequivalent on the Mössbauer timescale; the rate of electron transfer between them is slower than the rate of decay of excited ^{57}Fe $(7 \times 10^{-6}\,\mathrm{s^{-1}})$.

The proton NMR spectra of the three species show lines with contact and dipolar shifts, and the chemical shifts are different for Q^{3-}, Q^{4-} and Q^{5-}. The lines due to Q^{4-} remain as single peaks at 238 K, and do not split into resonances close to those of Q^{3-} and Q^{5-} so that in Q^{4-} the two cubane-type subclusters remain indistinguishable on the NMR timescale at this temperature. This implies an intramolecular rate constant for electron transfer of at least $5 \times 10^3\,\mathrm{s^{-1}}$. So by combining these

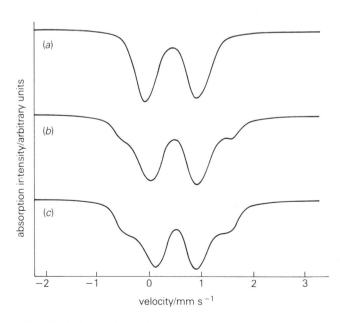

Fig. 10.34 Mössbauer spectra of $[Fe_6Mo_2S_8(SPh)_9]^{n-}$ in dimethylformamide solution: (a) n = 3, (b) n = 4 and (c) n = 5. Redrawn from Ref. [65].

results with the conclusions from the Mössbauer spectra we can put the electron-transfer rate constant as somewhere between about 10^3 and $10^6\,s^{-1}$.

Measurements of magnetic susceptibility indicate that spins are coupled within individual subclusters but that in Q^{4-}, where the two subclusters are formally different, there is little interaction between them.

It is easy to see why conclusions about the detailed electronic structures of complicated systems like these are hard to draw, and must depend on measurements of as many different kinds as possible. The systems are so important, however, that work of this type is likely to grow in significance, and that is why we have included this example in our set of case histories.

10.10 Methylidynephosphine, HCP

This compound, a surprisingly stable gas, is produced together with acetylene when phosphine is passed through a low-voltage arc between carbon electrodes [66]. It polymerizes slowly as a solid or liquid even at low temperatures, but can be handled as a gas without complete decomposition. The composition of the polymer was established as $(HCP)_x$ by analysis. Mass spectrometry of the gas showed the parent ion $[HCP]^+$ as the strongest peak, with only minor fragmentation, and no sign of the presence of oligomers. The IR spectrum of the monomeric solid condensed from the gas onto a surface at 77 K showed three strong bands, as expected for a triatomic molecule, at 3180 cm^{-1} (assigned to νCH), 1265 cm^{-1} (νC≡P) and 671 cm^{-1} (the deformation, ν_2). These frequencies indicate that the compound is HCP, and not the isomeric HPC. The fact that reaction with HCl at 160 K gave CH_3PCl_2 as the sole product supports this idea.

The suggested formulation was confirmed by a microwave spectrum [67], which showed the lines expected for a linear molecule. Detection of the signals from the 1.1% natural abundance of H^{13}CP and from DCP (prepared from PD$_3$) allowed the definition of the bond lengths rCH $= 1.0667 \pm 0.0005$ Å and rCP $= 1.5421 \pm 0.0005$ Å. The short CP bond is consistent with the formulation as a triple bond.

More recently the infrared spectrum of the gas has been studied in some detail [68], and the electronic excitation spectrum observed in the near-UV region (400-200 nm). Very precise values for the vibration frequencies and the rotation constant in the equilibrium configuration [69] are now known. Analysis of the UV spectrum [68] shows that some of the excited electronic states are linear, others bent. Other spectroscopic data, including a millimeter-wave spectrum and an NMR spectrum [70] have also been reported. The high value of $^1J_{CH}$ (211 Hz) is plausibly attributed to a high degree of s-character in the C—H bond. The photoelectron spectrum [71] shows the first two ionizations at 10.79 eV (CP π-bonding) and 12.86 eV (σ, lone pair on P). In all, a remarkably complete picture of the structure and bonding in HCP has been built up using a wide range of spectroscopic techniques.

10.11 Bipyridine complexes

Transition metal complexes containing 2,2-bipyridine (bpy) ligands (10.X) have been much studied in recent years. In particular, there has been a great deal of interest in their electrochemistry, and in this section we consider what happens to the electrons when complexes are oxidized or reduced, or when there is an internal electron shift on

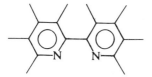

10.X

photochemical excitation. We also look at a by-product unexpectedly formed during the synthesis of an iridium bipyridine complex, and see how spectroscopic data enabled it to be identified.

10.11.1   Electrons in reduced or excited bipyridine complexes

Many transition metal bipyridine complexes show interesting and extended series of electroreductions [72]. $[Ru(bpy)_3]^{2+}$, for instance, can be reduced in three successive steps to give $[Ru(bpy)_3]^-$ which could be formulated as containing ruthenium(-1). This is nothing more than a formalism, though, and it is interesting to ask whether the electrons successively added on reduction are centered on the metal or on the ligands; their location is likely to affect the reactivity of the complex. Similarly, species such as $[Ru(bpy)_3]^{2+}$ can be excited photochemically, and the excited species may show important chemical properties; ruthenium derivatives, for instance, are effective in the water-splitting reaction. The excitation involves transferring an electron from the highest filled orbital to an empty level, and we can ask the same sort of questions about the electron in the excited species as we can about the electrons added on reduction.

Let us consider reduction first, and deal specifically with complexes of the platinum metals. Many of these show series of stepwise reductions like the one mentioned above. From the regularity of the differences between successive reduction potentials in complexes with different metals, and from the overall reversibility of the processes, it has been concluded that the electrons are added to ligand-based orbitals. There remains an important question. Are those electrons localized on the individual bipyridine ligands, or are they delocalized over all the ligands in the reduced complex, thus making the ligands equivalent? The question has been answered by careful use of a range of techniques, and some of the most significant experiments are described here. As $[Ru(bpy)_3]^{2+}$ is reduced electrochemically in steps, it has proved possible to record the electronic spectrum of each successive reduction product, using a special OTTLE (Optically Transparent Thin Layer Electrode) cell [73]. These spectra can be seen as superpositions of two sets of bands. One set is characteristic of $Ru^{II}–bpy^0$ groups like those in $[Ru(bpy)_3]^{2+}$ itself; the other is very like the bands observed in the spectrum of $M^+–bpy^-$, where M is an alkali metal. With each successive reduction, the bands associated with bpy^- become relatively stronger. This is powerful evidence in favor of the localized model, in which reduction should lead to distinct bpy^0 and bpy^- ligands; in the delocalized model each successive reduction-product should give a different spectrum. Similar work with Ir–bpy complexes has led to similar conclusions (see Fig. 10.35) [74]. The resonance Raman spectra of $[Ru(bpy)_3]^n$ (where $n = +2, +1, 0$ and -1) have also been recorded [75]. In $[Ru(bpy)_3]^-$, each ring should be represented as bpy^- if the localized model is correct. In keeping with this expectation, a 1:1 correspondence is

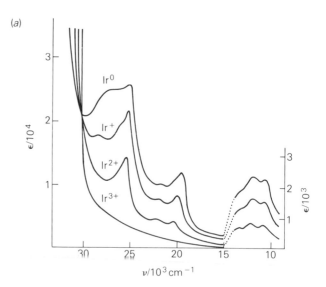

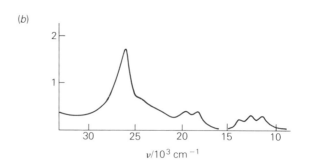

Fig. 10.35 (*a*) Absorption spectra for [Ir(bpy)$_3$]^n, where n is $3+$, $2+$, $1+$ and 0, recorded in acetonitrile. (*b*) Spectrum of Li$^+$bpy$^-$, presented for comparison. It is clear that bands corresponding closely to those of bpy$^-$ appear on reduction and grow in relative intensity with each successive reduction step, while those due to Ir^{III}—bpy become relatively weaker. Redrawn from Ref. [74].

quoted between the ligand-based resonance Raman bands in the spectra of this complex and of M$^+$—bpy$^-$. For [Ru(bpy)$_3$] and [Ru(bpy)$_3$]$^+$, the spectra can be analyzed as superpositions of bands due to Ru^{II}—bpy and to Ru^{II}—bpy$^-$ groups; all the results are consistent with the localized model.

ESR spectra also support the localized model: linewidths and **g**-values for reduced Ru complexes are consistent with this localized formulation. Unfortunately none of these spectra showed hyperfine structure, but the spectra of analogous complexes of iron were also recorded, and in the resonances from [Fe(bpy)$_3$]$^+$ and [Fe(bpy)$_3$]$^-$ it was possible to resolve five hyperfine lines due to coupling with two equivalent ^{14}N nuclei. This is the pattern required by the localized model; if the delocalized model were correct we should expect to see thirteen lines from coupling with six equivalent ^{14}N nuclei [76]. Details of the electronic and ESR spectra were interpreted in terms of electron-hopping from one ligand to another, but the overall conclusion was clear and unambiguous: [Ru(bpy)$_3$]$^+$ is best formulated as [Ru^{II}(bpy)$_2$(bpy$^-$)]$^{2+}$.

Optical excitation of [Ru(bpy)$_3$]$^{2+}$ at 265 nm or longer wavelengths produces an excited species which is emissive, photochemically active and relatively long-lived [77]. At this stage it is convenient to label the excited species [*Ru(bpy)$_3$]$^{2+}$. Its electronic spectrum has been recorded and assigned in terms of a structure in which the excited electron is localized on one bipyridine ligand, so that the species would be more exactly represented as [*Ru^{III}(bpy)$_2$(bpy$^-$)]$^{2+}$. The details of the assignment depend on spectra of bpy$^-$ and of Ru^{III}—bpy^0 complexes obtained from spectroelectrochemical studies of $[Ru(bpy)_3]^{3+}$ and related species. Some overlapping transitions are postulated, and the assignments are complex, though convincing [78]. Resonance Raman spectroscopy has provided independent and quite unambiguous evidence for the same conclusion for this and other related species [77, 79, 80]; indeed, the first results were published before the study of the electronic spectrum [77]. There is a strong ligand-based band in the electronic spectrum of $[*Ru(bpy)_3]^{2+}$ at 360 nm. By using the third harmonic of a Nd–YAG pulsed laser it proved possible both to excite $[Ru(bpy)_3]^{2+}$ to $[*Ru(bpy)_3]^{2+}$ and to obtain the time-resolved resonance Raman (TR^3) spectrum of $[*Ru(bpy)_3]^{2+}$. The new bands of the excited complex could be easily identified by the reduction in their intensities relative to those of $[Ru(bpy)_3]^{2+}$ as the laser pulse power was decreased. Seven symmetric modes were detected for both $[Ru(bpy)_3]^{2+}$ and $[*Ru(bpy)_3]^{2+}$. If we assume that the normal modes of the excited species correlate with those of $[Ru(bpy)_3]^{2+}$, we can

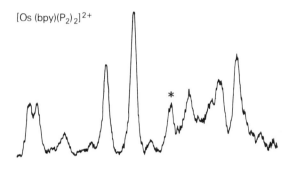

$[Os\,(bpy)(P_2)_2]^{2+}$

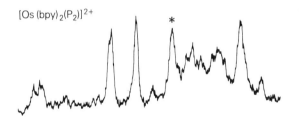

$[Os\,(bpy)_2(P_2)]^{2+}$

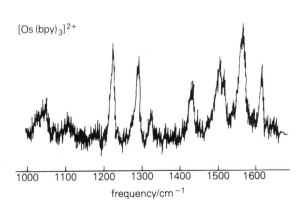

$[Os\,(bpy)_3]^{2+}$

1000   1100   1200   1300   1400   1500   1600

frequency/cm^{-1}

Fig. 10.36 Time-resolved resonance Raman spectra of the MLCT states of $[Os^{II}(bpy)_{3-n}(P_2)_n]^{2+}$, where n is 2, 1 and 0, and P_2 is $Ph_2PCH{=}CHPPh_2$. The middle spectrum can be seen as a superposition of the top and bottom spectra. Peaks marked * are due to CH_3CN. Adapted, with permission, from Ref. [80]. Copyright (1984) American Chemical Society.

conclude that the new modes show a general and marked reduction in frequency; indeed, the pattern of frequencies for the new bands of $[*Ru(bpy)_3]^{2+}$ is very like that for the fully-reduced bpy$^-$. Careful analysis of the TR^3 spectra of a series of ruthenium and osmium complexes confirms this conclusion. The bands due to Ru— or Os—bpy and —bpy$^-$ were observed together, with relative intensities varying as expected. Moreover, spectra of osmium complexes of the form $[Os(bpy)_{3-n}(P_2)_n]^{2+}$, ($n = 0$-2) were recorded [80]. (Here P$_2$ is short for $Ph_2PCH=CHPPh_2$.) These show a common electronic band due to charge transfer from metal to ligand (MLCT), and can all be excited by harmonics of the Nd–YAG pulsed laser. The TR^3 spectra of the excited species all show bands due to Os—bpy$^-$ at much the same frequencies (see Fig. 10.36). In addition, where $n = 2$ there were no bands due to Os—(bpy) modes. This is exactly what the localized model predicts. If the delocalized model were correct, then the excited electron should be delocalized over three bipy ligands when $n = 0$, over two when $n = 1$, and localized on one when $n = 2$. Thus the charge and electron density on the bpy ligands should differ markedly with n, and so should the vibrational spectra.

This extremely elegant work has also shown how it may prove possible to derive information about the geometries of the excited species from the resonance Raman and electronic emission spectra [80]. There is a rule called Badger's Rule which relates the length of a bond to its stretching frequency. It is an empirical rule and is not generally valid, but it can be used in comparing the vibrations of coordinated bpy and bpy$^-$, taking the reductions in the stretching frequencies to estimate how much the additional electron has lengthened the bonds in M—bpy$^-$ compared with those in M—bpy. This approach has been reinforced by results obtained by a completely different method, making use of a Franck–Condon analysis of the vibrational progressions in the emission spectra of the excited species. The details may be found in Ref. [80]. The important thing is that both approaches lead to the same general conclusions. While it is not yet possible to use either method to work out accurate geometries for ligands in charge-transfer excited complexes like these, we can now see how it may be possible in the future to obtain such unexpectedly detailed information.

10.11.2   An iridium bipyridine complex

As we have explained in Section 10.11.1, transition metal complexes containing 2,2'-bipyridine (bpy) ligands have been much studied. In the course of this work it has been necessary to prepare pure samples of compounds such as $[Ir(bpy)_3]^{3+}$ in fairly

10.XI

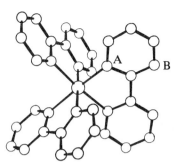

Fig. 10.37 Structure of a cation of an iridium tris(bipyridine) complex. Redrawn from Ref. [81].

large quantities. Unfortunately, the preparation of the iridium complex is not straightforward, and things are made complicated by the fact that at least three different complexes have been reported – all diamagnetic, and all containing an iridium(III) atom associated with three bipyridine ligands.

One of these compounds (compound X) has been shown by UV and NMR spectroscopy to be the expected complex, 10.XI, with three ligands each linked to the central atom by both nitrogen atoms. The identity of the others is therefore of interest as they cannot also contain simple, normal $[Ir(bpy)_3]^{3+}$. Fortunately, the second of the three compounds (compound Y) could be obtained in crystalline form, and an X-ray diffraction study was undertaken [81]. There were three similar cations in the asymmetric unit, and Fig. 10.37 shows one of them – with the atom identifications deliberately removed! At first sight this *is* a normal tris(bipyridine) complex, containing six Ir—N bonds, but careful analysis showed that the atoms labeled A and B had abnormal thermal parameters, and that the Ir—N bond *trans* to A was longer than the others. It was suggested that in this complex, atom A was carbon, and B was nitrogen. Refinements based on this unexpected structure gave a slightly better fit to the data (the *R* value improved from 0.061 to 0.059), and the thermal parameters became much more reasonable. This structure made chemical sense. The salt studied has only two perchlorate anions, instead of the three expected for a normal type of complex, and the third form (compound Z) could now also be accounted for: it could simply be compound Y, protonated at the uncoordinated nitrogen atom. But the crystallographic result is not clear-cut; distinguishing nitrogen and carbon in species like these is not decisive, and some supporting spectroscopic evidence would be valuable.

This evidence is provided by the ^{13}C and ^{1}H NMR spectra [82]. Fig. 10.29(*a*) shows the ^{13}C spectrum (with ^{1}H decoupling) for the normal complex, X. There are just five resonances, corresponding to the five carbon atoms in one ring, as all six rings are equivalent. The spectrum of the compound Y [Fig. 10.38(*b*)] is strikingly different. It has 30 resonances in 25 distinguishable peaks, showing that the ion no longer has any symmetry. Even more significantly, the spectrum showing quaternary carbon atoms only [Fig. 10.38(*c*)] contains *seven* resonances, one of which has an exceptionally low-frequency shift. This gives convincing evidence for the proposed carbon-metalated structure, and it is believed that it is the carbon atom bound to the metal that has the unusual chemical shift. The ^{1}H NMR spectrum is more complicated, and we will not discuss it here, but it has been assigned, with the aid of many double resonance experiments, and fully supports the structure. The protonated compound, Z, has also been studied by ^{13}C and ^{1}H NMR, with equal success [83]. Its ^{13}C spectrum shows 23 distinct peaks representing the 30 carbon atoms, and off-

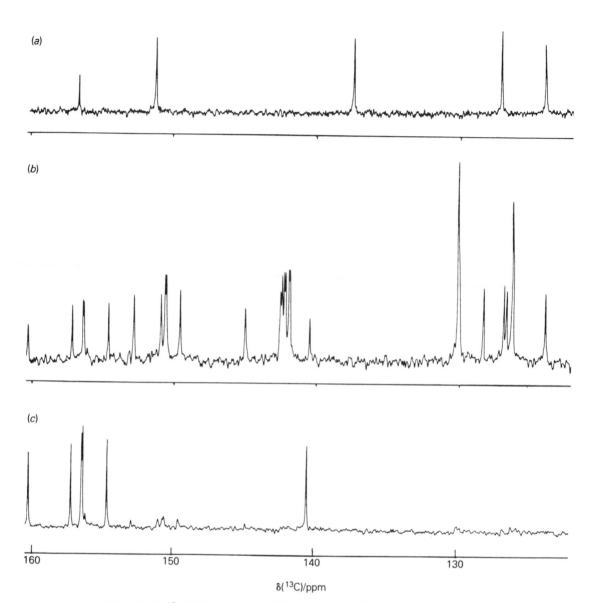

Fig. 10.38 ^{13}C NMR spectra of iridium tris(bipyridine) complexes: (*a*) the normal complex, with iridium coordinated to six nitrogen atoms, (*b*) the *ortho* C-metalated complex, and (*c*) the *ortho* C-metalated complex, but showing only quaternary carbon atoms.

resonance ^{1}H decoupling experiments again show that seven of the carbon atoms are not bound to hydrogen. One of these seven has a chemical shift which is at an unusually low frequency for a quaternary carbon atom.

10.12   A boron disulfide macrocycle

During studies of boron sulfides [84] a mixture of B_2S_3 and sulfur was heated to *ca.* 570 K in a sealed tube at a very low pressure, and a crystalline material was deposited in a cooler part of the tube. Analysis of the product for boron and sulfur showed that they were present in the ratio 1:2, and that it was therefore different from all known boron–sulfur compounds.

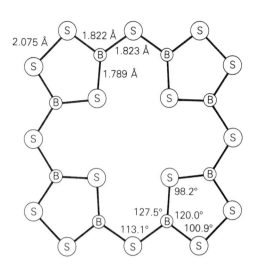

Fig. 10.39 Structure of the B_8S_{16} molecule, with parameters averaged over chemically equivalent bond lengths and angles. Redrawn from Ref. [84].

Characterization of such compounds by spectroscopic means is very difficult. NMR helps little, although [11]B spectra might be of some use. The mass spectrum shows various fragments, of which $B_2S_3^+$ is the most abundant, but nothing that can be confidently assigned as a molecular ion (although this might be seen if fast atom bombardment or field desorption was used to ionize the sample). The Raman spectrum shows 12 bands in the region associated with stretching vibrations, but that in itself does not tell us much, except that the compound must have a substantial number of atoms.

In this case, everything depends on a complete X-ray structure analysis, the results of which are shown in Fig. 10.39. The molecule is a remarkable porphine-like macro-cycle, with planar B_8S_{16} molecules, made up of four 1,2,4,3,5-trithiadiborolane rings linked by sulfur bridges. The symmetry in the crystal is C_i, but the molecule has D_{4h} symmetry within experimental error. It is a most unusual compound, and has exciting potential as a ligand. It was fortunate that it was formed as crystals large enough for X-ray diffraction: otherwise it would have been very difficult indeed to characterize.

10.13 A novel lead–selenium anion

Some interesting cluster anions have been prepared in recent years by dissolving solids, which have been made by fusing alkali metals with mixtures of more electro-negative elements. The solids themselves are usually amorphous and hard to charac-terize structurally, but once they are dissolved it is often possible to use spectroscopic methods to investigate the structures of the anions themselves. In favorable cases the anions may be crystallized with a suitable cation, and if X-ray diffraction data can be obtained and interpreted, then the anion structures can be determined.

In one particular case, a solid was formed by dissolving elemental selenium in molten KPb, a preformed alloy [85]. It was dissolved, after cooling, by standing for several days in liquid ethylenediamine with a stoichiometric amount of the cryptand ligand [2,2,2-crypt] (10.XII), which speeds dissolution by complexing strongly with potassium ions and also stabilizes the resulting solution by providing a large, very stable cation $[K(2,2,2\text{-crypt})]^+$. The original solid has the empirical formula KPb_xSe, where x is close to 0.5. Tellurium may be partially or totally substituted for the selenium but x varies with the proportion of tellurium, reaching about 0.65 when all

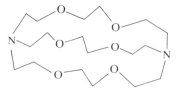

10.XII

the selenium has been replaced by tellurium. Clearly the solid is not a simple mixture of K^+ cations and anions containing lead and selenium (and/or tellurium).

The solution in ethylenediamine was studied using ^{77}Se and ^{207}Pb NMR. Both spectra showed signals having satellites due to the other nucleus, with the same coupling constant, $|J_{PbSe}|$ 194 Hz. Only one pair of satellites was visible in each spectrum, but the intensities of the satellites relative to the central peak were consistent with two equivalent lead atoms coupling to each ^{77}Se and three equivalent selenium atoms coupling to each ^{207}Pb. The other satellites would probably be too weak to be observed, arising from the small proportions of molecules containing two ^{207}Pb or two or three ^{77}Se nuclei. This suggests a highly symmetrical Pb_2Se_3 species, and the only plausible model seems to be that shown as structure 10.XIII. The charge depends on the oxidation state of the Pb atoms; their geometry suggests Pb^{II}, which would lead to a very reasonable charge of 2 – overall. This is also consistent with the observed coupling constant $|J_{PbSe}|$, which is much smaller than the measured value in $Me_3PbSeMe$, (–)1170 Hz, suggesting a low degree of s orbital participation in the Pb—Se bonds.

$$
\begin{bmatrix}
\text{Pb} & & & \\
 & \diagdown & \text{Se} & \\
\text{Se} & & \diagup & \\
 & \diagup & \text{Se} & \\
\text{Pb} & & &
\end{bmatrix}^{2-}
$$

10.XIII

Crystals were grown from the solution, and proved to contain the di-anion suggested by the NMR data, with two $[K(2,2,2\text{-crypt})]^+$ cations, confirming the net anion charge and oxidation state of lead. There is no sign of any bonding between anions, each of which is surrounded by ten cations. The bond angles at the lead atoms are about 90°, which is consistent with very little s orbital involvement in the bonding to selenium. The Pb$\cdots$Pb distance is very small, only 3.18 Å, which is less than that found in metallic lead (3.49 Å) so there may be some direct Pb—Pb bonding. However, this short Pb$\cdots$Pb distance may be enforced by the relatively short Pb—Se distances (2.70 Å) and the need to keep the non-bonded selenium atoms apart.

10.14   Structure of N_2O_3

The oxide N_2O_3 has been known since 1816 as formed from NO and NO_2; its blue color makes its presence noticeable. It is substantially dissociated at room temperature and atmospheric pressure into the components from which it is made; at lower temperatures, the dissociation equilibrium is complicated by the volatility of NO and

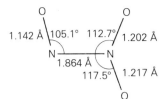

Fig. 10.40 Structure of N_2O_3, as determined by microwave spectroscopy.

the dimerization of NO_2, but N_2O_3 can be obtained as an effectively pure compound at around its melting point of 172 K.

The structure of gaseous N_2O_3 cannot be determined satisfactorily by electron diffraction, because the compound cannot be obtained in the vapor phase without substantial proportions of NO, NO_2 and (at lower temperatures) N_2O_4. There are problems with microwave spectroscopy. At the low pressures normally used to obtain microwave spectra, it is hard to get more than about 0.5% of the mixture as N_2O_3, but in an excellent study involving isotopic substitution the microwave spectrum has been obtained and analyzed [86] to give the structure shown in Fig. 10.40. There is a weak and long N—N bond, much longer than those in N_2H_4 (1.47 Å) or N_2O_4 (1.75 Å). The molecule is planar, and the two nitrogen atoms are not equivalent. The vibrational spectrum is consistent with this structure, though a full assignment has only recently been established by a modern Raman study [87]. The ^{14}N NMR spectrum [88] shows that the non-equivalence of the two nitrogen atoms is maintained in the liquid and in solutions at 240 K. There are two distinct ^{14}N resonances, one in the region associated with nitroso groups and the other in the region associated with nitro groups, and $^1J(NN)$ has been determined from the ^{15}N spectrum. The color of the liquid is also consistent with this structure, and the transition that gives rise to the blue color is believed to be of the $(n - \pi^*)$ type. The vibrational spectra are not grossly different for the gas, the liquid and the annealed solid; it therefore seems likely that the same structure persists through all three phases. An attempt to determine the crystal structure at low temperatures was frustrated by disorder, but there is no infrared band in the spectrum of the annealed solid at a high enough frequency to be assigned to NO^+. In contrast to N_2O_5, whose crystals are made up of NO_2^+ and NO_3^- (see Section 5.15), N_2O_3 does not appear to ionize in the crystal.

There are, however, persistent reports of the presence of small amounts of an isomeric form. As long ago as 1940, isotopic exchange between ^{15}NO and $^{14}NO_2$ was shown to occur in the gas phase at 240 K so fast that equilibrium was reached in less than 15 seconds. This led to the suggestion that the molecule had the symmetrical ONONO structure, sometimes referred to as s-N_2O_3. It is now certain that the major form is $ONNO_2$, or as-N_2O_3, but studies of the IR and Raman spectra of the solid and of N_2O_3 in matrices such as N_2, O_2 or NO [87] have led to the assignment of weak bands to s-N_2O_3, and calculations of the force-constants of s-N_2O_3 have been made. It has also been shown that irradiation of N_2O_3 in a matrix of NO with laser light with wavelengths between 568.2 and 752.6 nm (that is, under the envelope of the electronic band assigned to the $n - \pi^*$ transition) converts as-N_2O_3 into other species, whose spectra include the bands assigned to s-N_2O_3. The conversion is facile, since it is reversed by irradiation at 514.5 or 488.0 nm, and so it seems likely to involve a simple process like an isomerization; s-N_2O_3 is a plausible product, and a fairly complete assignment for the vibrations of this species has been made [87], using spectra from molecules containing ^{15}N and ^{18}O. Work with solutions of N_2O_3 in xenon at low temperatures has not fully confirmed these conclusions [89]. The bands of as-N_2O_3 were readily identified; the spectra were analyzed as showing that there was an equilibrium between two forms of N_2O_3, probably the symmetric and the asymmetric molecules, and the free energy of isomerization was calculated to be 7.6 ± 0.9 kJ mol^{-1}. However, it was only possible to identify some of the bands that had been assigned in previous studies to s-N_2O_3. Most of the missing bands were weak, and their absence is not significant, while some spectroscopic differences could be put down to the different environments of the molecules – xenon solvent in this study, matrices of N_2, O_2 and NO in the others. The isotopic exchange could easily be understood if there were an equilibrium in solution between symmetric and asymmetric forms, and the presence of only small amounts of the symmetric form would be enough to account for the observed exchange. But the NMR spectra make it clear that the two different nitrogen atoms in as-N_2O_3 have lifetimes in their different environments of at least 10^{-2} s in solution at 240 K. There is as yet no direct evidence to show that exchange involves s-N_2O_3, and some questions about this species will remain until it has been unambiguously identified in the vapor phase.

10.15 A radical monomer

It is well known that N—N single bonds are often long and weak, and N_2F_4 dissociates a little above room temperature to give the free radical $\cdot NF_2$. The replacement of fluorine by bulky electronegative substituents on the nitrogen may be expected to increase the tendency to dissociate, and this turned out to be the case with $(SF_5)_2N—N(SF_5)_2$.

The compound was prepared in the course of experiments intended to lead to the formation of $N(SF_5)_3$, which was not found in the products [90]. A gaseous mixture of $SClF_5$ (in excess) and SF_5NCl_2 was irradiated with a medium-pressure mercury arc lamp at room temperature, and the most interesting product, formed in 20% yield (based on N), was the tetra-substituted hydrazine. Its IR spectrum is unremarkable, containing strong bands due to internal vibrations of the SF_5 group. The ^{19}F NMR spectra is very complex because of the multiple possibilities of inequivalence and coupling between groups. The mass spectrum shows only fragments. The ESR spectrum of a solution in CCl_3F, at temperatures between 273 and 313 K, shows a strong signal, whose intensity increases and decreases reversibly as the temperature rises and falls, suggesting a dissociation/dimerization equilibrium.

$$(SF_5)_2N—N(SF_5) \rightleftharpoons 2 \cdot N(SF_5)_2$$

The spectrum [Fig. 10.41(a)] shows a clear eleven-line pattern due to coupling of the electron spin to nuclear spins, with a coupling constant of 1.35 mT. Rather than attribute this to equivalent coupling to all ten spin-$\frac{1}{2}$ ^{19}F nuclei, the authors suggest that it is due to an overlapping triplet of nine-line patterns (nonets) due to coupling respectively to a single spin-1 ^{14}N nucleus and the eight ^{19}F atoms in the equatorial planes of two SF_5 groups attached to the nitrogen. The remaining axial fluorine atom on each group appears to have no observable coupling to the electron spin.

This explanation requires the coupling constants to nitrogen and fluorine to be accidentally equal. This still leaves the electron residing more on nitrogen than on the equatorial fluorine atoms, as the magnetogyric ratio for ^{14}N is only 7.6% of that for ^{19}F, so equal couplings imply that the electron density at the nitrogen nucleus is 13

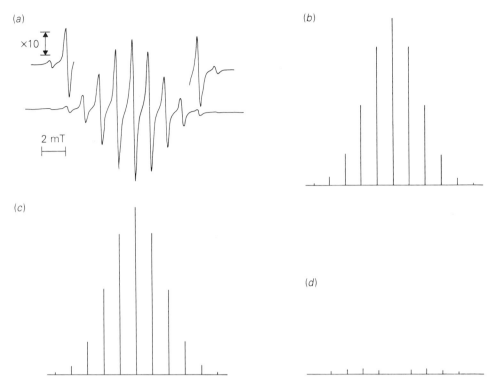

Fig. 10.41 (a) ESR spectrum of $\cdot N(SF_5)_2$, showing an eleven-line multiplet, with the outermost lines clearly seen in the vertical expansion of the outer regions of the spectrum. (b) The theoretical intensity pattern for a 1:1:1 triplet of nonets, (c) the theoretical pattern for coupling to ten equivalent spin-$\frac{1}{2}$ nuclei, and (d) the difference between (b) and (c), on the same scale. The spectrum is redrawn from Ref. [90].

times that at fluorine. The intensity pattern is also said to be consistent with this interpretation, though it is hard to be sure that the relative intensities of weak and strong peaks can be measured with sufficient precision to distinguish between patterns due to coupling to ten equivalent ^{19}F nuclei [Fig. 10.41(b)] and the suggested triplet of nonets [Fig. 10.41(c)]. The latter is certainly the more reasonable possibility on chemical grounds.

10.16   Fe(CO)$_4$

The simplest stable carbonyl of iron is $Fe(CO)_5$. When this molecule is photolyzed with UV radiation in a variety of inert matrices at 77 K or lower, CO is formed, with an unstable species $Fe(CO)_4$. The shape of this unusual 16-electron molecule was established using infrared spectroscopy, and the electronic structure was confirmed using magnetic circular dichroism.

In order to decide between possible alternative geometries, we have to work out what we would expect to see in the (CO) stretching region of the IR spectrum, and then we must compare what we observe with what is predicted from the different models. Of the reasonably plausible structures, tetrahedral $Fe(CO)_4$ (point group T_d) should give only one IR band in this region. A trigonal bipyramidal molecule with an axial lone pair of electrons (point group C_{3v}) should give three bands. A structure with D_{2d} symmetry should give two bands, while a trigonal bipyramid with an equatorial

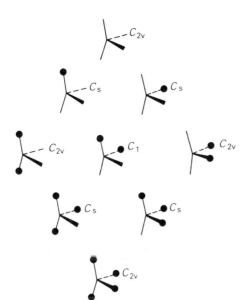

Fig. 10.42 Molecular symmetries for species containing ^{12}CO and/or ^{13}CO, derived from the form of Fe(CO)$_4$ having C_{2v} symmetry.

lone pair (C_{2v}) should give four bands. The IR spectrum of Fe(CO)$_4$ in various matrices did not allow distinction between these forms [91]. Though only three bands were observed, the C_{2v} structure could not be excluded: two of its four bands might overlap, or one might be too weak to be detected. Similarly, the more symmetrical structures might give more bands than predicted because of matrix splittings.

The question was answered [91] by recording the spectra of samples of Fe(CO)$_4$ substantially enriched with ^{13}C, which consisted of statistical mixtures of Fe(^{12}CO)$_4$, Fe(^{12}CO)$_3$(^{13}CO), Fe(^{12}CO)$_2$(^{13}CO)$_2$, Fe(^{12}CO)(^{13}CO)$_3$ and Fe(^{13}CO)$_4$. However, the number of different molecular species depends on the geometry of Fe(CO)$_4$. Tetrahedral Fe(CO)$_4$, for instance, would give five species: Fe(^{12}CO)$_4$ and Fe(^{13}CO)$_4$ (both T_d), Fe(^{12}CO)$_3$(^{13}CO) and Fe(^{12}CO)(^{13}CO)$_3$ (both C_{3v}), and Fe(^{12}CO)$_2$(^{13}CO)$_2$ (C_{2v}). The form of Fe(CO)$_4$ with point group C_{2v} would give nine species, as illustrated in Fig. 10.42. The symmetries for the species derived from some of the possible model structures for Fe(CO)$_4$ are given in Table 10.4. All the different species will give different IR spectra. In order to decide which structural model is correct, the infrared spectra of all the different species derived from one model are calculated using a CO-factored force field (which assumes that the CO stretching vibrations do not couple with any of the other molecular vibrations [92]), and bond angles and dipole derivatives are chosen to give the best fits to the observed band

Table 10.4 Point groups of molecular species for different forms of Fe(CO)$_4$ containing ^{13}CO

	Point group of parent molecule			
Species	T_d	D_{2d}	C_{3v}	C_{2v}
Fe(^{12}CO)$_4$	T_d	D_{2d}	C_{3v}	C_{2v}
Fe(^{12}CO)$_3$(^{13}CO)	C_{3v}	C_s	C_{3v}, C_s	C_s, C_s
Fe(^{12}CO)$_2$(^{13}CO)$_2$	C_{2v}	C_2, C_2	C_s, C_s	C_{2v}, C_{2v}, C_1
Fe(^{12}CO)(^{13}CO)$_3$	C_{3v}	C_s	C_{3v}, C_s	C_s, C_s
Fe(^{13}CO)$_4$	T_d	D_{2d}	C_{3v}	C_{2v}
Total	5	6	8	9

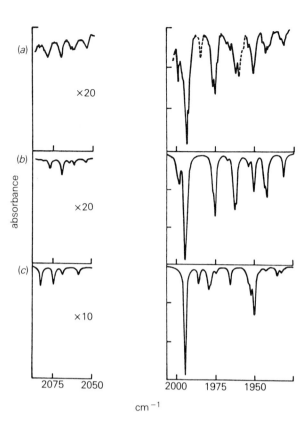

Fig. 10.43 (*a*) IR spectrum of Fe(CO)$_4$ with 40% ^{13}CO in a matrix of SF$_6$ at 35 K. The bands marked with a dashed line are due to Fe(CO)$_5$. (*b*) Calculated spectrum for a molecule with C_{2v} symmetry, with optimized structure and dipole moment derivatives. (*c*) Calculated spectrum for a molecule with C_{3v} symmetry, with optimized structure and dipole moment derivatives. Redrawn from M. Poliakoff, *Chem. Soc. Rev.*, **7**, 527 (1978).

intensities. The observed spectrum is compared with the one obtained by summing the spectra of all the species derived from one geometric model, and the fit is optimized as far as possible. The process is repeated for all the plausible geometries, and finally the resulting 'best fits' are compared. The comparison leaves no doubt: the spectrum calculated for the C_{2v} model fits the observed spectrum very well, and much better than any of the others. In Fig. 10.43 you can see the observed spectrum for Fe(CO)$_4$ containing 40% ^{13}C in a matrix of SF$_6$ compared with spectra calculated for C_{2v} and C_{3v} models. The detailed assignments based on the C_{2v} model have been confirmed by some very elegant photochemical experiments using IR laser irradiation to transform one partly-substituted species into another. The calculations also show that the fourth of the expected (CO) stretching modes is too weak to be detected.

An interesting conclusion from an analysis of the molecular orbitals of Fe(CO)$_4$ in its C_{2v} geometry is that the molecule should contain two unpaired electrons. This prediction was corroborated using magnetic circular dichrosim. Paramagnetic species give temperature-dependent MCD spectra, and so does Fe(CO)$_4$ between 5 and 25 K [93].

10.17   Gold ylide complexes

Gold complexes of phosphorus ylides show a range of remarkable reactions. In particular, the binuclear species 10.XIV show two different modes of oxidative addition, leading either to species of type 10.XV, which contains a bond between two Au^{II} atoms, or to type 10.XVI, in which there is a relatively weak interaction between Au^I and Au^{III} centers. In some cases, for example when the oxidizing species, XY, is Cl$_2$, Br$_2$ or I$_2$, the two isomers are formed simultaneously.

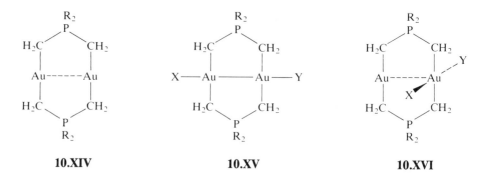

10.XIV   **10.XV**   **10.XVI**

A product of the former type (10.XV, X = Y = I) has been shown to react with excess of the phosphorus ylide, MePh$_2$PCH$_2$, to give a very unusual product [94]. The ^{13}C-{^{1}H} NMR spectrum of this new compound showed only two CH$_2$ environments; the ^{31}P-{^{1}H} NMR spectrum showed that all ^{31}P nuclei were equivalent. The NMR spectra of the corresponding chloride and bromide are almost identical to those of the iodide. The complex had electrical conductivity corresponding to a 1:1 electrolyte, and the ^{197}Au Mössbauer spectrum (Fig. 10.44) consists of two doublets, the isomer

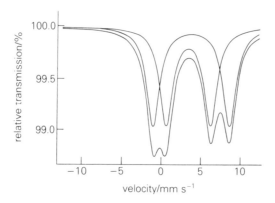

Fig. 10.44 ^{197}Au Mössbauer spectrum of the tri-gold complex 10.XVII, recorded at 4 K. Redrawn from Ref. [95].

shifts and quadrupole splittings of which indicate the presence of Au^I and Au^{III}. The doublets are of equal intensity, which would seem to imply that gold atoms in oxidation states I and III are present in equal numbers. However, the NMR spectra show that the complex contains a very symmetrical cation; if this cation contains both Au^I and Au^{III} and is based on the starting structure 10.XV, then it is plausible that the species contains two rings fused through Au^{III}. In that case, the Au^{III} Mössbauer doublet should be half as intense as that due to Au^I, and not of roughly the same

10.XVII

intensity. X-ray crystallography showed that the conclusion based on chemistry and NMR spectroscopy as set out above was correct: the complex does contain two rings fused through a square-planar gold (III) atom in a paddle-wheel structure (10.XVII). The problem with the Mössbauer spectrum lies in different recoil-free fractions for Au^{III} and Au^{I} [95]. Thus the Mössbauer lines for Au^{I} are relatively weaker than expected. It is clearly important to remember this possible complication when interpreting Mössbauer spectra, though it is unlikely to be signficant in most circumstances.

10.18   Chromium–chromium quadruple bonds

In the first half of the twentieth century, transition metal complexes were considered to consist simply of single metal atoms, each surrounded by a set of ligands. Metal–metal bonds were almost unknown. Since then, however, the availability of many spectroscopic and structural techniques, particularly X-ray crystallography, has led to enormous advances in synthetic transition metal chemistry, which have been paralleled by advances in our theoretical understanding of the principles involved. In particular, compounds containing metal–metal bonds have been intensively and increasingly studied, and very many compounds containing single, double, triple or quadruple metal–metal bonds are now known. The whole subject of multiple bonds between metal atoms has been reviewed in detail [96].

Perhaps the most remarkable feature of this story is that the advances in metal–metal multiple bond chemistry started so late. For the first compound with a quadruple bond [chromium(II) acetate] was prepared in 1844. This differed from the usual chromium(II) compounds in being red instead of blue, but it was not studied further for over 100 years, when it was observed that it was diamagnetic, in both hydrated and anhydrous forms. It was also noted that the hydrated compound was isomorphous with copper(II) acetate, which had a Cu–Cu distance of 2.64 Å; and in 1956 it was suggested that all these data implied that in the chromium compound there could be weak d–d interactions, one σ, two π and one δ, thus pairing the spins of the d electrons on the chromium atoms [97]. This is tantamount to postulating a quadruple bond, but it was not explicitly described as such, and apparently the report went unheeded until after the publication of a full description of the quadruple bond in $[Re_2Cl_8]^{2-}$.

Rather surprisingly, hydrated chromium(II) acetate was not studied by X-ray crystallography until 1970 [98]. The compound was shown to have the structure 10.XVIII, with four acetate groups bridging between the metal atoms, and the water molecules occupying the axial positions. The distance between the chromium atoms was 2.362 Å, which certainly supports the idea of a strong metal–metal interaction, but is nevertheless rather longer than expected for a quadruple bond. Indeed, the Mo≡Mo distance in crystalline anhydrous molybdenum(II) acetate is only 2.093 Å.

There are now many known compounds with structures similar to 10.XVIII, with different bridging or axial ligands. In complexes with bridging carboxylate groups, the Cr≡Cr distances vary enormously, between about 2.20 and 2.55 Å. The implication of this is that the energy changes on varying the distance between the metal atoms is very small, and that the distance is therefore very sensitive to the nature of the axial ligands. This idea is supported by the best molecular orbital calculations, and also by the observation that the variations in M≡M distances in molybdenum and tungsten

Me

Me

10.XVIII

complexes (which associate only weakly with axial ligands) are much smaller. Clearly, therefore, it would be valuable to determine the structure of a complex with *no* axial ligand. The Cr≡Cr distance would be expected to be less than the Mo≡Mo distance in molybdenum(II) acetate. Unsolvated chromium acetate is an obvious candidate, but in the crystalline state this forms infinite chains of molecules, as shown in Fig. 10.45, with the axial sites occupied by oxygen atoms of neighboring molecules. The Cr≡Cr

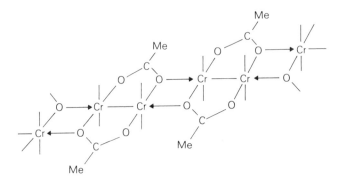

Fig. 10.45 Formation of infinite chains of chromium(II) acetate molecules in the crystalline phase. The oxygen bridge bonding makes the chromium atoms six-coordinate.

distance is 2.288 Å [99]. This is interesting enough, but leaves the question about the length of an undistorted chromium–chromium quadruple bond unanswered. One approach to solving this problem has been to try to block the axial sites or to screen the carboxylate oxygen atoms, and so prevent association. Various attempts to do this have been made, using carboxylate groups with sterically demanding substituents, but so far these have not been successful. But with other bridging ligands, compounds with unperturbed quadruple bonds can be prepared. For example, the acetate ligands in chromium acetate can be replaced by dimethoxyphenyl groups, so that one chromium atom is bound directly to a ring carbon atom, and the other to one oxygen atom [100]. With four such ligands, each chromium atom has two carbon and two oxygen substituents, and two of the four free methoxy groups are at each end of the molecule, effectively blocking approach to the axial sites. The structure is shown in Fig. 10.46. In this case, the chromium–chromium bond is strikingly short – 1.847 Å. There can be no doubt that this really is a quadruple bond. The structures of more

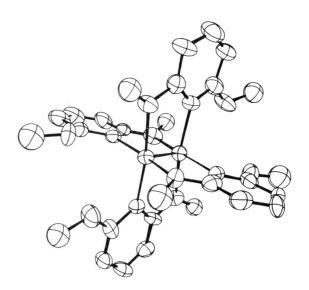

Fig. 10.46 Structure of tetrakis(dimethoxy-phenyl)dichromium in the crystalline phase. Redrawn from Ref. [100].

than a dozen similar compounds, with a range of ligands such as 10.XIX and 10.XX, are now known, and whenever the axial sites are unoccupied, the Cr≡Cr bond is very short, between 1.82 and 1.94 Å.

10.XIX

10.XX

At this point we have two groups of compounds. In one the complexes have no axial ligands, and bond lengths of 1.82-1.94 Å; in the other there are axial ligands, and the bonds are between 2.20 and 2.55 Å long. But to be sure that it really is the extra ligands that lengthen the bonds, rather than the electronic properties of the bridging ligands, we need to find sets of complexes that differ *only* in the presence or absence of axial ligands. Such systems are not easy to find, but with the ligand 10.XXI as the

10.XXI

bridging group a suitable series of complexes can be made. In the simple compound the Cr≡Cr distance is short, 1.937 Å. Addition of one axial THF group lengthens this to 2.023 Å, and with two such groups the distance goes into the normal 'long' range, at 2.221 Å. With two pyridine ligands in the axial positions the distance is 2.354 Å.

It is now quite clear that an unperturbed chromium–chromium bond is extremely short, but that it is easily lengthened, by up to 0.7 Å, by axial ligands. We therefore

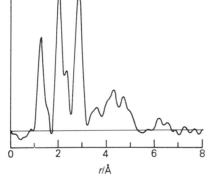

Fig. 10.47 Radial distribution curve for electron diffraction by gaseous chromium(II) acetate. Redrawn from Ref. [101].

expect that chromium(II) acetate should also have a very short bond, if only it could be obtained unassociated. For this reason it has been studied in the gas phase by electron diffraction [101]. This is a most difficult experiment, as the compound is not particularly volatile, and it is reactive and somewhat unstable. After years of work the data were collected, using a glass nozzle to minimize decomposition, and working at the lowest possible vapor pressure and temperature. But the effort was well worth while. The radial distribution curve (Fig. 10.47) shows a peak at 2.00 Å, which contains contributions from the Cr–O and Cr≡Cr atom pairs, and it is at once obvious that the Cr≡Cr distance cannot be around 2.3 Å, as it is in the solid phase. The refined distance is in fact 1.966 Å, slightly longer than in other unperturbed compounds, but nevertheless quite clearly in the expected 'short' range.

10.19 Dihydrogen and dihyride complexes

The proton is a unique chemical species. It is extremely small, extremely electron deficient, and extremely mobile. So when a complex molecule is protonated, it is by no means obvious just where the proton will end up. It is quite possible that the initial site of attack (perhaps guided largely by electrostatic attraction) may not determine the most stable product, which is then formed by subsequent rearrangement. The products formed by protonating some cyclopentadienyl (Cp) ruthenium complexes already containing hydride bound to the metal have been studied using ^{1}H NMR spectroscopy [102], which shows quite clearly that more than one type of product forms, depending to some extent on the other ligands present.

The protonation of $[Ru(C_5H_5)HL_2]$, where L is a phosphine ligand PR_3, with HBF_4 in ether led to species formulated as $[Ru(C_5H_5)H_2L_2]^+[BF_4]^-$. The ^{1}H and ^{31}P NMR spectra showed that one of the products in each case, the only one when the phosphine ligand was not too bulky, had a single type of hydrogen associated directly with the metal, with no observable coupling, or only a very small coupling, to ^{31}P. The key observation was that the relaxation time T_1 for this signal was extremely short, a few milliseconds, whereas T_1 values of up to a second are expected for hydrides bound directly to metal, and the protons in the other ligands show T_1 values of several seconds. Such short relaxation times appear to be characteristic of hydrogen bound as H_2 units, and lead to very broad lines (width at half-height of the order of 10–15 Hz), as found in these compounds. The short relaxation time is due to the intense dipole–dipole interaction between the two proton spins, which are uniquely close together in the H_2 unit.

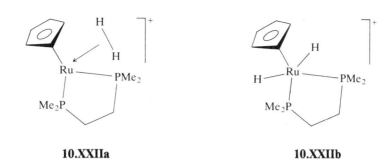

10.XXIIa   **10.XXIIb**

The identity of the product was confirmed by reaction of the neutral precursors with DBF_4, which gave HD complex cations. Now the ^1H spectra showed the one-bond HD coupling, whose magnitude, of the order of 30 Hz, is also characteristic of dihydrogen. The linewidths were much reduced, because the ^2H nucleus has a smaller magnetogyric ratio than ^1H, and so the dipole–dipole interaction between ^1H and ^2H nuclear spins is decreased proportionately. There was no evidence that the hydrogen atoms in the ligand occupied nonequivalent sites.

When the dibasic phosphine ligand $Me_2PCH_2CH_2PMe_2$ was used, the product of protonation contained both the η^2-H_2 cation 10.XXIIa and a conventional dihydride cation 10.XXIIb, which is formally seven-coordinate at ruthenium. The rate of interconversion was too slow to cause any broadening of the peaks in the NMR spectra (Fig. 10.48), but because of the very long relaxation times (10 s) associated with the cyclopentadienyl ligand protons it was possible to measure the effective rate of interconversion by a spin saturation transfer experiment, in which the decrease in intensity of the Cp resonance of one form was monitored as a function of time while the other Cp resonance was steadily irradiated. This showed a rate constant for the conversion of $9 \times 10^{-3}\,s^{-1}$ at 298 K, corresponding to a half-life for each form of the

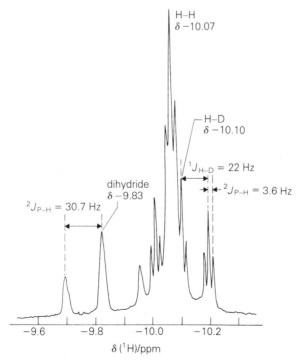

Fig. 10.48 ^1H NMR spectrum of a mixture of H—H and H—D isotopomers of the ruthenium complexes 10.XXIIa and 10.XXIIb. The η^2-dihydrogen complex shows a narrow triplet at –10.07 ppm, and on replacement of one proton by deuterium there is a small isotopic chemical shift, and the H—D coupling is also observed. The dihydride isomer appears simply as a triplet, coupling to the phosphines, and there is no evidence of coupling to deuterium or an isotopic shift. Taken, with permission, from Ref. [102].

order of 100 seconds. It turned out that if the protonation reaction was carried out at a low temperature only the η^2-dihydrogen species was formed, and its conversion to the equilibrium mixture could be followed on warming, giving a half-life of about 600 seconds at 283 K.

The dihydrogen ligand in this case seems to be very strongly bonded to the metal, and the H—H bond is correspondingly longer and weaker than in the products of protonation of complexes with two simple phosphine ligands; this results in a smaller $^1J_{HD}$, only about 22 Hz, and a measurable $^2J_{PH}$ about 4 Hz, in this case. The PH coupling constant in the dihydride complex is much larger, about 30 Hz, as expected.

Similar reactions of related species, giving a series of cations $[Ru(C_5H_5)H_2\{Ph_2P(CH_2)_nPPh_2\}]^+$ with $n = 1$, 2 or 3, have also been studied [103]. When $n = 1$, only the η^2-dihydrogen species was formed, while for $n = 3$ the sole product was the dihydride. In the case of $n = 2$, both species were found, but this time no interconversion could be observed, even using spin saturation transfer, and no variation of composition was seen on changing the temperature. The relaxation times, T_1, of the hydrogen ligand nuclei of both complexes were determined by the inversion-recovery method (Fig. 10.49). The value for the dihydrogen complex is about 40 times shorter than that for the dihydride, a result which is consistent with the earlier data for ruthenium compounds, and other dihydrogen complexes which can be identified unequivocally. These include the tungsten compound $[W(CO)_3(\eta^2\text{—}H_2)(PPr^i_3)_2]$, the first dihydrogen complex to be studied by neutron diffraction [104], and thereby shown to have a short H—H distance (0.84 Å) characteristic of a bonded pair of atoms. (The distance is 0.74 Å in free H_2.)

The criterion of short T_1 was naturally put to use, to identify possible dihydrogen ligands in the many polyhydrido derivatives of transition metals. In the course of a

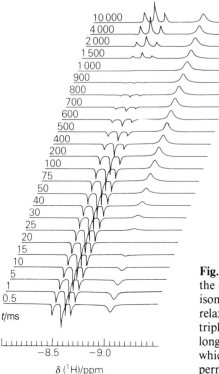

Fig. 10.49 Inversion recovery T_1 determination for the dihydrogen and hydride ligands in the mixture of isomers of $[Ru(C_5H_5)H_2\{Ph_2P(CH_2)_2PPh_2\}]^+$. The relaxation time for the dihydride, which gives the triplet near -8.5 ppm, is clearly about 40 times longer than that for the η^2-dihydrogen complex, which gives the resonance near -9 ppm. Taken, with permission, from Ref. [103].

detailed and careful study of a series of these complexes [105], it was concluded that some of the seven hydrogen atoms in [ReH$_7$(Ph$_2$PCH$_2$CH$_2$PPh$_2$)] were linked in pairs, with an H—H distance of not more than 1.25 Å. Yet an equally careful and detailed study of this same complex using neutron diffraction to reveal the exact positions of the hydrogen atoms showed that all seven were bound individually to the rhenium, with a minimum H$\cdots$H distance of 1.77 Å [106]. Only if the bonding arrangements in the crystal and in the solutions studied by NMR are different can these two conclusions be reconciled, yet a comparison of the IR spectra in the Re—H stretching region showed no clear changes from one phase to another that might confirm such a change of coordination. There is a similar apparent conflict between a short T_1 for some of the protons (the critical parameter is actually the minimum value attained as the sample is cooled below room temperature) in [ReH$_5$(PMePh$_2$)$_3$] [107] and a neutron diffraction study which shows no unusually short H$\cdots$H contact [108]. So it appears that although the ^{1}H nuclei in coordinated dihydrogen do have very short relaxation times, that parameter, of itself, does not necessarily indicate the presence of an η^2-H$_2$ ligand. Identification by NMR spectroscopy must depend on more substantial evidence.

Such evidence is available for some remarkable complexes which appear to have three hydrogens atoms bound to one another [109]. Protonation of [Ir(C$_5$H$_5$)H$_2$L] (L = PMe$_3$, PPh$_3$ or AsPh$_3$) yields [Ir(C$_5$H$_5$)H$_3$L]$^+$, complexes which have [AB$_2$X] (X = ^{31}P) or [AB$_2$] patterns in the region where metal hydride resonances are expected. The AB coupling constants are very large, ranging from 100 to over 400 Hz in the three compounds. There is also a large isotopic shift for the A nucleus when the B nuclei are replaced by deuterium (ca. 0.07 ppm per substitution), and a shift of about half this amount in the B resonances when A is substituted, but replacing one B by deuterium has no effect on the chemical shift of the remaining B. All this evidence indicates that there is a substantial bonding interaction between the A atom and both B's, but no direct link between the two B atoms. The ion would appear therefore to have structure 10.XXIII, which contains either H$_3^+$ bound to Ir^I or H$_3^-$ bound to Ir^{III}.

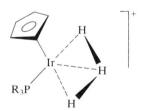

10.XXIII

However, even this conclusion is not so clear-cut as it appears to be; it has recently been shown [110, 111] that the large J_{AB} couplings in such complexes are ascribable to a further unique property of hydrogen, its propensity for quantum mechanical tunnelling. It appears to be unnecessary to postulate any direct bonding between the hydrogen atoms in [Ir(C$_5$H$_5$)H$_3$L]$^+$; all that is needed is that the unique (A) hydrogen nucleus is close in space to each of the two B nuclei. Then ^{1}H nuclei will then exchange *via* a tunnelling mechanism, which gives a contribution to the observed J_{AB} if both A and B are ^{1}H, but not if one is ^{2}D or ^{3}T. This has been confirmed [111] by observation of the ^{3}T NMR spectrum of partially-tritiated samples, where the quadrupolar relaxation effects associated with ^{2}D are absent, but ^{1}H—^{3}T J_{AB} coupling is small (20–30 Hz) even though the corresponding ^{1}H—^{1}H coupling is very large.

In this case it is the nonequivalence of ^{1}H and ^{3}T nuclei that removes the exchange contribution. In a fully tritiated tri-hydride the exchange contribution would also be small, because of the greater mass of ^{3}T, which reduces the tunnelling rate. The proposed tunnelling mechanism is consistent with the observation that J_{AB} *increases* with rising temperature when A and B are both ^{1}H, as the tunnelling rate is strongly influenced by the available thermal energy. The $H_A \cdots H_B$ distances in $[Ir(C_5H_5)H_3(PPh_3)]^+$ have been estimated [111] to be 1.65 ± 0.05 Å, similar to those determined in the neutron diffraction studies mentioned above; it therefore appears that some of the anomalous NMR parameters reported for hydride species of other types may be ascribed to similar tunnelling effects.

Problems

10.1 Measure the relative heights of the peaks in the ^{19}F and ^{129}Xe NMR spectra shown in Fig. 10.3, and compare them with the relative intensities for the central lines of the multiplets expected for Xe_4F_{24}.

10.2 In this chapter we have discussed only NMR evidence for the structure of xenon hexafluoride in solution, and X-ray diffraction evidence for the crystalline phase structure. What information about these phases could be obtained using other techniques? [For a short review of what has been done, see K. Seppelt and D. Lentz, *Prog. Inorg. Chem.*, **29**, 172-180 (1982).]

10.3 The ions $[HCN-XeF]^+$ and $[HCN-KrF]^+$ have been shown to exist, mainly by NMR spectroscopy. How would you determine whether the ions were linear or bent at the nitrogen atoms?

10.4 Read the description in Ref. [25] of the effects of temperature changes on the ^{1}H NMR spectrum of $Al(BH_4)_3$. Discuss the observations in the light of alternative theories (a) that the compound can exist in two distinct isomeric forms and (b) that at high temperatures a small amount of a new species is formed, irreversibly, and that this catalyzes exchange of BH_4 groups between molecules.

10.5 The compounds $Be(BH_4)(C_5H_5)$, $Be(BH_4)(B_5H_{10})$ and $Be(B_3H_8)_2$ are all known. What reasonable structures could these molecules have, and how could you distinguish the various possibilities? Which, if any, could show fluxional behavior? How would you recognize such behavior?

10.6 What would be the effect of replacing ^{14}N by ^{15}N on the vibrational spectra of Na_3NO_4? What would be the effect of replacing *ca.* 10% of ^{16}O by ^{18}O?

10.7 To what extent would it be possible to distinguish between D_{5h} and D_{5d} symmetries for ferrocene in gaseous and crystalline phases using (a) infrared and Raman spectroscopy or (b) NMR spectroscopy?

10.8 What would you expect the structure of $[Re(C_5Me_5)_2H]$ to look like? What methods could you use (a) to show whether the rings were η^5- or η^1-bonded, (b) to find out whether all ten carbon atoms in each ring system were co-planar and (c) to measure the Re—H bond distance?

10.9 The compound $(C_5H_5)BeC\equiv CH$ is believed to have C_{5v} symmetry. How could this be proved?

10.10 At low temperatures the ^{13}C NMR spectrum of $PF_2C_5H_5$ (10.IXa) has three doublet resonances, at 140 ppm (J_{PC} 5 Hz), 126 ppm (J_{PC} 11 Hz) and 66 ppm (J_{PC} 44 Hz). The last of these is half the intensity of the others. Describe what the high temperature spectrum will look like, for all possible combinations of signs for the three coupling constants.

10.11 How would you identify the isomers of $PF_2C_5H_5$, 10.IXb and 10.IXc? Would you expect them to show fluxional behavior by migration of a hydrogen atom round the ring? How would you recognize this behavior?

10.12 On addition of chloride to a solution of $[Au_{11}(PMe_2Ph)_{10}]^{3+}$ the cluster changes to $[Au_{13}Cl_2(PMe_2Ph)_{10}]^{3+}$. How could you monitor the course of this reaction in solution, and how could you identify the product?

10.13 In the photoelectron spectrum of HCP the first band, at 10.8 eV, shows vibrational structure, with lines separated by 1110 cm^{-1}. The second band, at 12.9 eV, has just one weak vibrational satellite, separated from the main line by 1250 cm^{-1}. The C—P stretching frequency of gaseous HCP is at 1265 cm^{-1}. How do these observations help in the assignment of the photoelectron spectrum? What would you expect to see in the photoelectron spectrum of HCN?

10.14 In Section 10.8.2 the preparation and characterization of large clusters containing 55 metal atoms are described. Which structural methods would you use to show that the product of such a preparation is a single compound, and not a mixture of clusters of different sizes?

10.15 The bipyridine complex $[Cr(bpy)_3]^{3+}$ can be reduced six times, and it is believed that the first three reductions are ligand-based but that the rest are metal-based. How could this be proved?

10.16 The Raman spectrum of B_8S_{16} (see Fig. 10.39) has 12 bands in the region associated with B—S and S—S stretching vibrations: the IR spectrum has not been described. How many bands should be expected in this region of IR and Raman spectra, assuming that the molecules have D_{4h} symmetry? How might the packing of molecules into the crystalline solid affect the number of bands observed?

10.17 Explain how (a) vibrational spectroscopy, (b) solid state NMR spectroscopy and (c) EXAFS might help in identifying $[Pb_2Se_3]^{2-}$ ions in the solid formed by dissolving selenium in molten KPb (Section 10.13).

10.18 In Section 10.14 a symmetrical form of N_2O_3 was discussed. What techniques, other than vibrational spectroscopy, could be used to characterize this species?

10.19 The ESR spectrum of $\cdot N(SF_5)_2$ (Fig. 10.41) may be interpreted as showing coupling to ten fluorine nuclei, or to one nitrogen and eight fluorine nuclei. How could the use of isotopically labeled samples help to resolve this uncertainty, and what extra information could also be obtained in this way?

10.20 If compound 10.44 is treated with the ylide Ph_3PCH_2, one halogen atom is replaced by $-CH_2PPh_3^+$. What would you expect to observe in the ^1H-decoupled ^{31}P and ^{13}C NMR and ^{197}Au Mössbauer spectra of the product? What other structural methods might be helpful in characterizing the new complex?

10.21 It has been suggested that it might be possible to study the unperturbed Cr≡Cr distance in chromium(II) acetate by EXAFS, with the molecules isolated in an inert matrix of methane or a noble gas. What would you expect to see in the radial distribution curve, using chromium EXAFS? How could you tell whether axial donation by atoms of the matrix was occurring?

10.22 It has been observed that some at least of the protons bound to the metal atom in $[ReH_7(Ph_2PCH_2CH_2PPh_2)]$ have short relaxation times, T_1, but that this cannot be attributed to coordinated H_2 ligands. What other factors could lead to rapid relaxation in this compound?

10.23 In many of the case histories included in this chapter isotopic substitution has been used to gain structural information. Discuss its application to vibrational, rotational, NMR, ESR and Mössbauer spectroscopy.

10.24 What structural methods would you use to characterize the following compounds and ions? $[Os_3(CO)_{10}(CH_3C{\equiv}CH)]$; $KFeS_2$; $B_2H_4(NMe_3)_2$; $[Cr_2Cl_6(PEt_3)_4]$; $[I(OSO_2F)_2]I$; FeI_3; $[Rh_2(CH_3COO)_4(PPh_3)_2]^-$; Se_5S_2; $[Ru_4(BH_2)(CO)_{12}H]$; CF_3SSF; $[Ru(C_5H_5)(NO)_2(PPh_3)_2]$; $[Fe_6S_6(OAr)_6\{Mo(CO)_3\}_2]^{3-}$, where Ar is an aryl group.
All these species are described in *Inorganic Chemistry*, Vol. 29 (1990).

Further reading

On xenon hexafluoride – K. Seppelt and Lentz, *Prog.Inorg.Chem.*, **29**, 172 (1982).

On tetrahydroborates – T.J. Marks and J.R. Kolb, *Chem.Rev.*, **77**, 263 (1977).

On gold clusters – Ref. [54].

On Fe(CO)$_4$ – M. Poliakoff, *Chem.Soc.Rev.*, **7**, 527 (1978).

On multiple bonds between metal atoms – Ref. [97].

By this stage almost any issue of any inorganic chemistry journal is suitable for further reading. Choose an article dealing with the structure or identification of inorganic compounds. Read it and criticize it. Identify the strong and weak points. Then look for ways in which the work can be confirmed, improved or extended.

References

1 R.M. Gavin and L.S. Bartell, *J.Chem.Phys.*, **48**, 2460 (1968).
2 L.S. Bartell and R.M. Gavin, *J.Chem.Phys.*, **48**, 2466 (1968).
3 G.L. Goodman, *J.Chem.Phys.*, **56**, 5038 (1972).
4 H.H. Classen, G.L. Goodman and H. Kim, *J.Chem.Phys.*, **56**, 5042 (1972).
5 U. Nielsen, R. Haensel and W.H.E. Schwarz, *J.Chem.Phys.*, **61**, 3581 (1974).
6 K.S. Pitzer and L.S. Bernstein, *J.Chem.Phys.*, **63**, 3849 (1975).
7 H. Rupp and K. Seppelt, *Angew.Chem.Int.Ed. Engl.*, **13**, 612 (1974).
8 G.J. Schrobilgen, J.H. Holloway, P. Granger and C. Brevard, *Inorg. Chem.*, **17**, 980 (1978).
9 R.D. Burbank and G.R. Jones, *J.Am.Chem.Soc.*, **96**, 43 (1974).
10 L. Stein, J.R. Norris, A.J. Downs and A.R. Minihan, *J.Chem.Soc., Chem. Commun.*, 502 (1978).
11 A.J. Downs and R.C. Fowler, unpublished results.
12 R.J.H. Clark, *Chem.Soc.Rev.*, **13**, 219 (1984).
13 A.A.A. Emara and G.J. Schrobilgen, *J.Chem.Soc., Chem.Commun.*, 1644 (1987).
14 G.J. Schrobilgen, *J.Chem.Soc., Chem.Commun.*, 863 (1988).
15 G.J. Schrobilgen, *J.Chem.Soc., Chem.Commun.*, 1506 (1988).
16 K.M. Melmed, D. Coucouvanis and S.J. Lippard, *Inorg.Chem.*, **12**, 232 (1973).
17 F. Takusagawa, A. Fumagalli, T.F. Koetzle, S.G. Shore, T. Schmitkons, A.V. Fratini, K.W. Morse, C.-Y. Wei and R. Bau, *J.Am.Chem.Soc.*, **103**, 5165 (1981).
18 D.G. Holah, A.N. Hughes, S. Maciaszek, V.R. Magnuson and K.O. Parker, *Inorg. Chem.*, **24**, 3956 (1985).
19 T.J. Marks, W.J. Kennelly, J.R. Kolb and L.A. Shimp, *Inorg. Chem.*, **11**, 2540 (1972).
20 S.W. Kirtley, M.A. Andrews, R. Bau, G.W. Grynkewich, T.J. Marks, D.L. Tipton and B.R. Whittlesey, *J.Am.Chem.Soc.*, **99**, 7154 (1977).
21 H.D. Empsall, E.M. Hyde, E. Mentzer, B.L. Shaw and M.F. Uttley, *J.Chem.Soc., Dalton Trans.*, 2069 (1976).
22 P.W. Frost, J.A.K. Howard and J.L. Spencer, *J.Chem.Soc., Chem.Commun.*, 1362 (1984).
23 A. Almenningen, G. Gundersen and A. Haaland, *Acta Chem.Scand.*, **22**, 328 (1968).
24 R.A. Ogg and J.D. Ray, *Disc. Faraday Soc.*, **19**, 239 (1955).
25 P.C. Maybury and J.E. Ahnell, *Inorg.Chem.*, **6**, 1286 (1967).
26 A. Almenningen, G. Gundersen and A. Haaland, *Acta Chem.Scand.*, **22**, 859 (1968).
27 G. Gundersen, L. Hedberg and K. Hedberg, *J.Chem.Phys.*, **59**, 3777 (1973).

28 D.S. Marynick and W.N. Lipscomb, *Inorg.Chem.*, **11**, 820 (1972).

29 J.W. Nibler, *J.Am.Chem.Soc.*, **94**, 3349 (1972).

30 K. Brendhaugen, A. Haaland and D.P. Novak, *Acta Chem.Scand., Ser. A*, **29**, 801 (1975).

31 D.F. Gaines, J.L. Walsh, J.H. Morris and D.F. Hillenbrand, *Inorg.Chem.*, **17**, 1516 (1978).

32 M. Jansen, *Angew.Chem.Int.Ed. Engl.*, **16**, 534 (1977).

33 M. Jansen, *Angew.Chem.Int.Ed. Engl.*, **18**, 698 (1979).

34 M. Jansen, *Z.Anorg.Allg.Chem.*, **491**, 175 (1982).

35 T.J. Kealy and P.L. Pauson, *Nature*, **168**, 1039 (1951).

36 G. Wilkinson, M. Rosenblum, M.C. Whiting and R.B. Woodward, *J.Am.Chem.Soc.*, **74**, 2125 (1952).

37 J.D. Dunitz, L.E. Orgel and A. Rich, *Acta Crystallogr.*, **9**, 373 (1956).

38 A. Haaland and J.E. Nilsson, *Acta Chem.Scand.*, **22**, 2653 (1968).

39 P. Seiler and J. Dunitz, *Acta Crystallogr., Sect. B*, **35**, 1068 (1979).

40 P. Seiler and J. Dunitz, *Acta Crystallogr., Sect. B*, **35**, 2020 (1979).

41 P. Seiler and J. Dunitz, *Acta Crystallogr., Sect. B*, **38**, 1741 (1982).

42 J.A. Bandy, F.G.N. Cloke, G. Cooper, J.P. Day, R.B. Girling, R.G. Graham, J.C. Green, R. Grinter and R.N. Perutz *J.Am.Chem.Soc.*, **110**, 5039 (1988).

43 G.L. Morgan and G.B. McVicar, *J.Am.Chem.Soc.*, **90**, 2789 (1968).

44 K.W. Nugent, J.K. Beattie, T.W. Hambly and M.R. Snow, *Austr. J. Chem.*, **37**, 1601 (1984).

45 C. Wong, T.Y. Lee, T.J. Lee, T.W. Chang and C.S. Liu, *Inorg. Nucl. Chem. Letters*, **9**, 667 (1973).

46 A. Almenningen, A. Haaland and J. Lusztyk, *J. Organometal Chem.*, **170**, 271 (1979).

47 K.W. Nugent and J.K. Beattie, *Inorg. Chem.*, **27**, 4269 (1988).

48 J. Lusztyk and K.B. Starowiesky, *J. Organometal. Chem.*, **170**, 293 (1979).

49 K.W Nugent, J.K. Beattie and L.D. Field, *J.Phys.Chem.*, **93**, 5371 (1989).

50 C. Wong and S. Wang, *Inorg. Nucl. Chem. Letters*, **11**, 677 (1975).

51 J.E. Bentham, E.A.V. Ebsworth, H. Moretto and D.W.H. Rankin, *Angew.Chem.Int.Ed. Engl.*, **11**, 640 (1972).

52 R.T. Paine, R.W. Light and D.E. Maier, *Inorg.Chem.*, **18**, 368 (1979).

53 P.L. Bellon, F. Cariati, M. Manassero, L. Naldini and M. Sansoni, *J.Chem.Soc., Chem. Commun.*, 1423 (1971).

54 K.P. Hall and D.M.P. Mingos, *Prog.Inorg.Chem.*, **32**, 237 (1984).

55 D.M.P. Mingos, *Proc.R.Soc. London, Ser. A*, **308**, 75 (1982).

56 J.M.M. Smits, B.T. Beurskens, J.J. Bour and F.A. Vollenbroek, *J.Cryst.Spectrosc.Res.*, **13**, 365 (1983).

57 K.P. Hall, B.R.C. Theobald, D.I. Gilmour, A.J. Welch and D.M.P. Mingos, *J.Chem.Soc., Chem.Commun.*, 528 (1982).

58 G. Schmid, R. Pfeil, R. Boese, F. Bandermann, S. Meyer, G.H.M. Calis and J.W.A. van der Velden, *Chem. Ber.*, **114**, 3634 (1981).

59 G. Schmid, U. Giebel, W. Huster and A. Schwenk, *Inorg. Chim. Acta.*, **85**, 97 (1984).

60 G. Schmid and W. Huster, *Z. Naturforsch., Teil B*, **41**, 1028 (1986).

61 G. Schmid and N. Klein, *Angew.Chem.Int.Ed. Engl.*, **25**, 922 (1986).

62 F. Scherbaum, A. Grohmann, B. Huber, C. Krüger and H. Schmidbaur, *Angew.Chem. Int.Ed. Engl.*, **27**, 1544 (1988).

63 D.M.P. Mingos and R.P.F. Kanters, *J. Organometal. Chem.*, **384**, 405 (1990).

64 A. Grohmann, J. Riede and H. Schmidbaur, *Nature*, **345**, 140 (1990).

65 G. Christou, P.K. Mascharak, W.H. Armstrong, G.C. Papaefthymiou, R.B. Frankel and R.H. Holm, *J.Am.Chem.Soc.*, **104**, 2820 (1982).

66 T.E. Gier, *J.Am.Chem.Soc.*, **83**, 1769 (1961).

67 J.K. Tyler, *J.Chem.Phys.*, **40**, 1170 (1964).

68 J.W.C. Johns, H.F. Shurvell and J.K. Tyler, *J.Chem.Phys.*, **47**, 893 (1969).

69 J.-M. Garneau and A. Cabana, *J.Mol.Spectrosc.*, **87**, 490 (1981).

70 S.P. Anderson, H. Goldwhite, D. Ko, A. Letsou and F. Esparza, *J.Chem.Soc. Chem. Commun.*, 744 (1975).

71 D.C. Frost, S.T. Lee and C.A. McDowell, *Chem.Phys.Letters*, **23**, 472 (1973).

72 Y. Ohsawa, M.K. DeArmond, K.W. Hanck, D.E. Morris, D.C. Whitten and P. Neveux, *J.Am.Chem.Soc.*, **105**, 6522 (1983).

REFERENCES

73 G.A. Heath, L.J. Yellowlees and P.S. Braterman, *J.Chem.Soc., Chem.Commun.*, 287 (1981).

74 V.T. Coombe, G.A. Heath, A.J. MacKenzie and L.J. Yellowlees, *Inorg.Chem.*, **23**, 3423 (1984).

75 S.M. Angel, M.K. DeArmond, R.J. Donohoe, K.W. Hanck and D.W. Wertz, *J.Am.Chem. Soc.*, **106**, 3688 (1984).

76 A.G. Motten, K. Hanck and M.K. DeArmond, *Chem.Phys.Letters*, **79**, 541 (1981).

77 R.F. Dallinger and W.H. Woodruff, *J.Am.Chem.Soc.*, **101**, 4391 (1979).

78 P.S. Braterman, A. Harriman, G.A. Heath and L.J. Yellowlees, *J.Chem.Soc., Dalton Trans.*, 1801 (1983).

79 P.G. Bradley, N. Kress, B.A. Hornberger, R.F. Dallinger and W.H. Woodruff, *J.Am.Chem. Soc.*, **103**, 7441 (1981).

80 J.V. Caspar, T.D. Westmoreland, G.H. Allen, P.C. Bradley, T.J. Meyer and W.H. Woodruff, *J.Am.Chem.Soc.*, **106**, 3492 (1984).

81 G. Nord, A.C. Hazell, R.G. Hazell and O. Farver, *Inorg.Chem.*, **22**, 3429 (1983).

82 P.S. Braterman, G.A. Heath, A.J. MacKenzie, B.C. Noble, R.D. Peacock and L.J. Yellowlees, *Inorg.Chem.*, **23**, 3425 (1984).

83 P.J. Spellane, R.J. Watts and C.J. Curtis, *Inorg.Chem.*, **22**, 4060 (1983).

84 B. Krebs and H.-U. Hürter, *Angew.Chem.Int.Ed. Engl.*, **19**, 481 (1980).

85 M. Björgvinsson, J.F. Sawyer and G.J. Schrobilgen, *Inorg. Chem.*, **26**, 741 (1987).

86 A.H. Brittain, A.P. Cox and R.L. Kuczkowski, *Trans. Faraday Soc.*, **65**, 1963 (1969).

87 E.M. Nour, L.-H. Chen and J. Laane, *J.Phys.Chem.*, **87**, 1113 (1983).

88 L.-O. Andersson and J. Mason, *J.Chem.Soc., Chem. Commun.*, 99 (1968).

89 R.F. Holland and W.B. Maier, *J.Chem.Phys.*, **78**, 2928 (1983).

90 J.S. Thrasher and J.B. Nielsen, *J.Am.Chem.Soc.*, **108**, 1108 (1986).

91 M. Poliakoff and J.J. Turner, *J.Chem.Soc., Dalton Trans.*, 2276 (1974).

92 G. Bor, *Inorg.Chem. Acta*, **1**, 81 (1967).

93 T.J. Barton, R. Grinter, A.J. Thomson, B. Davies and M. Poliakoff, *J.Chem.Soc., Chem. Commun.*, 841 (1977).

94 H. Schmidbaur, C. Hartmann, G. Reber and G. Müller, *Angew.Chem.Int.Ed. Engl.*, **26**, 1146 (1987).

95 H. Schmidbaur, C. Hartmann and F.E. Wagner, *Angew.Chem.Int.Ed. Engl.*, **26**, 1148 (1987).

96 F.A. Cotton and R.A. Walton, *Multiple Bonds Between Metal Atoms*, Wiley-Interscience, New York (1982).

97 B.N. Figgis and R.L. Martin, *J.Chem.Soc.*, 3837 (1956).

98 F.A. Cotton, B.G. DeBoer, M.D. LaPrade, J.R. Pipal and D.A. Ucko, *Acta Crystallogr., Sect. B*, **27**, 1664 (1971).

99 F.A. Cotton, C.E. Rice and G.W. Rice, *J.Am.Chem.Soc.*, **99**, 4704 (1977).

100 F.A. Cotton, S.A. Koch and M. Millar, *Inorg.Chem.*, **17**, 2087 (1978).

101 S.N. Ketkar and M. Fink, *J.Am.Chem.Soc.*, **107**, 338 (1985).

102 M.S. Chinn and D.M. Heinekey, *J.Am.Chem.Soc.*, **109**, 5865 (1987).

103 F.M. Conway-Lewis and S.J. Simpson, *J.Chem.Soc., Chem.Commun.*, 1675 (1987).

104 G.J. Kubas, R.R. Ryan, B.I. Swanson, P.J. Vergamini and H.J. Wasserman, *J.Am.Chem. Soc.*, **106**, 451, (1984).

105 D.C. Hamilton and R.H. Crabtree, *J.Am.Chem.Soc.*, **110**, 4126 (1988).

106 J.A.K. Howard, S.A. Mason, O. Johnson, I.C. Diamond, S. Crennell, P.A. Keller and J.L. Spencer, *J.Chem.Soc., Chem.Commun.*, 1502 (1988).

107 F.A. Cotton and R.L. Luck, *Inorg. Chem.*, **28**, 6 (1989).

108 T.J. Emge, T.F. Koetzle, J.W. Bruno and K.G. Caulton, *Inorg. Chem.*, **23**, 4012 (1984).

109 D.M. Heinekey, N.G. Payne and G.K. Schulte, *J.Am.Chem.Soc.*, **110**, 2303 (1988).

110 D.H. Jones, J.A. Labinger and D.P. Weitekamp, *J.Am.Chem.Soc.*, **111**, 3087 (1989).

111 K.W. Zilm, D.M. Heinekey, J.M. Millar, N.G. Payne and P. Demou, *J.Am.Chem.Soc.*, **111**, 3088 (1989).

Appendix: Symmetry

Knowledge of molecular symmetry is of central importance in the analysis of vibrations, electronic orbitals and other aspects of molecular properties and behavior. Many molecules have low symmetry overall, but even so some groups, such as methyl or phenyl substituents or triphenyl phosphine ligands, may have effective local symmetry, and we may find it helpful to consider this local symmetry in analysis of the whole molecule. The use of symmetry to classify molecular vibrations is explained in Chapter 5 of this book, and symmetry properties of molecular orbitals are discussed in Chapter 6. In this Appendix we simply state some of the basic ideas and rules needed so that we can use symmetry arguments properly. For a full treatment of symmetry for isolated molecules (point group symmetry) or crystals (space group symmetry) you should consult the specialized books, intended for chemists, which are listed at the end of the Appendix (Refs [1] – [10]). Ref. [11] is an invaluable compilation of information relating to the use of symmetry.

Symmetry operations

We begin with the idea of a symmetry operation, which is an operation that generates a configuration indistinguishable from the initial one. Thus for the molecule $SiBrCl_2H$ [Fig. A.1(a)] the reflection operation ('reflect in the yz plane') is a symmetry operation, and the plane specified is called a symmetry element (in this case a mirror plane). We cannot actually perform this operation, either on a real molecule or on a macroscopic model, but this does not matter; it is quite legitimate (and indeed necessary) to use some non-feasible symmetry operations. Fig. A.1 shows several molecules that have one or more mirror planes. Another non-feasible operation is the inversion, which operates through a point, called an inversion center, as in Fig. A.2.

The only feasible symmetry operations needed for isolated molecules are the rotation operations, 'rotate about an axis by $2\pi/n$ (*i.e.* $1/n$th of a revolution), where n is an integer between 1 and infinity'. The symmetry element is the axis of rotation. Some examples of rotational symmetry in molecules are shown in Fig. A.3.

Where only a single rotation axis occurs it is conventionally assigned as the z axis, while if only a single mirror plane occurs it is the yz plane, unless the plane is perpendicular to an axis which has already been defined as the z-axis.

Sometimes we need a compound operation to describe the symmetry of a molecule. For isolated molecules it is usual to choose a rotation–reflection operation, 'rotate about an axis by $2\pi/n$ and also reflect in the plane perpendicular to the axis'. The rotation–reflection operation may also be called an improper rotation. An example of a rotation–reflection operation may be seen in $SiCl_4$ (Fig. A.4). Remember that the reflection plane is perpendicular to the rotation axis, and note that we often find that a rotation–reflection axis of order $2n$ is associated with a pure-rotation axis of order n.

For completeness we must also define the identity operation ('leave alone') as a

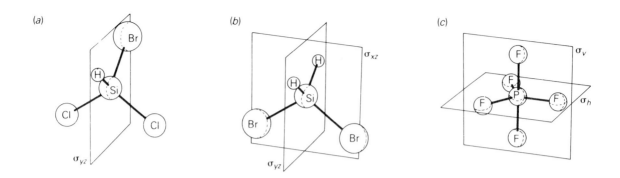

Fig. A.1 Mirror planes relating equivalent atoms in (*a*) $SiBrCl_2H$, (*b*) SiH_2Br_2 (two vertical planes, σ_v) and (*c*) PF_5 (one of three vertical planes, σ_v, and one horizontal plane, σ_h).

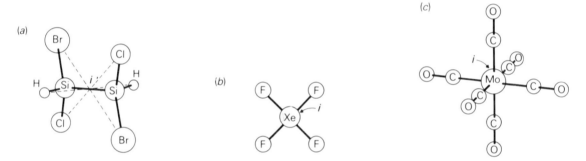

Fig. A.2 Inversion centers relating equivalent atoms in (*a*) $(SiBrClH)_2$, (*b*) XeF_4 and (*c*) $Mo(CO)_6$.

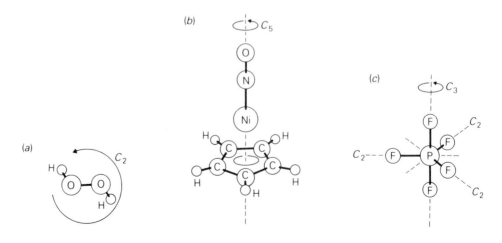

Fig. A.3 Rotation axes relating equivalent atoms in (*a*) H_2O_2 (C_2 axis), (*b*) $[Ni(C_5H_5)(NO)]$ (C_5 axis) and (*c*) PF_5 (C_3 and three C_2 axes).

APPENDIX: SYMMETRY

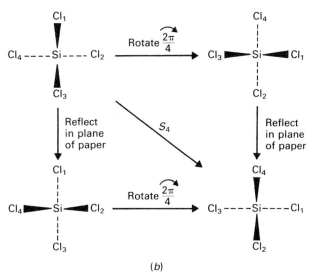

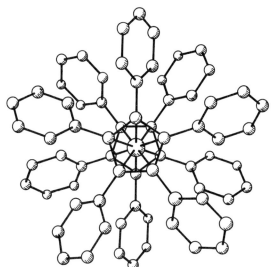

(b)

Fig. A.4 Rotation reflection axes in (a) SiCl$_4$, drawn looking down the S_4 axis; note that neither the rotation by $2\pi/4$ by itself nor the reflection in a plane perpendicular to this axis generates a configuration equivalent to the initial one, but the combined operation does; (b) Sn(η^5-C$_5$Ph$_5$)$_2$, showing an almost perfect S_{10} operation, looking down the axis; redrawn from M.J. Heeg, C. Janiak and J.J. Zuckerman, *J.Am.Chem. Soc.*, **106**, 4259 (1984).

symmetry operation, which applies to all molecules, so that strictly speaking we should never say that a molecule has no symmetry.

The various symmetry operations are each given a conventional symbol in the Schoenflies system as follows:

Operation	Symbol	Element
reflection	σ	mirror plane
inversion	i	inversion center
n-fold rotation	C_n	rotation axis
n-fold rotation–reflection	S_n	rotation–reflection axis
identity	E (or I)	none

This set of operations contains some hidden duplications. Thus for almost all purposes the operation C_1 (rotate by 2π) has exactly the same effect as the identity operation E.

A rotation–reflection operation S_1 is equivalent in its effect to a reflection alone, σ, and S_2 is equivalent to an inversion, i. In such cases the special symbol (E, σ, i) is used, and the general symbol is ignored.

The symmetry operations used in space group symmetry to describe regular arrays of atoms or molecules are not quite the same as those used for isolated molecules. The n-fold rotation, reflection in a mirror plane, inversion and identity operations are the same, but rotation–reflection is replaced by rotation–inversion ('rotate by $2\pi/n$ about the axis and then invert through a center lying on that axis'). The operations are given labels in the Herman–Mauguin system as follows:

n-fold rotation	n e.g. 2
n-fold rotation–inversion	$\bar{n}$ e.g. $\bar{3}$
reflection	m
inversion	$\bar{1}$
identity	1

The correlation between the Schoenflies system and the Herman–Mauguin system is simple for the rotation ($C_n \equiv n$), the reflection ($\sigma \equiv m$), the identity ($E \equiv 1$) and the inversion ($i \equiv \bar{1}$) operations. The rotation–reflection operation S_n correlates with a rotation–inversion operation, but the correlation is complex:

$$S_{2n}^{-1} \equiv \overline{2n} \text{ if } n \text{ is even } e.g. \ S_4^{-1} \equiv \bar{4}$$
$$S_{2n}^{-1} \equiv \bar{n} \text{ if } n \text{ is odd } \quad e.g. \ S_6^{-1} \equiv \bar{3}$$
$$S_{2n+1}^{-1} \equiv \overline{4n+2} \quad\quad e.g. \ S_3^{-1} \equiv \bar{6}$$

The symmetry of regularly repeating structures as found in crystals requires a more elaborate set of symmetry operations. In particular some translational motions are involved, as these can move one structural unit (molecule or ion) to the position previously occupied by another. They include the glide and screw operations illustrated in Fig. A.5. On the other hand, the set of symmetry operations is also more limited

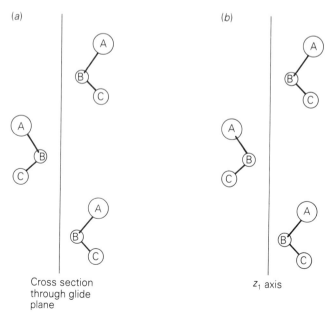

Fig. A.5 Symmetry operations involving translation: (a) a glide and (b) a 2-fold screw operation. In (a) the atoms A are all above the plane of the paper, and the symmetry operation involves translation and reflection. In (b) translation is followed by rotation: the A atoms are alternately above and below the plane of the paper.

(a) Cross section through glide plane

(b) z_1 axis

than that needed for isolated molecules; only axes of order 2, 3, 4 or 6 are possible as all others are incapable of leading to repeating patterns in 2 dimensions. The Herman–Mauguin system is used both for individual symmetry elements and for the space groups resulting from their combination. For example, the space group symbol $P2_1/m$ represents a primitive lattice having a two-fold screw axis (2_1) and a mirror plane perpendicular to the axis. A full description of the symmetry of crystals may be found in Refs [7] and [8].

Point groups

Many molecules have no symmetry elements except E, the identity. Many others have only one other symmetry operation, and for these it is generally sufficient for us to be able to decide whether a particular molecular property such as a vibration is symmetric with respect to this symmetry operation, *i.e.* whether the effect of the operation is to leave the property unchanged. If a property is reversed by the operation it is said to be antisymmetric with respect to the operation. All properties are of course left unchanged by the identity operation and are therefore symmetric with respect to it. Examples of vibrations symmetric and antisymmetric to a symmetry operation are shown in Fig. A.6.

More elaborate analysis is necessary if more than one symmetry operation can be performed on a molecule. A collection of symmetry operations forms a group, and for an individual object such as a molecule this is called a point group. All the symmetry elements associated with a point group pass through at least one common point. The total number of operations making up the group defines the order of the group, h. A

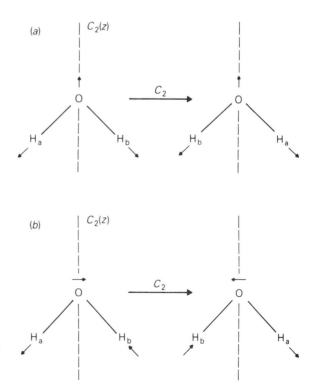

Fig. A.6 (*a*) A vibration symmetric to the $C_2(z)$ rotation axis of H_2O and (*b*) a vibration antisymmetric to this axis.

point group with only the identity operation has an order of 1; point groups each having only a single symmetry operation apart from the identity are of order 2.

When more than two symmetry operations are present, the effects of the various operations must be mutually consistent, and this limits the possible combinations. Thus there are only two possible orientations for the combination of a mirror plane and rotation axis. The plane may be perpendicular to the axis, as in *anti*-N_2F_2 [Fig. A.7(*a*)] or parallel to (and including) the axis, as in *syn*-N_2F_2 [Fig. A.7(*b*)]. The theory of groups expresses this by requiring that the product of two members of a group must itself be a member of the group. If we define 'product' in our point group as meaning successive application of the two operations we can then construct a multiplication table showing the results of the successive application of any pair of

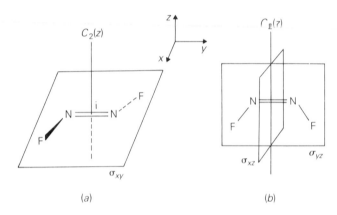

Fig. A.7 (*a*) *Anti*-N_2F_2, showing a mirror plane (σ_{xy}) perpendicular to the 2-fold axis $C_2(z)$; the point where these elements intersect is an inversion center, i; (*b*) *syn*-N_2F_2, showing the two mirror planes σ_{xz} and σ_{yz} parallel to the 2-fold axis $C_2(z)$.

operations. If we do this for the two molecules illustrated in Fig. A.7 we find in each case that the two symmetry operations we originally specified together imply a third operation as a necessary member of the group. In these two cases multiplication is commutative, but this is not always so, and we may have to specify which operation is to be performed first.

(*a*)

	E	$C_2(z)$	σ_{xy}	i
E	E	C_2	σ_{xy}	i
$C_2(z)$	C_2	E	i	σ_{xy}
σ_{xy}	σ_{xy}	i	E	C_2
i	i	σ_{xy}	C_2	E

(*b*)

	E	$C_2(z)$	σ_{xz}	σ_{yz}
E	E	C_2	σ_{xz}	σ_{yz}
$C_2(z)$	C_2	E	σ_{yz}	σ_{xz}
σ_{xz}	σ_{xz}	σ_{yz}	E	C_2
σ_{yz}	σ_{yz}	σ_{xz}	C_2	E

In each case constructing a partial group multiplication table using only E and the two operations originally specified introduces another operation, but including this in the table leads to a closed set of operations. We have in fact defined two different point

APPENDIX: SYMMETRY

groups of order 4; the multiplication table is the most detailed form of the definition of a group. In each case we could have chosen any two of the three symmetry operations as our defining operations, and would have found the third to be implied.

Point groups are distinguished by symbolic names, chosen to indicate the symmetry operations necessary to define the group. If only one symmetry operation (besides E) is present its symbol is also the group symbol (C_2, S_3, etc.), except that if the sole symmetry

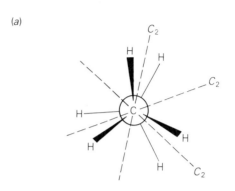

(a)

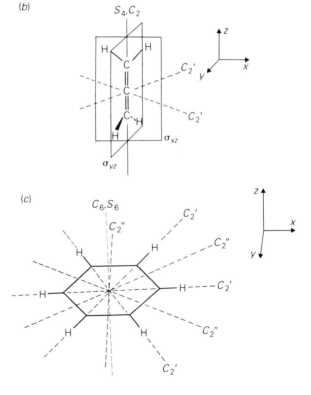

(b)

(c)

Fig. A.8 The symmetry elements comprising the groups (a) D_3, (b) D_{2d} and (c) D_{6h}. In (a) the drawing shows a twisted form of ethane, viewed along the C_3 axis; (b) shows the true structure of allene, where the planes xz and xy are σ_d planes and the C_2' axes lie at 45° to the x and y axes in the xy plane. In (c), which shows only the axes C_6 (which coincide with C_3, C_2, S_6 and S_3 axes), C_2' (one of which lies along x) and C_2'' (one of which lies along y), the xy plane is σ_h, and there are vertical mirror planes in-including each C_2' and C_2'' axis; the center of the molecules is an inversion center, i.

element is σ the point group is called C_s and if it is i the point group is called C_i. If E is the only symmetry operation the group is called C_1.

Combining a rotation operation C_n with a reflection through a plane perpendicular to the rotation axis leads to a point group called C_{nh}; case (a) above therefore defines the point group C_{2h}. The mirror plane is labeled σ_h in this case. Case (b) defines the point group C_{2v}, in which a C_2 operation is combined with σ_v (vertical, as opposed to horizontal). In general, combining C_n with σ_v leads to the point group C_{nv}, with a set of $n\sigma_v$ mirror planes.

Combining a rotation operation C_n with C_2 operations about axes at right angles to the C_n axis leads to the point group D_n; there are altogether n such C_2 axes [Fig. A.8(a)]. Adding a reflection operation parallel to the C_n axis between two of the C_2 axes leads to a point group D_{nd}. When vertical mirror planes are bisected by these subsidiary C_2 axes they are labeled σ_d rather than σ_v, so this point group involves n C_2 and n σ_d elements [Fig. A.8(b)]. Finally, adding a reflection operation through a plane perpendicular to the C_n axis leads to a point group D_{nh} [Fig. A.8(c)]. The multiplication table for the point group D_{4h} is shown in Table A.1. It has an order of 16, but only 10 distinct types of symmetry operation, so the table is 10×10. These types are called symmetry classes, and the number of operations in each class is shown as a prefix in the headings to the table.

It may be noted that the $C_4'(z)$ operation, if carried out twice, has the same effect as a C_2 operation about the z axis, so this latter operation has to be included in the group as implied by the C_4 operation. In the same way, applying the C_4 operation three times has a different effect from the C_4 operation itself (rotation by 3 times $2\pi/4$ rather than $2\pi/4$), but this operation (which can be written as C_4^3) is equivalent to C_4^{-1} (rotation by $-2\pi/4$) and is included in the same symmetry class as the operation C_4. Notice that the top left-hand corner of this table defines a self-consistent group of order 8, as only the operations C_4, C_2, C_2' and C_2'' are generated as products of these same operations. This is in fact the multiplication table for D_4, a sub-group of D_{4h}. Within this sub-group this is another sub-group, of order 4, which defines the point group C_4.

Table A.1   Multiplication table for the point group D_{4h}

	E	$2C_4(z)$	$C_2(z)$	$2C_2'$	$2C_2''$	i	$2S_4$	σ_h	$2\sigma_v$	$2\sigma_v'$
E	E	C_4	$C_2(z)$	C_2'	C_2''	i	S_4	σ_h	σ_v	σ_v'
$2C_4$	C_4	$C_2(z)$	C_4	C_2''	C_2'	S_4	C_2'	S_4	σ_v'	σ_v
$C_2(z)$	$C_2(z)$	C_4	$\cdot\,E$	C_2'	C_2''	σ_h	S_4	i	σ_v	σ_v'
$2C_2'$	C_2'	C_2''	C_2'	E	C_4	σ_v	σ_v'	σ_v	i	S_4
$2C_2''$	C_2''	C_2'	C_2''	C_4	E	σ_v'	σ_v	σ_v'	S_4	σ_h
i	i	S_4	σ_h	σ_v	σ_v'	E	C_4	$C_2(z)$	C_2'	C_2''
$2S_4$	S_4	C_2'	S_4	σ_v'	σ_v	C_4	$C_2(z)$	C_4	C_2''	i
σ_h	σ_h	S_4	i	σ_v	σ_v'	$C_2(z)$	C_4	E	C_2'	C_2''
$2\sigma_v$	σ_v	σ_v'	σ_v	i	S_4	C_2'	C_2''	C_2'	E	C_4
$2\sigma_v'$	σ_v'	σ_v	σ_v'	S_4	σ_h	C_2''	i	C_2''	C_4	E

There are two important point groups that cannot be defined through their multiplication tables, as they contain C_∞ rotation axes and have infinite order. They are $C_{\infty v}$, with an infinite number of reflection operations through planes parallel to the infinite-order rotation axis, and $D_{\infty h}$, which also has a mirror plane perpendicular to the axis, plus an inversion center and an infinite number of C_2 axes at right angles to the axis. They are applicable to linear molecules, those with unlike ends such as HCN belonging to the group $C_{\infty v}$ and those with identical ends such as CO_2 belonging to the group $D_{\infty h}$ (Fig. A.9).

All the point groups we have described above have an identifiable major axis, the single axis of highest order, except for D_2 and D_{2h}, which each have a set of three orthogonal 2-fold axes, and C_1, C_i and C_s, which have no axes. However, there is an

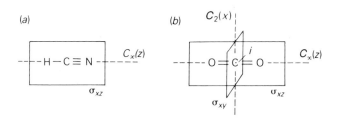

Fig. A.9 (a) HCN, with point group $C_{\infty v}$, showing the axis $C_\infty(z)$ and one of the infinite number of mirror planes σ_v, in this case σ_{xz}; (b) CO_2, with point group $D_{\infty h}$, showing $C_\infty(z)$ and the perpendicular mirror plane σ_{xy}, with one each of the infinite number of parallel mirror planes (σ_{xz}) and the infinite number of 2-fold rotation axes perpendicular to C_∞, in this case $C_2(x)$. The point at which all these elements intersect is an inversion center.

important set of point groups based on the underlying symmetry of a tetrahedron, which has four C_3 axes intersecting in a point, at a special angle (the 'tetrahedral angle', 109.47°). These are called the cubic point groups. The most important of these groups in chemistry are called T_d, which is the symmetry of a simple tetrahedron, and O_h, which is the symmetry of a cube or an octahedron. These are groups of high order ($h = 24$ for T_d, 48 for O_h). The icosahedral point groups are based on the six intersecting C_5 axes of an icosahedron; the full symmetry of an icosahedron gives the point group I_h with an order $h = 120$.

Characters and symmetry species

We must now return to our purpose of using symmetry to classify molecular properties such as vibrations or molecular orbitals. We define a number $\chi_v(R)$, called a character, which expresses the behavior of our property v when operated on by the symmetry operation R. A collection of characters, one for each class of symmetry operations of a group, forms a representation, Γ_v. The 'property' v is technically referred to as the basis vector of the representation Γ_v. We can define all sorts of basis vectors, some of which have very little apparent connection with our original object, such as the non-symmetry operation 'translate along the z-axis', often given the symbol z, or the non-symmetry operation 'rotate by an arbitrary amount about the x-axis', often referred to as R_x.

Other possible basis vectors include true spatial vectors such as a dipole-moment component (*e.g.* μ_y) or a displacement of an atom along a direction defined by a bond (a bond stretching motion). More elaborate basis vectors are also useful; one very important one consists of the three Cartesian-axis displacements of each of the N atoms in a molecule, and has a dimensionality of $3N$. We can also use the wavefunction of a molecular orbital as a basis vector, and classify the orbital according to the effects of the various symmetry operations on the wavefunction. Even a tensor such as the molecular polarizability can be used as a basis vector. We shall start by illustrating the use of basis vectors of the ordinary spatial sort, so that we only need to visualize their behavior in three dimensions.

The character $\chi_v(R)$ of a vector v for an operation R is related to the transformation matrix T that expresses the result of the operation in terms of the original vector.

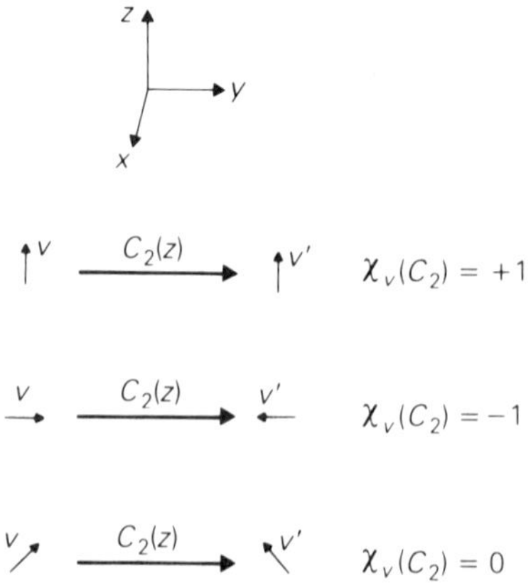

z

y

x

$\uparrow v \xrightarrow{C_2(z)} \uparrow v' \qquad \chi_v(C_2) = +1$

$\xrightarrow{v} \xrightarrow{C_2(z)} \xleftarrow{v'} \qquad \chi_v(C_2) = -1$

$\nearrow v \xrightarrow{C_2(z)} \nwarrow v' \qquad \chi_v(C_2) = 0$

Fig. A.10 Effects of a 2-fold rotation $C_2(z)$ on three different vectors, showing cases where $\chi_v(C_2)$ takes values of $+1$, -1 and 0.

$v' = R(v) = Tv$. Thus if v is a single spatial vector, which we can represent as $\vec{x}$, T is simply $+1$ if the operation leaves $\vec{x}$ unaltered in position, magnitude and direction, and -1 if the direction is reversed. If $\vec{x}$ is rotated by 90° by R, T is zero; these cases are illustrated in Fig. A.10.

In these examples T is simply a number, as we can handle the basis vector in one dimension; for a more general basis vector in N dimensions, T is a matrix of dimensions $N \times N$. The character χ is then equal to the trace of T, that is the sum of its diagonal elements. If we take a vector in three dimensions as a basis, which we can write as $\begin{pmatrix} x \\ y \\ z \end{pmatrix}$, and consider as an example the effect of a C_2 operation about the z axis on each component in turn, we can represent the effect as:

$$\begin{pmatrix} x' \\ y' \\ z' \end{pmatrix} = C_{2_z} \begin{pmatrix} x \\ y \\ z \end{pmatrix} = \begin{pmatrix} -1 & 0 & 0 \\ 0 & -1 & 0 \\ 0 & 0 & 1 \end{pmatrix} \begin{pmatrix} x \\ y \\ z \end{pmatrix}.$$

The character $\chi_{xyz}(C_2)$ is then equal to -1, the trace of the transformation matrix $\begin{pmatrix} -1 & 0 & 0 \\ 0 & -1 & 0 \\ 0 & 0 & 1 \end{pmatrix}$. In the same way we can determine the characters for all the classes of symmetry operations, to give the representation Γ_v. We must note that for any component of a basis vector the contribution to the character is zero if its origin is moved by the operation. This is most important in considering the motions of atoms [Section 5.7.4].

As we can in principle choose any basis vector v we like, we might think that there is an infinite number of possible representations Γ_v, each describing the behavior of one basis vector. In fact, each group has a set of irreducible representations Γ_i, and any basis vector we choose must generate either one of the irreducible representations *or* a reducible representation Γ_r, which can be reduced to a collection of irreducible representations. A table showing the representations Γ_i as rows of characters $\chi_i(R)$ therefore covers all possible modes of behavior under the symmetry operations of the groups. Such a table is called the character table for the group. An example is shown in Fig. A.11. The number of irreducible representations is equal to the number of symmetry

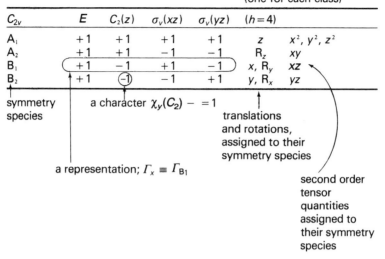

Character Table for C_{2v}

symmetry operations
(one for each class)

C_{2v}	E	$C_2(z)$	$\sigma_v(xz)$	$\sigma_v(yz)$	$(h=4)$	
A_1	$+1$	$+1$	$+1$	$+1$	z	x^2, y^2, z^2
A_2	$+1$	$+1$	-1	-1	R_z	xy
B_1	$+1$	-1	$+1$	-1	x, R_y	xz
B_2	$+1$	-1	-1	$+1$	y, R_x	yz

symmetry   a character $\chi_y(C_2) - = 1$
species

translations
and rotations,
assigned to their
symmetry species

a representation; $\Gamma_x \equiv \Gamma_{B_1}$

second order
tensor
quantities
assigned to
their symmetry
species

Fig. A.11 Character table for the point group C_{2v}, showing the location of various items of information.

classes, which is less than or equal to the order of the group. Each irreducible representation or row of the character table relates to a distinct symmetry species of the group, and the species are labeled at the left of the table. The character tables for a selection of point groups most likely to be of interest to chemists are included at the end of this Appendix.

At the right-hand side of the character table some common basis vectors are listed against the appropriate symmetry species labels and irreducible representations. They are x, y and z, representing translations along the three Cartesian axis, or any other spatial vectors of the same sort; R_x, R_y and R_z representing rotations about the three axes; and the six components of a second-rank tensor such as the inertia tensor or the polarizability. The use of these is described in some detail in Chapter 5.

For point groups with only two-fold symmetry operations, symmetry species are labeled A or B for $\chi(C_2) = +1$ or -1, respectively. An inversion center i generates symmetry species with subscript labels g or u for $\chi(i) = +1$ or -1, as in the C_{2h} point group, where the representation

$$
\begin{array}{cccc}
E & C_2 & \sigma_h & i \\
+1 & +1 & +1 & +1
\end{array}
$$

is labeled A_g, while

$$
\begin{array}{cccc}
+ & -1 & +1 & -1
\end{array}
$$

is labeled B_u. For a point group with a mirror plane we can distinguish $\chi(\sigma) = +1$ or -1 with the symbols $'$ or $''$, respectively.

When an axis of order higher than 2 is present, the x and y vectors become degenerate, as the effect of the C_n operation about the z axis is to convert each of them wholly or partially into the other. This is most easily seen for a C_4 axis (Fig. A.12), for which we

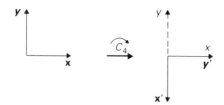

Fig. A.12 Effect of a $C_4(z)$ operation (rotate by $2\pi/4$ about z) on the **x** and **y** unit vectors. The resultant vectors are labeled **x'** and **y'**, respectively, and clearly point along the directions defined originally by the $-y$ and $+x$ axes, respectively.

can write $C_4(x) \equiv -y$, $C_4(y) \equiv x$. These two relations may be expanded slightly and then combined in matrix form as follows:

$$C_4(x) = 0.x - 1.y$$
$$C_4(y) = 1.x + 0.y$$

$$C_4 \begin{pmatrix} x \\ y \end{pmatrix} = \begin{pmatrix} 0 & -1 \\ 1 & 0 \end{pmatrix} \begin{pmatrix} x \\ y \end{pmatrix}$$

Here $\begin{pmatrix} 0 & -1 \\ 1 & 0 \end{pmatrix}$ is the transformation matrix that expresses the result of the operation C_4 on the basis vector made up of both x and y spatial vectors, and $\chi_{xy}(C_4)$ is zero. For the operation E, $\chi_{xy}(E) = 2$, as the transformation matrix is simply $\begin{pmatrix} 1 & 0 \\ 0 & 1 \end{pmatrix}$, meaning that the identity operation transforms x to x and y to y. *All* point groups with one rotation or rotation−reflection axis of order three or more have at least one symmetry species for which $\chi(E) = 2$. Such symmetry species are given the label E, with additional subscripts (g or u, 1 or 2) or superscripts ($'$ or $''$) as necessary. Different species of the same type are distinguished by subscript numbers (*e.g.* A_1, A_2; B_{1g}, B_{2g}, B_{3g}).

For cubic point groups there are also symmetry species for which $\chi(E) = 3$, corresponding to a triple-degeneracy of all the Cartesian axes. These are usually labeled T, but occasionally vibrations are labeled F instead. Even higher degeneracies (up to five-fold) are present in the icosahedral groups.

Permutation groups

We must mention that a completely different set of symmetry operations could in principle be selected. One such alternative set is used for non-rigid molecules, where two or more parts of the molecule can move fairly freely relative to one another. Examples are molecules with effectively unrestricted internal rotation, such as $SiH_3C \equiv CSiH_3$, or with a very high-amplitude bending motion connecting configurations of different formal symmetry, such as CH_3OH. The static or equilibrium geometry here is bent at oxygen (A.I), giving a maximum possible symmetry of C_s, but a bending motion leading to a bond angle at O of $180°$ generates a conformation with symmetry C_{3v} (A.II). For systems like these we may use permutation group symmetry, where the elementary symmetry operation is defined as the permutation (interchange) of two chemically-

A.I

A.II

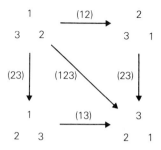

Fig. A.13 Effects of some permutation operations on an array of numbered points. (12) interchanges points 1 and 2, (23) interchanges points 2 and 3, while (123) moves 1 to 2, 2 to 3 and 3 to 1 cyclically. Note that successive operations interchanging two points can lead to the same result as a cyclic permutation; (13)(23) ≡ (123) and (23)(12) ≡ (123).

identical atoms, or a cyclic permutation in a larger set, such as $\{1 \rightarrow 2, 2 \rightarrow 3, 3 \rightarrow 1\}$ (Fig. A.13). The resulting groups can be handled in the same way as point groups, and have multiplication tables and character tables which allow us to classify molecular motions and other properties. A full treatment of permutation group symmetry will be found in Refs [9] and [10].

In some cases the order and structure of the molecular symmetry group defined in terms of permutation operations is the same as that of the point group for the most symmetrical possible structure. This is the case for CH_3OH, which is bent at oxygen, so that its static symmetry point group is C_s. Its rotation/internal rotation/vibration energy levels must be classified under a group labeled G_6, which is isomorphous with (or equivalent to) C_{3v} in its order and the structure of its multiplication table. This, of course, is the point group of CH_3OH with a bond angle of 180° at oxygen.

A.III

A.IV

In other cases the molecular symmetry group is larger than any possible point group, as for nitromethane, A.III, classified under the group G_{12}, or disilyl acetylene, A.IV, which requires a group of order 36, G_{36}. G_{12} is isomorphous with D_{3h}, though CH_3NO_2 cannot possibly adopt a structure of this symmetry, and there are *no* point groups of order 36 except D_{9h} and D_{9d}!

References

1 F.A. Cotton, *Chemical Applications of Group Theory*, Wiley-Interscience, New York and London (1974).
2 H.A. Jaffé and M. Orchin, *Symmetry in Chemistry*, Wiley, New York (1965).
3 D.S. Urch, *Orbitals and Symmetry*, Penguin Harmondsworth and Baltimore (1970).
4 A. Vincent, *Molecular Symmetry and Group Theory*, Wiley, New York (1977).
5 R. McWeeney, *Symmetry: An Introduction to Group Theory and its Applications*, Pergamon, Oxford (1963).
6 S. Califano, *Vibrational States* (Chaps 5 & 6), Wiley, New York (1976).
7 M.J. Buerger, *Elementary Crystallography*, Wiley, New York (1956).
8 *International Tables for X-ray Crystallography*, Kynoch Press, Birmingham (1974).
9 P.R. Bunker, *Molecular Symmetry and Spectroscopy*, Academic Press, London and New York (1979).
10 H.C. Longuet-Higgins, *Mol. Phys.*, **6**, 445 (1963).
11 J.A. Salthouse and M.J. Ware, *Point Group Character Tables and Related Data*, Cambridge University Press, Cambridge (1972).

Point group character tables

C_1	E	$(h = 1)$
A	$+1$	

C_s	E	σ_h	$(h = 2)$	
A$'$	$+1$	$+1$	x, y, R_z	$\begin{cases} x^2, y^2 \\ z^2, xy \end{cases}$
A$''$	$+1$	-1	z, R_x, R_y	yz, xz

C_i	E	i	$(h = 2)$	
A$_g$	$+1$	$+1$	R_x, R_y, R_z	$\begin{cases} x^2, y^2, z^2 \\ xy, xz, yz \end{cases}$
A$_u$	$+1$	-1	x, y, z	

C_2	E	$C_2(z)$	$(h = 2)$	
A	$+1$	$+1$	z, R_z	x^2, y^2, z^2, xy
B	$+1$	-1	x, y, R_x, R_y	yz, xz

C_3	E	$C_3(z)$	C_3^2	$(h = 3); \varepsilon = \exp(2\pi i/3)$	
A	$+1$	$+1$	$+1$	z, R_z	$x^2 + y^2, z^2$
E	$\begin{cases} +1 \\ +1 \end{cases}$	$\begin{matrix} \varepsilon \\ \varepsilon^* \end{matrix}$	$\begin{matrix} \varepsilon^* \\ \varepsilon \end{matrix}$	$\begin{matrix} x+iy; R_x+iR_y \\ x-iy; R_x-iR_y \end{matrix}$	$\begin{matrix} (x^2-y^2, xy) \\ (yz, xz) \end{matrix}$

D_2	E	$C_2(z)$	$C_2'(y)$	$C_2''(x)$	$(h = 4)$	
A	$+1$	$+1$	$+1$	$+1$		x^2, y^2, z^2
B$_1$	$+1$	$+1$	-1	-1	z, R_z	xy
B$_2$	$+1$	-1	$+1$	-1	y, R_y	xz
B$_3$	$+1$	-1	-1	$+1$	x, R_x	yz

Note: in all groups of types D_n, D_{nh} and D_{nd} the x axis is chosen to coincide with one of the C_2' axes.

D_3	E	$2C_3(z)$	$3C_2'$	$(h = 6)$	
A$_1$	$+1$	$+1$	$+1$		$x^2 + y^2, z^2$
A$_2$	$+1$	$+1$	-1	z, R_z	
E	$+1$	-1	0	$(x, y) (R_x, R_y)$	$(x^2 - y^2, xy) (xz, yz)$

D_4	E	$2C_4(z)$	$C_2(\equiv C_4^2)$	$2C_2'$	$2C_2''$	$(h = 8)$	
A_1	$+1$	$+1$	$+1$	$+1$	$+1$		x^2+y^2, z^2
A_2	$+1$	$+1$	$+1$	-1	-1	z, R_z	
B_1	$+1$	-1	$+1$	$+1$	-1		x^2-y^2
B_2	$+1$	-1	$+1$	-1	$+1$		xy
E	$+2$	0	-2	0	0	$(x, y)\ (R_x, R_y)$	(xz, yz)

Note: here and in the Tables for C_{5v}, D_{5h} and D_{5d} τ represents $2\cos(\pi/5)$.

D_5	E	$2C_5(z)$	$2C_5^2$	$5C_2'$	$(h = 10)$	
A_1	$+1$	$+1$	$+1$	$+1$		x^2+y^2, z^2
A_2	$+1$	$+1$	$+1$	-1	z, R_z	
E_1	$+2$	$\tau-1$	$-\tau$	0	$(x, y)\ (R_x, R_y)$	(xz, yz)
E_2	$+2$	$-\tau$	$\tau-1$	0		(x^2-y^2, xy)

D_6	E	$2C_6(z)$	$2C_3$	$C_2(z)$	$3C_2'$	$3C_2''$	$(h = 12)$	
A_1	$+1$	$+1$	$+1$	$+1$	$+1$	$+1$		x^2+y^2, z^2
A_2	$+1$	$+1$	$+1$	$+1$	-1	-1	z, R_z	
B_1	$+1$	-1	$+1$	-1	$+1$	-1		
B_2	$+1$	-1	$+1$	-1	-1	$+1$		
E_1	$+2$	$+1$	-1	-2	0	0	$(x, y)\ (R_x, R_y)$	(xz, yz)
E_2	$+2$	-1	-1	2	0	0		(x^2-y^2, xy)

C_{2v}	E	$C_2(z)$	$\sigma_v(xz)$	$\sigma_v'(yz)$	$(h = 4)$	
A_1	$+1$	$+1$	$+1$	$+1$	z	x^2, y^2, z^2
A_2	$+1$	$+1$	-1	-1	R_z	xy
B_1	$+1$	-1	$+1$	-1	x, R_y	xz
B_2	$+1$	-1	-1	$+1$	y, R_x	yz

C_{3v}	E	$2C_3(z)$	$3\sigma_v$	$(h = 6)$	
A_1	$+1$	$+1$	$+1$	z	x^2+y^2, z^2
A_2	$+1$	$+1$	-1	R_z	
E	$+2$	-1	0	$(x, y)\ (R_x, R_y)$	$(x^2-y^2, xy)\ (xz, yz)$

C_{4v}	E	$2C_4(z)$	C_2	$2\sigma_v$	$2\sigma_v'$	$(h = 8)$	$(x$ axis in σ_v plane$)$	
A_1	$+1$	$+1$	$+1$	$+1$	$+1$	z		$x^2 + y^2, z^2$
A_2	$+1$	$+1$	$+1$	-1	-1	R_z		
B_1	$+1$	-1	$+1$	$+1$	-1			x^2-y^2
B_2	$+1$	-1	$+1$	-1	$+1$			xy
E	$+2$	0	-2	0	0	$(x, y)\ (R_x, R_y)$		(xz, yz)

C_{5v}	E	$2C_5(z)$	$2C_5^2$	$5\sigma_v$	$(h = 10)$	
A_1	$+1$	$+1$	$+1$	$+1$	z	x^2+y^2, z^2
A_2	$+1$	$+1$	$+1$	-1	R_z	
E_1	$+2$	$\tau-1$	$-\tau$	0	$(x, y)\ (R_x, R_y)$	(xz, yz)
E_2	$+2$	$-\tau$	$\tau-1$	0		(x^2-y^2, xy)

C_{6v}	E	$2C_6(z)$	$2C_3(z)$	$C_2(z)$	$3\sigma_v$	$3\sigma_v'$	$(h = 12)$	$(x$ axis in σ_v plane$)$
A_1	$+1$	$+1$	$+1$	$+1$	$+1$	$+1$	z	x^2+y^2, z^2
A_2	$+1$	$+1$	$+1$	$+1$	-1	-1	R_z	
B_1	$+1$	-1	$+1$	-1	$+1$	-1		
B_2	$+1$	-1	$+1$	-1	-1	$+1$		
E_1	$+2$	$+1$	-1	-2	0	0	$(x, y)\ (R_x, R_y)$	(xz, yz)
E_2	$+2$	-1	-1	$+2$	0	0		(x^2-y^2, xy)

C_{2h}	E	$C_2(z)$	i	σ_h	$(h = 4)$	
A_g	$+1$	$+1$	$+1$	$+1$	R_z	x^2, y^2, z^2, xy
B_g	$+1$	-1	$+1$	-1	R_x, R_y	xz, yz
A_u	$+1$	$+1$	-1	-1	z	
B_u	$+1$	-1	-1	$+1$	x, y	

C_{3h}	E	$C_3(z)$	C_3^2	σ_h	S_3	S_3^5	$(h = 6);\quad \varepsilon = \exp(2\pi i/3)$	
A'	$+1$	$+1$	$+1$	$+1$	$+1$	$+1$	R_z	x^2+y^2, z^2
E'	$\begin{cases}+1\\+1\end{cases}$	$\begin{matrix}\varepsilon\\\varepsilon^*\end{matrix}$	$\begin{matrix}\varepsilon^*\\\varepsilon\end{matrix}$	$\begin{matrix}+1\\+1\end{matrix}$	$\begin{matrix}\varepsilon\\\varepsilon^*\end{matrix}$	$\begin{matrix}\varepsilon^*\\\varepsilon\end{matrix}$	$\begin{matrix}x+iy\\x-iy\end{matrix}\Big\}$	(x^2-y^2, xy)
A''	$+1$	$+1$	$+1$	-1	-1	-1	z	
E''	$\begin{cases}+1\\+1\end{cases}$	$\begin{matrix}\varepsilon\\\varepsilon^*\end{matrix}$	$\begin{matrix}\varepsilon^*\\\varepsilon\end{matrix}$	$\begin{matrix}-1\\-1\end{matrix}$	$\begin{matrix}-\varepsilon\\-\varepsilon^*\end{matrix}$	$\begin{matrix}-\varepsilon^*\\-\varepsilon\end{matrix}$	$\begin{matrix}R_x+iR_y\\R_x-iR_y\end{matrix}\Big\}$	(xz, yz)

D_{2h}	E	$C_2(z)$	$C_2'(y)$	$C_2''(x)$	i	$\sigma(xy)$	$\sigma(xz)$	$\sigma(yz)$	$(h = 8)$	
A_g	$+1$	$+1$	$+1$	$+1$	$+1$	$+1$	$+1$	$+1$		x^2, y^2, z^2
B_{1g}	$+1$	$+1$	-1	-1	$+1$	$+1$	-1	-1	R_z	xy
B_{2g}	$+1$	-1	$+1$	-1	$+1$	-1	$+1$	-1	R_y	xz
B_{3g}	$+1$	-1	-1	$+1$	$+1$	-1	-1	$+1$	R_x	yz
A_u	$+1$	$+1$	$+1$	$+1$	-1	-1	-1	-1		
B_{1u}	$+1$	$+1$	-1	-1	-1	-1	$+1$	$+1$	z	
B_{2u}	$+1$	-1	$+1$	-1	-1	$+1$	-1	$+1$	y	
B_{3u}	$+1$	-1	-1	$+1$	-1	$+1$	$+1$	-1	x	

D_{3h}	E	$2C_3(z)$	$3C'_2$	$\sigma_h(xy)$	$2S_3$	$3\sigma_v$	$(h = 12)$	
A'_1	+1	+1	+1	+1	+1	+1		x^2+y^2, z^2
A'_2	+1	+1	−1	+1	+1	−1	R_z	
E'	+2	−1	0	+2	−1	0	(x, y)	(x^2-y^2, xy)
A''_1	+1	+1	+1	−1	−1	−1		
A''_2	+1	+1	−1	−1	−1	+1	z	
E''	+2	−1	0	−2	+1	0	(R_x, R_y)	(xz, yz)

D_{4h}	E	$2C_4(z)$	C_2	$2C'_2$	$2C''_2$	i	$2S_4$	σ_h	$2\sigma_v$	$2\sigma'_v$	$(h = 16)$	
A_{1g}	+1	+1	+1	+1	+1	+1	+1	+1	+1	+1		x^2+y^2, z^2
A_{2g}	+1	+1	+1	−1	−1	+1	+1	+1	−1	−1	R_z	
B_{1g}	+1	−1	+1	+1	−1	+1	−1	+1	+1	−1		x^2-y^2
B_{2g}	+1	−1	+1	−1	+1	+1	−1	+1	−1	+1		xy
E_g	+2	0	−2	0	0	+2	0	−2	0	0	(R_x, R_y)	(xz, yz)
A_{1u}	+1	+1	+1	+1	+1	−1	−1	−1	−1	−1		
A_{2u}	+1	+1	+1	−1	−1	−1	−1	−1	+1	+1	z	
B_{1u}	+1	−1	+1	+1	−1	−1	+1	−1	−1	+1		
B_{2u}	+1	−1	+1	−1	+1	−1	+1	−1	+1	−1		
E_u	+2	0	−2	0	0	−2	0	+2	0	0	(x, y)	

D_{5h}	E	$2C_5$	$2C_5^2$	$5C'_2$	σ_h	$2S_5$	$2S_5^3$	$5\sigma_v$	$(h = 20)$	
A'_1	+1	+1	+1	+1	+1	+1	+1	+1		x^2+y^2, z^2
A'_2	+1	+1	+1	−1	+1	+1	+1	−1	R_z	
E'_1	+2	$\tau-1$	$-\tau$	0	+2	$\tau-1$	$-\tau$	0	(x, y)	
E'_2	+2	$-\tau$	$\tau-1$	0	+2	$-\tau$	$\tau-1$	0		(x^2-y^2, xy)
A''_1	+1	+1	+1	+1	−1	−1	−1	−1		
A''_2	+1	+1	+1	−1	−1	−1	−1	+1	z	
E''_1	+2	$\tau-1$	$-\tau$	0	−2	$1-\tau$	$+\tau$	0	(R_x, R_y)	(xz, yz)
E''_2	+2	$-\tau$	$\tau-1$	0	−2	$+\tau$	$1-\tau$	0		

D_{6h}	E	$2C_6(z)$	$2C_3$	C_2	$3C'_2$	$3C''_2$	i	$2S_3$	$2S_6$	$\sigma_h(xy)$	$3\sigma'_v$	$3\sigma_v$	$(h = 24)$	
A_{1g}	+1	+1	+1	+1	+1	+1	+1	+1	+1	+1	+1	+1		x^2+y^2, z^2
A_{2g}	+1	+1	+1	+1	−1	−1	+1	+1	+1	+1	−1	−1	R_z	
B_{1g}	+1	−1	+1	−1	+1	−1	+1	−1	+1	−1	+1	−1		
B_{2g}	+1	−1	+1	−1	−1	+1	+1	−1	+1	−1	−1	+1		
E_{1g}	+2	+1	−1	−2	0	0	+2	+1	−1	−2	0	0	(R_x, R_y)	(xz, yz)
E_{2g}	+2	−1	−1	+2	0	0	+2	−1	−1	+2	0	0		(x^2-y^2, xy)
A_{1u}	+1	+1	+1	+1	+1	+1	−1	−1	−1	−1	−1	−1		
A_{2u}	+1	+1	+1	+1	−1	−1	−1	−1	−1	−1	+1	+1	z	
B_{1u}	+1	−1	+1	−1	+1	−1	−1	+1	−1	+1	−1	+1		
B_{2u}	+1	−1	+1	−1	−1	+1	−1	+1	−1	+1	+1	−1		
E_{1u}	+2	+1	−1	−2	0	0	−2	−1	+1	+2	0	0	(x, y)	
E_{2u}	+2	−1	−1	+2	0	0	−2	+1	+1	−2	0	0		

D_{2d}	E	$2S_4$	$C_2(z)$	$2C_2'$	$2\sigma_d$	$(h = 8)$	
A_1	$+1$	$+1$	$+1$	$+1$	$+1$		$x^2+y^2,\ z^2$
A_2	$+1$	$+1$	$+1$	-1	-1	R_z	
B_1	$+1$	-1	$+1$	$+1$	-1		x^2-y^2
B_2	$+1$	-1	$+1$	-1	$+1$	z	xy
E	$+2$	0	-2	0	0	$(x, y)\ (R_x, R_y)$	(xz, yz)

D_{3d}	E	$2C_3$	$3C_3'$	i	$2S_6$	$3\sigma_d$	$(h = 12)$	
A_{1g}	$+1$	$+1$	$+1$	$+1$	$+1$	$+1$		$x^2+y^2,\ z^2$
A_{2g}	$+1$	$+1$	-1	$+1$	$+1$	-1	R_z	
E_g	$+2$	-1	0	$+2$	-1	0	(R_x, R_y)	$(x^2-y^2, xy)\ (xz, yz)$
A_{1u}	$+1$	$+1$	$+1$	-1	-1	-1		
A_{2u}	$+1$	$+1$	-1	-1	-1	$+1$	z	
E_u	$+2$	-1	0	-2	$+1$	0	(x, y)	

D_{4d}	E	$2S_8$	$2C_4$	$2S_8^3$	C_2	$4C_2'$	$4\sigma_d$	$(h = 16)$	
A_1	$+1$	$+1$	$+1$	$+1$	$+1$	$+1$	$+1$		$x^2+y^2,\ z^2$
A_2	$+1$	$+1$	$+1$	$+1$	$+1$	-1	-1	R_z	
B_1	$+1$	-1	$+1$	-1	$+1$	$+1$	-1		
B_2	$+1$	-1	$+1$	-1	$+1$	-1	$+1$	z	
E_1	$+2$	$+\sqrt{2}$	0	$-\sqrt{2}$	-2	0	0	(x, y)	
E_2	$+2$	0	-2	0	$+2$	0	0		(x^2-y^2, xy)
E_3	$+2$	$-\sqrt{2}$	0	$+\sqrt{2}$	-2	0	0	(R_x, R_y)	(xz, yz)

D_{5d}	E	$2C_5$	$2C_5^2$	$5C_2'$	i	$2S_{10}^3$	$2S_{10}$	$5\sigma_d$	$(h = 20)$	
A_{1g}	$+1$	$+1$	$+1$	$+1$	$+1$	$+1$	$+1$	$+1$		$x^2+y^2,\ z^2$
A_{2g}	$+1$	$+1$	$+1$	-1	$+1$	$+1$	$+1$	-1	R_z	
E_{1g}	$+2$	$\tau-1$	$-\tau$	0	$+2$	$\tau-1$	$-\tau$	0	(R_x, R_y)	(xz, yz)
E_{2g}	$+2$	$-\tau$	$\tau-1$	0	$+2$	$-\tau$	$\tau-1$	0		(x^2-y^2, xy)
A_{1u}	$+1$	$+1$	$+1$	$+1$	-1	-1	-1	-1		
A_{2u}	$+1$	$+1$	$+1$	-1	-1	-1	-1	$+1$	z	
E_{1u}	$+2$	$\tau-1$	$-\tau$	0	-2	$1-\tau$	$+\tau$	0	(x, y)	
E_{2u}	$+2$	$-\tau$	$\tau-1$	0	-2	$+\tau$	$1-\tau$	0		

D_{6d}	E	$2S_{12}$	$2C_6$	$2S_4$	$2C_3$	$2S_{12}^5$	C_2	$6C_2'$	$6\sigma_d$	$(h = 24)$	
A_1	$+1$	$+1$	$+1$	$+1$	$+1$	$+1$	$+1$	$+1$	$+1$		$x^2+y^2,\ z^2$
A_2	$+1$	$+1$	$+1$	$+1$	$+1$	$+1$	$+1$	-1	-1	R_z	
B_1	$+1$	-1	$+1$	-1	$+1$	-1	$+1$	$+1$	-1		
B_2	$+1$	-1	$+1$	-1	$+1$	-1	$+1$	-1	$+1$	z	
E_1	$+2$	$+\sqrt{3}$	$+1$	0	-1	$-\sqrt{3}$	-2	0	0	(x, y)	
E_2	$+2$	$+1$	-1	-2	-1	$+1$	$+2$	0	0		(x^2-y^2, xy)
E_3	$+2$	0	-2	0	$+2$	0	-2	0	0		
E_4	$+2$	-1	-1	$+2$	-1	-1	$+2$	0	0		
E_5	$+2$	$-\sqrt{3}$	$+1$	0	-1	$+\sqrt{3}$	-2	0	0	(R_x, R_y)	(xz, yz)

$C_{\infty v}$	E	$2C_\infty^\phi$	$\cdots$	$\infty\sigma_v$	$(h = \infty)$		
$A_1 \equiv \Sigma^+$	$+1$	$+1$	$\cdots$	$+1$	z		$x^2+y^2,\ z^2$
$A_2 \equiv \Sigma^-$	$+1$	$+1$	$\cdots$	-1	R_z		
$E_1 \equiv \Pi$	$+2$	$2\cos\phi$	$\cdots$	0	(x, y)	(R_x, R_y)	(xz, yz)
$E_2 \equiv \Delta$	$+2$	$2\cos 2\phi$	$\cdots$	0			$(x^2-y^2,\ xy)$
$E_3 \equiv \Phi$	$+2$	$2\cos 3\phi$	$\cdots$	0			
E_n	$+2$	$2\cos n\phi$	$\cdots$	0			
$\cdots$							

$D_{\infty h}$	E	$2C_\infty^\phi$	$\cdots$	$\infty\sigma_v$	i	$2S_\infty^\phi$	$\cdots$	$\infty C_2'$	$(h = \infty)$		
$A_{1g} \equiv \Sigma_g^+$	$+1$	$+1$	$\cdots$	$+1$	$+1$	$+1$	$\cdots$	$+1$			$x^2+y^2,\ z^2$
$A_{2g} \equiv \Sigma_g^-$	$+1$	$+1$	$\cdots$	-1	$+1$	$+1$	$\cdots$	-1	R_z		
$E_{1g} \equiv \Pi_g$	$+2$	$2\cos\phi$	$\cdots$	0	$+2$	$-2\cos\phi$	$\cdots$	0	(R_x, R_y)		(xz, yz)
$E_{2g} \equiv \Delta_g$	$+2$	$2\cos 2\phi$	$\cdots$	0	$+2$	$2\cos 2\phi$	$\cdots$	0			$(x^2-y^2,\ x$-
$E_{3g} \equiv \Phi_g$	$+2$	$2\cos 3\phi$	$\cdots$	0	$+2$	$-2\cos 3\phi$	$\cdots$	0			$y)$
E_{ng}	$+2$	$2\cos n\phi$	$\cdots$	0	$+2$	$(-1)^n 2\cos n\phi$	$\cdots$	0			
$\cdots$											
$A_{1u} \equiv \Sigma_u^+$	$+1$	$+1$	$\cdots$	$+1$	-1	-1	$\cdots$	-1			
$A_{2u} \equiv \Sigma_u^-$	$+1$	$+1$	$\cdots$	-1	-1	-1	$\cdots$	$+1$	z		
$E_{1u} \equiv \Pi_u$	$+2$	$2\cos\phi$	$\cdots$	0	-2	$2\cos\phi$	$\cdots$	0	(x, y)		
$E_{2u} \equiv \Delta_u$	$+2$	$2\cos 2\phi$	$\cdots$	0	-2	$-2\cos 2\phi$	$\cdots$	0			
$E_{3u} \equiv \Phi_u$	$+2$	$2\cos 3\phi$	$\cdots$	0	-2	$2\cos 3\phi$	$\cdots$	0			
E_{nu}	$+2$	$2\cos n\phi$	$\cdots$	0	-2	$(-1)^{n+1}2\cos n\phi$	$\cdots$	0			
$\cdots$											

T_d	E	$8C_3$	$3C_2$	$6S_4$	$6\sigma_d$	$(h = 24)$	
A_1	$+1$	$+1$	$+1$	$+1$	$+1$		$x^2+y^2+z^2$
A_2	$+1$	$+1$	$+1$	-1	-1		
E	$+2$	-1	$+2$	0	0		$(2z^2-x^2-y^2,\ x^2-y^2)$
T_1	$+3$	0	-1	$+1$	-1	(R_x, R_y, R_z)	
T_2	$+3$	0	-1	-1	$+1$	(x, y, z)	(xy, xz, yz)

O_h	E	$8C_3$	$6C_2$	$6C_4$	$3C_2(\equiv C_4^2)$	i	$6S_4$	$8S_6$	$3\sigma_h$	$6\sigma_d$	$(h = 48)$	
A_{1g}	$+1$	$+1$	$+1$	$+1$	$+1$	$+1$	$+1$	$+1$	$+1$	$+1$		$x^2+y^2+z^2$
A_{2g}	$+1$	$+1$	-1	-1	$+1$	$+1$	-1	$+1$	$+1$	-1		
E_g	$+2$	-1	0	0	$+2$	$+2$	0	-1	$+2$	0		$(2z^2-x^2-y^2,\ x^2-y^2)$
T_{1g}	$+3$	0	-1	$+1$	-1	$+3$	$+1$	0	-1	-1	(R_x, R_y, R_z)	
T_{2g}	$+3$	0	$+1$	-1	-1	$+3$	-1	0	-1	$+1$		(xz, yz, xy)
A_{1u}	$+1$	$+1$	$+1$	$+1$	$+1$	-1	-1	-1	-1	-1		
A_{2u}	$+1$	$+1$	-1	-1	$+1$	-1	$+1$	-1	-1	$+1$		
E_u	$+2$	-1	0	0	$+2$	-2	0	$+1$	-2	0		
T_{1u}	$+3$	0	-1	$+1$	-1	-3	-1	0	$+1$	$+1$	(x, y, z)	
T_{2u}	$+3$	0	$+1$	-1	-1	-3	$+1$	0	$+1$	-1		

Problems — Solutions and Comments

Solutions are given for *odd-numbered* problems only. Solutions for even-numbered problems are available from the publishers.

Chapter 2

2.1 (*a*) One line. (*b*) Doublet and quartet, relative areas 3:1. (*c*) Wide doublet of triplets. (*d*). Wide doublet of triplets. (*d*) Wide doublet of triplets of triplets. (*e*) Doublet of triplets and wide doublet of quartets, relative areas 3:2.

2.3 For $n = 2$, two lines of equal intensity; $n = 3$, three lines intensity ratio 2:2:1; $n = 4$, three equal lines; $n = 5$, four lines, ratio 2:2:2:1; $n = 6$, four equal lines. All terminal R_3SiO- chemical shifts should be similar, as should all $-OSiR_2O-$ shifts.

2.5 $[A_3B_2]$ $[AX_2]_3$ $[AB_4X]$ $[A_2B_3X]$ $[A_9X]_3$ $[[A_9X]_2B_9Y]$ $[A_2B_6]$ ($[A_2[B_3]_2X]$ if ^{29}Si at position 1, $[A_2B_6X]$ if ^{29}Si at position 2).

2.7 One resonance (with weak ^{57}Fe satellites); eight equal lines, most of which would be very close together; two doublets, coupling to ^{31}P; five-line multiplet, intensity ratio 1:2:3:2:1; two resonances of equal intensity.

2.9 Two 1:1:1:1 quartets centered at same frequency, relative intensities, 3:1, relative coupling constants in ratio 1.20:1.

2.11 The product is $IF_6^+ SbF_6^-$. Both ions are octahedral, so couplings to quadrupolar ^{127}I ($I = 5/2$, 100% abundant), ^{121}Sb ($I = 5/2$, 57% abundant) and ^{123}Sb ($I = 7/2$, 43% abundant) are seen.

2.13 TeF_5 is square pyramidal, and has a doublet and a quartet in its ^{19}F spectrum, $\delta(F_{ax})$ − 30.9, $\delta(F_{eq})$ − 41.9 ppm, $J(F_{ax}F_{eq})$ 50.8 Hz. The intensity pattern is distorted because the spectrum is slightly second order. The high frequency satellites give $J(^{125}TeF_{ax})$ 2886 and $J(^{125}TeF_{eq})$ 1376 Hz. The low frequency satellites all fall near − 3840 Hz, and give a second order subspectrum, thus showing that the signs of the two ^{125}TeF couplings are equal.

2.15 (*a*) A 1:1:1:1 quartet and 1:1:1:1:1:1:1 septet with the same chemical shifts, relative areas 4:1, relative B−H coupling constants 2.99:1, due to ^{11}B and ^{10}B coupling. (*b*) Six equal lines coupling to ^{27}Al (100%, $I = 5/2$). (*c*) Two 1:1:1:1 quartets with the same chemical shifts, relative areas 3:2 and relative coupling constants 0.79:1 due to ^{69}Ga and ^{71}Ga coupling.

2.17 There are five products, with relative amounts 1:4:6:4:1, appearing as a singlet, doublet, triplet, quartet and quintet, for $SiCl_4$, $SiCl_3F$, $SiCl_2F_2$, $SiClF_3$ and SiF_4 respectively.

2.19 The regular tetrahedral and octahedral species should have sharp lines, so the only broad resonance would be that for $CrCl_2O_2$. CrF_4, unlike SF_4, is tetrahedral.

2.21 (*a*) In $GeClH_3$ and $GeFH_3$ interaction of the quadrupolar ^{73}Ge nucleus with the local electric field gradient causes very rapid relaxation, and so coupling with other spinning nuclei is lost. In tetrahedral GeH_4 the electric field gradient at germanium is zero.

(b) At 200 K there are two σ-bonded allyl groups with three chemically distinct carbon sites. On warming there is an exchange process, the free CH_2 groups replacing the coordinated ones, so their resonances coalesce.

(c) The triplet due to ^{14}N coupling broadens as relaxation becomes more efficient on cooling. The lines of the doublet due to ^{15}N coupling broaden by exchange with free $^{15}NH_2^-$. This exchange is slowed on cooling.

(d) In a liquid crystal the fluorines couple to one another, giving a triplet, as well as to phosphorus.

2.23 (a) 4 couplings (FH, HH, SiF and SiH), 1 orientation parameter, 3 structural parameters.
(b) 4 couplings, 3 orientation parameters, so only 1 structural parameter.
(c) 7 couplings (including 3 to ^{77}Se), 3 orientation parameters, 4 structural parameters.
(d) 6 couplings involving ^2H and ^{13}C or ^{15}N (plus others involving ^{13}C and/or ^{15}N only), 1 orientation parameter, 5 structural parameters.
(e) 4 useful couplings (H to H, P, W and *trans* C) as the HC(*cis*) coupling is averaged over rotation about the W—P bond, so 3 structural parameters (1 orientation parameter).
(f) 5 couplings (HH, FeH and 3 different CH), 2 orientation parameters, 3 structural parameters.
(g) 13 couplings (5 different HH, 5 CH, 3 PH), 3 orientation parameters, so up to 10 structural parameters.

2.25 ^{31}P high temperature, a quartet; low temperature, a triplet of doublets; ^{19}F high temperature, a doublet; low temperature, a doublet of doublets plus a doublet of triplets.

2.27 (a) The wide doublet and narrow quintet couplings would be lost, leaving a triplet of triplets.
(b) The small triplet coupling would be collapsed, giving a doublet of triplets of quintets.

2.29

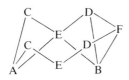

For full details see X.L.R. Fontaine, N.N. Greenwood, J.D. Kennedy, P. MacKinnon and M. Thornton-Pett, *J. Chem. Soc., Dalton Trans.*, 2809 (1988).

2.31 (a) A 1:6:15:20:15:6:1 septet. (b) A triplet of doublets. (c) A decet, 1:9:36:84:126:126:84:36:9:1 with ^{13}C satellites. (d) A triplet of quartets with widely spaced ^{77}Se and weaker ^{13}C satellites. (e) A 1:6:15:20:15:6:1 septet with ^{13}C satellites.

2.33 (a) Two quintets with 1:2:2:2:1 intensities, splittings 80 and 8 Hz. (b) A doublet of doublets (80 and 16 Hz). (c) A doublet of triplets (80 and 8 Hz).

2.35 Without any molecular symmetry in the crystal there would be 23 resonances, 9 for each indenyl ligand, 2 for the carbonyls and 3 for the allene. The carbonyl and two of the allene resonances would be split by ^{103}Rh into doublets, while the central allene carbon could be a triplet or doublet of doublets. If the molecule had C_s or C_2 symmetry there would be just 11 resonances corresponding to pairs of ^{13}C nuclei and one, a triplet, corresponding to the central carbon of the allene ligand.

2.37

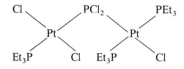

(Pr^{i}O)$_3$W⟨H⟩W(OPr^i)$_3$ with PH$_2$ bridge   [(Pr^{i}O)$_3$W⟨H⟩W(OPr^i)$_3$]$^-$ with PH bridge

A   **B**

The $^1J_{PH}$ coupling constants indicate 4- and 3-coordination in A and B respectively. The 26% intensity of the satellites shows that the phosphorus atom lies between and couples equally to two tungsten atoms, and that the complexes are therefore symmetrical. Each has another hydrogen, which must be bridging, to retain the molecular symmetry.

2.39 Both compounds must have η^5 C$_5$H$_5$ ligands, or be fluxional. The temperature invariance suggests that the former is true. The patterns of lines are those expected for the natural distribution of mercury and thallium isotopes. In fact the mercury compound is Hg(η^1-C$_5$H$_5$)$_2$ and the thallium compound is Tl(η^5-C$_5$H$_5$).

2.41 rC≡C 1.206 Å, rC—H 1.031 Å.

2.43 The ^{31}P spectrum would show two sets of resonances, each with couplings to the other phosphorus, to ^{19}F, ^{195}Pt, and one would include couplings to nine protons. The magnitudes of couplings, particularly J_{PP}, should indicate whether the phosphines are *cis* or *trans* to one another.

Showing the presence of the NCS ligands is not easy. ^{13}C or (even more difficult) ^{15}N spectrum could be run, and if the phosphines are mutually *trans*, only one chemically distinct nuclear environment would be observed in each case. Use of labeled samples might be necessary. Coupling constants should verify that the ligands are bound through nitrogen, not sulfur. The presence of sulfur is more or less impossible to deduce directly from NMR measurements.

2.45 Each BH$_4$ group has two bridging and two terminal hydrogen atoms. These couple to one another and to the quadrupolar ^{11}B. On warming they exchange positions by an *intra*molecular process. The mean lifetime at 180 K is 9.0×10^{-4} seconds.

2.47 The product is

Cl⟨Pt⟩PCl$_2$ ... Pt⟨PEt$_3$⟩; Et$_3$P / Cl ; Et$_3$P / Cl

The PCl$_2$ group has δ_P 196.7 ppm, and is a doublet ($^2J_{PP}$, 609.6 Hz) of triplets ($^2J_{PP}$, 10.6 Hz) with two sets of ^{195}Pt satellites ($^1J_{PtP}$ = 3197, 2312 Hz). The unique PEt$_3$ has δ_P 10.1 ppm, and is a doublet ($^2J_{PP}$, 610.7 Hz) of triplets ($^4J_{PP}$, 3.6 Hz) with two sets of ^{195}Pt satellites ($^1J_{PtP}$ = 2533 Hz, $^3J_{PtP}$ = 221 Hz). The other PEt$_3$ groups have δ_P 16.7 ppm and are a doublet ($^2J_{PP}$, 10.5 Hz) of doublets ($^4J_{PP}$, 3.6 Hz) with ^{195}Pt satellites ($^1J_{PtP}$ = 2454 Hz). $^3J_{PtP}$ is small in this case. Slight discrepancies between separate measurements of the same coupling constant arise from computer digitization errors.

Chapter 3

3.1 [AlH$_3$·]$^-$ – a sextet (equal intensities) of 1:3:3:1 quartets. [InH$_3$·]$^-$ – a decet (equal intensities) of 1:3:3:1 quartets from the 95% of the radicals which contain ^{115}In (I = 9/2), plus some weaker lines from radicals containing the remaining indium.

3.3 (*a*) The ten-line pattern arises from hyperfine coupling to ^{93}Nb ($I = 9/2$). The unpaired electron density must be mainly centered on the niobium atom.

(*b*) Nine lines are clearly visible, and must be the central part of a 13-line pattern, with coupling to twelve protons. Here the unpaired electron density must be at least in part ligand-based. Weak satellite lines due to ^{47}Ti and ^{49}Ti are expected, and have been observed outside the main pattern, so there is also coupling to the metal atom nuclei.

3.5 (*a*) Possible nuclear spin combinations for two nuclei with spin one-half each are 1 ($\frac{1}{2}, \frac{1}{2}$), 0 ($\frac{1}{2}, -\frac{1}{2}$ or $-\frac{1}{2}, \frac{1}{2}$, thus with statistical weight 2), and -1 ($-\frac{1}{2}, -\frac{1}{2}$). ESR transitions occur between electron energy levels but do not involve change in nuclear spin. Hence the hyperfine pattern observed is a 1:2:1 triplet.

(*b*) A randomly-oriented system with no axis of symmetry will give three lines [see Fig. 3.7(*d*), p. 121].

(*c*) The ground state of an octahedral high spin d^7 ion is $^4T_{1g}$. Jahn–Teller splitting will give $M_s = \pm 3/2$ and $\pm 1/2$ levels, and three lines result from $\Delta M_s = \pm 1$ transitions.

(*d*) Possible nuclear spin states here are 1, 0 and -1. These are of equal statistical weight, because they all derive from the spin of a single nucleus. The transitions allowed in the ESR spectrum are the same as those in (*a*) above, except for their relative intensities.

3.7 From Fig. 3.5: $A(^1H) = 1.6$, $A(^{11}B) = 2.1$ mT.
From Fig 3.6: $A(^{14}N) = 0.25$, $A(^{53}Cr) = 2.2$ mT.

3.9 The spectrum shows 15 lines, with an intensity pattern consistent with coupling equally to two ^{59}Co ($I = 7/2$) nuclei. The unpaired electron is therefore delocalized over the two metal centers, on the timescale of the ESR experiment. If $^{17}O_2$ was used, and there was significant unpaired electron density on the oxygen atoms, each line would be further split into a 1:2:3:4:5:6:5:4:3:2:1 multiplet.

3.11 The spectrum would be a wide doublet of quintets, with hyperfine coupling to ^{31}P and four equivalent ^{19}F nuclei. If the latter were not all equivalent, as in a trigonal bipyramidal structure, the quintet splitting would be replaced by a triplet of triplets.

3.13 The four-line pattern must arise from hyperfine coupling to ^{39}K, $I = 3/2$, and so at higher concentrations potassium atoms are probably present. The single resonance obviously arises from some other paramagnetic species, and it is most likely that this is a solvated electron.

3.15 The wide 1:1:1 triplet arises from coupling to the ^{14}N nucleus on which the unpaired electron is formally located. The remaining splitting must be ascribed to a 1:1:1 triplet of 1:3:3:1 quartets, the couplings being to the second ^{14}N nucleus and the three methyl protons. The observed intensity pattern shows that these couplings are (fortuitously) equal.

3.17 There are two isomers having one chlorine bound to each phosphorus atom. There should be only one signal in the ^{35}Cl NQR spectrum (assuming that there are no complications from crystal effects) from the form with all three Cl atoms on the same face of the ring; the isomer with two on one face and one on the other should show two resonances, one twice as strong as the other. The third isomer has one PCl_2, one PClR and one PR_2 group. This should give three resonances.

(*a*) There are three isomers. One contains a PR_2 group, and all of its chlorines are equivalent, so give just one resonance. In the other two isomers the R groups of PClR fragments may be on the same face or on opposite faces of the ring. Either way, there are two ^{35}Cl resonances of equal intensity, so these isomers cannot be distinguished this way.

(*b*) There are three isomers one with a PCl_2 group, and two with two PClR groups, with chlorines on the same or opposite faces. In each case the two chlorines are equivalent, so all three isomers give just one ^{35}Cl resonance, and so are indistinguishable by NQR spectroscopy.

3.19 Ignoring crystal effects, there would be two ^{35}Cl resonances, relative intensities 3:2, for $AsCl_5$, and for $AsCl_4^+ AsCl_6^-$, but the frequencies should enable the two to be differentiated, as the overall charge affects the electric field gradient at the nucleus. $AsCl_4^+ Cl^-$ would give just one resonance, as Cl^- has no field gradient at the nucleus. The bridged dimer should give three resonances, relative intensities 1:2:2, corresponding to bridging and two types of terminal chlorine sites.

3.21 The bromine atoms are equivalent, so we expect one ^{79}Br and one ^{81}Br resonance. The equivalent indium atoms give a set of four resonances for ^{115}In, $I = 9/2$. The cobalt atoms are of two types, and so two sets of three resonances should be seen for ^{59}Co, $I = 7/2$.

3.23 The eight crystallographically distinct chlorines each give rise to one ^{35}Cl and one ^{37}Cl resonance, the ^{127}I (100% abundant $I = 5/2$) gives two resonances, and ^{121}Sb (57%, $I = 5/2$) and ^{123}Sb (43%, $I = 7/2$) give two and three resonances respectively. The asymmetry parameters may be determined for those nuclei having $I \geqslant 5/2$.

Chapter 4

4.1 Only a linear molecule with a dipole moment will give such a simple spectrum; the obvious possibility is the monomer LiF. Using the given atom masses, and the conversion $I = 505\ 391/B$ amu Å^2 the line position implies $D = 44\ 870.23$ MHz (ignoring centrifugal distortion) so $I = 11.263\ 392$ amu Å^2, and the bond length is obtained from $I = \mu r^2$, where μ is the reduced mass, 4.567 636 amu, giving $r^2 = 2.465\ 913$ Å^2, $r = 1.570\ 323$ Å. This is a reasonable result, confirming that the spectrum is likely to be due to the simple monomer. As the bond distance is based on the ground state rotation B_0 it is called r_0.

4.3 The energy of each level is given by $BJ(J + 1)$, so the relative populations are:

0	10	20	30	40	50	60	70
1.0	18	22	16	7.7	2.6	0.6	0.1

4.5 The two $K = 0$ lines given may be used to define two equations in B_0 and D_J: $2B_0 - 4D_J = 20\ 023.21$ MHz; and $20B_0 - 4000D_J = 200\ 198.22$ MHz. Combining these gives $3960D_J = 200\ 232.1 - 200\ 198.2$ MHz $= 33.9$ MHz, so $D_J = 0.008\ 56$ MHz, or 8.56 kHz. Then B_0 is 10 011.62 MHz. The difference in frequency between the $K = 0$ and $K = 8$ lines is $2D_{JK}(J'' + 1)(8^2 - 0) = 1280D_{JK}$. This is 169.06 MHz, giving $D_{JK} = 132.08$ kHz.

4.7 CO_2: $I = 46.08$ amu Å^2, B = 10 968 MHz. Same for ^{12}C and ^{13}C.
CS_2: $I = 163.84$ amu Å^2, B = 3085 MHz. Same for ^{12}C and ^{13}C.
$O^{12}CS$: $I = 87.89$ amu Å^2 B = 5750 MHz.
$O^{13}CS$: $I = 88.17$ amu Å^2 B = 5732 MHz.

4.9 $B_0 = 57\ 635.9595$ MHz
$D_0 = 184.5$ kHz
$r_0 = 1.130\ 89$ Å.

4.11 SiF_4 is a spherical top and has no pure rotation spectrum. Its one moment of inertia could be measured from a high resolution vibrational spectrum. $XeOF_4$, PF_5 and $GeH_3C\equiv CH$ are symmetric tops, and each has two rotation constants. The others are asymmetric tops, and have three independent constants if they are non-planar, and two if they are planar (SCl_2). All except PF_5 and SiF_4 give pure rotation spectra in the microwave region, yielding one (symmetric tops) two (planar asymmetric tops) or three (other asymmetric tops) rotation constants. Rotational Raman spectra for PF_5 could give one rotation constant. Isotopic substitution increases the number of data obtainable, but it is not possible for P or F. Thus complete structures could be derived for all except PF_5. For this electron diffraction data are essential, and for all the others such data would be helpful.

4.13 The center of gravity lies on the N—Cl bond, 0.8667 Å from the nitrogen atom.

$I_A = 38.15$ amu Å^2, $A = 13\,249$ MHz.

$I_B = 97.68$ amu Å^2, $B = 5174$ MHz.

$I_C = 135.83$ amu Å^2, $C = 3721$ MHz.

Note that $I_C = I_A + I_B$ for a planar molecule.

4.15 Taking the first two species:

$I = 257.2179$ amu Å^2, $I* = 262.0987$ amu Å^2, $\Delta I = 4.8808$ amu Å^2.

$M = 240.9737$ amu, $M* = 241.9771$ amu, $\Delta m = 1.0034$ amu.

$z_C(\text{methyl})^2 = 4.8917$ Å^2, $|z_C| = 2.2117$ Å.

For the first and third species:

$I = 257.2179$ amu Å^2, $I* = 257.2650$ amu Å^2, $\Delta I = 0.0471$ amu Å^2.

$M = 240.9737$ amu, $M* = 242.9761$ amu, $\Delta m = 2.0024$ amu.

$z_{Hg}^2 = 0.0237$ Å^2, $|z_{Hg}| = 0.1540$ Å.

For the first and fourth species:

$I = 257.2179$ amu Å^2, $I* = 260.7604$ amu Å^2, $\Delta I = 3.5425$ amu Å^2.

$M = 240.9737$ amu, $M* = 241.9771$ amu, $\Delta m = 1.0034$ amu.

$z_C(\text{cyanide})^2 = 3.5352$ Å^2, $|z_C| = 1.8829$ Å.

The center of gravity is expected to be close to the mercury atom, but on the side of the cyanide group. Thus,

r_s (Hg—C)(methyl) $= 2.2117 - 0.1540 = 2.5077$ Å.

r_s (Hg—C)(cyanide) $= 1.8829 + 0.1540 = 2.0369$ Å.

The two bonds are therefore distinctly different in length.

For the first and final species:

$I = 257.2179$ amu Å^2, $I* = 266.3935$ amu Å^2, $\Delta I = 9.1756$ amu Å^2

$M = 240.9737$ amu, $M* = 241.9708$ amu, $\Delta m = 0.9971$ amu.

$z_N^2 = 9.2404$ Å^2, $|z_N| = 3.0398$ Å.

Thus $r_s(C{\equiv}N) = 3.0398 - 1.8829 = 1.1569$ Å.

Chapter 5

5.1 HBr: $k = 384$ N m^{-1} (3.84 mdyn Å^{-1})

NO: $k = 1549$ N m^{-1} (15.49 mdyn Å^{-1})

5.3 $v = 0, 1.000$; $v = 1, 0.056$; $v = 2, 0.003$.

Only about 5% of molecules are in even the first vibrationally excited state, so the anti-Stokes Raman lines and any hot bands arising from this state will be weak. Higher excited states have even lower populations.

5.5 SCl$_2$ has $3N - 6 = 3$ vibrations, 2 stretches and a bend. The stretches have symmetries a_1 (symmetric) and b_2 (antisymmetric), and the bend has symmetry a_1.

5.7 The six CO stretches combine in the O_h symmetry to give a_{1g}, e_g and t_{1u} components, of which only the t_{1u} mode is IR active, and only the a_{1g} and e_g modes are Raman active. Neglecting multiple ^{13}C labeling, the sample contains approximately 6% of M(^{12}CO)$_5$(^{13}CO), which has C_{4v} symmetry. This has $v(^{13}$CO) with a_1 symmetry and five (^{12}CO) stretching modes, whose symmetry species are $2a_1 + b_1 + e$. The three a_1 modes and the e mode are IR active, and all modes are Raman active. The ^{13}CO mode will be distinctive in having a lower frequency than the ^{12}CO modes, but the 'isotope shift' of about -2% (-40 cm^{-1}) may be not much larger than the separation between the ^{12}CO modes (due to mechanical coupling between the individual oscillators).

5.9 *Syn*-planar N_2F_2 has C_{2v} symmetry, while *anti*-planar N_2F_2 has C_{2h} symmetry. The latter point group contains an inversion center i, and the *anti* structure should therefore give fewer IR bands and fewer Raman bands than the *syn* structure. The numbers are:

	syn	*anti*
IR bands	5	3
Raman bands	6	3
coincidences	5	0

5.11 *cis*-SCl_2F_4 is C_{2v}; *trans*-SCl_2F_4 is D_{4h}.

For the *cis*-isomer, the number of modes with their symmetry species and activities is:
$$6a_1(\text{IR, Raman}) + 2a_2(\text{Raman}) + 3b_1(\text{IR, Raman}) + 4b_2(\text{IR, Raman}).$$

For the *trans*-isomer, the number of modes with their symmetry species and activities is:
$$2a_{1g}(\text{Raman}) + 1b_{1g}(\text{Raman}) + 1b_{2g}(\text{Raman}) + e_g(\text{Raman}) + 2a_{2u}(\text{IR})$$
$$+ 1b_{2u}(\text{forbidden}) + 3e_u(\text{IR})$$

5.13 Because BF_3 is planar the boron atom does not move in the totally symmetric stretch and so the frequency of the mode is not affected by the mass of the boron atom. It does move in the antisymmetric stretch, and so the frequency of that mode does show isotopic effects.

5.15 In frequency order:

$2v_3 + 2v_6$	2050	$a_1 + e$
$v_2 + 2v_6$	2052	$a_1 + e$
$v_5 + 2v_6$	2078	$a_1 + a_2 + 2e$
$5v_3$	2115	a_1
$v_2 + 3v_3$	2117	a_1
$2v_2 + v_3$	2119	a_1
$3v_3 + v_5$	2143	e
$v_2 + v_3 + v_5$	2145	e
$v_3 + 2v_5$	2171	$a_1 + e$
$v_3 + 3v_6$	2229	$a_1 + 2e$
$4v_3 + v_6$	2294	e
$v_2 + 2v_3 + v_6$	2296	e
$2v_2 + v_6$	2298	e
$2v_3 + v_5 + v_6$	2322	$a_1 + a_2 + e$
$v_2 + v_5 + v_6$	2324	$a_1 + a_2 + e$
$2v_5 + v_6$	2350	$a_1 + a_2 + 2e$
$4v_6$	2408	$a_1 + 2e$
$3v_3 + 2v_6$	2473	$a_1 + e$
$v_2 + v_3 + 2v_6$	2475	$a_1 + e$

5.17 This suggests a resonance Raman spectrum; the frequencies given can be assigned to two series of lines (154, 306, 456, 604, 750 . . and 240, 476, 708 . .) suggesting molecular vibration frequencies of about 156 cm^{-1} and 244 cm^{-1} (taking account of the anharmonicity shown in each series).

5.19 Irradiation in the envelope assigned to the $\delta \rightarrow \delta^*$ transition excited a vibrational frequency that does not depend much on the mass of the halogen; this is consistent with the assignment of the mode to Re—Re stretching, and so with the proposed assignment of the electronic band. Irradiation in the envelope of the electronic band assigned to the $\pi \rightarrow \delta^*$ transition excited the Re—Re stretch, but less strongly; the most strongly enhanced modes shift markedly to lower frequency as the mass of the halogen increases, suggesting that this mode is associated with Re—halogen stretching. This too is consistent with the proposed assignment of the electronic band.

5.21 The band is probably due to coordinated H_2, as this would account for its high frequency (far too high for a W—H stretch) and the fact that only one band appears when HD is used, with frequency midway between those of the H_2 and D_2 complexes.

5.23 The hydrogen atoms could both be axial (D_{3h}), both equatorial (C_{2v}) or one equatorial and one axial (C_s). In C_s there will be 12 modes, all active in both infrared and Raman. Eight of the Raman lines should be polarized. In C_{2v}, there will also be 12 modes; all will be active in the Raman, but only eleven in the infrared. Of the Raman lines, five should be polarized. For the D_{2h} model, there will be eight modes, $2a'$ (Raman; pol); $3e'$ (IR, Raman); $2a_2''$ (IR), and $1e''$ (IR, Raman). In principle it should be relatively easy from complete spectra with polarization measurements in the Raman to distinguish the structures, but the distinction between the C_{2v} and the C_s structures might be difficult.

5.25 The band at 169 cm^{-1} in aqueous $Hg_2(NO_3)_2$ is due to νHg—Hg in the Hg_2^{2+} cation and it seems most likely that the band at 167 cm^{-1} in solid Hg_2Cl_2 is also due to νHg—Hg. The weaker band at higher frequency in solid Hg_2Cl_2 must be due to an Hg—Cl stretching mode, which is at lower frequency than those of $HgCl_2$. Both chlorides must, however, contain Hg—Cl bonds, rather than separated anions and cations.

5.27 The species responsible have 6 IR bands each, suggesting 4-atom molecules, probably BF_2I and BFI_2 respectively. Both have C_{2v} symmetry, with $3a_1 + b_1 + 2b_2$ vibrations in each case, all IR active. The band contours quoted show that the C_2 axis is the A-axis for BF_2I, but the B-axis for BFI_2. The assignments are then:

Symmetry	BF_2I	BFI_2
a_1	1185(a) ν_{sym} BF_2	1295(b) ν BF
a_1	610(a) δ_{sym} BF_2	366(b) γ_{sym} BI_2
a_1	270(a) ν BI	125(b) δ_{sym} BI_2
b_2	1410(b) ν_{sym} BF_2	758(a) δ_{asym} IBF
b_2	318(b) δ_{asym} IBF	250(a) ν_{asym} BI_2
b_1	529(a) out of plane	403(c) δ out of plane

The two b_2 modes for BFI_2 may well be mixed to some extent, as they are further apart in frequency than might have been expected by comparison with the modes assigned in BF_2I.

5.29 Point group for NF_3Y is C_{3v}. Modes are:

	a_1	e
N—Y stretch	ν_1	
N—F stretch	ν_2	ν_4
NF_3 deformation	ν_3	ν_5
NF_3 rock		ν_6
Activity etc.	Raman pol	Raman dp
	IR parallel	IR perpendicular

Assignments are based on polarization data and order of frequencies:

	NF_3O/cm^{-1}	NF_3S/cm^{-1}
ν_1	1689	1512
ν_2	740	768
ν_3	548	520
ν_4	880	812
ν_5	513	430
ν_6	398	340

The very high values for the N—O and N—S stretching modes suggest strongly that the N—O and N—S bonds should be regarded as double; this seems inconsistent with the octet rule.

5.31 WF_4O might be square pyramidal (C_{4v}), trigonally bipyramidal with equatorial O (C_{2v}), or trigonally bipyramidal with axial O (C_{3v}). For the C_{4v} model we expect $3a_1$ (Raman, pol; IR), $2b_1$ (Raman, dp), $1b_2$ (Raman dp) and $3e$ (Raman dp; IR) modes – a total of 9 Raman bands (3 pol) and 6 IR bands. For the C_{3v} form we expect $4a_1$ (Raman pol; IR) and $4e$ (Raman dp; IR) modes, and for the C_{2v} structure we expect $5a_1$ (Raman pol; IR), $1a_2$ (Raman dp), $3b_1$ (Raman dp; IR) and $3b_2$ (Raman dp; IR) bands. We see 8 Raman bands, of which 3 are polarized, and 6 IR bands, of which one corresponds to a Raman band not detected. This fits well for the C_{4v} structure; the bands above 600 cm^{-1} should be assigned to stretching modes, and the others to deformations.

5.33 BiI_4^- could have T_d, C_{4v} or C_{2v} structures. A T_d ion would give 4 Raman bands (1 pol); C_{4v}, 7 Raman bands (2 pol); C_{2v}, 9 Raman bands (4 pol). We see 5 bands, of which 2 are polarized. This excludes the T_d structure. Moreover, the C_{4v} structure is unlikely. In that form, only one of the two polarized bands is due to a stretching mode; yet in the observed spectrum the two polarized lines that are observed are quite close together and seem likely to be due to the same sort of motion. In the C_{2v} form there are two a_1 stretching modes and so that seems to fit the data better. It is not very surprising with such heavy atoms that some low frequency bands have not been observed, and that some polarized bands have not been recognized.

Chapter 6

6.1 1.56, 1.57×10^7 m s^{-1}; 408, 403 eV. The more intense peak should correspond to ionization from the two terminal nitrogen atoms, which are equivalent. In valence bond structures for the azide ion the center nitrogen carries a positive charge, so its electrons are more tightly bound. The more intense peak should therefore be that at 1084 eV, corresponding to the lower binding energy.

6.3 For an octahedral d^3 complex there are two spin-allowed but Laporte-forbidden transitions, which are of moderate intensity. These are from the $^4A_{2g}$ ground state to $^4T_{1g}$ and $^4T_{2g}$ upper states, and each involves transfer of an electron from a t_{2g} orbital to an e_g orbital. The very weak band, at lowest energy, corresponds to a transition which is both spin- and Laporte-forbidden. This involves reversal of the spin of one electron, as the complex goes from its ground state to the ^2E_g state. The very intense band is a ligand-to-metal charge transfer band.

6.5 The molecular point group C_{3v}. Atomic orbitals are as follows: N $2p_z$, a_1; $2p_x$, $2p_y$, e; $2s$, a_1; H $1s$ (all three taken together), $a_1 + e$. The hydrogen and nitrogen orbitals of e symmetry combine to give a degenerate bonding pair and an anti-bonding pair of MOs. The three a_1 AOs give one bonding, one anti-bonding, and one approximately non-bonding MO. Four MOs are occupied, thus giving three bonds (a_1 and e orbitals) and a lone pair of electrons in a non-bonding orbital, which is derived primarily from the nitrogen $2s$ AO.

6.7 The spectrum shows four bands. The first is a doublet with no significant vibrational progression, and must represent ionization from an essentially non-bonding level that shows spin-orbit coupling, i.e. the π_g level of the valence shell. The second band, with a clear vibrational progression, corresponds to ionization from the bonding π_u level, and the separation between the component lines should therefore be less than v_1 for CS_2. The absence of spin-orbit splitting in the remaining two bands is consistent with their being associated with ionization from σ levels; the affectively non-bonding character must be due to mixing with other levels of the same symmetry.

6.9 0.0061 mol l^{-1}.

6.11 (a) one; (b) three; (c) one; (d) two.

6.13 In the oxyanions the band arises from charge transfer from ligand to metal, which is in a high oxidation state, The relative tightness of binding of the metal d orbitals increases with atomic number and positive charge, i.e. $V < Cr < Mn$, and so the transition energies decrease in this series. In the carbonyls the bands arise from metal-to-ligand charge transfer, and so the order of transition energies is reversed.

6.15 In the $^6A_1$ ground state of a high-spin octahedral d^5 complex there is one electron in each d orbital. The first d–d transition therefore involves bringing an electron from one of the e_g higher-energy orbitals to one of the lower-energy t_{2g} orbitals. The transition energy is the spin-pairing energy minus Δ, so decreasing Δ increases the energy of the transition.

6.17 The iron complex has one η^5-C_5H_5 ring and one η^1 ring, and so there are altogether five distinct chemical environments for carbon atoms, as well as one each for iron and oxygen. The XPS spectrum should therefore contain five peaks near 285 eV, and single peaks near 56, 95, 710, 723, 846 and 532 eV.

Chapter 7

7.1 The outer two lines are due to the two equivalent iron atoms with a large quadrupole splitting. The center line is due to the unique iron atom, with quadrupole splitting too small to resolve.

7.3 The parameters for the binuclear complex are close to a superposition of the parameters for the two model compounds. This implies that the binuclear species contains distinct low- and high-spin iron sites. Other techniques that could be used include measurements of magnetic susceptibility, ESR and UV-visible spectroscopy, and IR spectroscopy to explore the CN stretching frequencies.

7.5 (a) ^{197}Au Mössbauer spectroscopy could be used to explore the symmetry of the environment of the gold center and to see how far its electronic structure was consistent with the presence of Au$^-$.
(b) The ^{119}Sn Mössbauer spectrum would help to show the coordination number of the tin atoms. (They are actually five-coordinate in the solid phase, with CN groups bridging between tin atoms.)
(c) The ^{129}I Mössbauer spectrum should help to establish the extent to which the normal representation of the compound as $I(IO_3)_3$ was reasonable. The number of resonances and the chemical shifts and quadrupole coupling constants would tell us the relative numbers of iodine atoms in different environments, and their oxidation states.
(d) The ^{57}Fe Mössbauer spectrum should show whether the two iron centers are distinct, with a trapped-electron structure, or whether they share the odd electron; variable temperature studies would be necessary.

7.7 Fe^{II}, Mo^{VI}.

7.9 The formula of the complex shows that it contains one Fe(II) and two Fe(III) centers. The species with CH_3CN gives a spectrum that shows the two types of iron center are distinct at 300 K. On the other hand, the complex containing toluene shows that while the iron centers are distinct at low temperature on the Mössbauer timescale, they lose this distinction as the temperature increases until at room temperature and above the iron centers are all equivalent. X-ray crystallography shows that the two solvates have different crystal structures. The packing of the methylpyridine ligands is different, and in the CH_3CN solvate the inequivalence of the iron environments leads to appreciable potential energy barriers for intramolecular electron transfer. The molecular motion of the toluene molecules in the toluene solvate has also been studied by ^2D NMR in single crystals. The details of this fascinating study are given in S.M. Oh, S.R. Wilson, D.N. Hendrickson, S.E. Woehler, R.J. Wittebort, D. Inniss and C.E. Strouse, *J. Am Chem. Soc.*, **109**, 1073 (1987).

7.11 The Mössbauer spectrum of solid triphenylstannylpotassium [18]aneO$_6$ is consistent with the crystal structure; the environment of the tin nucleus is clearly asymmetric, and with the stereochemically active lone pair it bears some relation to Sn^{II}. In solid triphenylstannyl-potassium, the doublet shows that $SnPh_3^-$ ions are present. The singlet in the Sn^{IV} region implies that in this species the environment round the tin nucleus is pseudo-tetrahedral, presumably because of coordination of the lone pair of electrons to potassium.

Chapter 8

8.1 GeF_4: 2 peaks, Ge–F 1.80 Å area 640, F$\cdots$F 2.94 Å area 165. TeF_4: 4 peaks, Te–F 1.80 Å area 6.80, $F_{ax}\cdots F_{eq}$ 2.55 Å area 127, $F_{eq}\cdots F_{eq}$ 3.12 Å area 26, $F_{ax}\cdots F_{ax}$ 3.60 Å area 23. XeF_4: 3 peaks Xe–F 1.80 Å area 720, F$\cdots$F 2.55 Å area 127, F$\cdots$F 3.60 Å area 45.
(a) The Te–F peak would split, but the components would probably be unresolved, the $F_{ax}\cdots F_{eq}$ peak would shift to greater length, and the $F_{ax}\cdots F_{ax}$ peak would shift, by twice the change in the Te–F_{ax} distance.
(b) The $F_{eq}\cdots F_{eq}$ peak would be at a greater distance.
(c) The $F_{ax}\cdots F_{eq}$ peak would be at a shorter distance, and there would be a very small decrease in the $F_{ax}\cdots F_{ax}$ distance.

8.3 (a) Ge_2Cl_6 is ideal for electron diffraction. The eclipsed (D_{3h}) conformation could be easily distinguished from the staggered (D_{3d}) arrangement by the long Cl$\cdots$Cl distances.
(b) It would not be easy to determine the two different S–F distances, nor to detect any small deviation from regular octahedral coordination at sulfur. The S–H bond distance could not be measured very precisely.
(c) Good – a highly symmetrical molecule.
(d) Very difficult. The P–F, P–O, C–O and C–C peaks would all overlap, as would the various two-bond distances (F$\cdots$O, P$\cdots$C, O$\cdots$C etc.).
(e) The Sb–H peak is 36 times as intense as the H$\cdots$H peak, so the HSbH angle is not at all easy to measure accurately.
(f) Tetrahedral OsO_4 is an ideal molecule to study by electron diffraction.
(g) The Co–Co distance could be easily measured, but the terminal and bridging Co–C distances would be difficult to resolve, and the five angles needed to define the positions of carbonyl groups would probably also be very poorly determined.

8.5 (a) Electron diffraction or microwave spectroscopy.
(b) NMR spectroscopy in a liquid crystal solvent (see Section 2.13).
(c) X-ray diffraction, at low temperature because the compound is a volatile liquid at room temperature.
$SiH_3C\equiv CBr$: electron diffraction or microwave spectroscopy; virtually impossible, but X-ray diffraction or bromine EXAFS might give a sporting chance; X-ray diffraction.

8.7 There would be seven peaks, the first corresponding to the Fe–N bond distance, then to Fe$\cdots$C, Fe$\cdots$N and then to the four types of Fe$\cdots$C atom pair involving carbon atoms in the outer six-membered rings. There would also be weaker Fe$\cdots$H peaks. Moving the iron atom out of the ring plane increases all of these distances, the shorter ones by more than the longer ones.

The electron diffraction radial distribution curve contains all these Fe–C and Fe–N peaks, but also has many more peaks, corresponding to all the C–C, C–N and N–N atom pairs, as well as those involving carbon or nitrogen and hydrogen.

8.9 Each aluminum atom has three neighboring chlorine atoms in the plane above (solid circles) and three in the plane below, so it is octahedrally coordinated. The view shown is along a three-fold axis. Each oxygen has two nearest neighbors and has bent coordination. The Al–Cl distance is 2.46 Å, and the structure consists of infinite sheets – easily seen if four unit cells are shown, and all the bonds are drawn in.

8.11 If the chromium site has S_6 symmetry, the benzene ligands have only C_3 symmetry. The carbon (and hydrogen) atoms therefore are in two groups of three, and the bond lengths around the ring may alternate, as in the Kekulé structure. The ring angles may also alternate, the hydrogen atoms do not have to lie on the bisectors of the adjacent CCC angles, and there is no requirement for the benzene ligands to be planar.

8.13 Atoms 1 and 2 are related by inversion through the origin, 1 and 3 by reflection in the xz plane, and 1 and 4 by rotation around the y axis.
Coordinates:

Cu $0\,0\,0$; $\frac{1}{2}\frac{1}{2}0$; $\frac{1}{2}0\frac{1}{2}$; $0\frac{1}{2}\frac{1}{2}$

Cl $0.125\frac{1}{2}0.125$; $0.375\,0.5\,0.375$; $0.625\frac{1}{2}0.625$; $0.875\frac{1}{2}0.875$; $0.125\,0\,0.625$; $0.375\,0\,0.875$; $0.625\,0\,0.125$; $0.875\,0\,0.625$.

There are eight copper atoms at the corners of the cell, each $\frac{1}{8}$ in the cell, and six at the centers of the faces, each $1/2$ in the cell, giving a total of four atoms per cell. There are four chlorine atoms entirely in the cell, and four on each of two faces, giving a total of eight per cell.

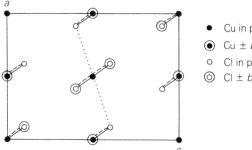

Fig. P.1

The projection along b (Fig. P.1) shows that there are square planar copper atoms linked in chains by double chlorine bridges. The view is along the axes of these chains. The Cu—Cl distance is 2.35 Å. There are also contacts, marked with dotted lines in the projection, completing the octahedral coordination of the copper atoms, with Cu$\cdots$Cl distances of 2.94 Å. The copper coordination is that expected for an octahedral d^9 metal with Jahn–Teller distortion.

8.15 (Molecular weight × 2/Avogadro's number) = (cell volume × density)
Left-hand side = $361.4 \times 2(6.02 \times 10^{23}) = 1.20 \times 10^{-21}$ g
Right-hand side = $4.29^2 \times 14.01 \times 4.6 \times 10^{-24} = 1.19 \times 10^{-21}$ g

Plotting a projection along b shows that all the atoms lie in planes, separated by $c/2$, i.e. more than 7 Å. A projection along c shows one of these planes, in which one type of atom forms a regular square array, with the second type of atom at the center of each square. The shortest Tl—I distance is therefore $\sqrt{2}a/2$, or 3.03 Å. The methyl groups must lie on the C_4 axis of one atom type, so the structure must be of the form XMe$_2$ Y. The most likely arrangement is $\text{TlMe}_2^+\text{I}^-$, with a linear CTlC configuration.

8.17 Two ligands can be accommodated, with the bipyridyl groups either *cis* or *trans* to one another. No further ligands could act as bidentate chelates if the first two were mutually *trans*, but if they were *cis* one or possibly two more groups could bind *via* their bipyridyl fragments. See *Inorg. Chem.*, **23**, 3254 (1984).

The ligand arrangment would be very difficult to determine by any method other than X-ray crystallography. The ^{15}N NMR spectrum would be useful, but it would not be easy to obtain it.

8.19 The Pd—S distances indicate that there are six bonds to the metal atom, but two of them are substantially longer than the remaining ones. This is expected for a d^7 palladium(III) complex, which should show a Jahn–Teller distortion, lengthening two bonds to give a tetragonal structure. See A.J. Blake, A.J. Holder, T.I. Hyde and M. Schröder, *J. Chem. Soc., Chem. Commun.*, 987 (1987).

8.21 A sulfur, B carbon and C nitrogen. The Hg$\cdots$Co distance is 5.66 Å. Measurement gives distances Co—N 1.93 Å, N—C 1.31 Å, C—S 1.65 Å and Hg—S 2.63 Å, and angles CoNC 161°, NCS 168° and HgSC 94°.

Chapter 9

9.1 [Rh$_6$C(CO)$_{16}$]

9.3 The heaviest observed fragment has a m/e value of 706, 134 less than that of the molecular ion M$^+$; the C$_5$Me$_5$ ligand has a mass of 135 amu. This suggests the loss of one C$_5$Me$_5$ ligand and addition of one H atom. Three other peaks (571, 435 and 300) also appear to be due to further losses of C$_5$Me$_5$ ligands, leaving the V$_4$O$_6^+$ core intact in each case. The other ions have masses 16 or 17 less than those in the first series, suggesting a second mode of fragmentation leading to ions with V$_4$O$_5^+$ cores, and even (403) V$_4$O$_4^+$.

$$[L_4V_4O_6]^+ \xrightarrow{-L+H} [L_3V_4O_6H]^+ \xrightarrow{-L} [L_2V_4O_6H]^+ \xrightarrow{-L-H} [LV_4O_6]^+ \xrightarrow{-L} [V_4O_6]^+$$

$$\xrightarrow{-L-O} \Big|{-OH?}$$

$$[L_3V_4O_5]^+ \xrightarrow{-L} [L_2V_4O_5]^+ \xrightarrow{-L} [LV_4O_5]^+ \xrightarrow{-O} [LV_4O_4]^+$$

9.5 [Re(CO)$_3$]$^+$ 269 37%: 271 63%. [ReBr]$^+$ 264 19%: 266 50%: 268 31%. Ratio [Re(CO)$_3$]$^+$: [ReBr]$^+$ = 0.39: 1.

9.7 (a) 412 5.8%, 413 11.0%, 414 30.3%, 415 24.0%, 416 27.0%.
(b) 241 18.2%, 242 17.5%, 243 45.3%, 244 5.5%, 245 12.5%.
(c) 276 13.9%, 277 13.3%, 278 38.9%, 279 8.4%, 280 20.4%, 281 1.3%, 282 3.0%.

9.9 9.4 [^{10}B$_2$H$_4$]$^+$, [^{10}B^{11}BH$_3$]$^+$, [^{11}B$_2$H$_2$]$^+$, [^{10}B^{12}CH$_2$]$^+$, [^{11}B^{12}CH]$^+$ and [^{12}C$_2$]$^+$. The observed peaks are due to [^{10}BCH$_2$]$^+$ and [^{11}BCH]$^+$.

9.11 *Either* exact mass measurement
or record mass spectrum of C$_2$D$_4$ complex. This was actually done, and only a weak peak [M-28]$^+$ was observed, with a strong peak for [M-32]$^+$, showing that at least the major part of the [M-28]$^+$ peak in the original compound probably derived from loss of C$_2$H$_4$.

9.13 The mass of the [M—H]$^+$ ion is 75, so the daughter ion has a mass of 41, and the neutral fragment has mass 34. The neutral fragment must be PH$_3$, leaving [C$_3$H$_5$]$^+$ as the daughter ion. The phosphorus atom is apparently extruded from the molecule, taking 3 hydrogen atoms with it, presumably leaving a cyclopropyl cation.

9.15 The mean dissociation energy of Mn—C in [Mn$_2$(CO)$_{10}$]$^+$ is 1.01 eV (97 kJ mol^{-1}) and for Re—C in [Re$_2$(CO)$_{10}$]$^+$ it is 2.07 eV (200 kJ mol^{-1}). The mean dissociation energy for the first five metal-carbon bonds of [MnRe(CO)$_{10}$]$^+$ to be broken is 0.78 eV, and for the remaining five it is 2.78 eV. This strongly suggests that the Mn—C bonds are all broken first.

Chapter 10

The questions in this chapter are intended to lead to discussion. No answers are given.

Compound Index

The ordering of the entries is determined by the following criteria in order of decreasing priority.
1 The central or most significant element. If this is ambiguous the compound is entered more than once.
2 The number of atoms of that element in the compound.
3 The coordination number of these atoms.
4 The substituent atoms bonded to the central atom, listed alphabetically.

Page numbers of main sections concerned with a particular compound are in italic.

The letters F and T refer to figures and tables.

Subject Index

Page numbers of major sections of text dealing with particular topics are indicated in italics. References to figures, tables and the glossary of techniques are indicated by the letters F, T and G after the page number.

DEPT 20G, 74–76, 75F
Detectors *15–17*
  position-sensitive 15
Diffraction 9, *331–370*
  of electrons, neutrons and X-rays *332–334*
  see also Electron, Neutron and X-ray
    diffraction
Dihydrogen and dihydrido complexes *452–456*
Dipole, and IR selection rule 190, 191F, 193
Direct methods in X-ray crystallography
    349–354
Disorder in crystallography 358–359, 358F,
    359F, 401–402, 418–419, 418F
Doppler effect, in electronic spectroscopy 283
  in Mössbauer spectroscopy 304–308
  in IR spectroscopy 159
Double resonance, see Multiple resonance

EDX 20G
ELDOR spectra 20G, 130
Electric field gradient, and Mössbauer
    spectroscopy 312, 314–316, 318
  and NMR relaxation 53–54, 64
  and NQR spectroscopy 131, 134–136
  sign of, from magnetic effects in Mössbauer
    spectra 316
Electron correlation 279
Electron diffraction 20G, *332–343*
  advantages for studying gases 333
  anharmonicity effects 336
  apparatus 334–335, 335F
  atomic number and scattering power 333F,
    334
  atomic scattering expression 335
  by gases *334–343*
  combining with other data 86, 169–170,
    169F, 340
  interatomic distances 6F
  interpretation of results *337–340*
  light atom location problem 340–341
  limitations *340–343*
  low-frequency vibration problem 342–343
  molecular scattering curve 336, 336F
  molecular scattering expression 336
  overlapping peak problem 341–342, 341F
  phase shifts 336
  rotating sector 335
  structure, of $Al(BH_4)_3$ 408–409
    of $Be(BH_4)_2$ 410–412, 411F
    of B_4Cl_4 339
    of Cr_2 acetate$_4$ 452, 452F
    of $Fe(C_5H_5)_2$ 415, 416F
    of XeF_6 396–397, 397F
  structure refinement 337
  symmetry effects 338
  vibrational effects 5, 336–337, 342–343,
    342F
  see also Amplitude of vibration, Low energy
    electron diffraction, Radial distribution
    curve, Shrinkage
Electron energy levels *256–259*
  in many-electron atoms 257–258
  in molecules 258–259
  in one-electron atoms 257
Electron energy-loss spectroscopy 20G,
    180–181, 195, 367
Electron impact mass spectrometry 20G
Electronics and computers *16–17*

Electronic spectroscopy 7, 8, 11, 20G,
    255–299
  electron delocalization in Fe-Mo-S clusters
    432–433, 432F
  excitation and ejection of electrons *255–256*,
    255F
  information in spectra *282–286*
  linewidths 282–283
  oriented samples and polarization of
    absorption 294
  rotational structure 156, 159, 282–284
  selection rules *284–286*
  spectra of lanthanides and actinides 299
  spectra of transition metal complexes 255,
    284–285, *286–294*
    assigning bands *293–294*, 294F
    charge-transfer bands *292–293*, 293F
    effects of cooling samples 293
    identification *296–299*
    intervalence transitions 293
    ligand field splitting 288–291, 289F
    ligand–ligand transitions 288
    measurement of concentration *296–299*
    metal, ligand, and metal–ligand bonding
      levels *287–288*, 287F
    metal–metal transitions *288–291*
    of HCP 434
    of $[Ir(bpy)_3]^z$ and Li^+bpy^- 435, 436F
    one-electron energy levels 288–289, 289F
    timescale 10T
  structures of gold clusters 425–426, 425F
  vibrational structure *179*, 282–283, 283F
  vibronic coupling 293
  see also Circular dichroism, Hot bands,
    Isosbestic point, Jahn–Teller distortion,
    Magnetic circular dichroism, UPS,
    UV/visible spectroscopy, Valence-electron
    photoelectron spectroscopy, Valence
    excitation spectroscopy
Electron impact ionization 378
Electron microscopy 21G
Electrons in reduced and excited bipyridine
    complexes *435–438*
Elemental analysis 1, 4
ENDOR spectra 21G, 129, 129F, 130
Energy loss processes in XPS 263–264
Energy spectrum 7, 8F
Equilibrium constants, measurement by NMR
    96
Equivalence, chemical 36, 44, 58
  chiral and pro-chiral non-equivalence 60–61
  full 84
  magnetic 44, 45, 58, 84
EPR spectroscopy, see ESR spectroscopy
ESCA 21G, 259
  see also XPS
ESR spectroscopy 3T, 7, 10, 21G, *115–130*
  anisotropic systems *120–122*
  electron–electron interactions 122–125,
    124F, 125F
  energy levels for electron/spin $\frac{1}{2}$ nucleus
    system 128F, 129
  experimental arrangement *115–116*
  Fermi contact 117
  fingerprints 128
  hyperfine coupling in isotropic systems
    117–120
  Jahn–Teller distortion 126–128
  magnetic field effect 122–124

spectrum of HCP 434
spin-orbit coupling *276–277*, 277F
structural information 3T
symmetry of molecular orbitals *267–269*
transition metal complexes *278–279*
vibrational structure of bands *270–276*,
271–276F
see also Electronic spectroscopy, Fingerprints,
XPS
UV/visible spectroscopy 26G, 255, 279–282,
286
spectra of $Xe_2{}^+$ 402–404, 403F
see also Electronic spectroscopy

Valence-electron photoelectron spectroscopy
269–279
see also UPS
Valence excitation spectroscopy *279–286*
calculation of energies *286*
experimental methods *279–282*
see also Electronic spectroscopy
Valence levels, and photoprocesses 257–258,
258F
Velocity analysis of ejected electrons 256, 259,
270
Vertical ionization energies in UPS 275
Vibrational effects, in crystallography 12, 346,
351, 359
in gas phase electron diffraction 5, 12,
336–338, 342–343
Vibrational coupling 221
Vibrational spectroscopy 3T, 27G, *173–248*
activities, of modes of $[M(CO)_4L_2]$ 225
of modes of $[M(CO)_4(PH_3)_2]$ 225–226,
226F
anharmonicity *175–176*
assignment of bands to vibrations *199–216*
characteristic frequencies *221–224*
CO stretching modes 222, 222T,
224–225, 227
deformation modes of germyl
halides 222–224, 223F, 223T
MH stretching modes 221–222
NO stretching modes 222, 222T
complete empirical assignments *233–237*,
235T
degenerate modes 195–196
effects of phase on spectra *187–189*, 187F,
189F
energy levels for IR, Raman and fluorescence
177F
experimental methods *176–182*
force fields 8, *237–241*
group frequencies *220–228*
impurities, detection and identification
218–219, 219F
inelastic scattering of electrons and neutrons
180–181
intensities of bands *210*
ligand binding types *227–228*
local modes 225–226
non-fundamental transitions *211–216*, 211F,
212T, 213F, 215F
normal modes 173–174
numbering of modes 210–211, 211T
number of normal modes 174–175
physical basis *173–176*
potential energy distribution 241

potential functions for diatomics 175–176
quantitative analysis 219–220, 220F
sample handling *182–187*, 184F, 186F
selection rules 177, 178, *190–193*, 196
spectrometers *181–182*
structural information *216–218*
symmetry *190–199*
symmetry of full set of normal modes
196–199
timescale 10T
use of isotopes *228–233*, 229T, 230–233F
vibrational coupling 222
vibrations of PF_5 197–202
see also Electronic spectroscopy, Fingerprints,
Infrared spectroscopy, Normal coordinate
analysis, Raman spectroscopy,
Vibration-rotation spectroscopy
Vibration-rotation spectroscopy 3T, 8,
151–156, 187, *202–209*
centrifugal distortion 153
Coriolis coupling 154, 156, 241
O, P, Q, R and S branches 153, 156,
203–206, 204F
parallel bands 153–154, 155F, 204F, 205
perpendicular bands 153–156, 155F, 204F,
205
Raman spectroscopy 158
selection rules 151, *202–209*
vibration-rotation constants 165–166
see also Infrared spectroscopy, Rotational
spectroscopy, Rotational Raman
spectroscopy
Virtual levels and photoprocesses 257–258,
258F
Virtual states in Raman spectroscopy 177F,
178
Visible spectroscopy, see UV/visible
spectroscopy

XANES 27G, 366
XPS 8, 27G, *259–267*
calibration 263–264
chemical shifts *263–267*
core binding energies 261T
core level photoionization processes
260–261
depth effects in solid samples *263*
electron spin complications 262
elemental analysis 259
experimental methods *259–260*
unhelpful in determining structures of gold
clusters 426
see also Electronic spectroscopy, UPS
X-ray diffraction 5, 27G, *332–334*, *345–361*
anomalous dispersion and absolute
configurations 333, 354–356
asymmetric unit 347–348, 347F, 348F
atomic number and scattering power 333F,
334, 354
direct methods 349–354
disorder 358–359, 358F, 359F
electron density distribution *363–366*,
363–365F
electron density maps 349–351, 352F
experimental arrangements 347
heavy atom method 349, 351
internuclear distances 7F, 361
limitations 10, *356–359*